T0295460

More than Nothing

More than Nothing

A History of the Vacuum in Theoretical Physics,
1925–1980

AARON SIDNEY WRIGHT

OXFORD
UNIVERSITY PRESS

OXFORD
UNIVERSITY PRESS

Oxford University Press is a department of the University of Oxford. It furthers
the University's objective of excellence in research, scholarship, and education
by publishing worldwide. Oxford is a registered trade mark of Oxford University
Press in the UK and in certain other countries.

Published in the United States of America by Oxford University Press
198 Madison Avenue, New York, NY 10016, United States of America.

© Oxford University Press 2024

CIP data is on file at the Library of Congress

ISBN 978-0-19-006280-4

DOI: 10.1093/oso/9780190062804.001.0001

Printed by Integrated Books International, United States of America

To Chloe

"Take some more tea," the March Hare said to Alice, very earnestly.

"I've had nothing yet," Alice replied in an offended tone, 'so I can't take more.'

"You mean you can't take *less*," said the Hatter: "it's very easy to take *more* than nothing."

"Nobody asked *your* opinion," said Alice.

Lewis Carroll, *Alice's Adventures in Wonderland*, illustrated by Sir John Tenniel, 1865.

Contents

Journal Abbreviations

AHES	*Archive for History of Exact Sciences*
AP	*Annalen der Physik*
AS	*Annals of Science*
BJHS	*The British Journal for the History of Science*
BJPS	*The British Journal for the Philosophy of Science*
CMP	*Communications in Mathematical Physics*
EJP	*European Journal of Physics*
GRG	*General Relativity and Gravitation*
HSNS	*Historical Studies in the Natural Sciences*
HSPBS	*Historical Studies in the Physical and Biological Sciences*
HSPS	*Historical Studies in the Physical Sciences*
INC	*Il Nuovo Cimento*
JMP	*Journal of Mathematical Physics*
MPCPS	*Mathematical Proceedings of the Cambridge Philosophical Society*
NP	*Nuclear Physics*
PL	*Physics Letters*
PM	*Philosophical Magazine*
PR	*Physical Review*
PRD	*Physical Review D*
PRL	*Physical Review Letters*
PRSLA	*Proceedings of the Royal Society of London A*
PT	*Physics Today*
RMP	*Reviews of Modern Physics*
SHPMP	*Studies in History and Philosophy of Modern Physics*
SSS	*Social Studies of Science*
ZP	*Zeitschrift für Physik*

Acknowledgments

I began asking questions about the wild picture of that reality physics presents, and its changes over time, as I shuttled between history classes and physics classes in Montreal. My mentors at McGill University, including James Delbourgo, Nick Dew, Keshav Dasgupta, and Brigitte Vachon, showed me how to continue asking thorny questions. I developed and refined my questions in graduate school in Toronto, with guidance and support from Chen-Pang Yeang, Ian Hacking, David Kaiser, Jim Brown, Chris Smeenk, and M. Murphy, and a close cohort of fellow students. As a postdoctoral fellow, I explored new archives and new arguments in conversation with Peter Galison and Michael Friedman. This book was made possible by extended visits to archives across the United States. I gratefully acknowledge the financial support of the Social Sciences and Humanities Research Council of Canada and Dalhousie University. The editorial staff at Oxford University Press and the critical comments I received from reviewers were unfailingly helpful. Writing this book has been a thoroughly iterative process of writing and rewriting. It would not have been possible without the support of good friends, colleagues, and family. I hope that all those who helped produce this book see their mark in it, but all errors are my own.

1
The Vacuum in Practice

Nothing is not always what it seems. Two of our best theories of physics take empty space to be full of strange fields and particles, or to be a substance itself. First, in quantum field theory (QFT): "According to present ideas," writes the physicist Peter Milonni, "there is no vacuum in the ordinary sense of tranquil nothingness. There is instead a fluctuating quantum vacuum."[1] Second, in general relativity (GR): instead of vast expanses of nothingness extending between the stars, instead of a passive *container* for matter and energy, in GR space is an active participant in physics. Here is the most succinct description of Einstein's theory of gravitation:

> Space acts on matter, telling it how to move. In turn, matter reacts back on space, telling it how to curve.[2]

Space *acts*. New measurements of the action of space—as gravitational waves—were awarded the 2017 Nobel Prize for Physics. In the earlier words of co-laureate Kip Thorne, "gravitational waves [are] ripples in the curvature of spacetime that propagate out through the Universe," which vibrate Earthbound experiments, and in fact the entire Earth itself.[3] Neither "quantum fluctuations" nor "ripples of curvature" sound much like a vacuum of "empty space." What sort of vacuum is this? A dictionary of physics gives little help:

> Vacuum. A space in which there is a low pressure of gas, i.e. relatively few atoms or molecules.[4]

But QFT and GR are paradigms of successful physical theories. Each has been the subject of decades of theoretical study and empirical confirmation. Despite the strangeness of these pictures, we have good cause to believe them. Have these great edifices of human reasoning only led to self-contradiction? On top of these puzzles, an additional problem immediately presents itself: the two pictures disagree. Fluctuating quantum fields are not a smooth rippling sheet. What are we to make of these competing paradoxes?

As an historian, my approach to this puzzle is to ask how it came to be. Where did these ideas come from? How did they change over time? *More than Nothing* is a history of the vacuum in QFT and GR. It examines the vacua at the center of physicists' research: as "seas" of electrons in space; as the virtual creation and annihilation of

---

1. Milonni, *Quantum Vacuum*, xiii, cf. 355, 378.
2. Overall italics removed, Misner, Thorne, and Wheeler, *Gravitation*, 5.
3. Thorne, *Black Holes*, 259; cf. Kennefick, *Speed of Thought*.
4. Law and Rennie, *A Dictionary of Physics*, s.v. "vacuum"; compare "vacuum state," under which heading we find: "A vacuum state does not mean a state of nothing," because of vacuum fluctuations.

More than Nothing. Aaron Sidney Wright, Oxford University Press. © Oxford University Press 2024.
DOI: 10.1093/oso/9780190062804.003.0001

particles; and as spacetime fabrics and bubbling foam. These vacua were represented in words, formalism, and diagrams; physicists regularly asserted that these vacua underlay experimental and observational phenomena and physical reality itself. This book focuses on the ways in which these vacua—these words, formalisms, and diagrams—constituted a large part of the day-to-day *practices* of theoretical physicists. Ontologies of the vacuum were tied together with physicists' pen-and-paper calculations and theoretical tools, such as perturbation theory and spacetime diagrams.

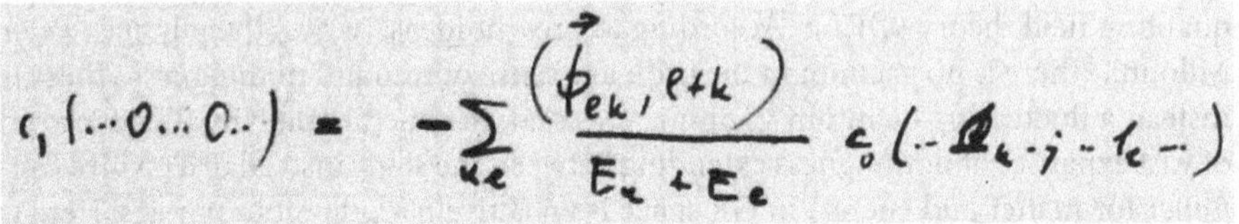

Figure 1.1. A fragment of Victor Weisskopf's vacuum calculations. Folder 2, Box 9.02, Weisskopf Papers, MIT Libraries collection MC-0572.

This focus on the practice of theory brings out a second important dimension of this book. It is not only the case that there is a history of strange pictures of the vacuum in QFT and GR; it is also the case that each subfield became increasingly *about* the vacuum. Physicists' arguments and analyses became centered on "vacuum states," "vacuum solutions," and "vacuum expectation values." This was a substantial change. In the nineteenth century, one physicist saw "a system consisting of no particles," as "a barren subject of study," to be addressed only for mathematical consistency.[5] Why was it that, in the 1930s, the young postdoctoral researcher Victor Weisskopf filled dozens of loose pages of calculations with expressions such as the one shown in figure 1.1? On the left side of the equation, he wrote an expression for a wavefunction, $c_1(\ldots 0 \ldots 0 \ldots)$, with zero positively charged and zero negatively charged particles. On the right side, he used perturbation theory to relate this to another wavefunction, $c_0(\ldots 1_k \ldots; \ldots 1_l \ldots)$, for a system with one of each kind of particle. Further details are given in chapter 3. Here I want to comment on the *strangeness* of Weisskopf's expression. Why would a physicist be writing down a representation of nothingness—a system with no particles, $c_1(\ldots 0 \ldots 0 \ldots)$—*at all*? And what is this business of it being equal to a representation of matter, of physical stuff? This book is a history of the vacuum as an entity in physics—as both a putative physical entity in the world and as pieces of formalism, text, and drawings. I want to understand not only physicists' *interpretations* of empty space; I want to understand physicists' *use* of empty space. The most important argument in the book is that these formal uses and ontologies were closely interrelated; they shaped each other over time. Sometimes a physicist's ontology shaped their choice of tools. Other times, a physicist's tools shaped their ontology. This is demonstrated in the chapters that follow through the analysis of physicists' published and unpublished papers, calculations, student notebooks, and letters. The history of vacuum physics is a history of making sense of nothing.

5. Gibbs, *Elementary Principles in Statistical Mechanics*, 192.

This book is the first history of the vacuum in modern physics.[6] It contributes to an established literature on the earlier history of concepts of space.[7] This book is different in scope and focus from histories of *space* in general. First, this book is not focused on *concepts*. The classic debate as unfurled from antiquity through eighteenth-century conflicts between Isaac Newton and Gottfried Wilhelm Leibniz has been seen as primarily a *philosophical* debate: a debate which was related to physics and mathematics, but which was argued with *meta*physical arguments.[8] In contrast, this book is concerned with the practice of physics, and how it shaped, and was shaped by, the vacuum. I am concerned with properly philosophical arguments only when they appear in the practice.[9] This book is closer to the approach taken by Steven Shapin and Simon Schaffer in their classic study of "the experimental life" of the vacuum in the seventeenth century, *Leviathan and the Air-Pump*.[10] Second, the scope of this book is narrower: the narrative is restricted to *active* pictures of the vacuum. In line with an older tradition, the mathematician Hermann Minkowski viewed spacetime as substantial—as full—in a famous 1909 contribution to Special Relativity:

> Not to leave a yawning void anywhere, we will imagine that everywhere and everywhen there is something perceptible.[11]

This spacetime was *physical* in some sense, but it was not *active*, in the Laser Interferometer Gravitational-Wave Observatory's sense that ripples of empty space *vibrated* the Earth. Minkowski's "world" contained regular massive electrons. In contrast, John Archibald Wheeler tried to build a formalism for electrons *out of* equations for empty space *alone* (chapter 5). This book is a history of *active* pictures of the vacuum. I see its first expression in Paul Dirac and Rudolf Peierls's reaction to the discovery of the positron in 1932 (chapter 2).[12] This focus is justified because—as subsequent chapters will show—this active vacuum has taken a central role in theoretical physics. This concentrated scope provides a stage for detailed analyses of the daily life of vacuum theory.

Following an entity through time provides a different historical perspective on physics than do traditional histories of maturing theories or crescendoing experiments. For example, it allows comparison and contrast between theories. Each of QFT

---

6. Popular accounts include: Thorne, *Black Holes*; Close, *Void*; Weatherall, *Void*.

7. Classic historical and philosophical studies include Grant, *Much Ado about Nothing*; Jammer, *Concepts of Space*; Earman, *World Enough and Space-time*. In this book, it is not my purpose to translate physicists' debates into the technical language of philosophical debate.

8. An exception is DiSalle, *Understanding Space-Time*.

9. For an example of this distinction, consider the discussion of Einstein and experiment in terms of practice and in terms of attitudes or philosophies in Dongen, *Einstein's Unification*, §§4.1–2.

10. Shapin and Schaffer, *Leviathan and the Air-pump*.

11. Minkowski, "Raum und Zeit," xvii; Galison, "Minkowski's Space-Time"; Walter, "Minkowski's Modern World."

12. It is possible to identify active pictures of the luminiferous aether as twentieth-century precursors, particularly as one retreats from scientific practice. Bromberg, "Concept of Particle Creation," §1; Kragh and Overduin, *Weight of the Vacuum*, §§2, 5. These active aethers were quite different from the dominant passive aether adumbrated by, for example, Hendrik Lorentz.

and GR espouses an active picture of the vacuum.[13] Did physicists consider the vacuum of quantum theory and the vacuum of relativity to be the same physical vacuum? Quantum theorists held a mixture of views. According to Dirac:

> There is no need to make the [relativistic quantum] theory [of the electron] conform to general relativity, since general relativity is required only when one is dealing with gravitation, and gravitational forces are quite unimportant in atomic phenomena.[14]

However, gravitational reasoning did appear in debates over Dirac's vacuum "sea," and in debates over zero-point energies (chapters 2 and 4, respectively). For example, Soviet physicist Vladimir Fock wrote to Dirac on February 12, 1930:

> An infinite number of negative energy electrons per unit volume [of empty space] entails an infinite negative mass density all over the world. This must have astronomical consequences and may stand in conflict with the general theory of relativity.[15]

Moreover, some physicists discussed in this book worked on quantum gravity directly. This work was a small minority within theoretical physics, but it demonstrates that mainstream quantum theory accepted, at least, the hypothesis that the two fields could be unified.[16] A fuller acceptance of the unity of quantum field theory and general relativity began when a rush of precariously employed postdoctoral fellows in particle physics embraced gravitation and cosmology in the late 1970s.[17]

What about from the GR side? Stephen Hawking's renowned work on particle creation by black holes is a useful point of departure for examining how "relativists" approached the QFT vacuum. Hawking pragmatically appropriated the quantum vacuum to discover his most famous results. In the 1960s, Hawking worked with his fellow research student at Cambridge, Roger Penrose, on black holes in GR and cosmology. Penrose's contributions to this work—particularly his diagrams— are the topic of chapter 6. In 1975, Hawking deployed a pragmatic sense of the unity of the two vacua: he acknowledged that a quantum theory of gravity would be "a deeper theory (still to be found) in which spacetime itself was quantized"— but went on to develop an approximate theory with a classical spacetime (fabric) and quantum particles (fluctuations). It was a "theory of quantum mechanics in curved space-time."[18] In his best-selling book, A Brief History of Time, Hawking explained

---

13. For example, two out of three of Tian Yu Cao and Sam Schweber's central tenets of QFT are about the vacuum. A formal tenet: operators are interpreted as creating and annihilating particles. "The creation of a particle is realized as a localized excitation of the vacuum." And a physical tenet: "The plenum assumption of the bare vacuum." The remaining tenet is that electrons are point-like objects without internal structure. Cao and Schweber, "Conceptual Foundations," 36–38.

14. Dirac, Principles (1967), 253.

15. Wright, "Beautiful Sea," 236.

16. See Stachel, "History of Quantum Gravity"; Rickles, "Pourparlers for Amalgamation"; Goenner, "History of Unified Field Theories"; Goenner, "History of Unified Field Theories, II"; Wesemael, "'Unaffected by Fortune.'"

17. D. Kaiser, "Whose Mass Is It?"

18. Hawking, "Particle Creation by Black Holes," 199.

his seemingly paradoxical result—in the chapter "Black Holes Ain't So Black"—as follows:

> How is it possible that a black hole appears to emit particles when we know that nothing can escape from within its event horizon? The answer, quantum theory tells us, is that the particles do not come from within the black hole, but from the "empty" space just outside the black hole's event horizon! We can understand this in the following way: what we think of as "empty" space cannot be completely empty because that would mean that all the fields, such as the gravitational and electromagnetic fields, would have to be exactly zero. However, the value of a field and its rate of change with time are like the position and velocity of a particle: the uncertainty principle implies that the more accurately one knows one of these quantities, the less accurately one can know the other. So in empty space the field cannot be fixed at exactly zero, because then it would have both a precise value (zero) and a precise rate of change (also zero). There must be a certain minimum amount of uncertainty, or quantum fluctuations, in the value of the field.[19]

Hawking layered the "quantum fluctuations" of the QFT vacuum on top of GR's fabric of spacetime.[20] *More than Nothing* explores just this kind of interaction between physicists' concepts of empty space.

1.1 Suprise, Paradox, and Practice

How should one react to these fantastic scientific statements? Some observers are incredulous. Physicist and philosopher James Cushing observed:

> consider the picture that quantum field theory gives us of the "vacuum" (aether?) as seething with particle-antiparticle pairs of every description [...]. Is nature *seriously* supposed to be like that?[21]

It is almost as if scientists are speaking a different language. This book is informed by an approach to technical language that has taken Ludwig Wittgenstein's philosophy as a touchstone. Its slogan is: look for the meaning of language in its uses.[22] This

---

19. Hawking, *Brief History of Time*, 111–12.

20. In his research, Hawking appealed to dynamic spacetimes, including the topological changes involved in black hole formation and evaporation. For example, he insisted that the stationary, and perhaps less active, Schwarzschild spacetime was insufficient to understand black hole radiation: "it is essential to consider not only the quasistationary final state of the black hole but also the time-dependent formation phase." Hawking, "Particle Creation by Black Holes," 205, cf. figs 1–3. I thank an anonymous referee for prompting this clarification.

21. Cushing, *Theory Construction and Selection*, 263.

22. This is only a gesture to a sophisticated literature. See, Winch, "Introduction"; essays by Michael Lynch and David Bloor in Pickering, *Science as Practice and Culture*, chs. 7–9; Pickering, *Mangle of Practice*, ch. 4; Warwick, "Laboratory of Theory"; for cautionary remarks, see Hacking, "Wittgenstein Rules"; for related historical approaches to scientific practice, see Buchwald, *Scientific Practice*, Galison and Warwick, "Cultures of Theory"; D. Kaiser, *Pedagogy and the Practice of Science*.

impulse motivates the close attention to scientists' arguments and practices I deploy in this book. How should we approach avowedly "paradoxical" descriptions of the power of empty space to describe "charge without charge" and "mass without mass" (chapter 5)? The definitions of these words are unlikely to bring clarity.

In 1939, Wittgenstein gave lectures on the foundations of mathematics at Cambridge. One of his central concerns was to distinguish between the multiple meanings of individual English words *inside* and *outside* mathematical discourse. He was particularly interested in the kind of self-contradictory phrase "which [set] the whole mind in a whirl, and gives the pleasant feeling of paradox." In a reference to Georg Cantor's transfinite numbers, he said to his class, "If you can show there are bigger numbers than the infinite, your head whirls." Wittgenstein's strategy for approaching "bigger numbers than the infinite" was to look for meaning in the use of this phrase within its technical discourse. "To understand a phrase, we might say, is to understand its use." He insisted that his students should *not even be surprised* by a scientific statement until they understood its use. His first attitude to paradoxical statements was simple incomprehension.[23] Before registering surprise, Wittgenstein wanted a scientist to explain how a given scientific phrase was used in the science.

Departing somewhat from his mathematical examples, Wittgenstein gave the example of a physicist with a new discovery. He enlisted his student Casimir Lewy:

> suppose that a physicist says "I have at last discovered how to see what people look like in the dark—which no one had ever before known."—Suppose Lewy says he is very surprised. I would say, "Lewy, don't be surprised," which would be to say, "Don't talk bosh."
>
> Suppose he goes on to explain that he had discovered how to photograph by infra-red rays. Then you have a right to be surprised if you feel like it, but about something entirely different. It is a different kind of surprise. Before, you felt a kind of mental whirl [...]—which whirl is a sign you haven't understood something.[24]

I would *like* to be able to say that this book aims to bring readers from the mental whirls of paradox about the vacuum to a more grounded sense of surprise. But before elaborating on that, here is a first example of finding meaning in use.

In the passage from *A Brief History of Time* quoted above, Hawking explains that black holes radiate particles because of the "uncertainty, or quantum fluctuations, in the value of the field." The existence of these fluctuations is "implied" by Werner Heisenberg's uncertainty principle. The uncertainty principle is often invoked in discussions of the QFT vacuum.[25] We can try to gain a better understanding of its role in Hawking's physics by following Hawking's texts from popular to research contexts. How was the uncertainty principle used in *Nature* (in 1974) and *Communications*

---

23. Wittgenstein, *Lectures on the Foundations of Mathematics*, 16, 19.
24. Wittgenstein, *Lectures on the Foundations of Mathematics*, 17.
25. E.g.: "From the energy-time uncertainty relation we know that virtual pairs of particles of mass m can exist for times $\Delta t \sim \hbar/mc^2$ [...]." Milonni, *Quantum Vacuum*, 315.

in Mathematical Physics (in 1975)?[26] There is a simple answer: as it turns out, the uncertainty principle is not mentioned in these papers.[27] Its non-use or absence from Hawking's initial arguments for particle creation from black holes suggests that this particle creation is *independent* of the uncertainty principle. It suggests that the uncertainty principle was a later addition to his understanding.

Looking for the meaning of uncertainty and vacuum fluctuations in Hawking's work has uncovered a surprise of its own. This (non-)use of the uncertainty principle is consistent with the analysis of vacuum fluctuations in the later chapters of this book. Beginning in the 1930s, physicists appealed to the uncertainty principle to *understand* a result that was derived by *other* means. These are the kind of questions with which this book is concerned: when were virtual particles, fluctuations, and uncertainty tied together in a picture of the vacuum? And how were these concepts shaped by the everyday practices—calculations, algorithms, diagrams—of theory?

There is an important difference between Wittgenstein's example of infrared physics and vacuum physics. Wittgenstein's example assumed that the concepts and practices of our ordinary life suffice to understand the meaning of technical phrases— at least those phrases which had any relevance for ordinary life. These concepts and practices served as a baseline from which to extend one's understanding. The seeming-paradox of the physicist who could see in the dark was dissolved by his appeal to photography and infrared rays. Photography was a well-known practice. Infrared rays might be beyond our experience, but rays of sunlight were common.[28] In his classroom, Wittgenstein was willing to simply reject as meaningless those mathematical claims which could not be brought out in terms of everyday concepts. (That is, meaningless *outside* mathematics.) This was more plausible for mathematics than for physics. Insofar as pure mathematics was not meant to be about the world of lived experience, such a rejection had limited consequences. The same is not true of modern physics. Physics not only purports to be about the physical world, physicists' technologies, institutions, and weaponry continue to shape everyday life. The question is, How much of physics, and vacuum physics, can be brought out in terms of the everyday? Despite the celebrated efforts of science popularizers, the consensus is that crucial elements of modern physics elide domestication. One might conclude that physics can only be understood by physicists, from within its technical discourse.

Here is one way to avoid this pessimistic conclusion. It follows the example of the infrared physicists' photography. Physics can be brought out in terms of everyday *practices*, like setting up cameras, adjusting lenses, and developing film.

26. Of course, there are many uses to consider within *A Brief History of Time*, but popularization as such is not my interest here. I take it that the strategy of looking at the underlying research is supported by the published intention (and Hawking's good faith) in presenting a popular account of such research.

27. Hawking does diagnose an "uncertainty" in the quantum field he studies, but this arises from the "ambiguity" in defining the vacuum states before and after the black hole is formed from a collapsing star. Hawking, "Black Hole Explosions?"; Hawking, "Particle Creation by Black Holes." It is unclear to me whether Hawking's paper actually applied quantum commutation relations to the field; the analysis proceeds on a classical level. Later work developed the QFT aspect of Hawking's calculation, such as Wald, *General Relativity*, 397, §14.3; Birrell and Davies, *Quantum Fields*, ch. 8.

28. Compare Pickering's analysis of science in terms of "cultural extension," not free invention: Pickering, *Mangle of Practice*, 3–4, 114–18.

8   MORE THAN NOTHING

In this book, I discuss how physicists read published papers; how they made notes; how they drew; and how they applied and extended methods of mathematical calculations on paper. No matter the difficulty of the content of physics, physi*cists* were always human beings undertaking a *practice* that can be understood. These practices manifested as individual actions; they were also part of the larger *discipline* of theoretical physics. Acceptable forms of argument, calculation, drawing, and note taking were shaped by social institutions such as pedagogy and publishing. Studying these institutions provides insight into physicists' practices, particularly when ephemeral actions are lost to the historical record.[29]

Another answer to this challenge follows the example of infrared beams, as concepts. Infrared beams are a scientific extension of everyday beams of light. If I were to continue digging at Hawking's black holes, could I ground his sentences in everyday concepts? This is one way to frame the well-known debates over the foundations of quantum theory between, among others, Werner Heisenberg and Erwin Schrödinger. Schrödinger's "*wave* mechanics" worked with well-known physical concepts; Heisenberg's "*matrix* mechanics" abandoned the concept of visualizable atomic motions in space and time. Physics was not kind to visualizable waves. Schrödinger renounced the hope that his wavefunction might stand in some direct relationship to physical reality. Still, he strove for *at least* a physics without paradoxical contradiction. He wrote to Niels Bohr on October 23, 1926:

> What hovers before my eyes is only the one thesis: even if a hundred attempts fail, one must not give up hope of reaching the goal—I say not by means of classical images, but by logically non-contradictory ideas, of the true nature of spatiotemporal events.[30]

Bohr himself developed a doctrine of the indispensability of classical concepts in communicating quantum phenomena.[31] My point here is that whether relativistic and quantum physics was expressible in intuitive—or at least consistent—terms cannot be taken for granted. It was debated by the physicists in this book. In physics, it may be that there is no escaping some of the "mental whirl" of paradox. This does not rule out the application of the broad strategy of looking for meaning in use. But it requires us to be open to the idea that an analysis that retains its air of paradox may be the best explanation possible.

The vacuum fascinates me. It is at the heart of modern theoretical physics, yet it is endlessly elusive. It has been the locus of some of the most interesting and challenging exercises of human reason about the natural world. And in its remoteness from experiment and observation, I believe its history reflects what is characteristically

29. On insights from Michel Foucault, Thomas Kuhn, and Ludwig Fleck on pedagogy and training: Warwick and Kaiser, "Kuhn, Foucault"; Olesko, "Science Pedagogy"; on coherent technical approaches to physics: Warwick, *Masters of Theory*, 278–80; Seth, *Crafting the Quantum*, 44.
30. Kalckar, "Correspondence Included (1985)," 460–61; Regt, "Schrödinger, Anschaulichkeit, and Quantum Theory," 478.
31. Beller, *Quantum Dialogue*, ch. 2, 160–67.

theoretical about theoretical physics.[32] This book is a history of reasoning about the physical void.

The next section lays out the scope of the narrative and historiographic framing of this book. Those readers who wish to dive into the narrative itself may skip to the Prologue in section 1.3.

1.2 Scope and Framing

As a history, this book shares features with several formations in the historiographic landscape. As a history of a scientific entity, it sits near Ian Hacking's "historical ontology" and Lorraine Daston's biographical approach to scientific objects.[33] Like historical ontology, this book is a *genealogical* investigation.[34] I am interested in how we got to now. This manifests in my choice of material. This book is not a comprehensive history of the vacuum, even within theoretical physics. I have chosen episodes that, in my view, mark important shifts in a larger series of events. (I discuss some paths not taken in section 1.2.4.) Among historians of science, these choices often evoke concern about "Whig history," roughly meaning a history of glorious victories for the current status quo (say, a Whig government).[35] Triumphalism would be a difficult attitude to maintain in the present study; while there are (arguably) widely agreed-upon pictures of the vacuum *within* QFT and GR, it is broadly acknowledged that there is no acceptable unified picture. According to the relativist P. C. W. Davies, "Empty space [...] holds the key to a full understanding of the forces of nature"; this was approvingly cited by the quantum theorist Milonni.[36] This was an expression of faith—there is no full understanding of the vacuum for physicists today. A purely anti-Whig history is impossible, like Leopold von Ranke's ideal representation of the past "as it actually occurred."[37] Within a genealogical framework, I have tried to bring out the physics of the vacuum in its own time, for example, by attending to the now-discredited concepts of Dirac's "hole" theory.

32. There was continued advance and interesting work in experiment and instrumentation for vacuum systems in the period covered by this book. Vacuum technologies enabled many of the experimental results that shaped vacuum theory. For example, Arthur Compton used a vacuum tube to create X-rays for his scattering experiments. However, the detailed quality and characteristics of this vacuum were not relevant to the resulting theoretical work. Instrumentalists considered a vacuum to be a region of low gas pressure. Kragh, *Quantum Generations*, ch. 9; in relation to particle physics experiments, Krige, "CERN from the Mid-1960s to the Late 1970s," §1.3.3; Galison, *Image and Logic*, §5.4.

33. Historical ontology aims at scientific entities and classifications that have come to structure the lives of non-scientists, such as psychological trauma. My study is narrower. The vacuum has become a central organizing principle of theoretical physics. Hacking, "Historical Ontology"; Daston, *Biographies of Scientific Objects*; cf. Arabatzis, *Representing Electrons*; D. Kaiser "Whose Mass Is It?"

34. Hacking, "Historical Ontology," 24.

35. Gordin, "Book Review: The Tory Interpretation of History."

36. Under the heading "The Living Vacuum," Davies, *Superforce*, 104; there are minor discrepancies in Milonni's text, Milonni, *Quantum Vacuum*, xiii.

37. Which is not to say it is not an ideal worth pursuing. Iggers, *Historiography in the Twentieth Century*, 2; cf. "all studies of local cultures will be coloured by the analyst's perspective and [...] the cultural characteristics identified as significant in a study will not necessarily be deemed important or similarly accounted for by the subjects themselves." Galison and Warwick, "Cultures of Theory," 290.

1.2.1 Theoretical Physics

This book is subtitled "a history of the vacuum in theoretical physics"; what, then, is theoretical physics? By the 1790s, mathematics was becoming the language of physical science. In John L. Heilbron's evocative terms:

> Not the rigorous, captiously logical, deadweight mortmain mathematics of the Greeks, however, or the difficult synthetic methods of Newton, but democratic easy-going algebra and creative slip-shod calculus.[38]

By 1900 it was generally recognized that physical science could be divided into disciplines such as chemistry and physics, and that these could in turn be subdivided into larger experimental and smaller, subsidiary, theoretical parts.[39] (A compact overview of physics in the *longue dureé* is John L. Heilbron's *Physics: A Short History from Quintessence to Quarks*.[40]) Certainly it has been commonplace to identify bits and pieces of the work of early modern instrument makers and courtiers, chymists and theologians, as constituting a prehistory of theoretical physics (Galileo and Newton, respectively[41]). But as either a cohesive body of work or a community of practitioners, theoretical physics was a product of the later nineteenth century. At this point, several German universities had established chairs and institutes of theoretical physics, such that about 10 percent of physicists practiced theory.[42] (While the analysis of physical observation and experiment required more mathematics over time, mathematics itself pulled away from the natural world, toward *purity* and self-subsistence.[43]) The institutional development of theoretical physics was uneven. In the United States, Edwin Kemble recalled that in 1917, Harvard refused to accept his doctoral thesis on quantum theory until he had completed additional experimental work.[44] When Enrico Fermi (1901–1954) set out on his career in the early 1920s, there were no chairs of theoretical physics in Italy and "a dissertation in theoretical physics as such would have been shocking."[45] But even if one could rarely *train* entirely in theory, increasingly, by World War I, physicists saw value in workers in theoretical physics *as such*.[46]

It is difficult, if not impossible, to make useful generalizations about the aims, accomplishments, or activities of theoretical physics in the early years of its history

---

38. Heilbron, "Weighing Imponderables," 1, quote: 6; Gingras, "What Did Mathematics Do to Physics?"; for the view from mathematics, see Garber, *Language of Physics*.

39. See Nye, *Chemical Philosophy to Theoretical Chemistry*.

40. Heilbron, *Physics*; for perspectives on disciplinary histories, see Graham, Lepenies, and Weingart, *Disciplinary Histories*.

41. Biagioli, *Galileo, Courtier*; Biagioli, *Galileo's Instruments*; W. Newman, *Newton the Alchemist*.

42. Forman, Heilbron, and Weart, "Physics circa 1900"; Pyenson and Skopp, "Educating Physicists in Germany"; Heilbron, "Fin-de-siècle Physics"; Jungnickel and McCormmach, *Intellectual Mastery of Nature*; Olesko, *Physics as a Calling*; Buchwald and Warwick, *Histories of the Electron*; Buchwald and Hong, "Physics"; Seth, "Crisis and Construction."

43. Heilbron, "Weighing Imponderables," 29–33; Gray, *Plato's Ghost*; Hacking, *Philosophy of Mathematics*, §5.A.

44. Galison, *Image and Logic*, 785; see Kevles, *Physicists*, 90, 168–69, 197–99.

45. Holton, "Fermi Group in Rome," 56; Reeves, "Einstein Politicized," 218–19.

46. If one draws the disciplinary boundaries to include more mathematical investigators, then this generalization breaks down. See Warwick, *Masters of Theory*; Garber, *Language of Physics*.

(say, 1875–1925). Historian Suman Seth has argued that this difficulty stems from the simple fact that physicists often disagreed with one another about these topics, at length, and in public. It seems that methodological and philosophical reflection was part of the developing job description, and a corresponding diversity of opinion flourished.[47] By the middle of the twentieth century, disciplinary identity had become more settled.[48] A useful point of departure is a report, *Physics: Survey and Outlook*, published by the US National Academy of Sciences in 1966 on "scientific manpower" in physics.[49] (Since World War II—and more intensely after the launch of the Soviet satellite Sputnik in 1957—funding for research in physics was seen as an American national security issue.[50]) The authors of the chapter on "Theoretical Physics" describe it as acquiring, over time, "a distinctive spirit."[51]

Before elucidating that "spirit," here is how the NAS report saw theoretical physics within physics. The table of contents of the report was a list of the sub-disciplines of physics, shown on the vertical axis of table 1.1. (Most of these sub-disciplines were newly formed. This list was very different from similar divisions made circa 1900.[52]) Each chapter was written by committee, normally a bane of readers, but here a useful feature, compared to idiosyncratic individual views. The authors of the chapter "Theoretical Physics" were established physicists at American universities, though some were immigrants and others were educated abroad: Chen N. Yang, Geoffrey F. Chew, Robert F. Christy, Marvin L. Goldberger, Robert G. Sachs, Leonard I. Schiff, John A. Wheeler, and Eugene P. Wigner. They were at the top of their field; for example, Yang and Wigner were recent Nobel laureates (in 1957 and 1963, respectively). Historian Peter Galison has usefully distinguished four divisions within twentieth-century physics, from instrumentation to mathematical physics, shown on the horizontal axis of table 1.1.[53] Obviously, theoretical physics as its own sub-field is not so divided, as Yang et al. recognized: the theoretical data tables "should *not* be additively combined with those of other panel reports, since most activities in theoretical physics are already included in the considerations by the other panels."[54] Instead, the authors of "Theoretical Physics" divided their speciality into categories similarly arrayed from

47. Seth. "Crisis and Construction." esp. 32; for programmatic statements by nineteenth-century German physicists, see Jungnickel and McCormmach. *The Second Physicist*. ch. 1; building from Jungnickel and McCormmach, some historians describe an overall Machian positivist view. E.g., Galison. *Image and Logic*. 785. In contrast, I would suggest that—to be brute with labels—ideological diversity was exemplified by neo-Kantians like Henri Poincaré; by Idealists like A. S. Eddington and James Jeans; and by realists like Lord Rayleigh and H. A. Lorentz.

48. On disciplinary identity, see Nye. *Chemical Philosophy to Theoretical Chemistry*. ch. 1; Galison. *Image and Logic*. ch. 9 pt. I; see also Carson. *Heisenberg in the Atomic Age*. ch. 3.

49. PSC. *Physics Survey*; this report is discussed in D. Kaiser, "Cold War Requisitions," 541; D. Kaiser "Whose Mass Is It?"; cf. Kevles. *Physicists*. chs. XXI–III.

50. Forman. "Behind Quantum Electronics"; Galison and Hevly. *Big Science*; Leslie. *Cold War and American Science*; Westfall. "Rethinking Big Science."

51. PSC. "Physics Survey." 165.

52. Staley, *Einstein's Generation*, pt. II.

53. Galison, "Theory Bound and Unbound"; Galison, *Image and Logic*, 800–2; cf. Hacking, "Disunities of Sciences."

54. PSC, "Physics Survey," 164; several later surveys, administered by the American Institute of Physics, did not include "theoretical physics" as a separate subdivision at all. Records of the American Institute of Physics, Education and Manpower Division, 1940–1973, AIP.

Table 1.1. Two schemes of physics. The horizontal division is primarily a set of *roles* that physicists undertake to create specific *products*: theoretical physicists produce theories (Galison 1995, 1997). The vertical division is the first item of each subfield of physics as listed by the US National Academy of Sciences in 1966. (A full listing was, e.g. "Solid-state and condensed matter.") This was primarily a division by *subject matter*. The two schemes can be interrelated as a matrix—with "theoretical" the odd-one-out (see text). Instead, theoretical physics could be divided into "levels." By population, a substantial plurality of modern physicists have been solid-state physicists. The NAS estimated that from 1957 to 1962, about 15 percent of US physics PhDs were granted in theoretical physicists; about half of those in elementary-particle physics.

	Instrumentation	Experimental	Theoretical	Mathematical
Astrophysics	-	-	-	-
Atomic and Molecular	-	-	-	-
Elementary-particle	-	-	-	-
Intermediate-energy	-	-	-	-
Nuclear	-	-	-	-
Plasma	-	-	-	-
Solid-state	-	-	-	-
Theoretical		Phenomenology	Laws	Principles

closest to a physical phenomena—say, a track in a cloud chamber—to most abstract. The order was: phenomenology, laws, principles.

Yang et al.'s chapter began with a "Description of Theoretical Physics":

> Theoretical physics deals with the concepts and the relations between those concepts that are useful for organizing, in the language of mathematics, an understanding of experimental information on physical phenomena.[55]

The authors repeatedly emphasized that theoretical physics was directed toward experiment. They acknowledged the "centuries of intimate contact" between theoretical physics and mathematics, but they not only distinguished sharply between them but also expressed anxiety about theoretical physics becoming too much like mathematics:

> The theoretical physicist, unlike the mathematician, is constrained by the realities of nature. It is vitally important for the future of physics that this experimental character be recognized at all levels in the education of physicists. In particular, theoretical physics students must be taught not to invent and manipulate formalism for its own sake, but rather to use formalism as a means for constructing theoretical interpretations of experimental observations. (165)

For Yang et al., mathematics was properly a *means* not an *end*.[56]

55. Subsequent citations will be given in text, PSC, "Physics Survey," 162.
56. This was boundary work, not a criticism of mathematics. Mathematically inclined physicists on the committee included Chew and Wigner; in 1967 Wheeler co-organized a conference to break down barriers of communication between physicists and mathematicians with Cécile DeWitt-Morette (ch. 5).

Returning to the introductory sentences, the report distinguished between the "levels" in table 1.1:

Activities in theoretical physics proceed on many different levels. Concepts applicable in the description of a few experiments are considered in phenomenological investigations. Concepts applicable to a general type of physical phenomena are considered in the formulation of laws of nature. Concepts with broad applicability are considered in the formulation of general principles of physics. (162)

Here the demarcation of levels was in terms of a spectrum of physical phenomena from most specific/least general to least specific/most general.

Rounding out their opening description, Yang et al. took a pluralistic approach to classic questions about scientific reasoning.

Activities in theoretical physics proceed in many different modes. In some, the emphasis is on inductive thinking in a search for concepts useful for describing known experimental information. In others, the emphasis is on deductive reasoning starting from accepted concepts to suggest new relevant experiments and to predict their results. In still others, the emphasis is on the construction of models that isolate from complex physical phenomena certain specific characteristics. (162)

Theoretical physicists used induction and deduction; they searched, accepted, and constructed; the immediate objects of these activities were concepts and models; their aim was experimental information, suggestions, and predictions and specific characteristics of physical phenomena. Yang et al.'s report to the NAS provides a good first-pass understanding of theoretical physics in the twentieth century, its place within physics, its intellectual activities, and its persistent anxieties.

1.2.2 Objects of Theory

Something is missing from Yang et al.'s description of "activities in theoretical physics": theories. Theories were neither built nor applied. This self-conception fits with recent historical and philosophical work that questions the role—or even existence—of scientific theories (particularly as described by analytic philosophers). For particular pedagogical or analytical purposes, it may be perfectly acceptable to refer to, say, Quantum Theory. But this is an idealization without reference to specific times, places, and people. Which QT is it? Circa 1919 or 1939? In Berlin or Cambridge? Expressed in texts by Max Planck or by Paul Dirac?[57] An idealized, homogeneous, contextless "Quantum Theory" is inadequate when discussing many difficult historical and philosophical questions, at least in my view. (This is the case because it is an empirical fact that expressions of QT in specific contexts do vary from one another in important ways.) Rich and contextualized theory-based studies—such as Sam Schweber's *QED and the Men Who Made It*—form

---

57. For an enlightening take on a classic question, see Badino and Navarro, *Research and Pedagogy*.

an important foundation for my analysis in this book. But there is more to the story.

David Kaiser's *Drawing Theories Apart* makes the negative case against homogeneous "theory," but also a positive case for developing alternative organizing principles. He demonstrates that attending to physicists *practices* and *tools*—even *theoretical* practices and *paper* tools—allows historians to tell stories and make arguments that do not respect the boundaries between one theory and another.[58] Kaiser's book is a history of diagrams in Quantum Electrodynamics, but also in meson (nuclear) theory, and abstract approaches to scattering problems, each of which could be considered a separate theory. Similarly, Andrew Pickering's *Constructing Quarks* ranges over multiple modes of experiment and theories.[59] Pickering's book is not a history of the *quark model* or of *quantum chromodynamics* or *quantum field theory*; it is focussed on the objects, *quarks*. *More than Nothing* ranges over non-relativistic quantum theory, "hole" theory, quantum electrodynamics, geometrodynamics, and general relativity; it is focused on an object, the vacuum.

What, then, is the vacuum? An object? A concept? In this book I take the vacuum to be an "object of theory." David Kaiser deployed the term to describe a field, ϕ, which was introduced in the 1960s in cosmology and in particle physics, *separately*, for the same purpose (to generate mass). Only in the mid-1970s did the "fields collide" and be recognized as the same object by practitioners of a new sub-discipline, particle astrophysics. Kaiser's object of theory has (at least) three distinguishing features. (1) It transcended boundaries between two theories (cosmology and particle theory). In fact, long past the decline of its original cosmological (Brans-Dicke) theory, the *object* ϕ remains central to particle cosmology today. Kaiser emphasizes that it remains central to the *practice* of particle cosmologists. (2) The object was shaped by physicists daily practices, including techniques of calculation. (3) Kaiser argues that the properties of ϕ were determined mostly by physicists' pedagogical inculcation. (In particular, for ϕ, pedagogy and institutional structure mattered more than either overarching theories or observational data.) Objects of theory are nodes of the intellectual, practical, and social arcs of theoretical physics. According to Kaiser, "we may use the objects of theory as tracers, mapping intertwined epistemic and social relations."[60] The history of the vacuum exhibits similar features: it was independent from individual theories, it was shaped by practices, and it was shaped (if not determined) by pedagogy and institutions. But there was more than this to the vacuum.

For the physicists in this book, the vacuum was an ambiguous object and it was a source of seemingly endless novelty. Its ambiguity came from its status as both an object of physical reality and as an invention of physicists. In a 1976 address, physicist Steven Weinberg expressed his sense of the physical *reality* of the mathematical models he *made*:

---

58. D. Kaiser, *Drawing Theories Apart*, esp. ch. 10; D. Kaiser, "Whose Mass Is It?"; he built on Galison, *Image and Logic*, ch. 9 pt. I; for a philosopher's view, see M. Morrison, "Where Have All the Theories Gone?"; a dissent is Smeenk, "Tools without Theories."

59. Pickering, *Constructing Quarks*.

60. D. Kaiser, "Whose Mass Is It?," 554; he built on earlier studies of "biographies of scientific objects" such as Daston, *Biographies of Scientific Objects*; Arabatzis, *Representing Electrons*.

We have all been working in what Husserl called the Galilean style; that is, we have all been making abstract mathematical models of the universe to which at least the physicists give a higher degree of reality than they accord the ordinary world of sensation.[61]

The vacuum was an object both in the sense of a goal, the object of their eyes, and in the sense of a material constituent of reality. Historians of science have struggled to develop what John Stachel describes as

a vocabulary for talking about human projects that does not imply that the outcome of such a project is *foreordained* by "objective reality," on the one hand; nor on the other, that "anything goes"—that there are no objective constraints on such projects.

He notes that Einstein's favorite word for advances in his own work was "invention [*Erfindung*]."[62] Stachel hopes that we can see the creative, human work of scientists as just that—creative and human—without relinquishing ourselves to two opposing views: naive realism and anti-realism. Naive realism holds that current scientific results are perfectly and necessarily true and so were *foreordained*; anti-realists hold that scientific results are unconstrained, as if literary fiction. The middle road is to represent some sense of contingency in scientific investigation.[63] In this book, physicists make creative leaps, which are then disciplined with mathematics, and buttressed or battered by experiment—or some other series of interactions between imagination, discipline, and the world.[64] In their report, Yang et al. wrote that

in the frontiers of theoretical physics the theorist is confronted with the formulation of new concepts, new mathematical schemes, rather than with solving precisely defined mathematical problems.[65]

This was surely the work of invention. Not only invention, but reinvention: the vacuum was shaped and reshaped over the course of the twentieth century.

I am left with five qualities for the vacuum as an object of theory: transcending theories and sub-disciplines; shaped by practices; shaped by institutions; ambiguity; and creativity. Perhaps the most compelling description of scientific practice that addresses these desiderata is Hans-Jörg Rheinberger's sustained analysis of the experimental life sciences. Rather than trace the history of (sub-)disciplines like cytology or embryology, Rheinberger organizes his analysis around "experimental systems," particularly test-tube systems for protein synthesis. Experimental systems are complexes of concepts, people, practices, institutions, and material that allow scientists to

61. Weinberg, "Forces of Nature," 28; cf. Hacking, "Language, Truth, and Reason," 162.

62. Stachel, "Scientific Discoveries," 140, 146; cf. "I shall try again and again to show that what is called a mathematical discovery had much better be called a mathematical invention." Wittgenstein, *Lectures on the Foundations of Mathematics*, 22.

63. This is of course a caricature, but some observers have defended these views, namely Steven Weinberg and Paul Feyerabend. They have since moderated their positions. Hacking, *Social Construction of What?*, Weinberg: 74–75; Feyerabend: 96–99; Contingency: ch. 3.

64. Pickering calls this a "mangle of practice." Pickering, *Mangle of Practice*.

65. PSC, "Physics Survey," 165.

pose and re-pose questions and to materialize answers, which suggest new questions, etc. At the heart of experimental systems are the objects of scientific research, which Rheinberger calls "epistemic things"—material entities such as chemical reactions. Epistemic things embody the ambiguity of human-made things and natural things, in part because they are objects of *ongoing* research, not clearly defined entities that are identified in retrospect. They are the object of epistemic activity but are (by construction) not completely known—scientists ask *new* questions and seek *new* answers. Completely known processes are the object not of science but of technology.[66]

Rheinberger's "epistemic things" provide a framework for extending Kaiser's "objects of theory." They overlap on the importance of disciplinary boundaries and institutions. (Epistemic things sit at the center of experimental systems; unfortunately, "theoretical system" has rather different connotations. If Rheinberger's experimental systems can be seen as a type of "research system," then objects of theory are found at the center of "theoretical research systems.") Rheinberger bites the bullet of ambiguity—he does not attempt to find a language that would untangle the human-made from the natural. This may be unsatisfying to some readers. But Rheinberger's choice pays dividends in his ability to position this ambiguity as an origin of scientific novelty and invention. Alongside the vacuum, objects of theory include the luminiferous aether and the electron. These examples immediately raise two questions, Are objects of theory real? and Are they the same as laboratory objects? These are good questions, but answers to them are not required to trace the history of physicists' engagement with these objects, including their claims about the nature of reality. I will defer my discussion of these questions to the Concluding chapter, after my central narrative. My central questions are: how and why did the vacuum become an object of theoretical research; how and why was it transformed and reinvented over and over; how was it shaped by institutional structures and physicists' philosophical commitments?

The answers to these questions often center on physicists' practice and tools. Where Rheinberger's epistemic things are manipulated with laboratory instruments, theorists manipulate their objects with "paper tools," as richly described by Ursula Klein.[67] These paper tools include formalisms, such as vacuum states in the hands of Furry, Heisenberg, and Weisskopf (chapter 3); ready-to-hand mathematical techniques, including perturbation analysis; and diagrams, including Hawking and Penrose's (figure 6.1). The ultracentrifuge transformed protein synthesis; perturbation analysis transformed the vacuum.

The goal of understanding physics through its practice is also underscored by physicists. Theory has been described as a tool, as a practice, and as an instrument. In 1933, Albert Einstein advised:

> If you wish to learn from the theoretical physicist anything about the methods which he [*sic*] uses, I would give you the following piece of advice: Don't listen to his words, examine his achievements.[68]

---

66. Rheinberger, *History of Epistemic Things*, esp. ch. 2; Rheinberger, "Cytoplasmic Particles"; Rheinberger, "Cultures of Experimentation."

67. U. Klein, *Experiments, Models, Paper Tools*; see also D. Kaiser, *Drawing Theories Apart*.

68. Einstein, "On the Method of Theoretical Physics," 163; Dongen, *Einstein's Unification*, §2.2.

Richard Feynman told his students at the California Institute of Technology that "You don't know anything until you have practiced."[69] In 1953, J. Robert Oppenheimer gave the BBC Reith Lectures. He said that the quantum theory

> has become not a subject of curiosity and an object of study but an instrument of the scientist to be taken for granted by him, to be used by him, to be taught to him as a mode of action, as we teach our children to spell and to add.[70]

A central thread of this book is the history of these "instruments," these paper tools, these uses, these lessons, these modes, and how they became taken for granted.

1.2.3 Narrative Overview

The central narrative of *More than Nothing* begins with chapter 2 at Cambridge University in the 1920s. It follows Paul Dirac's development of the quantum theory of the electron. Famously, Dirac proposed that empty space is actually filled with an infinite collection of electrons in strange, negative-energy states of motion. "Holes" or "bubbles" in this "sea" of electrons could represent positively charged particles. This is a well-studied story in the history of physics, but questions remain about the origin and reformulation of Dirac's radical reimagining of empty space. Chapter 2 shows in unprecedented detail how Dirac's "hole" theory emerged from his pedagogical training, his institutional context, and his research program. This program aimed for a *unified* quantum theory that described particles moving at relativistic velocities *and* that described collections of many particles (what would become "solid-state" theory). Simply put, in 1929 Dirac's filled vacuum was produced when he applied a many-particle framework to a relativistic problem.

Initially, Dirac's "hole" theory was a rearguard action. It was the most drastic in a series of attempts to justify Dirac's equation for relativistic electrons that he had introduced two years prior, in 1927. The equation was heralded as a heroic success, but its Achilles' heel was its negative-energy states. Dirac's vacuum sea filled the problematic states. It was a device to aid his research on his true object of theory, the electron. However, in the early 1930s physicists began to connect Dirac's "hole" theory to newly observed "positive electrons" or positrons. Dirac revisited his theory with a pair of young Jewish postdoctoral researchers—Rudolf Peierls and Victor Weisskopf—who found refuge from the rise of fascism in England. Under their pencils, the vacuum was a source of surprises. Dirac, Peierls, and Weisskopf discovered a new materiality in the vacuum—it became a polarizable medium—that might have novel physical effects. The vacuum took a central role as an object of investigation in its own right. (Rheinberger documents similar shifts—from device to object of research, and back again—in laboratory research.[71])

69. Series of Lectures on theory of fundamental processes 1959–1960, p. 141. Box 47, Folder 47.2, Papers of Richard Phillips Feynman, Archives, California Institute of Technology, hereafter Feynman Papers.
70. Oppenheimer, *Science and the Common Understanding*, 36–37.
71. Rheinberger, "Cytoplasmic Particles," 271–75.

Chapter 3 takes up the story of this material vacuum in the 1930s as it was represented and re-worked by three young postdoctoral researchers: Weisskopf (now working with Wolfgang Pauli), Hans Euler (working with Werner Heisenberg), and Wendell Furry (working with Oppenheimer). This chapter demonstrates the influence on the vacuum of high-minded philosophical commitments as well as practical techniques of calculation. When physicists applied the machinery of perturbation analysis and mathematical "operators" to Dirac's vacuum, they began to represent the creation and annihilation of particles on the page. Oppenheimer's legendary epistemological pessimism informed his and Furry's attempt use these operators to empty the vacuum of Dirac's sea. However, they were left with a vacuum of ineliminable "nascent" pairs of electrons and positrons. This operator machinery also made possible the concept of "virtual" particles. There is a suggestive link between the "virtual" language of Heisenberg's group at Leipzig and their simultaneous intensive discussion of neo-Kantian philosophy. Chapter 3 uses Weisskopf and Euler's correspondence to demonstrate how the new "electrodynamics of the vacuum" was related to experimental evidence. In one important case, I argue, the theorists attempted to leapfrog their arguments around evidence which contradicted theory.

Where did "virtual" particles and "nascent" pairs come from? Chapter 3 includes my argument that perturbation analysis (appendix A) led physicists to many of their conclusions about the vacuum. A typical question I am interested in is, Why did Heisenberg think that the "virtual possibility of pair creation" in empty space would cause deviations from Maxwell's equations? From where did this "virtual possibility" arise? The answer is that Heisenberg and his group at Leipzig applied a perturbation algorithm to the Hamiltonian function in Dirac's "hole" theory.

Weisskopf emigrated to the United States in 1937, became a citizen, and was a leader of the Manhattan Project at Los Alamos, New Mexico. After the war, he taught at the Massachusetts Institute of Technology, where he returned to fluctuations in quantum electrodynamics (QED), in collaboration with Julian Schwinger (down the road at Harvard). Chapter 4 begins by laying out Weisskopf's claim that quantum theory *requires* that there are electromagnetic field fluctuations in empty space—as an *independent* postulate. (This is distinct from, for example, Hawking's description in section 1.1, of innate fluctuations in empty space, which were rooted in the uncertainty principle.) Schwinger's view was close to Weisskopf, but not identical. I argue that for Schwinger, fluctuations were unavoidable consequences of measurements, not independent phenomena. While Weisskopf was more influential in his physical picture, Schwinger was far more influential in his mathematical treatment of quantum electrodynamics. The vacuum was at the center of Schwinger's paradigmatic techniques. The fundamental quantities of Schwinger's quantum electrodynamics were vacuum states and "vacuum expectation values" (his coinage). These became the central formal expressions of QED—for example, physicists used Feynman diagrams to evaluate vacuum expectation values.

Sam Schweber has argued that the postwar successes of QED were *conservative* successes: they did not reinvent the wheel assembled in the 1930s. I historicize this view by considering its expression by Oppenheimer in 1948. Chapter 4 also follows the story of this conservative strategy to its failure. Richard Feynman's attempt at quantum gravity in the late 1950s and early 1960s fit the mold of his earlier successes;

rather than following Einstein's geometrical path, Feynman followed his QED path integrals. Earlier, he had struggled to fit vacuum loops into his QED, but eventually succeeded. Feynman's attempt at quantum gravity came to center on efforts to "cut" open diagrammatic loops. But like Einstein before him, Feynman found the limits of his earlier strategy in the territory of quantum gravity. He remained tangled in knots of vacuum loops.

Chapter 5 follows John Archibald Wheeler's attempt, at Princeton University, to unify quantum theory with a general relativity of empty space, what he called "geometrodynamics." From the 1950s through the 1970s, Wheeler saw in GR's fabric of empty space a "magic" building material that could explain the disorderly diversity of new particles. This chapter follows Wheeler's work through abundant documentation of his research notebooks. In the course of two decades of work that aimed at a revolutionary new theory, Wheeler and his students made major contributions to the "renaissance" of GR that began in the mid-1950s. Wheeler proposed new models of matter, and coined "wormhole" and "black hole" to describe new geometries and singularities of spacetime. Eventually, Wheeler proposed that spacetime was, at the smallest scales, a *quantum foam* of bubbles in a constant froth. He followed Weisskopf's view of fluctuations as inherent in isolated matter. And his students looked for a formalism of quantum gravity and astrophysical grounds to test it. Wheeler exemplifies a scientist driven by what historian Gerald Holton calls "themata"—commitments about science and the natural world that are not derivable from axioms or evidence.[72] Wheeler was committed to simplicity and unity, and he held those commitments metaphysically, in which the universe was composed of a single single substance—empty space.

Chapter 6 takes up an example of the metaphysics of spacetime driven by paper tools: Penrose diagrams. At Cambridge, Roger Penrose's diagrams allowed him to represent an *entire* infinite universe—or even pairs of infinite universes—in a single finite diagram. "Paper tools" are not only conceptual, they are material; Penrose's new tool allowed theorists to do something that had been previously physically impossible.[73] In gaining this new representational power, Penrose paid a price in distortions: changes in topology and the interpretations of lengths on the page. Chapter 6 follows the development of Penrose's visual analysis from application in recreational mathematics, the psychology of perception, and Maurits C. Escher's art. To apply his techniques to black holes and "big bang" cosmology, in 1969 Penrose began a collaboration with another research student at Cambridge, Stephen Hawking. In this work, rigorous mathematics, visual invention, and interdisciplinary research reshaped our picture of the vacuum.

This book's central narrative concludes with a chapter on the particle physicist Sidney Coleman. He trained at Caltech with leaders of the application of symmetry to physics, Murray Gell-Mann and Sheldon Glashow. Coleman was a master of pedagogy, of explication, of explanation, and of technique in theory. Coleman's mastery of symmetry was in creative tension with another of his lifelong passions: narrative.

---

72. Holton, "Role of Themata."
73. In Pickering's terms, this was an overcoming of both *conceptual* and *material* resistances. Pickering, *Mangle of Practice*, 29–30.

Narrative occupied both his professional life (e.g., in narrating physical processes) and in his personal life (e.g., in 1956 he cofounded a publisher of science fiction—Advent). Chapter 7 shows how symmetry and narrative intersected in two episodes of Coleman's physics: his interpretation and research on Spontaneous Symmetry Breaking (ca. 1973) and his proposal that we might be living in a "false vacuum"—and that a bubble of true vacuum may be approaching us at almost the speed of light, with apocalyptic consequences (1977–1980). Coleman's work illuminates the role(s) of the vacuum in contemporary particle physics. The false vacuum was part of the developing new sub-discipline of particle astrophysics.

The narrative ends just before two major developments in theoretical physics. Coleman was an early champion of Alan Guth, who developed the false vacuum as part of his path-setting "inflationary universe" cosmology in 1981. Some observers see Coleman and Guth's work as marking a fundamental shift in our understanding of the laws of physics: that they change over cosmological time.[74] And in 1984, the "superstring revolution" ushered string theory in the mainstream of theoretical physics, bringing with it changes in standards of demonstration, evidence, and physical relevance. I sketch some modern developments in a brief Epilogue.

1.2.4 Paths Not Taken

There are several lines of thought about the vacuum that are not taken up in this book. Despite the sometimes wild nature of vacuum theory, this book is a study of the *mainstream*. It is about physicists who were in the prime of their careers, who were doing the work for which they are best remembered, who published in prestigious venues, and who held secure institutional positions. (A significant exception to the general institutional security were Jewish physicists, particularly in the 1930s.[75]) As a study of the mainstream, this book does not detail the work of "dissidents" or "hippies."[76] Astronomy and cosmology are not my central focus. Readers interested in cosmology should consult Helge Kragh's *Cosmology and Controversy* and *Conceptions of Cosmos*; in a recent volume with James Overduin he discusses the history of dark energy.[77] For histories of quantum field theory that hew closer to infinities and the renormalization schemes that eventually tamed them, Sam Schweber's *QED and the Men Who Made It* is the best source, which can be supplemented by other literature.[78] In topics covered, the book most similar to this one is philosopher Tian Yu Cao's *Conceptual Developments of 20th Century Field Theories*, which also discusses both quantum field theory and general relativity.[79]

---

74. Schweber, "Metaphysics of Science," 184; cf. Wise, "Forman Reformed, Again"; for an earlier example, see Kragh, "Cosmo-Physics."

75. Dobkowski, *Tarnished Dream*; Ash and Søllner, *Forced Migration*.

76. D. Kaiser, *Hippies Saved Physics*; Freire, *Quantum Dissidents*.

77. Kragh, *Cosmology and Controversy*; Kragh, *Conceptions of Cosmos*; Kragh and Overduin, *Weight of the Vacuum*.

78. Schweber, *QED*; Schweber, "Hacking the Quantum Revolution"; Darrigol, "Quantized Matter Waves"; Brown, *Renormalization*; A.I. Miller, *Early Quantum Electrodynamics*.

79. Cao, *Conceptual Developments*.

Spontaneous symmetry breaking (SSB) does not have its own chapter in this book, but it is well covered by the historical literature. As Laurie Brown and Tian Yu Cao write: "Essential to this [SSB] program was the conceptual shift from vacuum as void to vacuum as plenum."[80] My discussion in chapter 2 of Dirac's solid-state *and* relativistic theory sets up Yoichiro Nambu's SSB work in the 1950s. SSB was integrated into the Standard Model of particle physics in the 1960s and 1970s; the most famous example of SSB is the so-called Englert-Brout-Higgs-Guralnik-Hagen-Kibble mechanism.[81] My discussion in chapter 7 of Coleman's interpretation of SSB catches the topic, as it were, on the other side of its establishment. Two edited collections map out the history of particle physics from the 1950s to 1979 and contain recollections by physicists, including Nambu.[82] In the context of particle physics, another fundamental shift in physicists concept of spacetime was begun by Chien-Shiung Wu when in 1957 her laboratory at Columbia demonstrated that parity, or mirror symmetry, was broken in natural phenomena.[83]

The most attractive path not taken in this book leads to a chapter about Albert Einstein: any history of quantum theory and relativity is a history of his work and its development. The lack of a chapter on Einstein reflects both the historical trajectory of GR and the place of this book in the history of physics. Einstein is the subject of a vast literature. Since the 1980s, historians and philosophers of science have established a solid understanding of how Einstein moved from the special to the general theory of relativity in 1915.[84] Compared to this literature, little scholarly attention has been paid to GR research by others and in more recent decades.[85] *More than Nothing*'s chapters on Wheeler and Penrose are among the only discussions of their work by historians. There is also the question of whether Einstein's work was (always) considered within the mainstream of theoretical physics, and his view toward the active picture of empty space on which this book is centered. The final section of this introduction picks up the narrative of Einstein's views on the vacuum.

1.3 Prologue: The Aether, Gone and Back Again

Famously, aethers were a centerpiece of nineteenth-century studies of electricity, magnetism, and optics.[86] Proponents of aether theories argued that subtle fluids or particles provided the medium through which light shone and by which electric and

---

80. Brown and Cao, "Spontaneous Breakdown of Symmetry," 213; cf. D. Kaiser, *Drawing Theories Apart*, 93 fn90, 125–49, 238 fn58, 282–96

81. As Kibble maximally describes it, Kibble, "EBHGHK Mechanism"; Pickering, *Constructing Quarks*, ch. 6; Brown and Cao, "Spontaneous Breakdown of Symmetry"; Brown et al., "Rise of the Standard Model."

82. Brown, Dresden, and Hoddeson, *Pions to Quarks*; Hoddeson et al., *Rise of the Standard Model*, ch. 28, panel discussion; for a compact overview, see Schweber, "Quantum Field Theory."

83. See Jammer, *Concepts of Space*, 203; Benczer-Koller, "Chien-Shiung Wu"; Zhu, "Chien-Shiung Wu"; Crystal, "Quantum Times," ch. 2.

84. See the comprehensive four-volume collection: Janssen et al., *Genesis of General Relativity*.

85. For exceptions, see Blum, Lalli, and Renn, "Reinvention of General Relativity."

86. This and the next paragraph draw on: Whittaker, *Theories of Aether and Electricity*, 1953; Hunt, "Whittaker. A History of the Theories of Aether"; Cantor and Hodge, *Conceptions of Ether*; Hunt, *Maxwellians*; Buchwald, *Creation of Scientific Effects*; Darrigol, *Electrodynamics*; Navarro, *Ether and Modernity*.

magnetic forces acted. In Britain, James Clerk Maxwell (1831–1879) and his followers argued that optical, electrical, and magnetic phenomena could be described with a single *universal* aether. Maxwell's Continental European rivals rejected his aether and accepted the existence of true vacua. In Maxwell's laboratory, glass Geissler tubes could be evacuated of all "gross matter," but because of the aether, in Maxwell's eyes they were only one of the "so-called vacua" of electrical experiments.[87] Maxwell's unified aether theory was spectacularly confirmed by Heinrich Hertz's creation of electromagnetic waves in 1887. By the end of the nineteenth century, an array of aether theories spanned the discipline of physics.

The most celebrated aether theory was developed by Dutch physicist Hendrik Antoon Lorentz (1853–1928).[88] In 1865, Maxwell christened his "Dynamical Theory of the Electromagnetic Field" as *dynamical* because it treated aetherial matter in motion. But Maxwell's dynamic aether faced problems when physicists tried to use this picture to understand the interactions of light and matter. The speed of light changes depending on its medium of propagation—for example, air, or glass, or water. But it did not seem to change when a medium was moving in relation to an experiment, for example water moving with or against a beam of light. If the aether was a material substance, its motions were expected to affect the behaviour of light. Lorentz tried to address these problems with a stationary aether: it had zero velocity at every point in space. In his work on this theory, Lorentz supposed that matter itself would shrink, depending on its velocity relative to the aether, according to a formula we now call the FitzGerald-Lorentz contraction (co-developed by the Irish Maxwellian George Francis FitzGerald). The aether provided a material substrate from the propagation of electrodynamic waves, and it provided a fixed frame of reference for motion.

In 1905, Albert Einstein (1879–1955) reinterpreted "The Electrodynamics of Moving Bodies" in a way that made a true vacuum possible—a space with no aether.[89] This was a significant shift. Einstein argued that the phenomena of electrodynamics are the same, regardless of whether these results are observed from a frame of reference at rest with the aether or in constant linear motion. His example was the motion of a coil of wire and a magnet. If the coil was stationary and the magnet moved, a current flowed through the wire. If the wire was stationary and the coil moved, a current flowed through the wire. It made no measurable difference if the coil or the magnet was at rest with respect to the aether. This was an extension of the principle of relativity in mechanics: the description of a mechanical system aboard a ship is the same regardless of whether the ship is resting at port or under sail at a steady clip. Einstein extended this principle of relativity to electrodynamics.[90] This eliminated the need for the fixed frame of reference provided by Lorentz's immobile aether. At first, Einstein claimed that if the aether had no observable consequences, it was (merely)

87. Maxwell, "Dynamical Theory of the Electromagnetic Field," 460; on vacuum experiments, see Buchwald, "Hertz Was Right."

88. This and the next paragraph draw on: Hirosige, "Lorentz' Theory of Electrons"; Hirosige, "Ether Problem"; Hunt, *Maxwellians*, ch. 8; Stachel, *Einstein from "B" to "Z"*, pt. IV; Staley, *Einstein's Generation*, ch. 8.

89. Einstein, "Electrodynamics of Moving Bodies."

90. Stachel, "Introduction," 17–18.

unnecessary to physics. By 1910, his argument sharpened into a stronger empiricism: without observable consequences, physicists should deny the aether's existence. In order to reconcile both the principle of relativity and the empirical correctness of Lorentz's electromagnetic theory, Einstein argued that "the first step we must take is to *give up the ether*."[91] Unlike Maxwell, Einstein could look at a Geissler tube and see a true vacuum.

The aether was by no means vanquished by special relativity.[92] Some, particularly in Britain, maintained an aetherial physics—and as evidence for relativity grew throughout the 1920s, they adapted their aether rather than admit the existence of truly empty space.[93] Beginning with Emil Cohn around 1900, others attributed electromagnetic and optical characteristics directly to a newly independent electromagnetic field.[94]

Einstein's view of the aether continued to shift in the period surrounding his invention of *General* Relativity, in 1915.[95] He deeply admired Lorentz, but the senior physicist never embraced a relativistic viewpoint. In 1920, Einstein traveled from his home in Berlin, Germany, to Holland to take up an Extraordinary Professorship at Lorentz's University of Leiden. On May 5, he gave his inaugural lecture "Ether and the Theory of Relativity." There, Einstein returned to his original view of his 1905 work: that abandoning the aether was only *an option*. In fact, it was an option he would no longer take.[96]

Lorentz had recently reaffirmed his commitment to his aether theory by publishing a second edition of his *Theory of Electrons* in 1916.[97] In Leiden, Einstein might have had a triumphal attitude toward Lorentz's theory. He was fresh from the dramatic results of Arthur Eddington's 1919 eclipse expedition that confirmed his prediction of the sun bending starlight. But he was so cautious about harsh feelings from the just-ended world war that he held back from writing to Eddington, the scion of British academia. (For news, Einstein instead wrote to his friend, the Dutch physicist Paul Ehrenfest).[98] The eclipse occurred on May 29, 1919; the Treaty of Versailles was signed June 28, 1919. Both Einstein and Eddington were pacifists who took part in the ideology of scientific internationalism, and Einstein would have been doubly glad to speak at Lorentz's university as a scientist from Germany coming to neutral Holland

91. Einstein, "Relativity and Its Consequences," quote from the translation supplement: 124. See Canales, *Physicist and the Philosopher*.
92. Classic studies are in: Glick, *Comparative Reception of Relativity*; for a wider view including experimental details, see Kragh, *Quantum Generations*, chs. 7–8; Staley, *Einstein's Generation*.
93. Goldberg, "In Defense of Ether"; Warwick, *Masters of Theory*, chs. 7–9.
94. Previously, the "electromagnetic field" was a subsidiary entity, defined as a region near an electric or magnetic body filled with aether, and later redefined as the state of motion of the aether. Darrigol, *Electrodynamics*, 260–61, §9.4.1.
95. See Renn, "Third Way"; and others in this collection.
96. Einstein, "Ether and the Theory of Relativity"; by 1923, this lecture had been translated into French, English, and Russian, see *Barchas Collection Catalogue*.
97. Lorentz, *Theory of Electrons*.
98. There has been significant controversy and misunderstanding of Eddington's work. On the light bending calculation, see Pais, *Subtle Is the Lord*, 198–200, §§12b, 14c, 16b; on the eclipse, see Stanley, "Eddington as Quaker Adventurer"; the controversies are addressed in the authoritative analysis by Kennefick, "Not Only Because of Theory," Einstein to Ehrenfest letter on 208.

in an exchange with European scientists.[99] In this context, I read Einstein's lecture as conciliatory.[100]

At Leiden, Einstein argued that General Relativity *required* an aether.[101] This aether was, like Lorentz's, not mechanical matter. But where Lorentz's aether was stationary and rigid, Einstein's was dynamic. A picture of spacetime in General Relativity requires three mathematical objects: a set of spacetime points; a topographic metric, $g_{\mu\nu}$, that describes how measuring rods and clocks behave at each point; and a stress-energy tensor, $T_{\mu\nu}$, that describes the locations of matter and energy.[102] These objects were related by Einstein's Equations.[103] In John Archibald Wheeler's evocative terms, in Einstein's Equations the curvature ($g_{\mu\nu}$) tells matter ($T_{\mu\nu}$) how to move, and the matter tells spacetime how to curve (see chapter 5). In his 1920 lecture, Einstein put the aether in the center of his theory: the metric described the aether—it was space itself. For Einstein

> the fact that empty space in its physical relation is neither homogeneous nor isotropic [like Lorentz's aether], compel[s] us to describe its state by ten functions (the gravitation potentials $g_{\mu\nu}$) has, I think, finally disposed of the view that space is physically empty. But therewith the conception of the ether has again acquired an intelligible content.[104]

The metric, $g_{\mu\nu}$, encoded spacetime's function as a container and reference-structure—it described how meter sticks and clocks marked events. But it was also a physical *material* that, I think, Einstein hoped could fill the psychological or metaphysical role of the aether in the thought of his colleagues.[105] Many held that without an underlying mechanical medium, electrodynamics was *incomprehensible*.[106] If "the ether has again acquired an intelligible content," it could provide the desired *mechanical comprehensibility* of many of Einstein's peers, including Lorentz. Similarly, Eddington made a place for an aether within relativity:

99. Goenner and Castagnetti, "Einstein as Pacifist"; the classic reference on scientific internationalism is Forman, "Scientific Internationalism and the Weimar Physicists."

100. See also his scientific correspondence with Lorentz, in which he uses aether-language (e.g., "*der Zustand des Aethers*") E. to L., 15.11.1919, Einstein, *CPAE 9*, doc. 165; others judge Einstein's relationship with Lorentz to be unimportant: Janssen, "'No Success like Failure...'"

101. Einstein's relationship with Lorentz was not the only factor in this change. Crucially, from 1914 to 1918 Einstein shifted his view of spacetime such that it attained a role more on-a-par with matter. The classic work on this is Holton, "Mach, Einstein"; and the fruits of recent scholarship are synthesized in Janssen, "'No Success like Failure...'"

102. Einstein's own non-technical introduction to the theory is still the best. Einstein, *Relativity*.

103. They are: $R_{\mu\nu} - \frac{1}{2}g_{\mu\nu}R = \kappa T_{\mu\nu}$. $R_{\mu\nu}$ is the Ricci tensor, and R is the Ricci scalar; they are constructed out of the metric, $g_{\mu\nu}$. Each symbol with Greek indices can be thought of as a 4×4 matrix with rows and columns arrayed according to the four dimensions of spacetime.

104. Einstein, "Ether and the Theory of Relativity," 176–77.

105. For example, Tetu Hirosige located a distinguishing feature of Einstein compared to his colleagues around 1905: he appreciated Ernst Mach's critique of the mechanistic worldview, which required that light-as-vibrations must have a mechanical substrate. Hirosige, "Ether Problem"; insofar as Einstein moved away from Mach, as Holton and Janssen discuss, his 1920 aether can be seen as a rapprochement with mechanism

106. Comprehensibility was set as an organizing principle for physics in 1847 by Hermann von Helmholtz: Darrigol, *Electrodynamics*, §6.3.1, 311; Goldberg, "In Defense of Ether," 106; Goldberg, "New Wine in Old Bottles," 10; on Cunningham: Warwick, *Masters of Theory*, 426.

What we have here called the [four-dimensional] *world* might perhaps have been legitimately called the *aether*; at least it is the universal substratum of things which the relativity theory gives us in place of the aether.[107]

Those of Eddington's readers who might insist on the existence of an aether could, perhaps, be satisfied with this nomenclature. In his conclusion at Leiden, Einstein expressed uncertainty about how the aether of General Relativity would become integrated into physics. However, about his own theory he was clear: "According to the general theory of relativity space without ether is unthinkable."[108] It was a complete about-face.

The astronomer Karl Schwarzschild (1873–1916) worked out the first exact solution to Einstein's Equations for the curvature of spacetime, only one year after Einstein presented them to the Königlich Preußische Akademie der Wissenschaften in Berlin in 1915.[109] Schwarzschild found a metric, $g_{\mu\nu}$, that described the curvature of an empty spacetime surrounding a sphere of matter—like a lone star. A visualizable analog is a ball weighing down the center of a rubber sheet. A marble rolling around the ball on the sheet could not travel in a straight line, and were it not for friction on the sheet, it would find a stable orbit. This is how General Relativity transformed gravitation from a force (like electric attraction) into spacetime curvature. As long as Newton's first law—the law of inertia—is maintained, planets orbiting the Sun do not *need* a force to curve their motion.[110] Rather, they travel in the straightest path available in a curved environment. The materiality of empty space manifests in its power to change the motion of regular bodies.[111] This curvature also changed the motion of starlight, as in Eddington's eclipse observations, shown in schematic in figure 1.2.[112]

The word "aether" was taken up by other "relativists." The mathematician Hermann Weyl referred to both historical and contemporary aethers in his influential 1918 *Space-Time-Matter* (which went through five German editions in the subsequent five years).

We shall use the phrase "state of the world-æther" as synonymous with the word "metrical structure," [$g_{\mu\nu}$] in order to call attention to the character of reality appertaining to metrical structure;

though this came with a caveat,

107. Eddington, *Space, Time and Gravitation*, 187.
108. Einstein, "Ether and the Theory of Relativity," 181, English supplement.
109. Einstein's path to GR has been the subject of intense scholarship. Representative works are, Pais, *Subtle Is the Lord*, section IV; Janssen et al., *Genesis of General Relativity*; the papers and an excellent editorial introduction are available in Einstein, *CPAE 6*.
110. Inertia in relativity and its history have been closely debated. The classic text is Holton, "Mach, Einstein"; for recent commentary, see Barbour, "Einstein and Mach's Principle"; Renn, "Third Way."
111. Lehmkuhl, "Einstein Did Not Believe," esp. §2.
112. A freehand version of this diagram can be found in Folder: J. A. Wheeler notes for Spring 68, Charles W. Misner Papers, UMD.

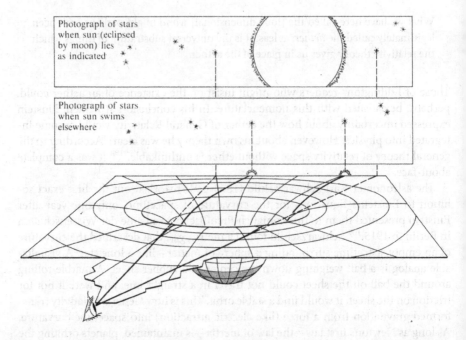

Figure 1.2. J. A. Wheeler's interpretation of Eddington's eclipse photographs in terms of the banding of spacetime by the Earth's mass. When the sun is behind the Earth, the viewer's line of sight falls across the Earth's gravitational well, and is distorted. Top caption: "Photograph of stars when sun (eclipsed by moon) lies as indicated"; middle caption: "Photograph of stars when sun swims elsewhere." Republished with permission of Princeton University Press, from Misner, Charles W., Kip S. Thorne, and John Archibald Wheeler, *Gravitation* (Princeton, NJ: Princeton University Press, 2017); permission conveyed through Copyright Clearance Center, Inc.

but we must beware of letting this expression tempt us to form misleading pictures.[113]

Weyl fully accepted Einstein's critique of Lorentz's physical, stationary, aether. He made aetherial arguments *within* relativistic physics. Similarly, in the late 1920s, the British mathematician Edmund Taylor Whittaker deployed a metric-as-aether that was similar to Einstein's 1920 picture. Whittaker used his aether, for both rhetorical and technical purposes, in his studies of relativistic astronomy:

> It is true that we do not speak much of the æther nowadays, and certainly do not regard it as a quasi-material medium filling all space; but when we endow space itself [...] with properties such as curvature, we are making it play the part of an æther.[114]

---

113. Weyl, *Space-Time-Matter*, 284; Ryckman, *Reign of Relativity*, §6.4.2.3.
114. Whittaker, "Influence of Gravitation," 137; Eisenstaedt, "Trajectoires et impasses," 305; Wright, "Paul Dirac's Æther."

This allowed him to celebrate geniuses of nineteenth-century British aether theorists, such as Michael Faraday. These examples show that the aether was invoked as part of arguments for the materiality or substantiality of curved spacetime.[115]

The history of the aether was political; it was particularly shaped by anti-Semitism in the Union of Soviet Socialist Republics (USSR) and Germany. Aether-talk was used in anti-Semitic denunciations of Einstein, as early as 1918.[116] His changed view on the aether in 1920 was not enough to deflect this line of argument. In the 1930s, Soviet philosophers and (some) physicists identified as "Mechanists." They attacked Einstein and modern physics. To these ideologues, Soviet dialectical materialism was incompatible with Einsteinian physics, which was alleged to be idealist, overly mathematical, bourgeois, and Jewish. The Mechanists proposed variations on a return to nineteenth-century ether theories. Internationally recognized Soviet physicists—including Yakov Frenkel, Igor Tamm, and Vladimir Fock—defended against these attacks, by turn rebutting and ridiculing the resurrected aether theories.[117] In Germany, while Philipp Lenard's aetherial *Deutsche Physik* gained prominence within National Socialism, the internationally renowned Arnold Sommerfeld illustrated a growing disdain for the aether. In 1935—Einstein by then a refugee in Princeton, NJ—Sommerfeld reviewed a volume of Grimsehl's *Textbook of Physics* (6th edition, the 27th edition now in print) on "*Materie und Äether.*" It was newly re-written by Rudolf Tomaschek, a student of Lenard's. Tomaschek's volume systematically inserted aether-language into modern physics, and downplayed relativity and Einstein.[118] Sommerfeld objected to sentences such as "The 'granular' structure of the aether, which manifests itself in the occurrence of the magnitude *h*…" (Planck's constant). He commented:

> Such an approach seems dangerous to us (despite the added attenuating note). Woe to you when a layman builds this granular structure into a new aether theory and gets it sent for review!

Sommerfeld accepted "the frequent use of the word 'aether' for historical reasons," but in his view—and his view of the community of physicists—the aether had no place in the creation of new physics.[119] When aether theories flourished under totalitarian regimes, physicists who resisted their ideological pressures took positions against the aether.

---

115. Absent an aether, others ascribed physical properties to empty space directly. Percy Bridgman argued that if light traveled through empty space with a characteristic velocity, then space must have physical properties such as electric and magnetic polarizability. Bridgman, *Dimensional Analysis*, 12; Macagno, "Historico-critical Review of Dimensional Analysis"; Noakes, "Making Space"; Whitworth, "Transformations of Knowledge"; Staley, "Ether and Aesthetics."
116. Goenner, "Reaction to Relativity Theory."
117. Josephson, *Physics and Politics*, chs. 7 and 8; cf. Graham, *Science, Philosophy, and Human Behavior*, ch. 11.
118. Hentschel, *Physics and National Socialism*, lxxiii fn255, 340 fn11, XLIX, docs. 93 and 110; Eckert, *Arnold Sommerfeld*, 363; Schirrmacher, "Lenard's Ether"; Hoffmann and Walker, *German Physical Society*.
119. Sommerfeld and Weisskopf, "Besprechungen," 71; this underscores the rhetorical force of Dirac's 1951 aether proposal. See Wright, "Paul Dirac's Æther."

In Princeton, Einstein continued his attempt to unify his relativistic theory with electromagnetism and atomic theory.[120] He gave empty space an increasingly prominent role. Schwarzschild's solution to the equations of general relativity was a *vacuum* solution. It described the curvature of the empty space *surrounding* the Sun. A unified theory of gravitation and electromagnetism was expected to account for the Sun's matter (atoms) and also its light (electromagnetic radiation). Physicists considered space filled with light to remain empty. This was not far from ordinary use of language. Consider the space between Earth and our nearest neighboring star, *Proxima Centauri*. This was usually considered empty, the vacuum of space—neglecting the odd speck of dust. Yet it is constantly full of light. Not just the light by which we see *Proxima Centauri* itself, but also transverse beams of light from every other neighboring star, including our Sun. This is the sense in which spacetime with electromagnetic fields can be "empty." In contrast, consider a region of spacetime containing electromagnetic fields *and* a star, or a lightbulb: the *sources* of electromagnetism. (In quantum theory, Tomonaga Sin-Itiro called equations without sources "vacuum equations"; in his evocative phrase they were "the field equations for the fields when they are left alone without interacting with each other."[121]) This sense of "vacuum equations" or zero-rest-mass fields as a vacuum is an important thread throughout the chapters of this book.

During the 1930s, Einstein himself was intellectually isolated. In 1933, Robert Oppenheimer wrote to his postdoctoral fellows Wendell Furry and John F. Carlson (see chapter 3) asking for updates about their calculations. Oppenheimer included a postscript about Einstein's unified field theory: "Einstein's 1/2 vector paper is out. It is pretty thin soup."[122] In addition to working toward a unified theory, Einstein also returned to the foundations of relativity. From the Institute for Advanced Study, Einstein and his assistants Leopold Infeld and Banesh Hoffmann argued in 1938 that

> the only equations of gravitation which follow without ambiguity from the fundamental assumptions of the general theory of relativity are the equations for empty space, and it is important to know whether they *alone* are capable of determining the motion of bodies.[123]

Empty space—whether as aether, as equations, or as physical medium—was again and again made the center of relativistic research. Here Einstein, Infeld, and Hoffmann explained that they insisted on understanding empty space because the path from the principles of relativity to equations of motion was certain, unambiguous, *only* for vacuum equations.[124] One might say that they were certain, at least, about nothing.

120. Dongen, *Einstein's Unification*.

121. See ch. 4, Tomonaga, "Relativistically Invariant Formulation. I," 30.

122. Folder: J. Robert Oppenheimer, Box 2, Furry Papers. Einstein visited Caltech, in 1933 and the American relativist R. C. Tolman was reportedly critical of this work. Dongen, *Einstein's Unification*, isolation: 186-89, semi-vectors: §5.3, Tolman: 125 fn75.

123. Einstein, Infeld, and Hoffmann, "Gravitational Equations I," 65.

124. Recently, Einstein's fuller motivations have been unearthed in Lehmkuhl, "General Relativity as a Hybrid Theory."

2

Paul Dirac's Seas and Bubbles

2.1 Introduction

In the summer of 1933, Paul Adrien Maurice Dirac (1902–1984) faced a dilemma. He had agreed to speak at a conference in Leningrad in September, and had just been given a last-minute invitation to speak at the Solvay conference in Brussels in October.[1] He was to speak on his electron theory. But this—his most-recognized contribution to physics—had suffered a steady decline since he introduced it at the turn of the new year of 1927/28. Then, he succeeded in integrating quantum mechanics with Einstein's special relativity into "The Quantum Theory of the Electron."[2] Astoundingly, without assumptions about spinning spheres or angular momentum, Dirac's theory produced electrons with spin—an intrinsic angular momentum that had been proposed to explain experimental puzzles. Dirac's work transformed spin from an ad hoc hypothesis to a necessary consequence of his combination of quantum theory and relativity.[3] In February 1928, Wolfgang Pauli (1900–1958) wrote to a friend and colleague: "Well Dirac's work has been published. It's wonderful how everything is right!"[4] But by 1932, those same colleagues were satirizing Dirac's "donkey-electrons"—alongside many other targets—in a mock-performance of Goethe's *Faust* at Niels Bohr's Institute in Copenhagen (figure 2.2).[5] Pauli was cast as Mephistopheles. By then, Dirac had almost abandoned his "hole" theory.

In 1930, as an effort to save his theory, Dirac had integrated protons into his electron picture—but he described protons as "holes" in an infinite distribution of electrons in strange, "negative-energy" states. Electrons in these negative-energy states filled empty space. Immediately, in personal correspondence and in print, Dirac's colleagues argued that his "holes" could not be protons. And so, in 1931, Dirac made a qualified prediction of the existence of a positive particle with the same mass as an electron, the "anti-electron." This was a prediction, of a sort; but it was not directly connected to a specific possible measurement.[6] At that time, Dirac was not inclined to develop his work further. He recalled in an interview: "I thought it was

---

1. Dirac, "Teoriya pozitrona"; Dirac, "Théorie."
2. Dirac, "The Quantum Theory of the Electron"; Dirac, "The Quantum Theory of the Electron. Part II."
3. Robotti, "Quantum Numbers"; Kragh, "Dirac's Relativistic Theory"; Moyer, "Origins of Dirac's Electron."
4. Pauli to Ralph Kronig, quoted in Kragh, "Dirac's Relativistic Theory," 61.
5. Kragh, *Dirac*, 106–17; a partial translation is included in Gamow, "Blegdamsvej."
6. Roqué, "Manufacture of the Positron," 76–77.

More than Nothing. Aaron Sidney Wright, Oxford University Press. © Oxford University Press 2024.
DOI: 10.1093/oso/9780190062804.003.0002

Figure 2.1. Photograph of Paul Dirac, Wolfgang Pauli, and Rudolf Peierls, taken in Birmingham, UK, in 1949. Photographer unknown. Copyright CERN, reused with permission from the Wolfgang Pauli Archives, photograph 071.

rather sick, yes. [...] I wouldn't say it had to be given up altogether, but it was rather sick."[7]

Then in 1932 at Caltech, Carl Anderson announced the observation of a never-before-seen particle—lightweight like an electron, but with *positive* electric charge. At Cambridge's Cavendish Laboratory, Dirac's friend Patrick Blackett and Giuseppe Occhialini took photographs of a cloud chamber which recorded the traces of electrically charged particles.[8] When high-energy cosmic rays hit a lead plate inserted across the diameter of the chamber, Blackett and Occhialini recorded traces, as in figure 2.3, that bent in the opposite direction that a negatively charged electron would travel. (The chamber was placed in a magnetic field that curved the particles' paths.) Blackett and Occhialini argued that these traces were left by particles described by

7. Interview of P. A. M. Dirac by Thomas S. Kuhn on May 14, 1963, Niels Bohr Library & Archives, American Institute of Physics, College Park, MD USA, www.aip.org/history-programs/niels-bohr-library/oral-histories/4575-5. Kragh, *Dirac*, 96–105; Moyer, "Evaluations of Dirac's Electron," §IV.
8. See Galison, *Image and Logic*, ch. 2.

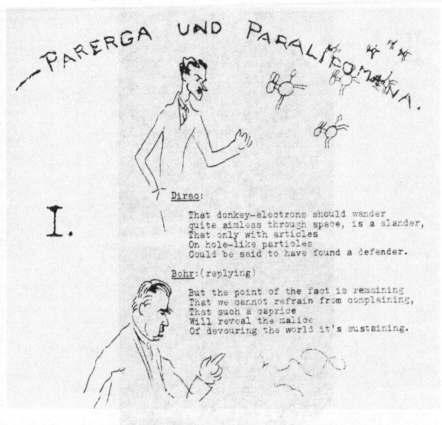

Figure 2.2. Caricature of Dirac from the Copenhagen *Faust*, with donkeys approaching at right. *Parerga und Paralipomena* [Appendices and Omissions] was the title of a 1851 book by Arthur Schopenhauer, surely the type of philosophy Dirac did not appreciate. The title was a double pun: first, on Dirac's "holes" as an appendix to his successful electron theory; second on "holes" and "omissions." *Blegdamsvej Faust*, Folder 8, Box 4, Series 2, Weisskopf Papers, MIT Libraries collection MC-0572.

Dirac's theory.[9] This gave new life to Dirac's anti-electron theory, and prompted his invitation to Brussels.

Theoretical physicists had mixed reactions to Anderson's "positive electron" and Blackett and Occhialini's "penetrating radiation." The Soviet physicist, and Dirac's friend, Igor Tamm wrote from Moscow on June 5, 1933:

> Many times this year I was about to write you, especially after Blackett's and Occhialini's paper appeared. I got used to say, that your prediction of the antielectron has no parallel in the history of science. One can think about Leverrier and Adams [who predicted the existence of the planet Neptune], but your case is different.

9. Anderson, "Apparent Existence"; Blackett and Occhialini, "Photographs"; Kragh, *Dirac*, 109–11; Roqué, "Manufacture of the Positron."

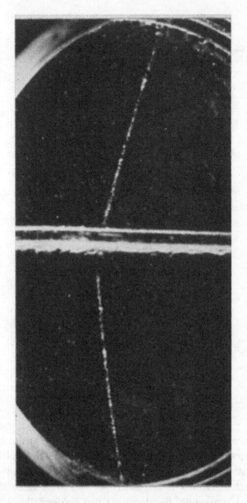

Figure 2.3. Photograph of track in a 13cm cloud chamber. "A track showing deflection at plate and a greater positive curvature [rightward] below than above. The direction must therefore be downwards, and hence the charge positive" (Blackett and Occhialini 1933, Plate 22). Republished with permission of The Royal Society (UK), from Blackett, Patrick M. S., and G. P. S. Occhialini. "Some Photographs of the Tracks of Penetrating Radiation." *Proceedings of the Royal Society of London A* 139, no. 839 (1933): 699–726; permission conveyed through Copyright Clearance Center, Inc.

Your theoretical prediction about the existence of the antielectron [...], seemed so extravagant and totally new, that you yourself dared not to cling to it and preferred rather to abandon the theory. And now the experiment unexpectedly proved you to be right and even ~~gave you~~ presented you with the neutron,[10] to make the "hole" stable with and to form a proton!

10. Discovered by James Chadwick in 1932. See, Bromberg, "Impact of the Neutron."

Nevertheless, even an admiring friend such as Tamm did not think Dirac's negative energy states would survive.

> One often hears people to say, that the unobservable negative electrification of the world, produced by the negative energy electrons, is a metaphysical notion. I personally am also inclined to think, that this notion in its present form will find no place in the future structure of physics, but nevertheless I think one has to use this notion and work with it.[11]

As Hans-Jörg Rheinberger has observed about researchers in the experimental life sciences, Tamm had a sophisticated view of the transient nature of the objects of theoretical physics.[12] Others were far more skeptical. Dirac's colleague and competitor Pauli responded: "I do not believe on your perception of 'holes,' even if the existence of the 'antielectron' is proved."[13] Much more observational and theoretical work would be required before the positron was broadly accepted.[14]

Looking forward to his fall 1933 presentations, Dirac had to consider more than just these issues. He collaborated with a young Rockefeller Fellow at Cambridge, Rudolf Peierls (1907–1995, figure 2.1). Dirac's dilemma was this: in the course of his calculations with Peierls to check if these discoveries truly vindicated the "hole" theory, they had discovered a *new* puzzle. His infinite distribution of electrons in empty space was not smooth and orderly (as he initially assumed). Instead, Dirac's vacuum could be "polarized." Polarization was a process in electricity and optics. Optical polarization was interpreted in classical electromagnetism as a preferred direction of vibration of the electromagnetic field; it played an important part in Dirac's radiation research. However, Dirac's polarized vacuum appealed to the electrical meaning of the term. One of Dirac's undergraduate engineering textbooks from University of Bristol discussed the polarization of liquid electrolytic cells. Solutions, for example dilute sulphuric acid, became polarized when an electric current was applied; when the applied current was cut, the energy used in decomposing the molecules appeared in a reverse current flowing out of the cell; it was a battery.[15] Polarized media *stored energy*. Much as in liquids, in solid-state insulators, polarization acted in opposition to applied electric fields. Polarized media *shielded charge*. In 1933, Dirac and Peierls found that the vacuum had exactly these physical characteristics. Just as an insulator shields the field of electric charges, they discovered that the vacuum of "hole" theory *screened* the "true" charge of electrons, which was hidden from everyday observations.[16] Renewed attention to "hole" theory revealed striking new phenomena.

---

11. Kojevnikov, *Dirac and Tamm*, 64, struck-out text in original.

12. Rheinberger, "Cytoplasmic Particles," 271.

13. Pauli to Dirac, May 1, 1933. Quoted in Kragh, *Dirac*, 112; for "competitor," see Dirac, "Recollections," 130.

14. Moyer, "Vindications of Dirac's Electron"; Roqué, "Manufacture of the Positron."

15. Thomälen, *Text-book of Electrical Engineering*, §15, cf. 100; Sommerfeld, *Atomic Structure*, I§5; the texts are nicely drawn out in, Farmelo, *Strangest Man*, 31, 72, 447 fn53-54; at Cambridge's informal Kapitza Club, Dirac likely found his training useful for understanding experimental papers; Kapitza himself devised new batteries; Kapitza, "Strong Magnetic Fields."

16. Dirac, "Théorie," 206; Dirac, "Teoriya pozitrona," 747.

Part of the mythology surrounding Dirac—of a solitary genius, ascetic, and remote—is a depiction of his writing process. In the recollections of physicist Christian Møller: "He could spend a whole day in the same position [in the library], writing an entire article, slowly and without ever crossing anything out." This story is at best inconsistent. Møller could hardly come close enough to Dirac's manuscript to see a perfect, correction-free document from his emotially heightened distance, "we hardly dared to creep into the room, afraid as we were to disturb him."[17] In fact, Dirac's archive includes drafts and re-drafts of his writing, including the 1933 presentations in the USSR and Belgium. In a pencil manuscript with many crossings and corrections, he began:

> The recent discovery of the positively charged electron or positron has revived interest in ~~some~~ an old ~~speculations~~ theory about particles with −ve kinetic energy, as the experimental results that have been obtained so far are in agreement with the predictions of the theory.

Where "−ve" was shorthand for "negative." Here Dirac wavered over the confidence he should express in his theory, or "speculations," but was firm on how experimentalists had made a "discovery" which had confirmed his "predictions." He cast his 1930–1931 work as temporally distant, "an old theory," from his present time (1933). He went on to lay out his picture of space filled with electrons in negative-energy states:

> We can then assume that in the world as we know it, nearly all the −ve energy states are filled, with one electron in each state, and that the vacuities or holes in the −ve energy distribution are the positrons.
>   A rough picture is to compare it with the sea, a bubble in the sea corresponding to a positron, just as a drop of water in the air corresponds to an ordinary electron.
>   The sea is a bottomless one, and we do not consider at all how it is supported from below, but are interested only in events near its surface.

After further arguments, Dirac concluded that "the hole is quite a rational particle and it is legitimate to interpret it as a positron."[18] As far as I know, this was his first use of the "sea" metaphor. (Indeed, water is electrically polarizable.) In Dirac's "rough" imagery, empty space in "the world as we know it" was like a bottomless sea. Matter was but bubbles and droplets near its surface.

A second manuscript, in ballpoint pen, is clearly a draft of Dirac's Solvay conference report. Most of the corrected pencil manuscript survived: "The recent discovery of the positively charged electron or <u>positron</u> has revived interest in the old theory about the states of negative energy of an electron, as the experimental results that

17. Kragh, *Dirac*, 38, cf. 13; Kragh, "Dirac's Relativistic Theory," 53; Mehra and Rechenberg, *HDQT*, vol. 4, 87; exceptions include Galison, "Suppressed Drawing"; Farmelo, *Strangest Man*, 138, other examples of extensive re-drafting include the introduction to his 1926 doctoral dissertation and his contributions to the 1927 Solvay report. Dirac Papers, Series 2, Box 50, Folder 3.

18. The order of composition is not clear, with the possibility of some marks and pages being added at different times in the 1930s. They appear to be in approximately chronological order. The re-drafting, rather than the order of drafts, is the relevant feature here. DP, Series II, Box 36, Folder 22.

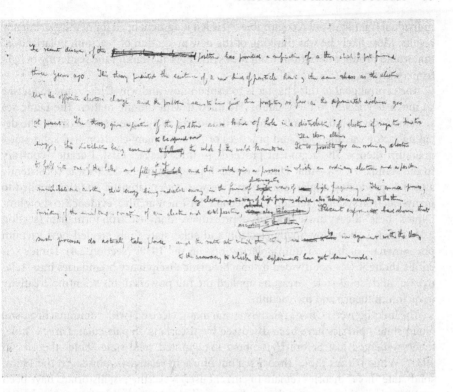

Figure 2.4. Verso of page from a draft of 1933 Solvay report. DP, Series 2, Box 36, Folder 22. Paul A. M. Dirac Papers, Special Collections & Archives, Florida State University.

have been obtained so far are in agreement with the predictions of the theory."[19] On the back of the fifth page of his Solvay draft he began again in pencil (figure 2.4). He showed deference to ongoing experiments, but a pronounced confidence. Rather than "the old theory," Dirac expressed personal ownership:

> The recent discovery of the positron has provided a confirmation of a theory which I put forward three years ago. This theory predicted the existence of a new kind of particle having the same mass as the electron but the opposite electric charge and the positron seems to have just these properties, so far as the experimental evidence goes at present.[20]

Stronger than his earlier description of experiments reviving interest in his work, here Dirac expressed his confidence in "a confirmation" of his theory. This personal ownership and confidence did not appear *tout court* in Dirac's published lectures, though

19. Dirac, "Théorie"; a translation without the discussion is included in Dirac, *Collected Works*; see Roqué, "Manufacture of the Positron," 106–7.
20. DP: Series 2, Box 36, Folder 22.

individual fragments are recognizable.[21] He left judgment open for new experimental results. Most likely he was thinking of the new prediction of "vacuum polarization" that would produce deviations from the existing predictions for the scattering of light from electrons.

The central goal of this chapter is to explain how and why Dirac came to ascribe so much *physicality* to the vacuum. He filled it with infinite matter. The Dirac sea sustained waves and currents, and was the source of strange new particles. The deciding factor was the tools and practices of Dirac's "many-electron" (or solid-state) quantum theory. The argument proceeds in three stages. First, I argue—contrary to the standard view—that Dirac's research program in the 1920s was consistently aimed at producing a quantum theory that was *both* relativistic *and* that applied to many-electron systems (section 2.2). Second, I use new archival evidence to show how Dirac's "hole" theory arose from his approximate, many-electron formalism (section 2.3). Third, I show how many-electron and solid-state tools created new vacuum phenomena—such as vacuum polarization—in the 1930s (section 2.4). Dirac's scientific thought was not divided by anachronistic disciplinary boundaries into "relativistic" and "solid-state" areas; he applied the full powers of his scientific creativity to quantum theory and the vacuum.

The links between Dirac's relativistic and many-electron (with some anachronism, "solid-state") physics have been discussed by historians. In particular, Dirac's "hole" theory emerged just before Heisenberg inaugurated solid-state "hole" theories in 1931. (While Dirac's "hole" theory fell out of use in relativistic physics in the 1950s, solid-state "hole" theories remain in current use.) Nevertheless, historians have been cautious about the amount influence between them. Alexei Kojevnikov notes the scientific and personal connections between and Dirac and Yakov Frenkel's solid-state "hole" theory. But Kojevnikov does not argue for a causal connection.[22] In his biography of Dirac, Helge Kragh lays out possible inspirations for the hole theory: chemical affinities, X-ray orbitals, Dirac's earlier radiation theory, and an (apocryphal) dream. In his excellent survey of twentieth-century physics, Kragh discusses both Dirac's and solid-state "hole" theories, but writes that "the two ideas were independent."[23] From the opposite view, historians of solid-state physics Lillian Hoddeson, Gordon Baym, and Michael Eckert note the temptation of integrating Dirac's "hole" picture into the history of solid-state physics. They acknowledge sympathetic views some physicists have presented in retrospect. But Hoddeson, Baym, and Eckert reject the connection. For them, there simply is not enough evidence to connect Dirac's hole theory to solid-state physics. The primary evidence is that Heisenberg's paper that inaugurated many-electron "hole" theories did not cite Dirac's work.[24] However, throughout

21. His lecture in Leningrad had a more personal tone; see Dirac, "Teoriya pozitrona," ex: 729.
22. Kojevnikov, "Freedom, Collectivism, and Quasiparticles"; cf. Kojevnikov, "Dirac's Quantum Electrodynamics."
23. Kragh, *Quantum Generations*, 369; cf. Kragh, *Dirac*, 105.
24. Heisenberg, "Zum Paulischen Ausschließungsprinzip"; Hoddeson, Baym, and Eckert, "Quantum Mechanical Electron Theory of Metals," 112–14.

this period Heisenberg was actively calculating with Dirac's theory of electrons and "holes," so a lack of citation cannot be attributed to ignorance.[25]

In this chapter, my goal is not to establish the influence of relativistic physics on many-particle (or solid-state) physics, or vice versa. Rather, I argue that Dirac's relativistic "hole" theory *was always* also a many-particle theory.[26] When are two ostensibly separate theories actually part of a single research program? Here are some indicators:

- programmatic statements connect them,
- single publications discuss them,
- common formal methods are applied to them,
- they share a common physical referent.

All of these indications apply to relativistic and many-particle physics in Dirac's research.

This depiction contrasts with the common presentation of Dirac as a formalist. It is easy to see how some of his statements fit into a broader pattern of theorists' denigration of "applied" problems. Wolfgang Pauli derided solid-state physics as filth (*Dreck*) and in 1931 wrote to his assistant Peierls: "On semiconductors one should not work, that's an obscenity (*Schweinerei*) [...]."[27] However, Dirac valued his engineering training and did not express such views.[28] Here is an often-quoted sentence about the state of quantum theory, from the introductory paragraph of a 1929 paper by Dirac: "The underlying physical laws necessary for the mathematical theory of a large part of physics and the whole of chemistry are thus completely known, and the difficulty is only that the exact application of these laws leads to equations much too complicated to be soluble."[29] How might this paper have continued? A formalist might have pursued the missing solution to the exact equations. A purist might have further analyzed the underlying physical laws, or rejected this dirty topic entirely. Dirac's direction was signaled by his following sentence: "It therefore becomes desirable that approximate practical methods of applying quantum mechanics should be developed [...]."[30] The development of "approximate practical methods" occupied him for the rest of his landmark twenty-page paper. As Albert Einstein remarked (sec. 1.2.2), theoretical physicists are best measured by their accomplishments.

25. Heisenberg to Dirac, 16.1.1930, DP, Series II, Box 2, Folder 1. Kragh, *Dirac*, 94–95; Pais, "Dirac Theory of the Electron"; relatedly, Enz claims that in 1931 Heisenberg was unaware of Hartree's work because he did not cite Hartree. Enz, "Heisenberg's Applications of QM," 999; however, Heisenberg discussed Hartree's work in his 1929 Chicago lectures: Heisenberg, *Physical Principles*, 164–65.

26. This contributes to a developing area of research on allegedly subsidiary physical topics. James and Joas, "Subsequent and Subsidiary?"; Martin, "Fundamental Disputations."

27. Enz, *No Time to Be Brief*, 157; cf. Hoddeson, Baym, and Eckert, "Quantum Mechanical Electron Theory of Metals," 159, fn438 and fn458.

28. At least in the period under consideration. See Kragh, *Dirac*, 280–82.

29. Dirac, "Quantum Mechanics of Many-electron Systems," 714; e.g., Simões, "Chemical Physics," 402.

30. Dirac, "Quantum Mechanics of Many-electron Systems," 714.

Dirac is well known for his appreciation of mathematical beauty, but he did not see approximations as ugly. In October 1931, Dirac lectured at Princeton. He explained:

> As soon as we have interaction between the particles of our system the problem becomes extremely difficult and in general involves equations that are so complicated that exact solutions cannot be found by present analytical methods; to attack these problems we must therefore make use of approximate methods; approximate methods do not necessarily have to be ugly and those we shall describe are based on general laws; all of physics is only approximation.[31]

Even Dirac's commitment to mathematical beauty was not in conflict with the use of approximations. In the 1920s and 1930s, Dirac was not the most mathematical or the purest theorist.[32] This chapter follows Dirac through his engagement with experiment; with the quantum theory of metals, molecules, and complex atoms; with approximations; and, with relativity. This unified view of relativistic physics and the solid state reveals how Dirac arrived at his polarizable sea.

2.2 Relativistic and Many-Electron Theory

An important continuity in Dirac's early research career was his repeated efforts to understand the Compton effect—the striking deviations from classical expectations Arthur Compton found when he scattered X-rays off of graphite in the early 1920s. Dirac published directly on Compton scattering in 1925, 1926, 1927, 1930, and 1934. The story of Dirac's relativistic quantum theory and the Compton effect has been told before.[33] In standard accounts, the Compton effect is considered in a very purified form: a single light quantum scattering off of a single electron. Here, I emphasize the Compton effect's material contexts and its diverse representations as a phenomena of light quanta or classical light waves; of classical or quantum electrons; and of single electrons or statistical assemblies of electrons. Most important, this chapter takes Dirac's various representations of the Compton effect seriously. This reveals deep connections between his many-electron quantum theory and his relativistic research.

2.2.1 Compton Effects

In the early 1920s, the American physicist Arthur H. Compton undertook a groundbreaking series of experiments with X-rays. After a postdoctoral year at the Cavendish Laboratory in Cambridge, UK, in 1920–1921, he became an assistant professor at

---

31. Notes by Banesh Hoffmann, which Dirac corrected. Dirac Papers, Series II, Box 26, Folder 15, page 72. See, Wright, "Beautiful Sea," §2.3; cf. Galison, "Suppressed Drawing," 159.

32. My analysis departs from Kragh, *Dirac*, 142, 340 fn77; Peierls was perhaps best able to compare Pauli and Dirac. He remarked on Dirac's broad appreciation of physics in Eden and Polkinghorne, "Dirac in Cambridge," 3–4.

33. Kragh, "Dirac's Relativistic Theory," 35, 41–43; Kragh, *Dirac*, ch. 3; Brown, "Compton Effect"; Mehra and Rechenberg, *HDQT*, vol. 6 §IV.3.

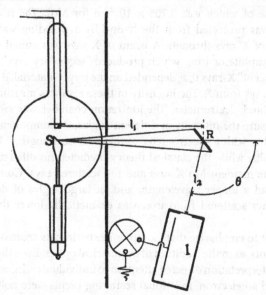

FIGURE 1. X-rays from the target S of the X-ray tube excite in the radiator R secondary rays which are measured by the ionization chamber I.

Figure 2.5. One of Compton's experimental arrangements.

Washington University in Saint Louis, MO. As Roger Stuewer has detailed, in 1922 Compton's X-ray studies led him to reject the classical theory of X-rays (as electromagnetic waves) in favor of a picture of X-rays as discrete light quanta, in accord with Einstein's 1905 analysis of black-body radiation.[34] Compton's results were a sensation among physicists, crystallizing what had been a diffuse field of mixed experimental results. They convinced many that light was not made up of macroscopic beams that experienced matter as an average of many particles.[35] Rather, light came in discrete quanta with energy, E, according to Einstein's formula $E = h\nu$, with h Planck's constant and ν the frequency. These quanta interacted with electrons or atomic nuclei, not as an entire beam, but as individuals.

In order to understand the physical phenomena Dirac was trying to analyze, it is necessary to understand Compton's experiment. One of Compton's experimental arrangement is depicted in figure 2.5. It included an evacuated Coolidge X-ray tube (at left), within which electrons streamed from cathode at the top of the tube toward the anode and molybdenum target at S. The electron beam excited X-rays in S, which radiated in all directions at its characteristic wavelengths

34. Compton, "Quantum Theory of the Scattering of X-rays"; Stuewer, *Compton Effect: Turning Point*, ch. 6; on prior interactions between X-ray experimentalists and theorists, see Katzir, "Theoretical Challenges."
35. Wheaton, *Tiger and the Shark*, 286; cf. Stuewer, *Compton Effect: Turning Point*, ch. 7.

(the most intense of which was 0.708×10^{-10}m for Mo). The rest of each apparatus, at right, was protected from the X-rays by a shielding wall; a narrow slit allowed a beam of X-rays through. A beam of X-rays was aimed at a second target, R, such as graphite or iron, which produced "secondary rays." Secondary rays included "fluorescent" X-rays that depended on the target material, β rays (electrons), and scattered X-rays from S. The intensity of these rays was measured with an ionization chamber and electrometer. The ionization chamber was swung around the target, R, to measure the intensity at different angles, θ. Compton also replaced the ionization chamber with a spectrometer to measure wavelength.[36] Here are Compton's central results: while the classical theory predicted no difference between the wavelength of the incident Mo X-rays and the scattered rays, Compton found the scattered rays had a longer wavelength; and, at large angles of deflection, θ, the intensity of X-rays scattered from iron was dramatically lower than the classical prediction.[37]

It is important to emphasize that Compton's experiments created a *steady state* of beams and currents, as in the tradition of geometrical optics. Even though Compton's light quantum interpretation directed attention to individual collisions between an X-ray quantum and an electron, individual scattering events were not legible by these methods.[38] Compton in fact measured a continuous electrical current that was produced by a continuous beam of X-rays. He argued that this beam should be considered as a stream of discrete light quanta. Compton was also clear about the limitation of his results to "A Quantum Theory of the Scattering of X-rays by *Light Elements*," such as the carbon graphite target, atomic weight 12 (compared to, say, platinum, with atomic weight 195). He assumed that the electrons scattering the X-rays were *free*, like idealized molecules in gas theory, not bound to atoms.[39] In December 1922, Compton concluded that "[t]he conditions of scattering in [heavier elements] remain to be investigated."[40] That is to say, he suggested that other physicists take up the challenge to analyze scattering by assemblies of electrons, interacting with each other and a nucleus, like those in heavy atoms. Indeed, an early application of Compton's scattering theory to statistical assemblies of electrons was completed in June 1923, by Compton himself.

At the time of Compton's experiments, Dirac was a newly minted electrical engineer with a first-class honors degree from the University of Bristol. With

36. Compton, "Secondary Radiations Produced by X-Rays," 2; Compton, *X-rays and Electrons*, 6, 10; Stuewer, *Compton Effect: Turning Point*, tubes: 77–78, 239, chambers: 138–39, 186, Compton's apparatus: fig. 5, 201–2; Kragh, *Quantum Generations*, 128–30, 177.

37. Compton, "Quantum Theory of the Scattering of X-rays," wavelengths: 494–96 and fig. 4; angles: 501 and fig. 7.

38. Compton, "Quantum Theory of the Scattering of X-rays," 496; I have drawn from: Compton, *X-rays and Electrons*, ch. 1; Stuewer, *Compton Effect: Turning Point*, 200–6; Stuewer, "Compton Effect: Transition," §3; by 1925, Compton and others studied the secondary β-rays, which were caused by individual X-ray quanta ejecting an electron from the target. Kragh, *Quantum Generations*, 161; Bonolis, "Walther Bothe and Bruno Rossi," individual events were studied via coincidence methods, beginning in 1925. Galison, *Image and Logic*, ch. 1.

39. W. Kaiser, "Electron Gas."

40. Compton, "Quantum Theory of the Scattering of X-rays," 502.

soldiers returning from the World War, he was unable to find an engineering job, and he lacked the funds to attend postgraduate studies. Dirac spent a further year studying mathematics at Bristol. In 1923, he won a Research Fellowship from the Royal Commission for the Exhibition of 1851 and a grant from the Department of Scientific and Industrial Research. He left Bristol and became a research student at St. John's College, Cambridge.[41] Dirac had originally hoped to study under Ebenezer Cunningham, a mathematician who published on special and general relativity and electron theory. While Dirac attended Cunningham's lectures, he was assigned to be supervised by Ralph Fowler (1889–1944), an applied mathematician who became Cambridge's premier theoretical physicist.

In 1924, Fowler was in the midst of a research program to merge classical statistical mechanics with the Bohr-Sommerfeld atomic theory. A central physical puzzle was to understand the Sun—both physicists and astronomers used spectroscopy to infer from the colors of radiant light the inner structure of radiating bodies. Fowler's astrophysical research connected the quantum theory of spectral lines with the thermodynamics of stars. In the early nineteenth century, Joseph Fraunhofer had observed that when sunlight is passed through a prism, hundreds of dark lines cross the spectrum. By the middle nineteenth century, these lines were being systematically compared to the inverse patterns which were emitted by chemical elements.[42] Another of Dirac's mentors was Edward Arthur Milne (1896–1950), the assistant director of the Solar Physics Observatory in Cambridge. In 1921, Milne considered a long-standing solar puzzle: why did the Sun appear darker and redder at its edge, or "limb"? Why was it that the whole series of Fraunhofer lines appeared shifted downward, toward the red, when the source of the light was from the limb of the Sun? Milne advanced a theoretical explanation that relied on a picture of the motions of the atoms in the Sun, as if the star was a box of gas molecules in equilibrium with velocities distributed according to Maxwell-Boltzmann statistical mechanics. This was seen as an excellent piece of work, for which Milne won Cambridge's Smith Prize in 1922. In 1923, Milne and Fowler collaborated on a further development of the theory.[43]

In 1924, this agreeable state of affairs was disturbed by an unusual article in the *Philosophical Magazine*, written and communicated from Saint Louis by Compton. Compton's terrestrial X-ray studies had shown that scattering by electrons made the wavelength of the scattered rays longer, or redder. He then made two assumptions: that what was true of X-rays was true of all wavelengths of light, and that light from

---

41. In this section I draw on: Mehra and Rechenberg, *HDQT*, vol. 4, §II.1; Kragh, *Dirac*, ch. 1; Darrigol, *c-Numbers to q-Numbers*, ch. XI; Farmelo, *Strangest Man*; Warwick, *Masters of Theory*, Cunningham: chs. 8–9; Kragh and Rebsdorf, "Before Cosmophysics"; Gavroglu and Simões, "Preparing the Ground."

42. The solar lines were first observed by William Wollaston. See the marvelous analysis in Hentschel, *Mapping the Spectrum*, §§2.2–4, 3.1.

43. Milne, "Radiative Equilibrium"; Fowler and Milne, "Absorption Lines in Stellar Spectra"; on Fowler and Milne, see Gavroglu and Simões, "Preparing the Ground"; Rebsdorf and Kragh, "Edward Arthur Milne."

the outer limb of the Sun underwent more scattering than light from the Sun's center.[44] Compton posed a challenge to Milne.

Under Milne's guidance, Dirac set out to discover whether the "limb effect" was due to Milne's velocities or to Compton's scattering.[45] Dirac's paper worked from general physical principles, through specific physical models, to specific observational consequences. Both Milne and Compton used the free-electron gas approximation. However, Compton made the additional idealization that the "the scattering electron is at rest [...]."[46] By contrast, Milne allowed electrons to fly about in any direction, with velocities described by a statistical distribution that depended on the temperature of the Sun. Compton's redshift arose from the scattering and recoil of electrons at rest; Milne's redshift (and blueshift) arose from a Doppler effect created by the electrons' thermal motions.

In order to compare the two theories, Dirac needed to bring together the two physical pictures of electrons in the Sun. He did this by re-deriving Compton's results within Milne and Fowler's statistical framework. He "generalise[d] [Compton's scattering rate equation] for the case when there are N scattering electrons in statistical equilibrium at the temperature T." Where Compton calculated the rate of scattering from electrons *at rest*, Dirac calculated the rate of scattering from electrons with *any* momentum. Increasing the generality of a theoretical expression—widening the array of phenomena it described—is often associated with increasing idealization and abstraction.[47] This was different. Dirac's more general expression for electron motion was a *more* realistic description of electrons (which were, of course, thought to move about). It was a *de*-idealization of Compton's physical picture. Dirac's paper concluded from a "numerical estimate" of the observable effects, that the thermal motion of the electrons was the dominant effect. "Compton scattering cannot therefore account for the limb effect in the sun."[48] Dirac's "The Effect of Compton Scattering by Free Electrons in a Stellar Atmosphere" used statistical mechanics to critique Compton's application of his scattering theory. Despite its negative conclusion

44. The light from the Sun's limb would scatter within the Sun and be deflected toward the Earth, as the X-rays were deflected toward the ionization chamber in figure 2.5. Compton, "Absorption Measurements," 909; Stuewer claims that the limb effect was "never before understood," but Milne's work shows that this was hardly the case. Stuewer, *Compton Effect: Turning Point*, 242; this impression may have been caused by Compton's lack of citations to the astrophysical literature.

45. This redshift caused some to doubt Einstein's relativistic prediction of the gravitational redshift of light. Hentschel, "Conversion of St. John"; Mehra and Rechenberg conjecture that Dirac's work "promised to provide support for general relativity" by securing a gravitational origin for the redshift. Mehra and Rechenberg, *HDQT*, vol. 4, 114. However, neither Compton nor Dirac mentioned the gravitational redshift in the articles under discussion. Fowler and Milne do briefly mention the "Einstein effect," Fowler and Milne, "Absorption Lines in Stellar Spectra," 416; but Dirac was narrowly "concerned with the effect of the electrons on light of particular wave-lengths near an absorption line, and not with any general effect produced throughout the spectrum." Dirac, "Effect of Compton Scattering by Free Electrons in a Stellar Atmosphere," 826; cf. Kragh, *Dirac*, 12.

46. Dirac cited both Compton's astrophysical paper and his landmark, Compton, "Quantum Theory of the Scattering of X-rays," 493.

47. An example from this paper: Dirac discussed the rate of scattering events independently of "some special theory of the process of scattering"; in order to accomplish this general discussion he employed a simple mathematical abstraction: he represented any "special theory" by an algebraic variable, F. Dirac, "Effect of Compton Scattering by Free Electrons in a Stellar Atmosphere," 828.

48. Dirac, "Effect of Compton Scattering by Free Electrons in a Stellar Atmosphere," 832.

about Compton's explanation of the limb effect, Dirac's general analysis was also an improvement of Compton's work. The "Stellar Atmospheres" paper is central to the argument of this chapter because it shows how Dirac thought of the Compton effect *within* statistical mechanics, a physics of many-body systems.

2.2.2 (Relativity) Quantum Mechanics

At Cambridge in 1925, Fowler received the proofs of Werner Heisenberg's landmark paper "Quantum-theoretical Re-interpretation of Kinematic Mechanical Relations" and, in August 1925, passed them to Dirac to study. Olivier Darrigol has detailed Dirac's initial response to Heisenberg's paper: an attempt to construct its relativistic extension, "Heisenberg's quantum mechanics and the principle of relativity." In this unpublished manuscript, he applied a theory of "the uni-directional emission of radiation"—analogous to Compton's light-quantum theory. A second unpublished attempt, "Virtual oscillators," followed the lines of Niels Bohr, Hendrik Kramers, and John Slater's 1924 quantum theory of radiation. These lines of inquiry petered out. A third attempt, which led to publication, took for a starting point Kramers and Heisenberg's dispersion theory, on which Fowler had recently lectured. In October, Dirac wrote out "The Fundamental Equations of Quantum Mechanics." This paper became a founding document of quantum mechanics in which he introduced his non-commuting q-numbers (quantum- or queer-numbers which disobeyed the regular commutative law of arithmetic).[49] Dirac's next publications on the new quantum theory aimed at the target Heisenberg had set: atomic and molecular structure.[50] But Dirac's quantum mechanics began as attempts at relativistic, light quantum, radiation theory—just the characteristics of Compton's theory. And he soon returned to these questions.

In April 1926, Fowler sent a copy of Dirac's latest article to the Royal Society, "Relativity Quantum Mechanics with an Application to Compton Scattering." In this paper, Dirac showed his stunning ambition. In rapid succession, he deployed his analysis across the five sections of the paper. (§1) Introduced Dirac's non-cummutative q-numbers, and argued that advancing the quantum theory in new relativistic domains required first generalizing the underlying classical mechanics. (§2) Systems for which the Hamiltonian function (i.e. energy) changed in time. Even advanced texts on dynamics focused on time-independent Hamiltonians (roughly, those cases in which energy was conserved).[51] Dirac argued that time-dependence was necessary for a theory of "Quantum Time." (§3) Systems which were composed of many particles *in motion*, a "Quantum Mechanics of Moving Systems." (§4) Systems characterized by the combination of §2 and §3, forming "Relativity Quantum Mechanics." And not least: (§5) Application of this formalism to the "Theory of Compton Scattering." This paper shows both continuity with Dirac's earlier work on the Compton effect

---

49. Darrigol, *c-Numbers to q-Numbers*, 309–15; Kragh, *Dirac*, 14–20; Dirac, "Fundamental Equations"; cf. Mehra and Rechenberg, *HDQT*, vol. 4 §IV.I; Hendry, "Bohr-Kramers-Slater."

50. E.g., Dirac, "Quantum Mechanics and a Preliminary Investigation of the Hydrogen Atom."

51. E.g., Whittaker, *Analytical Dynamics*.

and new perspectives from his fresh experience with Heisenberg's atomic mechanics. It demonstrates Dirac's unified approach to relativistic and statistical problems. As Helge Kragh has shown, this was one of the most influential of Dirac's early papers— it garnered notice, citations, and an appreciative letter from Compton, by then in Chicago.[52] Without question, "Relativity Quantum Mechanics" demonstrates Dirac integrating the new quantum mechanics with relativistic and many-electron physics.

Compton apparently did not notice, but Dirac departed from the light-quantum view of scattering. Dirac consistently compared and contrasted his analysis—"the present theory"—to Compton's "light-quantum theory."[53] Instead of particle ballistics, Dirac represented the Compton effect as an atom reacting to a classical electromagnetic wave. (This was consistent with Dirac's discomfort with Einstein's light-quantum hypothesis.[54]) Instead, Dirac applied a holistic analysis taken from atomic theory: "The electron and incident radiation together may be considered to form a dynamical system whose emission spectrum can be determined [...], although it is usually called not an emission spectrum but a scattered radiation."[55] Dirac considered this radiation and electron system one of a type of "systems acted upon by external fields of force," which required sophisticated mathematics.[56]

In order to describe an electron's motion, Dirac needed to develop a formalism in which the "Hamiltonian function" of a system could change in time. This was an additional layer of complication, from kinematics→dynamics→dynamics with time-varying forces. There were two interrelated reasons for this. For most of the dynamical systems physicists considered, the Hamiltonian represented the system's total energy. In particularly simple cases, the Hamiltonian, H, was the sum of the kinetic energy, W, and the potential energy, V: $H = W + V$. Allowing the Hamiltonian to change in time meant allowing for a class of systems in which the energy of the system changed, such as "systems acted upon by external fields of force."[57] The pertinent example was a charged body in a rapidly varying electromagnetic field—or, on Maxwell's theory, a beam of light. Dirac was not alone in studying the quantum mechanics of electrons exposed to light waves. In their famous "three-man paper," Max Born, Heisenberg, and Pascual Jordan wrote that this question was of "especial interest" because it illustrated the differences between classical and quantum mechanics. They developed an approximation scheme using perturbation theory (see appendix A).[58]

52. Kragh, *Quantum Generations*, 28–29, 296 table III; Kragh has denied that this early work "attempted directly to relativize quantum mechanics," Kragh, "Dirac's Relativistic Theory," 40; in my judgment, these papers show that Dirac did indeed make such attempts. Unsuccessful attempts were attempts nonetheless.

53. Dirac, "Relativity Quantum Mechanics with an Application to Compton Scattering," 414, 418–19, figure p. 422; contra Kragh, *Dirac*, 28.

54. See, Dirac, "Conditions for Statistical Equilibrium," 594, cf. 583; Darrigol, "Quantized Matter Waves," 216; he may have followed Bohr: Bohr, "Application of the Quantum Theory to Atomic Structure," 35.

55. Dirac, "Relativity Quantum Mechanics with an Application to Compton Scattering," 415; cf. Dirac, "Conditions for Statistical Equilibrium," §5.

56. Dirac, "Relativity Quantum Mechanics with an Application to Compton Scattering," 410.

57. Dirac, "Relativity Quantum Mechanics with an Application to Compton Scattering," 410.

58. Dirac, "Relativity Quantum Mechanics with an Application to Compton Scattering," 409; Born, Heisenberg, and Jordan, "On Quantum Mechanics. II.," 333, §1.5.

However, Born, Heisenberg, and Jordan's analysis did not satisfy Dirac. He had another motivation: to make quantum theory relativistic. "The principle of relativity demands that the time shall be treated on the same footing as the other variables, and so it must therefore be a q-number."[59] Dirac learned this version of the principle of relativity from Hermann Minkowski's famous words: "From henceforth, space by itself, and time by itself, are mere shadows, and only a kind of blend of the two exists in its own right."[60] To Dirac this meant that—if spatial position was to be a quantum number—*time* must be a q-number.[61] (Chapter 5 discusses John Wheeler's struggle with the same requirement.) Dirac's analysis was a muddled effort on this score. The relativistic dynamics of electrons had been investigated by Dirac's mentors Cunningham and another former Wrangler, Charles Galton Darwin, but the quantum question was unanswered.[62] Dirac admitted that "the requirements of the restricted principle of relativity will still not be completely satisfied," because of the demands of the formalism of Hamiltonian mechanics. Dirac was reduced to a defensive appeal to "immediately obvious" parallels to classical mechanics to maintain in his calculations a semblance of coherence.[63]

The third section of Dirac's "Relativity Quantum Mechanics" concerned many-particle systems in motion. It offers further direct evidence that Dirac thought of relativistic physics and many-particle physics together. First, recall that his "Stellar Atmosphere" paper allowed for (non-relativistic) electron motions, which were distributed according to statistical mechanics. Dirac may also have hoped to apply a result from a 1924 paper on radiation statistics. There, he found that the statistical distribution of a system's momentum was invariant under a relativistic change of reference frame.[64] In "Relativity Quantum Mechanics," Dirac considered a general system of many particles, with a center of gravity. It turned out that the motion of the center of gravity was "ambiguous on the quantum theory."[65] To calculate Compton scattering, Dirac also needed an expression for the center of charge of the many-electron system. This was not forthcoming, either. Dirac's formalism only allowed calculations for a single charged particle, with center of charge and center of gravity the same.[66] Though he established a general formalism for relativistic, many-electron quantum mechanics, he was unable to use it for calculations.

However, Dirac made the necessary approximations to connect his formalism with Compton's results, and with experiment. In the sixth section of the paper, Dirac

59. Dirac, "Relativity Quantum Mechanics with an Application to Compton Scattering," 407, eq. 8; Kragh, "Dirac's Relativistic Theory," 41.

60. Quoted in: Cunningham, *Relativity and the Electron Theory*, 70; Cunningham, *Relativity, The Electron Theory, and Gravitation*, 6, repeated on 72; cf. Cunningham, *Principle of Relativity*, 113; cf. Farmelo, *Strangest Man*, 36–42.

61. This was not a universal position. For example, Fock argued that "time plays a role quite different from the role of coordinates both in the relativity theory and in quantum relativity theory." Fock, "On the Notion of Velocity," 97.

62. Darwin, "Dynamical Motions"; Cunningham, *Relativity, The Electron Theory, and Gravitation*, §§51–54, 62–66; Navarro, "'Dedicated Missionary.'"

63. Dirac, "Relativity Quantum Mechanics with an Application to Compton Scattering," 412, 408.

64. Dirac, "Conditions for Statistical Equilibrium," §2.

65. Dirac, "Relativity Quantum Mechanics with an Application to Compton Scattering," 413. He noted that his quadratic Hamiltonian was not "in the standard form" $H - W = 0$, 409, eqs. 4, 9, 13; cf. Kragh, "Dirac's Relativistic Theory," 42; for centers of gravity in atomic systems, see Darwin, "Free Motion."

66. Dirac, "Relativity Quantum Mechanics with an Application to Compton Scattering," 414.

showed a small difference between his and Compton's predictions of the intensity (brightness) of the emitted radiation. In the seventh and final section of the paper, Dirac compared this difference with experiment. He proudly announced: "This is the first physical result obtained from the new mechanics that had not been previously known." Though a note, added in proof, acknowledged that the better part of Dirac's result had been found independently by Gregory Breit using Kramers's dispersion theory.[67] Consider Compton's experimental set-up, figure 2.5. On Compton's theory, scattered radiation measured with the ionization chamber at 90 degrees should not be polarized. Dirac predicted, in agreement with Bohr's correspondence principle, that it would in fact be polarized.[68] On August 21, 1926, Compton wrote to Dirac that new measurements had confirmed that maximum polarization was at 90 degrees, "in good accord with your theory."[69] "Relativity Quantum Mechanics" was a success, and yet it fell short of its lofty aims.

From spring to summer 1926, Dirac prepared another sprawling manuscript, "On the Theory of Quantum Mechanics," which Fowler sent to the Royal Society on August 26, 1926. This paper is most remembered for its contribution to the "statistics" obeyed by assemblies of electrons, protons and other "Fermions," soon called Fermi-Dirac statistics. (Other particles, such as light quanta, were called "bosons," and obeyed Bose-Einstein statistics, after Satyendra Nath Bose). But "On the Theory of QM" was very much a continuation of "Relativity Quantum Mechanics." It also ranged over relativity, many-body physics, and radiation theory. Again Dirac aimed for a relativistic "slightly more general point of view, in which the time t and its conjugate momentum $-W$ are treated from the beginning on the same footing as the other variables."[70] However, he was not able to carry through a relativistic analysis. The third section of "On the Theory" discussed the wavefunctions for "systems containing several similar particles." His fourth section discussed the statistical mechanics of an ideal gas of molecules in quantum mechanics. In the final section he applied his formalism to recover one of the two pillars of Einstein's radiation theory: the B coefficient (see section 3.2).[71] "On the Theory of QM" solved some of the key problems Dirac had identified in "Relativity Quantum Mechanics"—particularly a description of many particles—but he still had no practical method for relativistic calculations.[72] Dirac's holistic atomic-system analysis provided the energy levels of an ideal gas, but it said nothing about the positions of the molecules in space. Dirac's conclusion focused on improving the many-electron results by suggesting that future calculations should consider atoms "distributed at random," as in a gas "or arranged in a crystal lattice."[73] Here Dirac was likely trying for new connections to experiment. Such position information would give the phases of vibrations of the radiating molecules. At the

---

67. Dirac, "Relativity Quantum Mechanics with an Application to Compton Scattering," recoil: 413, Hamiltonian: eq. 26, frequency: 419, result: 421.

68. Dirac, "Relativity Quantum Mechanics with an Application to Compton Scattering," 420–23; on polarization, see Bohr, "Application of the Quantum Theory to Atomic Structure," 26–29.

69. Kragh, *Dirac*, 29.

70. Dirac, "On the Theory of Quantum Mechanics," 662, 666; see Kragh, *Dirac*, 33–38; Dirac's intentional framing of his formulae in "view of relativity generalisation" was recognized in Darwin, "Electron as a Vector Wave," 283.

71. This was a continuation of 1924's Dirac, "Conditions for Statistical Equilibrium."

72. Dirac, "On the Theory of Quantum Mechanics," §5.

73. Dirac, "On the Theory of Quantum Mechanics," 677.

1927 Solvay Congress, Dirac asked W. L. Bragg about phase relationships in his X-ray scattering experiments.[74]

2.2.3 The Quantum Theory of the Electron

As the summer of 1926 turned to fall, the recently graduated Dr. Dirac began a series of visits to major European physics centers. His advisor Fowler wrote to Bohr on Dirac's behalf, and Dirac began his visit in Copenhagen in September. Bohr's Institute was a hive of activity, with physicists visiting from across the continent and beyond.[75] During this time Dirac completed the development of his quantum formalism, from q-numbers to "uniformitising variables" into his new "transformation theory," which could answer a *far* greater range of physicists' questions about atomic systems. He framed the transformation theory as an alternative to Heisenberg's interpretation of matrices of radiation intensities, or of Born's interpretation of wavefunctions as probabilities. He even suggested it "can replace all the special assumptions previously used, and perhaps go further."[76] The transformation theory mobilized an even greater set of tools from classical Hamiltonian mechanics in the quantum theory. This work became his "darling."[77] (Jordan independently developed an equivalent formalism.[78]) According to Dirac, he developed this new approach from Heisenberg's version of statistical fluctuations in many-electron systems. On November 23, 1926, Heisenberg wrote to Pauli that "Dirac has managed an extraordinarily grandiose generalization of the assumption from my note on fluctuations." In place of energy levels or probabilities, Heisenberg's fluctuation theory considered the amount of time an atomic system spent within a certain range of a quantity, for example the time spent with a position between two boundaries.[79] Through Heisenberg's "exchange" formalism, Dirac again built many-particle physics into his fundamental "Physical Interpretation of the Quantum Dynamics."[80]

Once Dirac returned from his ten months abroad, he settled back into St. John's College, Cambridge, now as a fellow. From July to December 1927 he was quietly working on a relativistic theory of the electron, except for his participation in the Solvay Conference in October at Brussels. At Cambridge, he lectured on "Modern Quantum Mechanics" and worked toward an unsolved problem from his "Relativity Quantum Mechanics" (section 2.2): a relativistic description of the motion of a single electron in quantum theory. By December 1927, he found the solution. Around Christmas 1927, Darwin and Fowler each sent word of Dirac's new work to Bohr's Institute in Copenhagen.[81] Bohr's flattering letter to Dirac, of February 27, 1928,

74. Bacciagaluppi and Valentini, *Quantum Theory at the Crossroads*, 293.

75. Aaserud, *Redirecting Science*.

76. Dirac, "Physical Interpretation of the Quantum Dynamics," 621–22; Kragh, *Dirac*, 41–42, 44.

77. Cited in Pais, "Playing with Equations," 96, 97.

78. Duncan and Janssen, "Canonical Transformations."

79. Pauli, *Wissenschaftlicher Briefwechsel*, vol. 1, doc. 148, 357; Heisenberg, "Schwankungserscheinungen und Quantenmechanik," 504; Jammer, *Conceptual Development*, 305; partially quoted in: Kragh, *Dirac*, 39.

80. On exchanges: Carson, "Exchange Forces—II"; compare to the multiperiodic systems of the old quantum theory: Darrigol, *c-Numbers to q-Numbers*, 102–18.

81. Kragh, *Dirac*, 57–63.

indicates how eager he was to have Dirac return to Copenhagen: Bohr called it a "most wonderful paper."[82] There have been many scholarly discussions of Dirac's successful relativistic theory.[83] It was the source of the "negative-energy solutions" that would be filled by his "sea."

On January 2, 1928, the Royal Society received an envelope from Fowler containing Dirac's paper "The Quantum Theory of the Electron." The paper started with a Hamiltonian equation for the energy of a single electron with no surrounding electromagnetic field:[84]

$$W^2/c^2 - p_1^2 - p_2^2 - p_3^2 = m^2c^2, \qquad (2.1)$$

$$\text{or written as a wave equation: } (W^2/c^2 - p_1^2 - p_2^2 - p_3^2 - m^2c^2)\psi = 0. \qquad (2.2)$$

Following Schrödinger's quantization algorithm, the wave-equation symbols were reinterpreted as derivatives, $p_1 \rightarrow \partial/\partial x$, etc. Equation (2.2) is known as the Klein-Gordon equation, after Oskar Klein and Walter Gordon, though it was independently discovered by several others. Klein, Gordon, and Schrödinger each applied equation (2.2) to calculate the energy levels (Ws) of electrons in Compton scattering.[85] To make practical calculations with equation (2.1), physicists isolated W and took the square root of both sides, $W = \pm\sqrt{|H|}$. The complicated square root was then expanded in a series approximation.[86] Dirac argued that while equation (2.2) could be used to answer *some* questions about atomic systems, it was more limited than the general scope of his and Jordan's transformation theory. The "general interpretation" was "made possible by the wave equation being of the form $(H - W)\psi = 0$, *i.e.* being linear in W or $\partial/\partial t$."[87] In 1926, he had been unable to find a linear version of equation (2.1).[88]

82. Dirac papers, Series II, Box 1, Folder 6. See also letters from Nishina Yoshio and Heisenberg.

83. Dirac, "Recollections"; Kragh, "Dirac's Relativistic Theory"; Moyer, "Origins of Dirac's Electron"; Moyer, "Evaluations of Dirac's Electron"; Mehra and Rechenberg, HDQT, vol. 6, §II.7; Wright, "Beautiful Sea"; cf. Darrigol, "Quantized Matter Waves."

84. With variations from 1926–1928: Dirac, "Relativity Quantum Mechanics with an Application to Compton Scattering," eq. 9; Dirac, "Compton Effect in Wave Mechanics," 502; Dirac, "Quantum Theory of Dispersion," eq. 9; Dirac, "The Quantum Theory of the Electron," eq. 3; cf. Dirac, "Effect of Compton Scattering by Free Electrons in a Stellar Atmosphere," eq. 2.

85. Kragh, "Dirac's Relativistic Theory," 41–42.

86. Pauli, "Zur Quantenmechanik des magnetischen Elektrons," 616; cf. Dirac, "Quantum Theory of Dispersion," eq. 10.

87. Dirac, "The Quantum Theory of the Electron," 612; cf. Dirac, "Zur Quantentheorie des Elektrons," 371; in 1926, he noted that his quadratic Hamiltonian was not "in the standard form," Dirac, "Relativity Quantum Mechanics with an Application to Compton Scattering," 409.

88. Today, physicists may object that the operator $\frac{\partial^2}{\partial^2 t} = \partial_t^2$ is linear, because $\partial_t^2 c(f + g) = c\partial_t^2 f + c\partial_t^2 g$, where f and g are functions and c is a constant. This was not what Dirac was thinking, but because of the ambiguity between derivation and multiplication in quantum mechanics, there are two approaches to see why. First, take W algebraically. Then the operator acting on W is just squaring, $f \times f$. This operator, it is perhaps needless to say, is not linear because $(f+g)^2 \neq f^2 + g^2$. The second option is to follow Schrödinger's replacement of W or p_0 with ∂_t. Sometimes Dirac wrote ∂_t^2 and other times $p_0 \times p_0$, but neither is in general linear. To see this, consider the definition of the derivative: $\partial_t \phi(t) = \lim_{h \to 0} \frac{\phi(t+h)-\phi(t)}{h}$. Ask whether it is linear in the substitution $t \rightarrow f + g$. It is, for linear ϕ, but the product $p_0 \times p_0$ and second derivative are not, in general. Much more could be said about the relevant domains and functions involved, but even this level of detail is beyond the scope of historical discussion.

Dirac recalled that in late 1927, he was "playing around" with mathematics, including 2 × 2 rotation matrices Pauli had introduced to quantum theory earlier that year.[89] Dirac saw that Pauli's 2 × 2 matrices could be used to linearize quadratic equations with three terms. He wanted a linearization for the four terms on the left-hand side of equation (2.1), or between the brackets in equation (2.2). He succeeded. Here is Dirac's eponymous equation:

$$(W/c + \rho_1[\sigma_1 p_1 + \sigma_2 p_2 + \sigma_3 p_3] + \rho_3 mc)\, \psi = 0. \tag{2.3}$$

The linearization was possible because of the six matrices, σ_r and ρ_r where ($r = 1, 2, 3$). This equation contained W/c instead of the quadratic W^2/c^2 of equations (2.1) and (2.2). Linearity was acknowledged to be the key to his success.[90] In this way, Dirac's formalism satisfied two of his goals: its linearity was consistent with standard Hamiltonian mechanics and his general interpretation of quantum theory. And it treated time, W/c or $\partial/\partial t$, equivalently to the other coordinates. Equation (2.3) was genuinely quantum and genuinely relativistic.[91]

But this success came with a price. The matrices σ_r and ρ_r were 4 × 4 grids of numbers. Dirac explained: "[o]n account of the matrices ρ and σ containing four rows and columns, [equation (2.3)] will have four times as many solutions as the non-relativity wave equation." For example, one matrix, ρ_1, and a fourfold wavefunction can be written:

$$\rho_1 = \begin{pmatrix} 0 & 0 & 1 & 0 \\ 0 & 0 & 0 & 1 \\ 1 & 0 & 0 & 0 \\ 0 & 1 & 0 & 0 \end{pmatrix}, \quad \psi = \begin{bmatrix} \psi_1(p_0, p_x, p_y, p_z) \\ \psi_2(p_0, p_x, p_y, p_z) \\ \psi_3(p_0, p_x, p_y, p_z) \\ \psi_4(p_0, p_x, p_y, p_z) \end{bmatrix}. \tag{2.4}$$

The first two of these wavefunctions were welcome. In Dirac's formalism, ψ_1 and ψ_2 represented the two possible "spin angular momentum" states of an electron, which explained the "duplexity phenomena" of atomic spectral line-splitting.[92] More than this, Dirac's result came from "the point-charge electron," without additional assumptions about spinning spheres.[93] But the remaining two wavefunctions, ψ_3 and ψ_4, were problematic. The relation between their velocity and kinetic energy was deeply strange: it was negative. No particle like that had ever been observed. Negative kinetic energy would mean that as the electrons in ψ_3 and ψ_4 states *lost* energy they would *gain* velocity. Moreover, the theory allowed electrons to transition from a regular, positive energy, state to a negative energy state, say ($\psi_1 \rightarrow \psi_3$). Dirac explained: "Such a transition would appear experimentally as the electron suddenly changing its charge from $-e$ to e, a phenomenon which has not been observed." Faced with

89. Pauli, "Zur Quantenmechanik des magnetischen Elektrons"; Dirac, "Recollections," 142; Pais, "Playing with Equations."

90. Breit, "Interpretation of Dirac's Theory"; Darwin, "Wave Equations"; see a secondhand report of Jordan's reaction: "I cannot forgive myself that I didn't see that the point was linearization," in Kragh, *Dirac*, 62.

91. Dirac, "The Quantum Theory of the Electron," 615.

92. For an overview of "spin," see Jammer, *Conceptual Development*, §§3.4, 8.1.

93. Dirac, "The Quantum Theory of the Electron," 618, 610.

these absurdities, Dirac concluded that "half the solutions must be rejected." But Dirac understood that this was somewhat arbitrary, and that "[t]he resulting theory is therefore still only an approximation" to an unknown "true relativity" equation.[94] Dirac maintained this view into the 1930s, and in fact he viewed physical theories as approximations, in general.[95] This attitude may have contributed to the view that the negative-energy states were optional appendices (*Parerga*, section 2.1) to Dirac's relativistic theory.

"The Quantum Theory of the Electron" tested the limits of Dirac's empiricist strictures. In "On the Theory of QM" (1926), he argued that quantum mechanics "shall enable one to calculate only observable quantities" (section 1.2.2). In his notes for his 1927 "Lectures on Modern Quantum Mechanics" at Cambridge, he introduced some mathematical leeway:

> The main feature of the new theory is that it deals essentially only with observable quantities, a very satisfactory feature. One may introduce auxiliary quantities not directly observable for the purposes of math[ematica]l calculation, but variables not observable should not be introduced merely because they are required for the description of the phenomena according to ordinary classical notions, e.g. orbital frequencies in Bohr's theory.

Dirac also expressed a characteristic feature of many interpretations of quantum theory: that epistemology and ontology were intertwined. For Dirac, the restrictions on the questions quantum theory could answer were not motivated by philosophical scepticism. Rather these restrictions *themselves* told him something about the physical world. Or, so he presumed.

> It may at first sight appear unsatisfactory that the theory gives, in answer to certain problems, much less information than the classical theory, but actually the information that the theory can give is separated from what it can't give by simple clear-cut rules, which presumably show fundamental properties of nature.[96]

This was not so much a change in orientation as a consolidation of words and actions. In 1926, Dirac had expressed strong empiricist views, and he had also made extensive use of "auxiliary quantities not directly observable for the purposes of math[ematica]l calculation." When, in 1927, he was tasked with giving lectures on the quantum theory, he consolidated his position.[97]

In his electron theory, Dirac's negative-energy wavefunctions were certainly *not* "required for the description of the phenomena according to ordinary classical notions." One can infer that Dirac did not view his negative-energy states as eqivalent to

94. Dirac, "The Quantum Theory of the Electron," 612, 618.

95. "There are at present fundamental problems in theoretical physics awaiting solution, *e.g.*, the relativistic formulation of quantum mechanics," Dirac, "Quantised Singularities in the Electromagnetic Field," 60; cf. Kragh, *Dirac*, 170.

96. Lectures, pages 1–1A, Folder 2 Box 26, Dirac Papers.

97. I do not quite see Dirac as using empiricist remarks as only after-the-fact justifications, but see Beller, *Quantum Dialogue*, 51–59.

Bohr's orbits. They *were* required for equation (2.3) to be mathematically meaningful. But they were different from earlier mathematical "quantities not directly observable" that appeared in Dirac's papers. Earlier "auxiliary quantities" were directly related (by equations) to observable quantities. For example, the rate of occurrence of an atomic process—such as Einstein's A and B coefficients—were themselves unobservable; but they were related to the intensity of radiation at a given location. Intensities *were* observable, for example with Compton's ionization chamber (section 2.2.1).[98] Dirac's negative energy states neither were *in principle* unobservable, like Bohr's orbits, nor were they directly related to observation. At least, Dirac did not relate them explicitly in an equation. Dirac *needed* ψ_3 and ψ_4, but he also "rejected [them] as referring to the charge $+e$ on the electron," because that implied the existence of never-before-observed phenomena. This was a much weaker argument than the in-principle rejection of Bohr's orbits. When did this novel scientific result count as a prediction? Not yet.

By rejecting the negative energy states, Dirac was able to show that his relativistic theory matched the latest studies of the spectral lines of hydrogen. But his published paper contained only a partial and approximate analysis.[99] In 1928, Dirac wrote follow-up papers that further developed the connection to hydrogen spectroscopy and the hydrogen-like alkali metals. Darwin immediately picked up Dirac's equation (2.3) and worked through in detail how Dirac's theory could produce the hydrogen spectrum.[100] Darwin, Pauli, Gordon, and others transformed Dirac's simple rejection of the negative energy states into a more rigorous approximation procedure (which assumed that the negative energy solutions are non-zero, but small). The approximation had lasting successes, but also failed in studies of the nucleus.[101] In radiation theory, Dirac's relativistic equation was applied to the Compton effect with great success by Oskar Klein and Nishina Yoshio in 1928.[102] But in radiation theory, the negative energy solutions could not be so easily dismissed.[103]

2.3 "Hole" Theory

In two years, "The Quantum Theory of the Electron" (1928) was transformed into "A Theory of Electrons and Protons" (1930), in which protons were "holes" in a distribution of negative-energy electrons. During this time Dirac alternated between

---

98. Dirac, "Relativity Quantum Mechanics with an Application to Compton Scattering," 413; cf. Dirac, "Relativistic Quantum Mechanics," 456–57.

99. See Kragh, "Dirac's Relativistic Theory," 61; Kragh, *Dirac*, 61–62.

100. Dirac, "The Quantum Theory of the Electron," §4; Dirac, "The Quantum Theory of the Electron. Part II," §§2–3; Dirac, "Quantentheorie des Elektrons"; Darwin, "Wave Equations."

101. Technically, Darwin first wrote down an exact solution, before approximating. See Moyer, "Evaluations of Dirac's Electron"; successes: Bethe and Salpeter, *Quantum Mechanics of One- and Two-electron Atoms*, §§10–14; Weinberg, *QTF* vol. 1, ch. 14, cf. 6–14.

102. Ito, "Superposing Dynamos and Electrons."

103. The distinction between atomic applications and radiation applications reflects "actors categories," e.g., Pauli, "Théories quantiques du magnétism," 237.

relatively solitary stints at St. John's College and wide-ranging travels and lecturing.[104] Dirac spent another six months in Cambridge before crossing the Atlantic in March 1929 to take up a two-month position at the University of Wisconsin, Madison. Heisenberg was simultaneously visiting Compton at the University of Chicago. Heisenberg had been invited by Nishina to travel to Japan, and he invited Dirac to join him. They met at Berkeley, California, and set sail. During their Japanese travels they played go. A key element in go strategy is constructing vacant areas—called eyes—surrounded by one's own stones (pieces). Only formations with at least two eyes can survive to the end of the game and count for points. It seems Dirac was better than Heisenberg. In a 1930 letter about Dirac's new "hole" theory, Heisenberg mentioned that he recently "played 'Go' with a friend in Leipzig. I hope to beat you in Brussels in Go."[105]

After lecturing and touring in Japan, they parted company. Heisenberg returned to Europe by steamer; Dirac completed his circumnavigation of the globe. He traveled from Japan to the Eastern terminus of the Trans-Siberian Railway at Vladivostok, and then by rail to Moscow and Leningrad to visit again with his Soviet colleagues. Showing his taste for adventure, Dirac *flew* from Leningrad to Berlin—since 1928 the German-Soviet airline Deutsch-Russische Luftverkehrs operated flights with five- and eight-person monoplanes. He wrote to Tamm that the "air journey from Leningrad to Berlin was very enjoyable."[106] By October 1929, Dirac was back in his spare rooms at St. John's College, Cambridge, and back working on his relativistic electron theory.[107] In November, he wrote to Bohr with his new theory of electrons and protons.

Was all of space filled with electrons? Were protons not, in fact, *material* at all? Were protons instead negative spaces—*holes* in their surroundings? Dirac made a fantastic proposal. And yet Dirac described this to Bohr as a "simple way of avoiding the difficulty of electrons having negative kinetic energy."[108] In retrospect, he considered "A Theory of Electrons and Protons" to be only "a small step forward."[109] Is it possible to see Dirac's theory not as a strange leap, but as a small, simple, step? It is a major goal of cultural analysts of science to reveal seemingly inhuman leaps of genius to be composed of smaller steps taken in a specific context, by grounded human beings. Historian Andrew Warwick called for just such an analysis of Paul Dirac at the end of his study of Cambridge mathematical physics, *Masters of Theory*.[110] Indeed, close attention to Dirac's many-electron work reveals that when he developed his theory, he was not wearing seven-league boots, even if the path he walked was winding.

104. See Dirac to Tamm, October 4, 1928 in Kojevnikov, *Dirac and Tamm*, 10–11; Kragh, *Dirac*, travels: 67–69, politics: 154–61; Farmelo, *Strangest Man*, 227–29.

105. Heisenberg to Dirac, July 14, 1930. Dirac papers, Series II, Box 2, Folder 3. Heisenberg's friend was likely either Fujioka Yoshio or Kikuchi Seishi, young Japanese physicists who visited Leipzig in this period. Ito, "Making Sense of Ryôshiron," 285.

106. Dirac to Tamm, 21.2.1930. Kojevnikov, *Dirac and Tamm*, 35.

107. The preceding two paragraphs draw from Kragh, *Dirac*, 64–77; see also Brown and Rechenberg, "Paul Dirac."

108. Dirac papers, Series II, Box 14, Folder 2. Kragh, *Dirac*, 90–91.

109. Dirac, AIP Interview, session 5.

110. Warwick, *Masters of Theory*, 506–11; cf. Pickering, *Mangle of Practice*.

It is well known that Dirac drew on many-particle quantum mechanics to create his "hole" theory, but this was not only a metaphor or "visual inspiration."[111] This section shows how Dirac's many-electron theory provided the specific formal elements that he extended to form his new theory of electrons and protons. In particular, he drew on approximation strategies from the theory of complex atoms. This allowed him to give meaning to "a definite specifiable *place* in the [wave] function."[112] The formalism of quantum theory needed to be secure enough to hold its shape when a hole was punched through it.[113]

2.3.1 Radiation Theory and Waller

On July 24, 1928, Dirac wrote to Oskar Klein in Copenhagen: "I have not met with any success in my attempts to solve the $\pm e$ difficulty. Heisenberg (whom I met in Leipzig) thinks the problem will not be solved until one has a theory of the proton and electron together."[114] In the spring of 1929, while Dirac was in the United States, the mathematician Hermann Weyl put forth the first proposal to take the negative-energy states seriously: he proposed that they represented protons, the only positively charged particle known to physics.[115] Faced with the 2,000-fold difference between the masses of the electron and the proton, Weyl concluded that science's best theory of mass must have a role in electron-theory. He made perhaps the first attempt to integrate Dirac's theory into *general* relativity.[116] (Weyl's formalism became very important for Roger Penrose; see chapter 6.) But Dirac was not in favor.

In my judgment, Dirac was finally *pushed* into finding a physical interpretation for the negative energy states by the Swedish physicist Ivar Waller (1898–1991). This occurred between Dirac's return to Cambridge, in early October, and his letters to Bohr in late November 1929. Generally, Waller's research had de-idealized much of non-relativistic radiation theory. Where Compton and Dirac had worked with unrealistic individual electrons, Waller represented gases of atoms. Where they took the electrons to be free of any electric field, Waller represented electrons bound within atoms. Waller was realizing some of the suggestions Dirac made about statistical mechanics and radiation in 1926 (section 2.2.2). Much of this work was done in collaboration with Douglas R. Hartree, another of Fowler's students at the same time as Dirac.[117] Dirac learned of Waller's work in discussion at the 1927 Solvay conference. There, W. Lawrence Bragg reported on Waller and Hartree's collaboration

---

111. Mehra and Rechenberg, *HDQT*, vol. 6, 777, but see fn875.
112. Emphasis added, Dirac, *Principles* (1930), 209. This was another change in his thinking. In 1924, Dirac disliked Bose's "cells" in phase space, Darrigol, "Quantized Matter Waves," 215–16. Similarly, Dirac's 1926 "On the Theory of QM" followed de Broglie in assigning particles to waves instead. But in 1930, Dirac explicitly identified a "hole" with a "vacant cell of phase space." Dirac, "On the Annihilation of Electrons and Protons," 375.
113. In Pickering's terms, this formal "place" was Dirac's "bridgehead" between X-ray atomic levels and his relativistic theory. Pickering, *Mangle of Practice*, 116.
114. Quoted in Pais, "Playing with Equations," 99.
115. Kragh, *Dirac*, 89–90, 95.
116. Weyl's paper arrived in Leningrad while a paper by Fock was in press. Fock, "Geometrisierung der Diracschen Theorie des Elektrons."
117. Waller, "Transition from Ordinary Dispersion into Compton Effect"; Waller, "Memories," 123.

with the experimentalist Reginald W. James at Manchester. The trio provided strong evidence, from X-ray scattering off rock salt, that atoms retained a zero-point energy: a half quantum of energy per atom even at absolute zero temperature.[118] Both Waller and Dirac attended the first conference at Bohr's Institute in Copenhagen in April 1929.[119] It is not clear, however, if Dirac learned of Waller's work in person or through correspondence in late 1929.

In their winter correspondence, Bohr identified the central problem for Dirac's relativistic electron theory. It was not the possibility of never-before-observed transitions from one to the opposite charge. It was not the seeming paradoxes elaborated by Bohr's assistant, Klein. On December 5, 1929, Bohr wrote: "The question is, how much those features of the theory which claim the transitions [to negative energy] are involved in the problems, where your theory has been found in so wonderful agreement with experiments."[120] Compton scattering was the prime example.[121] If the states could be ignored, or downplayed *consistently*, so that connection to experiment was maintained, there would be little worry. Dirac replied negatively:

> There is one case where transitions of electrons from positive to negative energy levels does give rise to serious practical difficulties, as has been pointed out to me by Waller. This is the case of the scattering of radiation by an electron, free or bound. [...] Detailed calculations of this have been made by Waller.[122]

Darwin's atomic spectroscopy was not in question. Radiation theory was the proving ground for the negative energy states.

Almost certainly the calculations to which Dirac referred in his letter to Bohr were published by Waller as "The Scattering of Radiation by Bound and Free Electrons in Dirac's Relativistic Mechanics."[123] However, the chronology is somewhat delicate. Waller's paper was received by the *Zeitschrift für Physik* on February 12, 1930, and published in the November issue, *after* Dirac had explained his "hole" theory to Bohr. However, it is possible to infer which part of the work was discussed by Waller and Dirac *before* the "hole" theory. Waller's paper was divided in two parts according to the case of electrons bound to an atom, or for the free electron gas approximation.

118. James, Waller, and Hartree, "Zero-point Energy"; Mehra and Rechenberg, "Planck's Half-quanta," §2.3.2; Kragh, "Preludes to Dark Energy," 211; the question of a zero-point energy of matter was open somewhat later than Kragh's dating, to Robert Mulliken's 1924 experiments. Pauli's attitude was negative in 1924–1925 and accepting in 1927–1928. Bacciagaluppi and Valentini, *Quantum Theory at the Crossroads*, 272–79, 291–95.

119. P. Halpern, "Quantum Humor," 283–84; Pais, *Inward Bound*, 348–50.

120. Bohr's letter is reproduced in full in Moyer, "Evaluations of Dirac's Electron," 1057–58.

121. In 1928–1929 Oskar Klein and Nishina Yoshio successfully updated Dirac's 1926 Compton-effect equation with the new relativistic basis. In 1928, Dirac's colleague Louis Harald Gray's experiments suggested that the Klein-Nishina formula best fit the available scattering data. Mehra and Rechenberg, *HDQT*, vol. 6, 307–9, fn351; Roqué, "Manufacture of the Positron," fn53; Ito, "Superposing Dynamos and Electrons"; cf. Kragh, *Dirac*, 88–89.

122. Kragh, *Dirac*, 93.

123. Waller, "Die Streuung von Strahlung durch Gebundene und freie Elektronen nach der Diracschen relativistischen Mechanik"; Kojevnikov, *Dirac and Tamm*, 33; some historians refer to a different paper, which is about the self-energy of the electron. Waller, "Bemerkungen über die Rolle der Eigenenergie des Elektrons in der Quantentheorie der Strahlung"; e.g., Kragh, *Dirac*, 331 fn17.

Waller credited Pauli with the suggestion to extend his initial bound electron results to the free electron case.[124] Concomitantly, on February 21, 1930, Dirac told Tamm that "Waller, so far as I know, has not been dealing with free electrons, but only bound ones (where the same difficulty of intermediate states of negative energy occurs)."[125] From this, it is reasonable to conclude that the calculations which Dirac saw correspond to the first part of Waller's paper, which discussed bound electrons.

Waller considered an electron in an atom, first, absorbing an X-ray and rising to a higher energy level; then, emitting an X-ray and falling back into its normal energy level within the atom. Recall that solutions to Dirac's equation (2.3) had four components, with the third and fourth ψs representing negative-energy states. Waller found that, "[i]n the calculation of [the scattered electromagnetic field] only the 3- and 4-components of the wave functions ψ_n, ψ_m, and ψ_s, substantially come into consideration," in which n, m, and s label the initial, final, and intermediate stationary states of the electron, respectively (see section 3.2). Waller's next paragraph pointed toward Reginald James's X-ray crystallography: according to his analysis, the sum of negative-energy states,

Σ^- is that portion of [the emitted radiation] which is the most important for describing the scattering of X-rays by atoms. This is most clearly [schärften] demonstrated by R. W. James, in Manchester, and his associates. This shows in compelling ways the need to take the states of negative energy into account.[126]

Waller also cited unpublished calculations by Heisenberg, who had found that negative energy states were required by Dirac's theory.[127]

This was the conclusion that likely set Dirac on the course for a physical interpretation of the negative energy states. Waller set a choice before Dirac: accept the negative energy states or lose your theory's grasp on radiation experiments. He chose agreement with experiment and accepted the states.[128] There is a parallel here between Dirac's decision to give a physical interpretation of the negative energy states—as a last resort—and historian Peter Galison's analysis of experimentalists' declaration that they had observed weak neutral currents. They had to be convinced that they could *not* make the novel phenomenon disappear.[129] Once he was forced to accept the negative energy states, the question for Dirac was what to do with them.

124. Waller, "Die Streuung von Strahlung durch Gebundene und freie Elektronen nach der Diracschen relativistischen Mechanik," 844.

125. Kojevnikov, *Dirac and Tamm*, 35.

126. Waller, "Die Streuung von Strahlung durch Gebundene und freie Elektronen nach der Diracschen relativistischen Mechanik," 843.

127. See Heisenberg to Pauli, 31.7.1928, Pauli, *Wissenschaftlicher Briefwechsel*, vol. 1, doc. 204; Kojevnikov, *Dirac and Tamm*, 33 fn3.

128. Compare a physicist's view: Gottfried, "Quantum Electrodynamics," 117.

129. Galison, *How Experiments End*, §4.17.

2.3.2 Permutations of Hartree's Approximation

Douglas R. Hartree (1897–1958) introduced an approximation procedure in 1928 that allowed *quantitative* calculations of atomic properties, far beyond the Bohr-Sommerfeld theory. Following on Hartree's heels, yet another of Fowler's students stepped in to put Hartree's empirical work on a firmer mathematical footing: John Arthur Gaunt (1904–1944).[130] Dirac, Hartree, and Gaunt were colleagues at Cambridge, and within Fowler's circle. Dirac enthusiastically took up Hartree's method in Gaunt's form in his many-electron calculations.[131] Hartree and Gaunt's 1928 non-relativistic work was a centerpeice of Dirac's 1929 "Quantum Mechanics of Many-Electron Systems," a paper that set the path for solid-state physics to explain phenomena such as magnetism.[132] This many-electron formalism lay the foundation for the "hole" theory.

In December 1929, Dirac gave four days of lectures at the Institute Henri Poincaré in Paris, with the generic title "Several Problems of Quantum Mechanics"; it was his first public airing of the "hole" theory.[133] Much of the material was taken from his many-electron publications from earlier in 1929 (with some new results). Though he did not reflect on it, Dirac's analysis of many-electron systems had shifted since 1926. Earlier, Dirac rejected "degenerate" energy levels for many-electron systems (energy levels that corresponded to more than one stationary state). By contrast, Heisenberg's 1926 many-electron work allowed for degenerate levels. This made room for an "exchange" (*Austausch*) effect, which resulted in an attractive force between the particles. Heisenberg offered this physical picture:

> If one wanted to make oneself an intuitive [*anschauliches*] picture of the motions of the electrons in the atom corresponding somewhat to the quantum mechanical so-lution, then one would perhaps have to imagine that the two electrons periodically change places in a continuous manner.[134]

By 1929, Dirac had adopted Heisenberg's exchange idea, and developed an influential formalism to express it. Dirac made contributions to ferromagnetism, many-electron atomic theory, and Walter Heitler and Fritz London's theory of molecular bonding.

Dirac's new mechanics of many-electron systems rested on Hartree's approximation and Gaunt's formalism. In Paris, Dirac explained:

> The old [i.e. Bohr's] theory of orbits is now replaced by Hartree's self-consistent field method, based on quantum mechanics. This retains the simplifying hypothesis of the old theory, that each electron has its own individual orbit; but the "orbit" is now a quantum state of the electron under consideration, represented by

130. Hartree, "Wave Mechanics. I"; Hartree, "Wave Mechanics. II"; Gaunt, "Hartree's Atomic Fields."
131. Jeffreys, "John Arthur Gaunt (1904–1944)"; Simões and Gavroglu, "Quantum Chemistry," §3.
132. Dirac, "Quantum Mechanics of Many-electron Systems," 715; Kragh, *Dirac*, 76.
133. Dirac, "Quelques problèmes de mécanique quantique"; Kragh discusses this in, Kragh, *Dirac*, 94, 96.
134. Carson, "Exchange Forces—II," quote: 30, 42–45; Gavroglu and Simões, *Neither Physics nor Chemistry*, ch. 1.

a wave function in three dimensions. [...] Gaunt gives a theoretical justification for Hartree's hypothesis.

Hartree's approximation fundamentally contradicted Dirac's (and Heisenberg's) approach to many-electron systems, which insisted that only *holistic* questions could be asked of them (section 2.2.2). Hartree did not contest the validity of Dirac and Heisenberg's work, per se. But nevertheless, he represented each electron in an atom as an individual—as in hydrogen—that made his numerical calculations possible.[135] With the acknowledgment that it was approximate, Dirac made extensive use of Hartree's work.

However, Dirac explained that there was a deficiency in Hartree's picture. Like the old quantum theory, it had to simply *assume* that there was a large interaction between the electrons in an atom. The solution came from Dirac's realization that Heisenberg's "exchange" interaction could be integrated with Hartree's approximation. "The resolution of this difficulty is given by the *exchange* (*Austausch*) interaction of the electrons, which originates from the fact that the electrons are identical and indistinguishable from each other. Two electrons can permute themselves without our noticing." If the exchange interaction could be synthesized with Hartree's approximation, Dirac would have a theory of complex atoms with fewer arbitrary assumptions.[136] And the resulting theory would be formally closer to his holistic statistics.

In pictorial terms, Dirac's formalism for orbits and exchanges created the solid ground that was necessary for a "hole" to be seen. He began by giving each wavefunction, usually written $\psi(q_r)$, an extra place on the page, in terms of its position variables, q_r, and orbit α. He wrote, "each of the identical particles has its own 'orbit' represented by a wavefunction $(q_r|\alpha)$ depending only on the coordinates q_r of the particle." The q_rs could be position coordinates, x, y, z, or atomic quantum numbers, and the αs were the orbits of atomic theory (two 1s-orbits, six 1p-orbits, etc.).[137] In 1926, Dirac had considered exchanges, or permutations, of the coordinates only (section 2.2.2). With two labels for each wavefunction, Dirac explained,

it is necessary to distinguish between two sorts of permutations: those of the q's and those of the α's. [A permutation] may be regarded either as the transposition of the objects labelled 2 and 3, or as a transposition of the objects in the places 2 and 3, the objects in question being the q's.[138]

Figure 2.6 shows the two types of permutations. He took permutations very seriously, arguing that an operator for performing a permutation could be a constant of the motion of a dynamical system and a quantum observable, like position or

135. Hartree, "Wave Mechanics. II," 112–14.
136. Dirac, "Quelques problèmes (Eng.)," 541–43; in 1930, Fock made a fuller analysis, and it is now called the Hartree-Fock approximation. Fock, "Approximate Method"; Schweber, "Young John Clarke Slater," 378.
137. Dirac, "Quelques problèmes (Eng.)," 567; this was the beginning of the influential bra-ket notation, introduced a decade later, Dirac, "New Notation."
138. Dirac, "Quelques problèmes (Eng.)," 569.

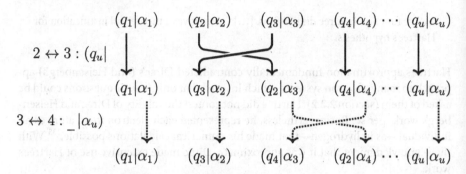

Figure 2.6. Dirac's two types of permutations. At top, the electron coordinates $(q_2|$ and $(q_3|$ are interchanged. At bottom, the coordinates at the orbits $|\alpha_3)$ and $|\alpha_4)$ are interchanged. The αs remain fixed in place on each line. Credit: author.

momentum.[139] Dirac was perhaps following A. S. Eddington, who called for "a wider principle of relativity which admits, in addition to a rotation of space and time a 'rotation' which will interchange the identity of two electrons."[140]

Dirac emphasized that permutation of the orbits, α, required a well-defined mathematical apparatus. The overall wavefunction for a system with u electrons was written:[141]

$$(q_1|\alpha_1) \times (q_2|\alpha_2) \times \cdots (q_u|\alpha_u) = (q|\alpha). \tag{2.5}$$

Then, "a permutation interpreted in the second way has a meaning only when applied to a function of the q's in which each of the q's *has a well-determined and labelled place*. [emphasis added]"[142] It would do no good to refer to permutations of the coordinates at $|\alpha_4)$ if $|\alpha_4)$ lacked a clear meaning. Close attention to *all* the orbits available in a given atom was a defining characteristic of Gaunt's re-presentation of Hartree's work. Where Hartree calculated the energy of a given orbit, say, the 6th 4p-orbit of rubidium, Gaunt realized that the analysis must include solutions to the wave equation for the remaining p-orbits, and even orbits for other stationary states.[143] Gaunt represented these orbits, even when there was no electron occupying them.

139. Dirac, "Quantum Mechanics of Many-electron Systems," §§3–4; Dirac, "Quelques problèmes (Eng.)," 547–67; on permutations, see Mehra and Rechenberg, *HDQT*, vol. 6, §III.4(c); Scholz, "Introducing Groups"; Kragh, *Dirac*, 77.

140. Eddington, "Charge of an Electron," 360.

141. He worked in the approximation of ignoring the Coulomb interaction between electrons.

142. Dirac, "Quelques problèmes (Eng.)," 569; similarly: Dirac, "Quantum Mechanics of Many-electron Systems," 727; Dirac, *Principles* (1930), 209.

143. Gaunt, "Hartree's Atomic Fields," eqs. 3.01–3.03.

PAUL DIRAC'S SEAS AND BUBBLES 59

2.3.3 Protons and X-ray Levels

In his last lecture in Paris, Dirac introduced "A theory of electrons and protons." It may have surprised his audience that his first topic was a "Theory of X-ray levels." Above, I suggested that Waller's calculations may have directed Dirac's attention to X-ray absorption and emission in radiation theory. X-rays were, of course, at the center of Dirac's own work on the Compton effect. They also featured strongly Hartree's and Gaunt's research. X-rays were already part of a discourse about vacant spaces within atoms. In the Bohr-Sommerfeld atomic theory, the basic picture of X-ray emission was drawn in 1914–1916. Bohr had proposed that atomic *optical* radiation was caused by electrons transitioning between stationary states of the atom. Walther Kossel proposed that *X-ray* radiation was caused by "inner" electrons, close to the nucleus of large atoms, being kicked up to a higher energy and then falling back to fill the space left in the "core" of central electrons (n.b., not the nucleus). When the inner gap was filled, the atom transitioned from an excited state back to its ground state, and released radiation equivalent to what the electron had lost in its fall; this was an X-ray. In 1924 at Cambridge, Edmund Stoner proposed a reorganization of Bohr's scheme for how electrons produced a shell structure in each atom which mirrored the structure of the periodic table (the *Aufbauprinzip*). Stoner noticed that as atoms filled in a d-shell—across the ten elements from potassium to copper—the magnetic properties of the first atoms equaled those of the last. He proposed that "the presence of x electrons [...] may be expected (by a sort of Babinet principle!) to produce the same paramagnetic properties as $(10 - x)$ electrons." Babinet's principle was drawn from optics. It stated that the diffraction pattern caused by a circular disk was exactly equal to the diffraction pattern caused by an opaque sheet with a hole of the same radius as the disk. In 1925, Pauli elevated Stoner's observation into a "general reciprocity law" between a stationary state with x electrons and a stationary state with $m - x$ electrons, where m was the degeneracy of the state—the number of electrons with the same energy. (This was the same paper in which he introduced the exclusion "rule.")[144] Before the advent of quantum mechanics, there was a quite developed treatment of X-ray "gaps" (*Lücken*) and X-ray emission by "an electron falling from outside the atom into a vacant space."[145]

This was all well and good, but after four years of work on matrix mechanics, wave mechanics, and q-numbers, Dirac was unlikely to fall deeply back into Bohr's orbits. Hartree's approximation offered a middle ground. In Paris, Dirac "consider[ed] a heavy atom thus having many electrons." He first appealed to an idealization developed by Enrico Fermi and independently by Llewellyn Hilleth Thomas (another of Fowler's students). They took the electrons near a large nucleus to be "distributed exactly as in a saturated gas." This was meant to capture, in statistical quantum theory, Bohr's picture of the ground state of an atom, in which each electron was in its state of

---

144. Stoner, "Distribution of Electrons among Atomic Levels," 733; Pauli, "Über den Zusammenhang des Abschlusses der Elektronengruppen im Atom mit der Komplexstruktur der Spektren," 778; Jammer, *Conceptual Development*, 138–44; Heilbron, "Kossel-Sommerfeld Theory"; Wheaton, *Tiger and the Shark*, 227-28, 264-70; Enz, *No Time to Be Brief*, 123; Peierls recalled Heisenberg telling him of this result, Hoddeson, Baym, and Eckert, "Quantum Mechanical Electron Theory of Metals," fn152.
145. Foote and Mohler, *Origin of Spectra*, 22.

lowest possible energy.[146] Constructing the Fermi-Thomas wavefunction was a step in Hartree's more elaborate approximation procedure.[147] Dirac continued: "If we take one electron away we will have an ion, whose energy will be equal to that of an X-ray [energy] level." In his fluent French, Dirac explained his technical requirements for further analysis: "To construct a theory of these levels, we must first show that the empty place [*la place vide*] left after the departure of an electron *can be described by an ordinary wave function* (in three-dimensional space) exactly as for a real electron [*un électron réel*] [emphasis added]."[148] Could a gap in a valence structure have a wavefunction?

Hartree had discussed Pauli's reciprocity law, but wrote that a detailed analysis "involves developments of quantum mechanics beyond those applied in this paper."[149] Dirac listened to a better-developed proposal from Enrico Fermi at the Leipzig University week lectures in July 1928. Fermi applied approximation techniques to iodine. Iodine was one column to the left of the noble gases in the periodic table; according to the valence theory of chemistry, this meant that it was one electron short of a saturated valence shell. Fermi was trying to understand why electrically neutral iodine formed a stable ion with one extra electron (which filled the shell). He reported: "I have numerically calculated the eigenfunction of this space [*Platzes*] ... one finds as it should be, that its energy is negative, i.e. that the electron is bound to the atom."[150] However, Gaunt most fully expressed the result Dirac required. He analyzed the addition of a "series electron," s, to an atom with otherwise complete inner shells. Hartree's example was the restoration of an electron to ionized rubidium, $e + Rb^+ \rightarrow Rb$. Here, Gaunt's equation is less important than his remark: "The same equations apply to the X-ray terms, if s refers to the orbit whose electron is removed from the core, and the series electron is included in the set of orbits α."[151] This was just Pauli's reciprocity law, but Gaunt showed that the mathematics of wave mechanics could meaningfully apply to an empty orbit. According to Dirac's lecture, this technical result made possible his theory of X-ray energy levels.

Dirac's X-ray theory proceeded by considering a neutral atom with n regularly occupied states, $\alpha_1, \alpha_2, \ldots, \alpha_n$, which had been ionized such that there were only u electrons actually present. Schematically:

$$u = n - 1; \quad (q_1|\alpha_1), (q_2|\alpha_2), \ldots (q_u|\alpha_u), \_|\alpha_n). \qquad (2.6)$$

He showed that, because of the properties of permutations, the combined wavefunction for the u electrons could be expressed in terms of a single wavefunction for the empty orbit, α_n.[152]

146. Mehra and Rechenberg, *HDQT*, vol. 6, 460.
147. Hartree, "Wave Mechanics. II," 114.
148. Dirac, "Quelques problèmes (Eng.)," 589–91; cf. Dirac, "Theory of Electrons and Protons," 363; Kragh, *Dirac*, 96.
149. Hartree, "Wave Mechanics. II," 112; he cited: Hund, *Linienspektren und periodisches System der Elemente*, 144ff.
150. E. Fermi, "Über die Anwendung der statistischen Methode auf die Probleme des Atombaues," 303.
151. Gaunt, "Hartree's Atomic Fields," 342.
152. Dirac, "Quelques problèmes (Eng.)," 591–92.

Following his analysis of X-ray levels, Dirac presented his relativistic equation and its negative energy states. He asked: "How might such states appear in our experiments?" As he noted in 1927, they would behave as if they had a positive charge. But Dirac rejected Weyl's proposal that negative energy states were protons (sec. 2.3.1). It implied protons would behave in strange, never observed ways. Instead, he introduced his "hole" theory:

> The connection between negative-energy electrons and protons can be established in a different way. We will make the following hypothesis: almost all the negative-energy states in the universe are occupied; it is the empty places which constitute the protons.
>
> From the general theory of X-ray levels which we gave at the beginning of this lecture, one of these empty places behaves rather like a true particle, and can be described by a wavefunction.[153]

Here, Dirac relied on the technical results of his theory of X-ray levels to substantiate his relativistic "hole" theory. (I discuss archival evidence of Dirac employing this formalism in section 2.3.4.)

This proposal had several difficulties. The one Dirac himself most emphasized was that from the theory he presented, one would expect that the electron and "hole" should have the same mass. The proton, however, was known to be about 2,000 times heavier than the electron. In his November 26 letter to Bohr, he explained:

> I have not yet worked out mathematically the consequences of the interaction. It is the "Austausch" effect that is important and I have not yet been able to get a relativistic formulation of this. One can hope, however, the a proper theory of this will enable one to calculate the ratio of the masses of proton and electron.[154]

This indicates that Dirac thought that a characteristically many-particle effect was the most important area of investigation for his new relativistic theory.[155] In "A Theory of Electrons and Protons" (received December 9), he referred to Eddington:

> The consequences of this dissymmetry are not very easy to calculate on relativistic lines, but we may hope it will lead eventually to an explanation of the different masses of proton and electron. Possibly some more perfect theory of the interaction, based perhaps on Eddington's calculation of the fine structure constant e^2/hc, is necessary before this result can be obtained.[156]

153. Dirac, "Quelques problèmes (Eng.)," 605; cf. Dirac, "On the Annihilation of Electrons and Protons"; Moyer, "Evaluations of Dirac's Electron."

154. A facsimile of the letter is found in Dirac papers, Series II, Box 14, Folder 2. It is also reproduced in Kragh, *Dirac*, 90–91.

155. Dirac was not alone here. Heisenberg and Pauli noted that exchange-style integrals appeared in their QED: Heisenberg and Pauli, "Zur Quantendynamik der Wellenfelder," 53.

156. Dirac, "Theory of Electrons and Protons," 364.

Table 2.1. The analogy between Dirac's "hole" theory of electrons and protons and many-electron quantum mechanics.

"Hole" theory	Many-electron QM
particles go to lowest energy	Bohr's *Aufbau* principle
saturated *e* distribution	Thomas-Fermi-Dirac atom
+ve energy hole	−ve energy valence gap
electron falls into hole, emits radiation	Kossel's X-ray theory
uniform saturation=unobservable	complete valence shell=unreactive
e-p mass difference	ortho-para- helium levels (?)

Eddington's major innovation, in the paper Dirac cited, was his integration of permutation symmetry into the discussion of Dirac's equation.[157] In his last lecture at Paris, on December 20, his final lines were:

> The interaction between the electrons, treated according to the general method of the beginning of this lecture may well be able to give rise to an asymmetry between electrons and protons, but I think that it will be necessary to improve our theory of the interaction before it will be possible to calculate the ratio of the masses with precision.[158]

This general method was, of course, his theory of X-ray levels.

My analysis goes some way toward re-scaling Dirac's proposed "hole" theory from a seven-leagued stride to a mountaineer's considered step. Within the work Dirac was reading, citing, and writing in 1929 was: X-ray absorption and emission; wavefunctions for orbits; the equivalence of systems of $(n-1)$ electrons and systems of one electron; saturated distributions of electrons; and even wavefunctions for *empty* orbits. Table 2.1 sets out the main features of Dirac's theory of electrons and protons, and an analogous feature in many-particle theory. Not all of these parallels were explicitly drawn by Dirac. How might the exchange interaction have addressed the mass difference between electron and proton? One possibility was that Dirac hoped to replicate Heisenberg's exchange analysis of the helium spectrum. According to Heisenberg, the electrons in the two species of helium molecules (ortho- and para-helium) obeyed different statistics (Bose-Einstein and Fermi-Dirac, respectively). Heisenberg was able to show how the exchange interaction reproduced the different *energy* levels of the helium spectra.[159] Of course, via $E = mc^2$, an energy difference was equivalent to a mass difference.

The framework that enabled Dirac's detailed analogy was provided by Hartree and Gaunt's approximation work. Where might Dirac have seen this approximation applied to relativistic electrons? From none other than Hartree and Gaunt. In April 1929, Hartree published "The Distribution of Charge and Current in an Atom Consisting of

---

157. Eddington, "Charge of an Electron"; Kragh, *Dirac*, 331 fn12; see also Breit, "The Effect of Retardation on the Interaction of Two Electrons," fn1 and fn3.
158. Dirac, "Quelques problèmes (Eng.)," 607.
159. Carson, "Exchange Forces—I," 27–31.

Many Electrons Obeying Dirac's equations." True to form, in May, Gaunt published "The Relativistic Theory of an Atom with Many Electrons," which put clearer mathematics under Hartree's work. Gaunt thanked Dirac for criticisms of the paper.[160] (This work formed a basis for relativistic quantum chemistry.[161])

An obvious *dis*analogy between the "hole" theory and atomic theory is that scientists saw good reasons to suppose that there were electrons outside an atomic nucleus. Even if the Thomas-Fermi picture was less than accurate, few questioned the late nineteenth-century idea that atoms contained electrons. The idea that the vacuum was in fact a plenum of electric charges had rather less support. As Fock wrote to Dirac on March 1, 1930, "infinite but unobservable masses and charges seems to me a serious disadvantage of the theory."[162] Was Dirac's argument for the unobservable sea an inexplicable leap after all?

Dirac argued that his negative energy states were acceptable because they were unobservable. This was a daring inversion of the empiricist-inspired foundation of quantum mechanics, that it would only calculate observable quantities. Here Dirac gave himself permission to assert the existence of almost any particle, so long as its consequences were *un*observable. But this maneuver, too, was rooted in Dirac's philosophical environment and his practice of theory. He claimed: "Our theory leads us to an infinite density of electrons, but completely uniform and regular, and hence unobservable. The only things accessible to experiment are electrons of positive energy and the empty places in the distribution of negative-energy electrons."[163] This argument was hardly accepted without comment. For example, Bohr reported that he and Klein could not see how electrostatics would bend to Dirac's cause.[164]

As was often the case, Dirac did not indicate whether he had drawn on the work of others in developing this claim.[165] Dirac studied Eddington's *The Mathematical Theory of Relativity*, and it included a parallel discussion. Eddington considered whether there was enough mass in the universe for Einstein's cosmological model to be correct:[166] "It has been suggested that since the amount of matter necessary for Einstein's world greatly exceeds that known to astronomers, most of it is spread uniformly through space and is undetectable by its uniformity." This concerned mass and gravitational fields, not charge and electric fields, but it was close to Dirac's case of uniform charge. In fact, the analogy went farther. Eddington continued:

> This is dangerously like restoring a crudely material aether—regulated, however, by the strict injunction that it must on no account perform any useful function lest it

160. Hartree, "Distribution of Charge and Current"; Gaunt, "Relativistic Theory of an Atom with Many Electrons," 176; cf. Fock, "On Quantum Exchange Energy," §4.

161. Bertha Swirles, Lady Jeffries, also a student of Fowler, made foundational contributions beginning in the 1930s. Byers and Williams, *Out of the Shadows*, ch. 16; Gavroglu and Simões, *Neither Physics nor Chemistry*, 154.

162. Dirac papers, Series 2, Box 2, Folder 1.

163. Dirac, "Quelques problèmes (Eng.)," 607; cf. Dirac, "Theory of Electrons and Protons," 362; with variations into positron theory: Dirac, "Quantised Singularities in the Electromagnetic Field," 61; Dirac, "Théorie," 205.

164. Moyer, "Evaluations of Dirac's Electron," 1057.

165. Bromberg, "Concept of Particle Creation," 187–88, fn82 and fn83.

166. Pais, *Subtle Is the Lord*, §15e; Wright, "Bringing Infinity to a Finite Place."

upset the principle of relativity. We may leave aside this suggestion, which creates unnecessary difficulties.[167]

Indeed, champion of the aether Oliver Lodge greeted Dirac's "hole" theory with enthusiasm.[168] Dirac accepted the difficulties of such a distribution in order to represent protons.

Formally, to do away with the infinite charge throughout space, Dirac redefined the vacuum. "A perfect vacuum is now to be considered as a region in which all the states of negative energy and none of those of positive energy are occupied."[169] This meant redefining the density of electric charge, ρ. The density featured in Maxwell's equation for the field, E, produced by a charge distribution:

$$\mathrm{div} E = -4\pi\rho. \tag{2.7}$$

Dirac argued: "It seems natural, however, to interpret the ρ in Maxwell's equation [(2.7)] as the departure from the normal state of electrification of the world," i.e. from a filled vacuum with no holes or free electrons.[170] An expression for the new density, ρ_N, might read:

$$\rho_N = \mathrm{const.} - \rho, \tag{2.8}$$

where the constant could be infinite, from the negative energy electrons. This is similar to Pauli's reciprocity law. Equation (2.8) has just the same form as a central result from Dirac's theory of X-ray levels for the Hamiltonian of the system of u electrons and one free orbit, n:[171]

$$H_{nn} = \mathrm{const.} - (r_1, r_2, \dots r_u | H | r_1, r_2, \dots r_u). \tag{2.9}$$

While in his X-ray theory, the constant was finite, in the relativistic theory, Dirac redefined the observable as a difference from an infinite constant. Chapters 3 and 4 discuss how this featured in "renormalized" quantum theory.[172] A first step to demystify Dirac's redefinition of the vacuum is to observe that redefining ρ was part of his many-electron calculations.

Dirac's striking redefinition of the charge density, ρ, was also rooted in the practice of interpreting quantum mechanics. Dirac's work was characterized by a particular coordination of formalism and interpretation: he strove for clear and unambiguous calculational rules which could be connected to experimental observation by

167. Eddington, *Mathematical Theory of Relativity*, 168; Bromberg, "Concept of Particle Creation," 189–90, fn90; see also Kragh, *Dirac*, 108.

168. Lodge to Dirac 9.9.1930, Folder 3, Box 2, Series II, Dirac Papers; Wright, "Beautiful Sea," 234.

169. Dirac, "The Proton," 606.

170. Dirac, "Theory of Electrons and Protons," 363; Dirac, "Quelques problèmes (Eng.)," 607.

171. I have used "n" for Dirac's "r" and rearranged the terms. Dirac, "Quelques problèmes (Eng.)," 593.

172. In his 1931 Princeton lectures, upon redefining ρ, Dirac remarked: "If one wanted actually to write down equations for this theory one may get infinities coming in but similar infinities have already appeared in the theory of radiation and it seems reasonable to suppose that one should first get over these infinities in the radiation theory and then employ the ideas thus discovered to overcome any infinities that occur in the theory of the negative energy electrons." Dirac Papers, Series II, Box 26, Folder 15, page 132.

clear principles.[173] In 1927's "The Physical Interpretation of the Quantum Dynamics," Dirac considered several "special assumptions" that connected formalism to observation, such as Born's rule that $|\psi|^2$ was a probability (section 2.2.3). Dirac's 1929 lectures in Paris gave several interpretations of ρ. As a density matrix, ρ had a meaning in classical statistical mechanics; in quantum theory, as the density of states of a Gibbs ensemble of systems; and as "an ordinary variable pertaining to an individual system." Dirac considered integrating his ρ method with a wavefunction approach by defining $\rho = \psi\psi*$.[174] This demonstrates Dirac's ease in redefining the central elements of his formalisms. With a well-defined formalism, interpretations could be tailored to fit the physical phenomena in question. In a famous 1931 passage, Dirac recommended that physicists

> employ all the resources of pure mathematics in attempts to perfect and generalise the mathematical formalism that forms the existing basis of theoretical physics, and *after* each success in this direction, to try to interpret the new mathematical features in terms of physical entities.[175]

It may never be easy to understand how there could be an infinite electric charge in the vacuum. But Dirac's self-understanding of his own methods makes it easier to understand how *Dirac* could see redefining ρ as a "natural" procedure.[176]

How did Dirac's new interpretation relate to its radiation and atomic theory applications? In the published version of "A Theory of Electrons and Protons," Dirac remarked only on the "application to scattering." He credited Waller with drawing his attention to the severity of the involvement of negative energy states in radiation problems. Without mathematical details, Dirac explained that his new theory replaced the transitions to negative energy states with transitions of electrons into and out of holes in the distribution of negative energy electrons. These new transitions fit exactly where the old were removed, leaving the calculations untouched (see section 3.2). Initially, he did not address atomic questions. However, Heisenberg, Tamm, Oppenheimer, and others quickly turned to the consequences of "hole" theory for atoms, in particular hydrogen. Dirac's new theory opened up a new process: the collision of an electron and proton, resulting in their annihilation and the release of radiation. (Particle creation and annihilation itself was not new in 1930.[177]) Tamm and Oppenheimer each calculated that Dirac's theory predicted that hydrogen

---

173. Heisenberg appreciated this in a letter to Dirac on April 9, 1926. Kragh, *Dirac*, 23.

174. Dirac, "Quelques problèmes (Eng.)," 523–39; cf. Dirac, "Basis of Statistical Quantum Mechanics"; Dirac, "On the Annihilation of Electrons and Protons," §2.

175. Dirac, "Quantised Singularities in the Electromagnetic Field," 60.

176. I emphasize self-understanding here, because not all of Dirac's work seems to me to actually fit his description. E.g., in the paper under discussion, Dirac's formal results relied heavily on the mathematical properties of objects which obey Born's probabilistic interpretation. He argued *from* the fact that the square modulus of an integral of a wavefunction is a probability: "In order that the integral may have a definite modulus." He did not first establish a result with wavefunctions and then seek their physical interpretation. Insofar as the probability interpretation was an "interpretation" or "special assumption" in Dirac's sense, his actual argument reversed the order of operations which he recommended in the same paper. Dirac, "Quantised Singularities in the Electromagnetic Field," 63.

177. Bromberg, "Concept of Particle Creation"; Galison, "Discovery of the Muon," §2.

atoms would almost instantaneously collapse as a collision of electron and "hole."[178] In March 1930, Dirac submitted a paper in which he further developed his Gibbs ensemble "density matrix" formalism; applied the density matrix to re-derive the Klein-Nishina formula for Compton scattering; and gave his own calculation of the unhappy result on the stability of hydrogen. In a prelude to his later "bra-ket" notation, he began writing wavefunctions as $(q|)$. The many-electron formalism was interwoven with scattering theory: "We choose our ψ_0 to represent an electron, (or rather a distribution of electrons,) at rest [...]."[179] On May 19, the paper was read before the Cambridge Philosophical Society. That day Dirac also read a "Note on Exchange Phenomena in the Thomas Atom," in which he added exchange effects to Fermi and Thomas's model.[180] In 1931 and 1932 he gave two seminar talks on the London-Heitler theory of molecular binding, which he had derived from his permutation formalism in 1929.[181] Dirac's integrated program of many-particle and relativistic research that began in 1924 continued into the 1930s.

2.3.4 A New Kind of Particle

The bulk of Dirac's drafts, calculations, and notes, which are preserved in his archive at Florida State University, were written in pencil on loose sheets of paper. Pages are rarely dated. Throughout his career, he also rearranged and commented on some pages. In 1963–1964, Dirac lectured at Yeshiva University on quantum field theory. In the lectures, he discussed developments since the 1940s: vacuum polarization and the Lamb shift in the energy levels of hydrogen (see section 4.3). On the verso side of the proofs of these lectures, Dirac returned to his calculation of vacuum polarization. Dirac inserted a note in this file on March 31, 1972, after which follow several more pages of calculations. The note reads: "Original calculation of vacuum polarization produced by an electrostatic field." He may have been prompted to return to this material by an upcoming conference with a historical theme.[182] Dirac noted that "There is an error of a factor of 2, appearing halfway down page 8," which caused "an error in formula (10) of the proofs." Comparison with Dirac's publications indicates that his 1972 comments referred to calculations that informed his report to the 1933 Solvay Congress.[183] I will return to these pages in the next section. Here I will discuss a further three undated pages, which were grouped with the Solvay calculations, but which do not follow a numbering scheme. A rough dating—independent of Dirac's filing—can be established by the notation. In 1939, Dirac introduced his "bra-ket" notation: for $(q|a)$ he wrote $\langle q|a\rangle$.[184] These pages use the older notation. The calculations themselves suggest that they may have been written *before* the vacuum polarization

178. Kragh, *Dirac*, 94, 99–103; Wright, "Beautiful Sea," §2.2.
179. Dirac, "On the Annihilation of Electrons and Protons," 367.
180. Dirac, "Exchange Phenomena in the Thomas Atom"; on notation: Dirac, "New Notation," 417.
181. Dirac, "Quantum Mechanics of Many-electron Systems," §6; Kragh, *Dirac*, 328 fn31.
182. Dirac, "The Development of Quantum Mechanics."
183. Dirac, "Théorie," eq. 10; apparently Dirac noticed the error and communicated it to Heisenberg: Heisenberg, "Remarks on the Dirac Theory," 179 fn10.
184. Dirac, "New Notation," 417; Kragh, *Dirac*, 177–78.

pages, perhaps in 1929–1931. They do not mention vacuum polarization or use the more sophisticated formalism developed by Dirac, Peierls, and Heisenberg in 1933–1934 (see chapter 3). They do record Dirac's attempts to calculate the interactions among the electrons in negative energy states, and the interactions between electrons and holes.[185] These calculations appear to record Dirac's attempts at many-electron interactions to produce different masses of electrons and "holes"; his attempts to save his "hole" theory as a theory of *protons*.

Figure 2.7 shows a page of Dirac's calculations. At the top left, he began with "states labelled $1, 2, 3 \ldots n$" and considered the eigenvalues of a dynamical variable u for the wavefunction $\Psi(u)$. (This followed his theory of X-ray levels, equation (2.6).) He then wrote the "<u>full</u> wave f[unctio]n" as a determinant with n rows and n columns.[186] He proceeded to investigate expressions for "<u>One hole</u>. Each point in domain of wave fn can be labelled by a single variable u. For Coulomb forces

$$(u_1 u_2 \ldots u_{n-1} | \mathbf{V} | u_1' u_2' \ldots u_{n-1}') \text{ "}$$

where $\mathbf{V}$ was the sum of the Coulomb interaction between the pairs of particles. As in his theory of X-ray levels, Dirac was working with $n - 1$ states. Next, he wrote down a version of Schrödinger's eigenvalue equation for a general operator $A(\rho)$: $A(\rho)\Psi_r(u_\rho) = \sum_t \Psi_t(u_\rho) A_{\rho r}$. He manipulated this expression for the eigenvalues, substituting in the determinants for the wavefunction Ψ. However, the full wavefunction at the top of the page had n rows and columns, at the center of the page, the determinants had only $n - 1$ columns. He denoted this missing nth state as Ψ_r, and wrote that it was "omitted" from the new expression. Dirac collected terms and used the simple property of exchanges in quantum theory: for antisymmetric wavefunctions (that follow Fermi-Dirac statistics), exchanging the positions of two particles (figure 2.6) multiplies the wavefunction by -1. The total effect of r permutations is simply $(-1)^r$, or in Dirac's notation $(-)^r$. As in his X-ray theory, this allowed him to find simple expressions for the properties of "holes":

$$(\textcircled{r} | A | \textcircled{r}) = A_{11} + A_{22} + \cdots A_{r-1,r-1} + A_{r+1,r+1} + \cdots = \sum_\rho A_{\rho\rho} - A_{rr}$$

The sum over A skips from $A_{r-1,r-1}$ to $A_{r+1,r+1}$ over the "hole" at A_{rr}. His result had the same form as the X-ray case, equation (2.9). Dirac then wrote down the contribution between to the interaction from occupied $\textcircled{p}$ and unoccupied $\textcircled{r}$ states. At the bottom left, he wrote:

Take new wave fn = old one $\times(-)^r$
Extra energy of hole is $-A_{rr}, -A_{rp}$

Dirac was clearly using his X-ray theory as more than a qualitative analogy or visual inspiration. This "hole" theory calculation just *was* a many-electron calculation.

---

185. Folder 2, Box 25, Series II, Dirac Papers.
186. Here he may have followed Slater. Schweber, "Young John Clarke Slater."

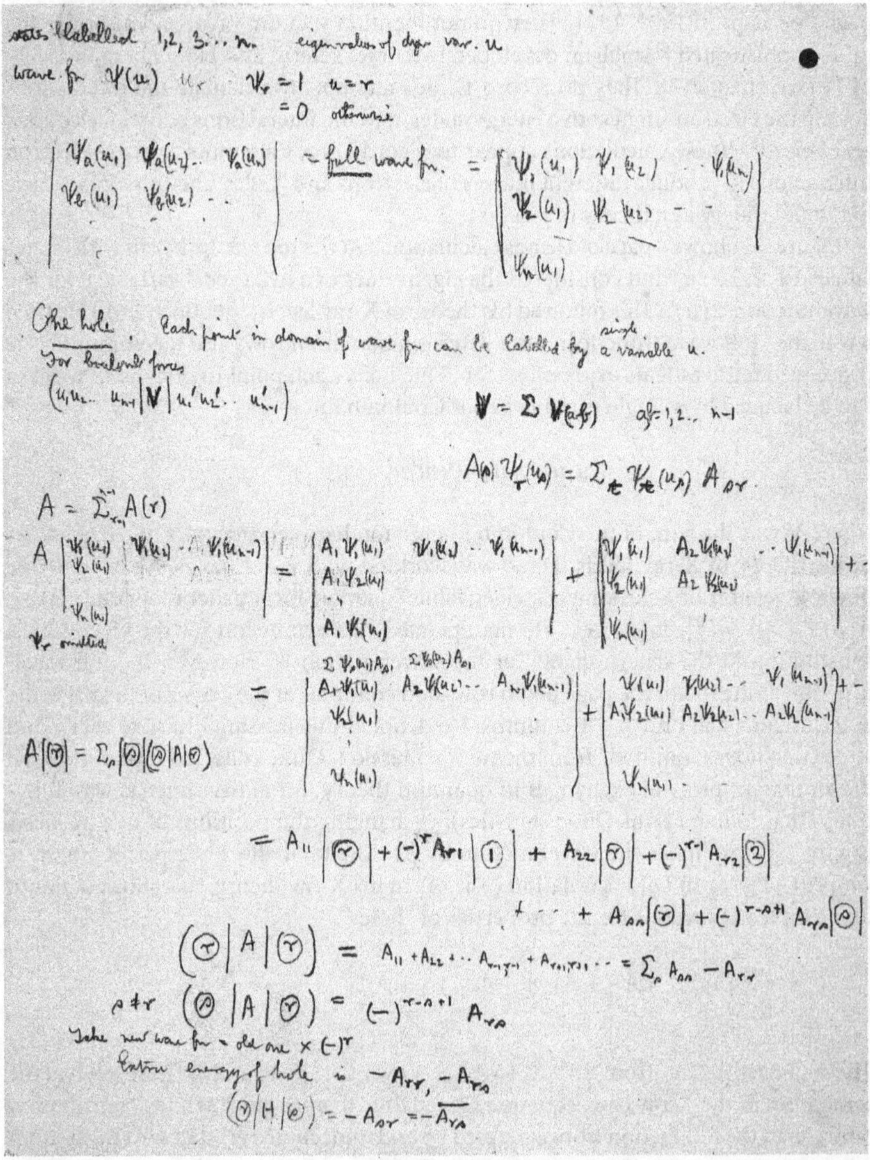

Figure 2.7. Dirac's calculations of the interaction between electrons and holes. Folder 2, Box 25, Series II, Paul A. M. Dirac Papers, Special Collections & Archives, Florida State University.

On the verso to figure 2.7, Dirac continued his calculation for a more complicated case than the eigenfunction $A(\rho)$, which was a function of a single set of states ρ. He turned to the Coulomb interaction, $V(u_1, u_2)$, which referred to the interaction between pairs of particles. He constructed expressions for the contributions of V to the Hamiltonian with the same (const. $-\sum_r$) form. One of these expressions was

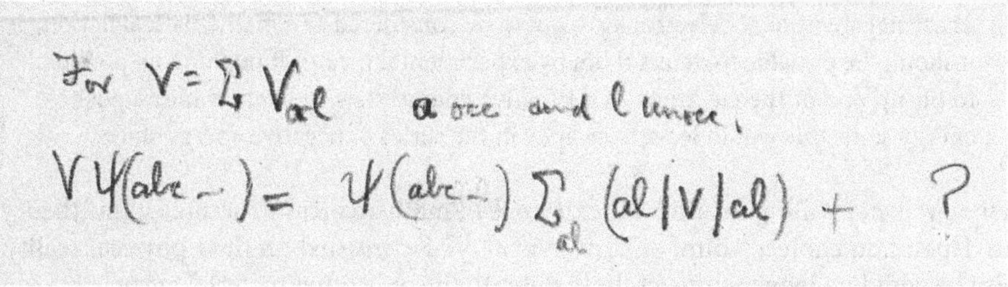

$$\text{For } V = \sum_{\alpha l} V_{\alpha l} \quad a \text{ occ and } l \text{ unocc,}$$
$$V\Psi(abc\cdots) = \Psi(abc\cdots)\sum_{al}(al|V|al) + \quad ?$$

Figure 2.8. Two lines from Dirac's calculations of the Coulomb interaction, V, between "a occ[upied] and l unocc[upied]" states, that is, between electrons and holes. Folder 2, Box 25, Series II, Paul A. M. Dirac Papers, Special Collections & Archives, Florida State University

"=energy of charge distribution $\overline{\Psi}_r(u_1)\Psi_\rho(u_2)$ in Thomas field." Dirac was making some progress, as he considered more complicated situations. On the final page, he moved from considering a single "hole" to the more general case of a wavefunction for multiple electrons, $\Psi(abc\ldots)$ and a wavefunction for multiple holes, $\chi(lmn\ldots)$. For the Coulomb interaction among *occupied* orbits, he wrote $V = \sum_{r,\rho} V_{r,\rho}$. Dirac was able to write out expressions for this operator applied to Ψ and χ. If he was to find some mechanism to explain the mass difference between electrons and protons in his new theory, there would have to be some difference between the electric interaction among the negative energy electrons and the interaction between these electrons and "holes." He wrote down an expression for the Coulomb interaction "For $V = \sum V_{al}$" with the labels "a occ[upied] and l unocc[upied]." What was the Coulomb interaction between an electron and a "hole"? Two-thirds of the way down the page, his calculations ended with a question mark, figure 2.8. Dirac had reached an impasse.

Neither in these nor any other calculation did Dirac succeed in calculating the mass difference between electrons and protons from his many-electron quantum mechanics. Neither the exchange effect nor the standard Coulomb interaction between charges produced vacuum electrons with a distinct mass from "holes." Tamm and Oppenheimer's problem of collapsing hydrogen atoms remained. Nevertheless, Dirac argued that the negative energy states could not be ignored. Waller had convinced him that the negative energy states were critical to the empirical success of his theory. Heisenberg, Pauli, and Weyl argued that the formalism of quantum theory would not give different masses for electrons and holes.[187] He had been unable to prove them wrong.

Dirac was convinced that to keep his relativistic theory he needed to keep the negative energy states, even if that meant giving up the proton. In Princeton in 1931, Dirac argued that "there is a small probability of a transition from a positive to a negative energy state and we can therefore no longer ignore the negative energy states." But he insisted that these states were not empty formalism.

187. Kragh, *Dirac*, 102–3.

These negative energy electrons are not to be considered as a mathematical fiction; it should be possible to detect them by experimental means. It ought to be possible to lift up one of the electrons in a negative energy state to bring it into a positive energy state; this would leave a vacancy in the series of negative energy states.

Dirac was not willing to allow the existence of mathematical structures in his theory to depend on choice, whim, or convenience.[188] He insisted on their physical reality. But he could no longer connect these states to the only other material particle known to exist, the proton.[189] Instead,

we shall refer to such a hole as an anti-electron; an anti-electron ought to have the same mass as an electron and this appears to be unavoidable; we should prefer to get a much larger mass so as to identify the anti-electrons with the protons, but this does not seem to work even if we take a Coulomb interaction into account.

His failed calculations testified to this. The closing lines of the lectures emphasized that, in principle at least, the reality of electrons in negative energy states could be established. "This idea of the anti-electrons doesn't seem to be capable of experimental test at the moment; it could be settled by experiment if we could obtain two beams of high frequency radiation of a sufficiently great intensity and let them interact."[190] Dirac wanted to preserve his equation (2.3). It required four wavefunctions as solutions. Ignoring the two negative energy solutions put the theory in contradiction with experiment (section 2.3.1) and was theoretically awkward. Dirac's position was contorted. He insisted on the observable consequences of an unobservable plenum.

These experimental and theoretical motivations were present in both 1929 and in 1931, but Dirac's affect differed. In 1929/30, Dirac was enthusiastic. His letters to Bohr show his enthusiasm and willingness to disagree with the master.[191] He had found a theory of two known particles, electrons and protons. His 1931 anti-electron remarks came at the end of a long series of lectures and in a methodological aside to a paper about another topic. Judging by the standard of Dirac's earlier work, and by the reactions of other physicists, Dirac's anti-electron remarks were less a central prediction than an appendix. He did not calculate the consequences of the anti-electron for a specific measurement, as he had for the Compton effect in 1926 (section 2.2). Neither did he connect the anti-electron to a well-verified experimental relation, as he had for Sommerfeld's formula in 1927 (section 2.2.3). In his 1931 Princeton lectures, and in his brief remarks in the *PRSL*, he presented hypotheticals and generalities in a pessimistic tone, expecting "negligible" results.[192] He was forced to admit that his theory implied almost unbelievable consequences. While a some physicists took up the anti-electron, these were a scattered few.[193] Historian Xavier Roqué has documented the

188. As in classical statistical mechanics: Darrigol, "Statistics and Combinatorics II," 243, 269.
189. Kragh, *Dirac*, 97.
190. "Lecture on Quantum Mechanics," notes by B. Hoffmann with corrections by Dirac, pages 130, 132, 134. Dirac Papers, Series II, Box 26, Folder 15.
191. Moyer, "Evaluations of Dirac's Electron."
192. Dirac, "Quantised Singularities in the Electromagnetic Field," 62.
193. Kragh, *Dirac*, 104.

extended and extensive process through which Dirac's anti-electron was connected to Carl Anderson's experimental *positrons* and certified as a genuine constituent of physical reality. This began with Anderson's cosmic ray experiments at Caltech in 1932 and continued, intertwining with controversies in nuclear physics, into the mid-1930s.[194] This brings the narrative back to the beginning of this chapter, with Dirac's maligned donkey electrons (section 2.1) and a newfound reason to return to the "hole" theory.

2.4 Theory of the Positron

This chapter began with Dirac's drafting and re-drafting of the presentation of his "hole"/positron theory. It was newly invigorated by evidence from cloud-chamber experiments. But it was also subject to a new puzzle: Dirac's infinite distribution of strange electrons throughout space was no longer uniform. Rather, it could be "polarized" like a dielectric medium (section 2.1).[195] This metaphor fit with the solid-state context of Dirac's relativistic research—which he carried through from his early studies of the Compton effect, the quantum mechanics of many-electron systems, and (non-relativistic) quantum electrodynamics. This chapter has discussed four theoretical treatments of the Compton effect, each with a physical description of X-ray scattering and with subtly differing quantitative predictions: Compton's 1923 single-light-quantum and single-electron theory; Dirac's 1925 light quanta and assembly of electrons approach; Dirac's 1926 classical electromagnetism and quantum mechanical electron approach; and, Klein and Nishina's 1928 classical electromagnetism and relativistic quantum electron approach.[196] In 1932–1933, as he returned to the "hole" theory, Dirac discovered that his theory allowed electromagnetic fields to create concentrations of charge and currents in the vacuum. This was a serious challenge. In 1929–1930, the unobservability of his infinitely charged space was a consequence of the *uniformity* of the infinite distribution (section 2.3.3). But concentrations and currents of negative-energy electrons spoiled this uniformity. Dirac and Rudolf Peierls debated whether vacuum polarization spoiled the agreement of the "hole" theory with special relativity.[197] On the other hand, if Dirac's negative-energy distribution had observable effects, perhaps he could finally respond to his doubters and force them to accept his "donkey-electrons" after all. Scattering was once again front of mind. At the 1933 Solvay Congress, Dirac argued that vacuum polarization "would result in deviations of order 1 in 100 in expressions such as the Klein-Nishina formula or the

194. Anderson, "Apparent Existence"; Roqué, "Manufacture of the Positron."

195. In inverted commas in Dirac, "Teoriya pozitrona," 747; without inverted commas, in Dirac, "Théorie," 212.

196. Klein and Nishina treated the incident X-ray as a classical perturbation. They also drew heavily on the correspondence principle. Klein and Nishina, "Scattering of Radiation," 113; Dirac's critique of Klein and Nishina's work was based on the correspondence principle, i.e., their "use of classical analogies not rigorously proved to be consequences of general quantum mechanics." Dirac himself continued to represent the incident X-ray as a classical perturbation in his 1930 re-derivation of the K-N formula: "we do not need to quantise the field, but can treat it simply as an external perturbation having a given value at each point in space-time." Dirac, "On the Annihilation of Electrons and Protons," 361–62; this was applied in the positron context: Dirac, "Teoriya pozitrona," 745.

197. Peierls to Dirac, 8.2.1934, 17.2.1934, and 20.3.1934. Peierls, *Correspondence*, Letters 176, 178, and 180; Peierls, "Vacuum in Dirac's Theory," 439; Rueger, "Attitudes towards Infinities," 322–24.

Rutherford scattering formula" for alpha particles scattered from atomic nuclei. "This also indicates a way to experimental verification of the theory."[198]

During 1932–1935, Dirac often visited the Cavendish Laboratory to observe the observers of anti-electrons and the Mond Laboratory for an experimental collaboration with Peter Kapitza.[199] In July 1932, Dirac was twenty-nine years old, living in his modest quarters in St. John's College, and had just been awarded his first permanent academic position—the Lucasian Chair of Mathematics, once held by Isaac Newton.[200] At Cambridge, he had two new, *slightly* younger, colleagues with whom to discuss his theory, visitors on fellowships from the Rockefeller Foundation. The first was Victor "Viki" Weisskopf (1908–2002), who had written his PhD under Max Born at the University of Göttingen, and then worked with Heisenberg (at Leipzig), then Schrödinger (at Berlin), and then Bohr (at Copenhagen), before arriving in Cambridge in 1933. The second visitor was similarly well-connected to the élite of European theoretical physics. Rudolf Peierls (1907–1995) was a former student of Heisenberg's, and assistant to Pauli (at Zurich). Peierls came to Cambridge in spring 1933 to complete his fellowship. He had spent the first half of the fellowship in Fascist Rome with Enrico Fermi, and almost took a position in his native Germany, when Adolf Hitler was appointed chancellor and German politics lurched further rightward. Peierls was Jewish, and saw little prospect of a good life for him and his pregnant wife Genia in Germany.[201] Weisskopf, a socialist and also Jewish, had similar concerns. Right-wing antisemites had been denouncing Albert Einstein and his work since the 1920s. However, the persecution of Jewish academics intensified dramatically in the spring of 1933, when the German Reichstag passed "reforms" that excluded "non-Aryans" from the civil service—including professors. Einstein was in the process of settling in Princeton, NJ. In addition, Ernest Rutherford managed to get a temporary position for Max Born in Cambridge, and Dirac helped him obtain an honorary position at St. John's. Dirac made few political statements, but he had a strong reaction to these events: he vowed to never again speak in German.[202]

Peierls and Weisskopf would soon publish influential treatments of the vacuum of quantum theory, and vacuum polarization in particular. But Peierls took to Dirac's relativistic mechanics more easily than Weisskopf. Weisskopf recalls that it was Peierls who taught him the "alpha gymnastics" needed to make use of Dirac's theory.[203] The two young men certainly had different strengths and weaknesses, but their different training may have determined their early reactions to "hole" theory. At Göttingen, Weisskopf's strongest work was on atomic theory and fluorescence: with Eugene Wigner he worked out a theory of how long an atom remained in an excited state before "spontaneously" jumping back to its ground state (see section 4.2.1). In 1930–1931 Peierls had collaborated with the young Soviet firebrand Lev Landau

198. Dirac, "Théorie," 212; see also Dirac, "Teoriya pozitrona," 749.
199. Roqué, "Manufacture of the Positron," 107; Kragh, *Dirac*, 142–43.
200. Farmelo, *Strangest Man*, 204–17.
201. Peierls, *Bird of Passage*, 89–90.
202. On this migration, see Ash and Søllner, *Forced Migration*, on Einstein, see: Pais, *Subtle Is the Lord*, 449–53; on Dirac, see: Farmelo, *Strangest Man*, 218–36, speak German: 220 see also L. Halpern "Observations."
203. Weisskopf, *Joy of Insight*, 73.

on quantum electrodynamics, but the bulk of his scientific career had been spent working on many-electron quantum theory: on magnetism and the conduction of heat and electricity in metals. His training in many-electron quantum mechanics surely contributed to his facility with positron theory.[204]

In particular, Peierls's 1929 study of magnetism gave a "peculiar" result: electrons could have a *negative* velocity, compared to their expected direction.[205] He used this to provide the first explanation for the anomalous, or inverse, Hall effect, in which electric currents seemed to flow in the wrong direction. Peierls's negative-velocity electrons were analogous to Dirac's negative-energy electrons, but were not quite analogous to Dirac's "holes."[206] Peierls's description of electrons occupying negative-velocity states was closer to Weyl's electron-proton proposal, or Heisenberg and Pauli's 1929 description of electron jumps from positive to negative *mass*.[207] These multiple interpretations gave similar results because of the following characteristic of Dirac's theory. The kinetic energy was written:

$$\frac{W}{c} = m\vec{v} + e\vec{A}, \tag{2.10}$$

with mass, m, velocity, $\vec{v}$, electric charge, e, electromagnetic potential, $\vec{A}$, and neglecting Dirac's 4×4 matrices. Reversing the sign of any variable could result in an overall negative energy. (If one writes velocity as the change in position over time, $\Delta x/\Delta t$, then reversing time performs the same function, as E. C. G. Stueckelberg, John Wheeler, and Richard Feynman later explained.) In 1931, Heisenberg reframed Peierls's work in terms of the same "holes" Dirac saw in X-ray orbits (without citing Dirac). Arnold Sommerfeld and Hans Bethe's canonical review 1933 article on the theory of metals firmly established the "hole" interpretation of Peierls's work.[208] In a draft for his 1933 "Theory of the Positron" presentations, Dirac included metals in his many-electron analogy: "Holes in distributions of electrons have occurred previously in atomic physics e.g. in the X-ray levels and in the theory of metals, where they are useful for describing the inverse Hall effect."[209] When fascism brought Peierls to Cambridge, he was well prepared to begin calculating with the "hole" theory.

In the Copenhagen *Faust* (section 2.1), a group of American physicists complained that Mephistopheles had brought them "no X-ings [*keine Ixerei*]!"[210] Peierls wrote to his friend Bethe on June 14, 1933: "I have given up my metals for a while

204. For their bibliographies, see Stefan, *Physics and Society*, 215–30; Peierls, *Selected Scientific Papers*; Weisskopf's solid-state-free background may have led Pauli to hire him as an assistant in 1933: Hoddeson, Baym, and Eckert, "Quantum Mechanical Electron Theory of Metals," 156.
205. Peierls, "Galvano-magnetic Effects," 1.
206. Here I differ from Peierls's own brief retrospective commentary. Peierls claimed that the hole idea was present in his earlier work. Peierls, "Galvano-magnetic Effects," 10; Peierls, "Hall Effect"; Hoddeson et al., *Crystal Maze*, 112–14.
207. Heisenberg and Pauli, "Zur Quantendynamik der Wellenfelder," 61; see also "states of negative mass": Landau and Peierls, "Extension of the Uncertainty Principle," 40.
208. Heisenberg, "Zum Paulischen Ausschließungsprinzip"; Peierls, "Elektronentheorie der Metalle," 308–9; Hoddeson, Baym, and Eckert, "Quantum Mechanical Electron Theory of Metals," 112–14.
209. Folder 22, Box 36, Series II, Dirac Papers.
210. Folder 6, Box 2.04, Weisskopf Papers; Gamow attributes X-ing to Einstein, Gamow, "Blegdamsvej," 191, 216 fn11.

and am now xing with Diracholes=Blackett-electrons [*ixe mit den Dirac-Löchern*]. Maybe this theory of Dirac's, in some approximation, will make some sense after all." Peierls and Bethe's "xing" was slang, I suggest, for crossing out terms in long pages of calculations.[211] He had productive work to do.

At the 1933 Solvay congress, Dirac presented his "Théorie du positron," including the new modifications to scattering processes. These results were done "in collaboration with Peierls."[212] Peierls jumped in with the first comment after Dirac's presentation with the addition that he had treated "the problem of the polarization of the vacuum [*la polarisation du vide*]" in a different way.[213] The published version of Peierls's work was titled "The Vacuum in Dirac's Theory of the Positive Electron."[214] There were parallels to many-electron quantum mechanics in both physicists' discussions. Dirac described the production of an electron-positron pair near an atomic nucleus "as a kind of photoelectric effect" in which "the electron is lifted into a positive-energy state and thus appears as an ordinary electron, leaving behind a hole which behaves as a positron." Dirac used the atomic language of his theory of X-ray levels, that is, he described "a negative-energy electron in a hyperbolic orbit."[215] Though Peierls also used Hartree's approximation and the density matrix, he did not mention orbits. Instead, his work had parallels in the photoelectric effect in metals.[216] This can be seen in Peierls's "schematic energy-co-ordinate diagram."[217] Peierls's diagram, figure 2.9b, included two horizontal energy levels for the positive and negative energy states. From left to right: the line B marks a location where the distributions of electrons could be compared if no external electromagnetic fields were present. External fields bent the energy levels, resulting in a new arrangement for comparison, labeled by the line A. The line E represented the vertical axis, energy, with key values marked, such as $\pm mc^2$. The horizontal line labeled x was the generalized coordinate being considered. Figure 2.9a is an energy-coordinate diagram from Peierls's 1932 review article on the theory of metals. It represented the gap between the energy bands of electrons in the metal. The gap defined the correct energy, $h\omega$, for light to eject

---

211. The crossing aspect is my inference. Sabine Lee uses "ixing." See also Peierls to Bethe 28.6.1933. Bethe and Peierls, *Bethe–Peierls Correspondence*, 99, 104, ixing: 80 fn133; Roqué, "Manufacture of the Positron," 99–102.

212. Dirac, "Teoriya pozitrona," 741.

213. Dirac, "Théorie," 213.

214. Peierls, "Vacuum in Dirac's Theory."

215. He also referred to charge screening—an atomic process modeled on ionic solutions. Dirac, "Théorie," 206; see also Dirac, "Teoriya pozitrona," 737.

216. Roqué credits Peierls with the photoelectric effect analogy, in July 1933. However, many physicists could have come to this view, independently. Max Delbrück visited Cambridge in April 1933 and his photoelectric note was communicated in May. Roqué, "Manufacture of the Positron," 99–100, fn107–fn10;9; Meitner and Kösters, "Über die Streuung Kurzwelliger γ-strahlen"; Heisenberg's assistant Guido Beck made a photoelectric analogy in April 1933: Beck, "Hat das negative Energiespektrum einen Einfluß auf Kernphänomene?," 502, 507; cf. this June 6 note, published July 1, which Peierls cites in his letters to Bethe: Oppenheimer and Plesset, "Production of the Positive Electron," 54; at the 1933 Solvay congress, Dirac credited both Peierls and Oppenheimer with independently calculating the photoelectric process: Dirac, "Théorie," 206; finally, note that the photoelectric absorption of X-rays by electrons near a nucleus was a well-developed topic: Hall and Oppenheimer, "Relativistic Theory of the Photoelectric Effect."

217. Peierls, "Vacuum in Dirac's Theory," 422; cf. Dirac, "Teoriya pozitrona," 723.

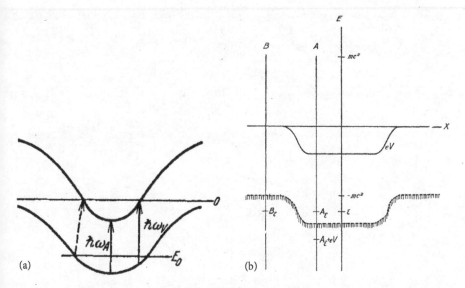

(a)   (b)

Figure 2.9. Peierls's energy-coordinate diagrams: the photoelectric effect in metals (1932) and the energy gap between negative- and positive-energy electrons (1934). (a) reprinted by permission from Springer Nature: *Elektronentheorie der Metalle*, copyright 1932. (b) republished with permission of The Royal Society (UK), from Peierls, Rudolf. "The Vacuum in Dirac's Theory of the Positive Electron." *Proceedings of the Royal Society of London A* 146, no. 857 (1934): 420–41; permission conveyed through Copyright Clearance Center, Inc.

an electron from the metal—the photoelectric effect.[218] The photoelectric picture of electron-positron pair production was a focal point of the developing "hole" theory, both in atomic and metallic frameworks.

In the mid-1930s, physicists argued that vacuum polarization had more consequences and greater significance than Dirac had depicted in his first presentations. Peierls wrote his "Vacuum" paper after leaving Cambridge for Manchester, but he kept in touch with Dirac by mail. In February 1934, Dirac sent Peierls the manuscript of his "Discussion of the Infinite Distribution of Electrons in the Theory of the Positron." Peierls thought that Dirac's analysis was too narrow. "For if we make a theory in which the vacuum has a polarisability, this obviously means that we modify the electrodynamics."[219] Chapter 3 discusses the new electrodynamics of the vacuum in detail.

Section 2.3.4 discussed Dirac's attempts to calculate the interaction between electrons and "holes," which had been filed with Dirac's original calculation of vacuum polarization. These calculations provide some insight into this work, though they

218. Peierls, "Elektronentheorie der Metalle," 292, see also the *Umklapp* diagram, 270; Roqué, "Manufacture of the Positron," 99–102, fn107; cf. a related diagram in Beck, "Hat das negative Energiespektrum einen Einfluß auf Kernphänomene?," fig. 1.
219. He gave Born's nonlinear electromagnetism as an example of a modification. Peierls to Dirac 8.2.1934. Peierls, *Correspondence*, vol. 1, 432; Dirac, "Infinite Distribution of Electrons," received 2.2.1934.

Figure 2.10. Dirac's pencil sketch, top left, of the sea of electrons and negative-energy electrons and holes. It reads: "Introduce the density ρ classically first," "ord[inary] electrons" and "holes." Series II, Box 36, Folder 22, Paul A. M. Dirac Papers, Special Collections & Archives, Florida State University.

follow his Solvay presentation reasonably closely. This chapter concludes with a slightly later sheet of calculations, which show Dirac reacting to Heisenberg's substantial "Remarks on the Dirac Theory of the Positron." Heisenberg's paper was received on June 21, 1934, and it acknowledges comments from Dirac, who likely saw a draft.[220] Dirac's manuscript calculations relate Dirac's notation to Heisenberg's, and Dirac copied out one of Heisenberg's equations (2.13).[221] At the top of the page, figure 2.10, Dirac instructed: "Introduce the density ρ classically first."[222] Underneath this sentence, Dirac sketched the distribution of "ord[inary] electrons" and "holes" in mathematical phase space separated by a horizon. Below, he sketched a charge density, R_S, with two equal-and-opposite regions separated by another horizon. R_S

220. Heisenberg, "Remarks on the Dirac Theory."
221. Heisenberg's equation (13). The full page is reproduced in Wright, "Beautiful Sea," 254.
222. Dirac Papers, Series II, Box 36, Folder 22.

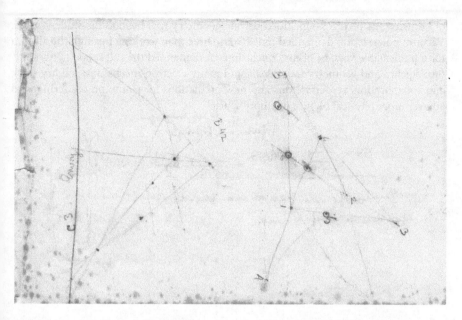

Figure 2.11. Detail of verso page from Dirac's archive. The leftmost vertical line is labelled "Energy." Series II, Box 36, Folder 22, Paul A. M. Dirac Papers, Special Collections & Archives, Florida State University.

was Heisenberg's notation for a relativistic density matrix with the form of equation (2.8).[223] In Dirac's "normal distribution," all the negative energy states were full, and all the positive energy were empty. Dirac labeled this new horizon "density $-\frac{1}{2}$" and lower horizon "density $\frac{1}{2}$." Two circles mark out regions—particles? bubbles?—of the opposite density. These sketches were not the only graphical method Dirac used. Peter Galison has explored Dirac's "suppressed drawing"—particularly his early training in technical drawing, engineering diagrams, and projective geometry.[224] Figure 2.11 is the verso page to figure 2.10, and it shows the characteristic form of projective geometry drawings. In faint pencil, Dirac wrote a general question: "What are proj[ective] coord[inates]s of (xyz) in a coordinate system in which $xx' + yy' + zz' = 0$ is line at infinity?"[225] Dirac never published such figures, and publicly argued with algebra. But this sheet demonstrates that Dirac's personal epistemology—his way

223. Heisenberg, "Remarks on the Dirac Theory," 171; this was Dirac's R_1: Dirac, "Infinite Distribution of Electrons," 152.

224. Galison, "Suppressed Drawing."

225. In his discussion of Dirac's mathematical background, Darrigol presumes that one of Dirac's resources was A. N. Whitehead's abstract approach to projective geometry. However, Barrow-Green and Gray write that even at Cambridge, Whitehead was exceptional in this regard. We do not know whether Dirac actually read Whitehead. Graham Farmelo notes that Dirac kept G. B. Mathews's 1914 *Projective Geometry* until the end of his life. Mathews was Senior Wrangler and associated with St. John's College. Mathews's book was not an abstract work of modern mathematics. It was pedagogical and declared that "mathematics is an inductive science, and not a set of rules and formulæ." Mathews, *Projective Geometry*, vii; Darrigol, *c-Numbers to q-Numbers*, 291, 306–8; Barrow-Green and Gray, "Geometry at Cambridge," §2.4; Farmelo, *Strangest Man*, 448, fn5.

of knowing his theories—included private geometry well beyond his student years. Vacuum polarization demanded more from Dirac than working through the algebra for a publishable version of his conclusions. It demanded the full scope of his abilities: algebra and geometry; relativistic and many-electron mechanics; solitary work and collaboration; re-derivations and new predictions for Compton scattering; and an ever more physical picture of empty space.

3
Nascent Pairs and Virtual Possibilities

3.1 Introduction

Chapter 2 showed how the vacuum was transformed when Paul Dirac and Rudolf Peierls applied the formalism and approximations of non-relativistic, many-electron quantum theory to the relativistic theory of electrons. With paper tools, Dirac first filled the vacuum with infinities of negative-energy electrons in 1929; and after 1932, with Peierls, he demonstrated the vacuum to be *dynamical*, a sea that experienced and exerted forces of nature. In Chapter 1, I positioned the vacuum as an "object of theory," akin to Hans-Jörg Rheinberger's experimental systems and "epistemic things." A key characteristic of Rheinberger's experimental epistemic things is that they oscillate between being the primary object of scientific investigation and being part of the "experimental system" of research. In cellular biology, cytoplasmic particles were at one point a primary object of investigation in their own right; later, they were a tool for unraveling the puzzles of the genetic code.[1] This chapter shows how new paper tools were applied to the vacuum—most importantly perturbation methods—to produce new theories and new vacuum phenomena. In the process, the vacuum became a *resource* for theorists looking to investigate new phenomena, build new theories, and build their careers. In parallel to the researchers in the life sciences, theoretical physicists began to *assume* the properties of the complex vacuum as a way to get on with studying *other* objects of theory.

This chapter takes up Dirac and Peierls's "hole" theory as it was developed, outside Cambridge, through the work of young postdoctoral fellows or assistants: Wendell Furry, Hans Euler, and Victor Weisskopf. Together with their mentors—J. Robert Oppenheimer, Werner Heisenberg, and Wolfgang Pauli, respectively—from 1932 to 1936, these physicists extended, critiqued, modified, and applied "hole" theory. They built competitor theories of the positron, based on "pairs," using complementary wave-formalisms. All these competitors shared with "hole" theory a startling picture of empty space: that the vacuum was, in some sense, full. Dirac transformed the vacuum from the least interesting entity in quantum theory into a resource for solving theoretical problems. In their own ways, Furry, Euler, and Weisskopf followed suit, *using* the vacuum positively to build formalisms, answer existing questions, and pose new ones. One reason—likely as important as the experimental reasons—these three physicists built positron theory into a foundation of Inter-War physics was the *fecundity* of the quantum vacuum.[2]

---

1. Rheinberger, "Cytoplasmic Particles," 271.
2. Cf. "Scientific objects [...] remain objects of research as long as they have the power to manifest themselves in yet unthought-of ways in the future." Rheinberger, "Cytoplasmic Particles," 273.

More than Nothing. Aaron Sidney Wright, Oxford University Press. © Oxford University Press 2024.
DOI: 10.1093/oso/9780190062804.003.0003

I argue that, to a large extent, there was a single cause for the fecundity of the quantum vacuum. It was the reason for the agreement among Peierls's, Furry's, Euler's, and Weisskopf's positron theories that the vacuum was more than nothing. The cause was their use of a paradigmatic "paper tool" theorists used to connect their formalism to experiment: perturbation analysis. (Often called "perturbation theory," I use "analysis" to distinguish this bundle of applied mathematical techniques from *physical* theories. See appendix A.)

This chapter continues with a discussion of the formalisms of radiation theory, section 3.2. Section 3.3 examines Wendell Furry's (1907–1984) attempt to master both the particle and wave formalisms. Furry's notes and calculations show what it meant to read, calculate, and write in radiation theory. Section 3.3.2 takes up Furry's best known work: his development of electron-positron "pair theory" with his postdoctoral supervisor, J. Robert Oppenheimer. An unpublished draft of their paper in Furry's archives gives insight into the impact of Oppenheimer's pessimism about theoretical physics on his published work. Furry and Oppenheimer broke from Dirac's many-electron formalism and used instead a representation of an electron-positron *field*. Despite this change, they independently discovered the same vacuum polarization phenomena with which Dirac and Peierls's were wrestling (section 2.4).[3] In 1934, Furry and Oppenheimer described the vacuum as containing "nascent pairs" of electrons and positrons.

This new theory of positrons—whether expressed in a many-electron formalism or in the language of waves and matter fields—proposed new physical processes in the vacuum. Vacuum phenomena were often described with "virtual" concepts. Section 3.4.2 discusses Werner Heisenberg's group at Leipzig, in particular Hans Euler. It focuses on their use of "virtual possibilities" to describe a new vacuum phenomenon: the scattering of light by light. But effects with less-energetic radiation bore no fruit; Weisskopf and others developed arguments that leapfrogged this failure to connect to experiments. By 1936, Furry and Oppenheimer's picture of "nascent" electron-positron pair energy in the vacuum was increasingly accepted. Euler had a correspondence with Weisskopf; section 3.5 uses their letters to show how vacuum processes became more important in theoretical physics. For example, new vacuum processes were a resource which the Leipzig group used to build new theories of the interaction of radiation and matter. This importance strengthened despite the weakening connection between vacuum processes and experimental observation. This chapter exhibits the qualities theoretical physicists appreciated in new positron theory papers that were theory for theory's sake.

3.2 Radiation Theory

Scattering theory was one of the central physical processes for quantum physics discussed in chapter 2. The concept of scattering was rooted in the tradition of geometrical optics, where steady beams of light reflected off of, say, crystals.

---

3. This was a general feature of wave theories. See, Wentzel, *Einführung in die Quantentheorie der Wellenfelder*, 183–91; Darrigol, "Quantized Matter Waves," 252.

Radiation theory was just as important to relativistic quantum theory but was rooted in the tradition of statistical mechanics. The paradigmatic physical example was "blackbody" radiation: a white hot ceramic chamber with a hollowed-out interior within which (in theory) the radiation in the chamber was in thermodynamic equilibrium with the ceramic walls. Around the turn of the nineteenth century, Max Planck and Albert Einstein used the statistics of radiation to establish quantum theory.[4] From 1916 to 1925, Einstein reconciled this statistical theory with the Bohr-Sommerfeld quantum theory of the atom (e.g., as applied to the molecules in the walls of a blackbody chamber).

In his radiation theory, Einstein considered blackbody radiation—radiation in thermodynamic equilibrium with a surrounding cavity, and possibly in equilibrium with gas molecules as well. Einstein considered that there were three types of "elementary process" by which a molecule could transition from one state, Z_n, to another, Z_m, with the kinetic energy of Z_n less than that of Z_m. First, a molecule could spontaneously emit a quantum of radiation and transition $Z_m \rightarrow Z_n$ ("spontaneous emission"). This was analogous to the spontaneous decay of a radioactive element. If the gas was stimulated by radiation with a frequency that matched the energy difference between the two states, $E = h(v_n - v_m)$, *two* elementary processes could occur. A molecule could absorb a quantum of radiation and jump from a lower-energy state to a higher-energy state, $Z_n \rightarrow Z_m$ ("stimulated absorption"). This was a normal feature of atomic theory. But Einstein proposed that there was a another elementary process, in which a molecule was *stimulated* by radiation but transitioned to a *lower* energy level: stimulated emission. These three processes were required for the radiation to be in thermodynamic equilibrium. The rates of occurrence for the processes was expressed by the "A_{mn}" and "B_{mn}" coefficients, which represented the probability of emitting (A) and absorbing (B) radiation at the frequency corresponding to the energy difference between states $E = h(v_n - v_m)$.[5] Historians Anthony Duncan and Michel Janssen consider Einstein's radiation theory as the most important work for the development of quantum mechanics.[6]

Considered an empirical and theoretical success, Einstein's A and B coefficients became a yardstick by which future theories would be measured: was each new theory consistent (or not)? New theories which failed to reproduce such success were unlikely to succeed.[7] In the new year 1927, Dirac was visiting Bohr's Institute in Copenhagen. He wrote a landmark paper, "The Quantum Theory of the Emission and Absorption of Radiation" which was not only consistent with Einstein's coefficients—it derived them. It began with observations on the state of quantum theory that were a direct parallel to his later remarks on many-electron quantum theory—like

4. Kuhn, *Black-Body Theory*.

5. Einstein, "Quantum Theory of Radiation," Pais, *Subtle Is the Lord*, ch. 21.

6. Duncan and Janssen, "Umdeutung in Minnesota. I," 572–73; For more on early scattering and dispersion, see Darrigol, *c-Numbers to q-Numbers*, pt. A; Taltavull, "Challenging the Boundaries."

7. E.g., Fowler to Lewis 6.6.1925: "my impression is that the Einstein coefficients are essentially correct & give the right picture of what the atoms are doing & that insofar as you regard this as denying the law of [detailed balancing] then I would take the view of denying that law for radiative processes." Stuewer, "Lewis on Detailed Balancing," 479.

an engineer, he pushed through with approximations, even at the cost of violating principles. A truly satisfactory Quantum Electrodynamics (QED) would be consistent with the special theory of relativity, but this was beyond Dirac's powers. Instead, his "main object" was "to build up a fairly satisfactory theory of the emission of radiation and of the reaction of the radiation field on the emitting system on the basis of a kinematics and dynamics which are not strictly relativistic."[8] Though Dirac presented his work as only "fairly satisfactory," it was greeted enthusiastically by his peers. Dirac combined a quantum theory of statistical ensembles of electrons with electrodynamics.[9] While in 1926 he had calculated Einstein's "B" coefficient (section 2.2.2), in 1927 he was able to derive its partner, the "A" coefficient. Until this point, theorists had to rely in their calculations on experimentally measured values of A. Dirac allowed them to calculate A from quantum theory and *compare* its value with experiment.

The next development of Dirac's radiation theory established a paradigm of radiation processes that profoundly shaped quantum theory ever since. In 1927, after his visit in Copenhagen, Dirac moved to visit Max Born's institute at Göttingen. Collision processes were Born's specialty, and during his visit Dirac both discovered errors in his just-published article and worked out corrections. At Göttingen, Dirac replaced Einstein's elementary processes with a picture from Hendrik Kramers and Heisenberg's 1925 radiation theory. While Kramers and Heisenberg considered statistical averages that included strings of absorptions and emissions, Dirac gave a micro-level description of individual scattering processes. Dirac arranged Einstein's regular emission and absorption processes in a series in time, as shown in table 3.1. An electron in an initial state m jumped to a new *intermediate* state, n, and then jumped to a final state m'. This intermediate state could have any energy at all; "in neither of" the jumps into and out of the intermediate state, "is the total proper energy even approximately conserved."[10] There were two critical elements. One has been well studied, particularly in histories of non-relativistic physics: the non-conservation of energy. The second has received less attention: the tripartite description of scattering events.[11]

Dirac's scattering theory was more than a shift in his thinking about energy conservation; it was a reversal. In 1924, Ralph Fowler—then Dirac's advisor—published a paper about the ionization of gas molecules. He considered processes in which an electron collided with a molecule and ejected an electron ($e + M \rightarrow e + e + M^+$), and the reverse process where two electrons collided with an ionized molecule to form an electron and a neutral molecule ($e + e + M^+ \rightarrow e + M$). In the tradition

8. Dirac, "Quantum Theory of the Emission and Absorption of Radiation," 243–44.

9. He applied perturbation methods to study an ensemble of quantized atoms and quantized electromagnetic radiation under the perturbing influence of a classical electromagnetic field.

10. Dirac, "Quantum Theory of Dispersion," errors: 715, double process and quote: 712, unimportant: 718; cf. Dirac, "Über die Quantenmechanik der Stoßvorgänge"; in 1930: Dirac, "Theory of Electrons and Protons," 364; Kramers and Heisenberg, "Dispersion of Radiation by Atoms," §4; Mehra and Rechenberg, *HDQT*, vol. 2, 179-85; Dresden, *H. A. Kramers*, §13.IV; on Born: Beller, *Quantum Dialogue*, 39–49.

11. Predecessor discussions of "intermediate" entities or states included HA Lorentz in 1921 and the BKS paper in 1924, Dresden, *H. A. Kramers*, 147.

Table 3.1. "[T]he problem of the scattering of radiation [γ] by an electron, [e,] free or bound," in Dirac's 1928 relativistic theory. Reading bottom to top, in this case, the intermediate states of an electron can have positive or negative energy, E.

Final State:	e, γ	e, γ
	—emission—	—absorption—
Intermediate State:	e	e, γ, γ
	—absorption—	—emission—
Initial State:	e, γ	e, γ

of Planck and Einstein's statistical radiation theory, systems of molecules and radiation were taken to be in thermodynamic equilibrium. This made it natural to pair processes such as Einstein's emission and absorption with Fowler's ionization, such that they balanced one another out. If each process was matched one-by-one to its opposite, this was called "detailed balancing." Fowler argued that in molecular processes one could not know whether a process was balanced, and whether conservation of energy held. A standard cycle of absorption and emission—in which a molecule transitions between stationary states 1 and 2—could be represented as $1 \rightarrow 2, 2 \rightarrow 1$. But, Fowler noted, a balance could occur by "the cyclic process $1 \rightarrow 2$, $2 \rightarrow 3, 3 \rightarrow 1$." *Overall* balancing was established because the molecule started and ended in the same state, 1. However, the nature of the internal transitions would be completely unknown. *Detailed* balancing for each individual process was no longer possible.[12]

One of Dirac's first publications challenged Fowler's pessimistic argument. He argued—surely to Fowler's delight—that detailed balancing indeed held for each individual process. This restored the conviction that conservation of energy holds universally.[13] Both quantum theory and experimental studies had advanced since 1924, but Dirac still appealed to detailed balancing in his first QED paper, which was communicated to the Royal Society by Bohr, and arrived on February 2, 1927.[14] Things changed during his stay with Born in Göttingen, and Fowler communicated Dirac's "The Quantum Theory of Dispersion," which arrived on April 4, 1927. In that paper he related his quantum electrodynamics to the dispersion analyses of Schrödinger, Klein, Jordan, and especially Heisenberg and Kramers. In 1925, Heisenberg and Kramers analyzed dispersion in terms of an atom jumping between three stationary states and for which conservation of energy was only statistically conserved.[15] (Kramers may also have modeled his intermediate states on his earlier study of chemical chain reactions.[16]) Between February and April, Dirac's picture of scattering changed to incorporate jumps into- and out of- an intermediate state, "in neither

12. Detailed balancing was taken up by a wide variety of authors, beginning in 1911. See Fowler and Milne, "Principle of Detailed Balancing"; Fowler, "Statistical Equilibrium," 263–64; these were surely the "cyclical processes" to which Tolman referred in Stuewer, "Lewis on Detailed Balancing," 477.
13. Dirac, "Conditions for Statistical Equilibrium"; cf. Fowler, *Statistical Mechanics*, §§17, 19.
14. Dirac, "Quantum Theory of the Emission and Absorption of Radiation," 260.
15. Kramers and Heisenberg, "Dispersion of Radiation by Atoms," §4.
16. Christiansen and Kramers, "Über die Geschwindigkeit chemischer Reaktionen."

[transition] is the total proper energy even approximately conserved." It was a substantial reversal, back to Fowler's 1924 pessimism. But it further connected radiation theory and scattering theory. This paper, too, was greeted as a major advance, and taken up by Dirac's peers.[17]

Dirac used his three-component picture, table 3.1, to describe scattering in his 1928 relativistic theory of the electron (section 2.3.1). When Dirac presented his "hole" theory of electron and protons in 1929/30, he revised the picture. He presented a new scheme that gave equivalent quantitative results, but which had a very different physical interpretation. In a scattering experiment, the sea of negative energy electrons made it "absolutely forbidden, by the exclusion principle, for the [scattering] electron to jump into a state of negative energy, so that the double transition processes with intermediate states of negative energy for the electron must be excluded." That is, those intermediate states in table 3.1 with negative energy, $e_{E<0}$, were now excluded. However, Dirac's "hole" theory proposed a new processes: a quantum of radiation could knock an electron out of the vacuum sea, into a positive-energy state, accompanied by a "hole."

[F]irst one of the distribution of negative-energy electrons, jumps up into the required final state for the electron, with absorption (or emission) of a photon, and then the original positive-energy electron drops into the hole formed by the first transition with emission (or absorption) of a photon.

In 1930, Dirac emphasized that these new processes exactly made up for the exclusion of the old transitions to negative energy states.[18] The creation and annihilation of an electron-hole pair occurred *inside* the radiation process. These intermediate states played a crucial role in the development of radiation theory and QED. They were in a sense hidden from physicists' knowledge. Peierls called transitions into and out of intermediate states "virtual transitions."

3.3 Furry and Oppenheimer

Wendell Furry (1907–1984) was one physicist who took up the challenges of Dirac's theory. Furry was an American from the Midwest. He studied at DePauw University in Indiana, and received his PhD in Physics from the University of Illinois in 1932. In 1931, he attended the summer school at the University of Michigan, and wrote up the lectures by Arnold Sommerfeld and Wolfgang Pauli. In 1932, in the midst of the great depression, Furry was awarded a National Research Council fellowship to work with J. Robert Oppenheimer at the University of California at Berkeley. Oppenheimer was

---

17. Bethe and Heitler, "Stopping of Fast Particles and on the Creation of Positive Electrons," 84; Bethe and Salpeter, *Quantum Mechanics of One- and Two-electron Atoms*, 90; Heitler, *Quantum Theory of Radiation* (1944), 89–90, 131ff, 147ff; Wentzel, "Recent Research in Meson Theory," 16; in his canonical treatment, Fermi wrote that the tripartite structure was a consequence of the linearity of Dirac's relativistic wave equation. E. Fermi, "Quantum Theory of Radiation," 123.

18. Dirac, "Theory of Electrons and Protons," 365.

the most dynamic figure in physics in the United States, shuffling between simultaneous appointments at Berkeley and the California Institute of Technology. At Berkeley, Ernest Lawrence was perfecting his cyclotron particle accelerator and at Caltech, Carl Anderson was photographing cosmic rays. The Michigan summer school and Oppenheimer's position marked a new era of American independence from European institutions.[19]

On May 5, 1932, Oppenheimer wrote to Furry:

> I am very happy indeed that you have been granted a fellowship, and am looking forward with great pleasure to our work together next year. [...] I think that we shall have a fine year. Several of the theoretical students are staying on, so that you will not want for good theoretical company; and Lawrence and I have already planned an intensive seminar in the physics of nuclei.[20]

One Californian colleague was Felix Bloch, an early member of the group of Jewish refugee physicists who fled continental Europe and settled in the United States. He was Heisenberg's former student, and Peierls's colleague, at Leipzig. In 1933, while a Rockefeller Fellow with Enrico Fermi in Rome, he was offered a position at the little-known physics department at Stanford University. Bloch already knew Oppenheimer, and upon Bloch's arrival they established a joint seminar in theoretical physics.[21]

One of Oppenheimer's central goals in this period was to develop his version of Dirac, Heisenberg, and Pauli's QED, and particularly to understand the production of "pairs" (of electrons and positrons). This occurred in the intermediate state of scattering processes in table 3.1, but in that case the electron half of the new pair fell back into the "hole"/positron and disappeared. Oppenheimer worked closely with experimentalists, and he had seen tantalizing cloud chamber photographs that appeared to show visible paths of pairs of oppositely charged particles exiting the same point in a sheet of lead. More quantitatively, both Anderson at Caltech and Patrick Blackett at Cambridge used the decay of radioactive thorium to measure the absorption of gamma rays in lead. (X-rays emanated from valence shells of electrons, gamma rays emanated from atomic nuclei.) But experiment and radiation theory were misaligned. Blackett and Occhialini suggested that this excess absorption could be the result of a new elementary process: real pair creation, figure 3.1.[22]

In 1933, Oppenheimer and Milton S. Plesset (1908–1991), an NRC Fellow at Caltech, worked on a description of pairs produced by high-energy gamma rays. Niels Bohr was scheduled to visit California in the spring of 1933.[23] In anticipation, Oppenheimer delegated the task of describing the production of pairs by high-energy

19. Kevles, *Physicists*, 216ff.

20. Folder: J Robert Oppenheimer, 105.8 Box 2, Furry Papers.

21. Interview of Wendell Furry by Charles Weiner on August 9, 1971, Niels Bohr Library & Archives, American Institute of Physics, College Park, MD USA, www.aip.org/history-programs/niels-bohr-library/oral-histories/24324; Brown and Hoddeson, "Birth of Elementary Particle"; Kevles, *Physicists*, 282; see also Uehling's impressions: Schweber, *Shadow of the Bomb*, 209 fn152.

22. Brown and Hoddeson, "Birth of Elementary Particle"; Galison, "Discovery of the Muon," 133–34; Mehra and Rechenberg, *HDQT*, vol. 6, 821; Roqué, "Manufacture of the Positron," 78.

23. See Oppenheimer to Bohr, 14.6.1933 in Smith and Weiner, *Robert Oppenheimer*, 161–62.

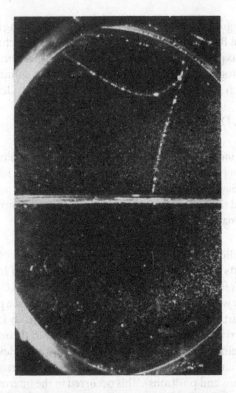

Figure 3.1. Chadwick, Blackett, and Occhialini's cloud chamber photograph with an electron-positron pair at top right (1934, fig. 5). Republished with permission of The Royal Society (UK), from Chadwick, J., P. M. S. Blackett, and G. P. S. Occhialini. "Some Experiments on the Production of Positive Electrons." *Proceedings of the Royal Society of London A* 144, no. 851 (1934): 235-249; permission conveyed through Copyright Clearance Center, Inc.

electrons to his students in Berkeley, John F. Carlson and Furry. He wrote from Pasadena:

> Dear Furry + Carlson,
>   You will know that I am eager to hear about the problem: whether you have run into any grave difficulties, whether you can yet tell anything of the results. I have an extremely crude estimate of the ratio σ_{rad}/σ_{el} for large ϵ; it seems to me suggestive, but it may very well be wrong.

He went on to describe his crude estimate of this ratio—the amount of scattering due to gamma radiation, σ_{rad}, to the amount of radiation due to incident electrons, σ_{el}—and the conditions in which it might hold. He signed off with news that he would travel up to Berkeley before Bohr arrived in May; with best wishes to Furry's and Carlson's families; and with another request for news about their calculation, signed

"Oppie." (Neither Oppenheimer's nor his students' paper were ready before Bohr arrived.)[24]

3.3.1 Pedagogy between the Lines

Furry prepared for this work by reading the papers on QED and scattering by Dirac and Waller (section 2.3.1) and writing between their lines.[25] Synthesizing these papers was a major goal of physicists—Waller's papers followed Dirac's relativistic quantum theory of electrons and "holes," which described radiation as a classical electromagnetic perturbation. Dirac's radiation theory was a quantum theory of assemblies of electrons and light quanta but did not allow for electrons with relativistic motion.[26] Synthesizing the two pictures would produce a theory of relativistic electrons and quantized electromagnetism. But simply *reading* these published papers would not prepare Furry for doing original work. Physicists, like all readers, engaged with texts with varying levels of intensity, from scanning or skimming to perusing. But theoretical physics texts had something of the character of a hiking guidebook: they began from a well-known location, and proceeded to give a series of instructions and landmarks. Rather than literal step-by-step instructions, in two lines of text a reader might be informed that they have 300 meters to hike before cresting a ridge. Much as a hiker would have to actually walk the 300 meters to fully know the route of the hike, Furry sat down with Dirac's 1927 "The Quantum Theory of the Emission and Absorption of Radiation" and in neat pencil on unlined paper, figure 3.2., he worked his way through Dirac's mathematics. The published, numbered, mathematical expressions served as guideposts. He reproduced Dirac's conclusion that, in Furry's words "double quantization" of an assembly of particles "is the same as imposing E[instein].B[ose]. statistics on the assembly." Furry worked through the equivalence Dirac established between a description of radiation in terms of "double quantization" of wavefunctions and a description in terms of a gas of particles which obeyed Bose-Einstein statistics (light quanta).

Once he was comfortable with this equivalence, Furry turned to Waller's treatment of relativistic scattering.[27] (The same calculations which I argue convinced Dirac of the necessity of providing a physical interpretation for the negative energy

---

24. Folder: J Robert Oppenheimer, 105.8 Box 2, Furry Papers; see also Oppenheimer to Bohr 14.6.1933, Niels Bohr Scientific Correspondence, 1903–1962, Niels Bohr Archive; see Pais, *J. Robert Oppenheimer*; ch. 6; Oppenheimer and Plesset, "Production of the Positive Electron"; Furry and Carlson, "Production of Positive Electrons by Electrons"; Plesset continued this work during a postdoctoral year at Bohr's Institute in Copenhagen with another young NRC Fellow, John Archibald Wheeler; Plesset and Wheeler, "Ineslatic Scattering." Furry recalled the dubious claims of these notes: "It was stylish in those days to say things that people couldn't possibly understand and put them in the Letters to the Editor so you could be sufficiently brief." AIP Interview.

25. Furry interview, AIP.

26. Note that Dirac's radiation theory also appealed to classical fields to perturb the quantum electrons and light quanta. Waller, "Über eine Verallgemeinerte Streuungsformel"; Waller, "Die Streuung von Strahlung durch Gebundene und freie Elektronen nach der Diracschen relativistischen Mechanik"; cf. Oppenheimer, "Note on Light Quanta and the Electromagnetic Field."

27. Waller, "Die Streuung von Strahlung durch Gebundene und freie Elektronen nach der Diracschen relativistischen Mechanik."

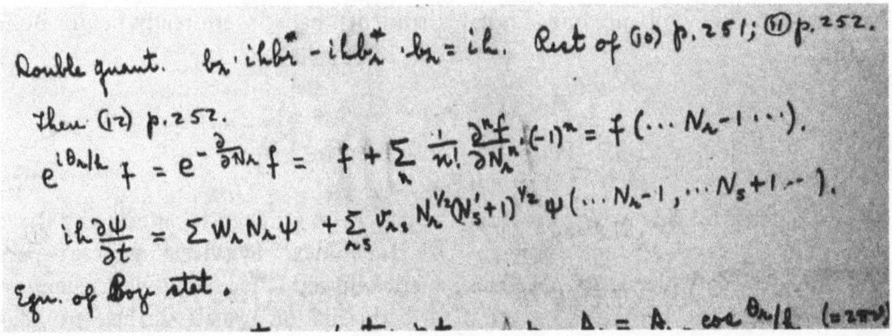

Double quant. $b_\lambda \cdot i h b_\lambda^* - i h b_\lambda^* \cdot b_\lambda = i h.$ Rest of (30) p.251; (31) p.252.

Then (32) p.252.

$e^{i\theta_\lambda/h} f = e^{-\frac{\partial}{\partial N_\lambda}} f = f + \sum_r \frac{1}{r!} \frac{\partial^r f}{\partial N_\lambda^r} (-1)^r = f(\cdots N_\lambda - 1 \cdots).$

$i h \frac{\partial \psi}{\partial t} = \sum_\lambda W_\lambda N_\lambda \psi + \sum_{\lambda s} v_{\lambda s} N_\lambda^{1/2} (N_s' + 1)^{1/2} \psi(\cdots N_\lambda - 1, \cdots N_s + 1 \cdots).$

Eqn. of Boze stat.

$A_r = A_r \cos \theta_\lambda/h$ (=254)

The wave equation written in terms of the variables N_r, is[†]

$$ih \frac{\partial}{\partial t} \psi (N_1', N_2', N_s' \ldots) = F \psi (N_1', N_2', N_3' \ldots), \tag{12}$$

where F is an operator, each θ_r occurring in F being interpreted to mean $ih \, \partial/\partial N_r'$. If we apply the operator $e^{\pm i\theta_r/h}$ to any function $f(N_1', N_2', \ldots N_r', \ldots)$ of the variables $N_1', N_2', \ldots$ the result is

$$e^{\pm i\theta_r/h} f(N_1', N_2', \ldots N_r', \ldots) = e^{\mp \partial/\partial N_r'} f(N_1', N_2', \ldots N_r' \ldots)$$
$$= f(N_1', N_2', \ldots N_r' \mp 1, \ldots).$$

If we use this rule in equation (12) and use the expression (11) for F we obtain[‡]

$$ih \frac{\partial}{\partial t} \psi(N_1', N_2', N_3' \ldots)$$
$$= \Sigma_{rs} H_{rs} N_r'^{\frac{1}{2}} (N_s' + 1 - \delta_{rs})^{\frac{1}{2}} \psi(N_1', N_2' \ldots N_r' - 1, \ldots N_s' + 1, \ldots). \tag{13}$$

Figure 3.2. Comparison of page 252 of Dirac's "The Quantum Theory of the Emission and Absorption of Radiation" (1927) and a page from Furry's calculations. Equation 12 is a multi-particle version of Schrödinger's wave equation. Dirac shows how the operator $e^{\pm i\theta_r/h}$ acts on the eigenfunction f by changing the number of particles in the state N_r by plus or minus one (according to the opposite of the sign chosen in the operator e). Equation 13 shows how the wavefunction of the assembly, ψ, changes in time, with changes in the number of particles N in a given state r. Furry notes that this is the "Eq[uatio]n of Boze stat[istics]" (meaning Bose). Box 8, Folder: Notes on Dirac and Waller, Furry Papers. Courtesy of the Harvard University Archives. Republished with permission of The Royal Society (UK), from Dirac, P. A. M. "The Quantum Theory of the Emission and Absorption of Radiation." *Proceedings of the Royal Society of London A* 114, no. 767 (1927): 243–65; permission conveyed through Copyright Clearance Center, Inc.

states, section 2.3.1). Waller had only used one of Dirac's two approaches. Furry tried to reproduce Waller's results separately from *each* method. This analytical strategy was well suited to advanced pedagogy. In a classroom setting, students worked through physics problems, and their answers were compared to correct solutions. Dirac's "complete harmony between the wave and light-quantum descriptions of

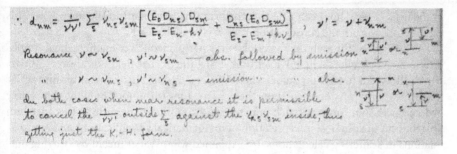

Figure 3.3. Furry's notes on Waller's calculation of scattering. At right, he drew scattering processes as absorption and emission transitions between atomic stationary states. Such atomic diagrams were common in the Bohr-Sommerfeld theory, as found, for example, in Sommerfeld's *Atomic Structure and Spectral Lines* (146). Furry wrote that when the frequency of the incident radiation was near the difference in energy between the states, these formula reduce to those of "K.-H."—Kramers and Heisenberg's dispersion theory. Box 8, Folder: Notes on Dirac and Waller, Furry Papers. Courtesy of the Harvard University Archives.

the interaction" provided an opportunity for Furry to check his own calculations against one another.[28] This was, perhaps, a practical use of wave-particle duality. If all went well, he would find that the paths taken in two separate calculations converged on the same endpoint. Historian Mara Beller has remarked on the *productive* lack of consensus—and lack of feeling of paradox—physicists expressed toward wave-particle duality.[29] Furry's calculations elucidate the pragmatic use of an apparent paradox.

Furry's recalculations of Waller's results were far more involved than those referring to Dirac's article. He drew a diagram of radiation processes. In contrast to the time-series representations of scattering in tables 3.1, Furry drew scattering as if it was an atomic transition process, figure 3.3.[30] Furry encountered differences between his results and Waller's. Once he found an additional set of terms, but discarded them to continue his investigation: "Omit the terms 1 for present, in order to get Waller's result. Then, [...]. This gives just 1/2 of Waller's formula (10)."[31] "Reason (10) is twice above result is that (9) contains twice the real parts of I_{nm}," the intensity of scattered radiation. The notes used an impersonal voice, perhaps modeled on the tone of published articles.[32] "We have given a reason different from Waller's for omitting the term in k." But more colloquial language entered when Furry encountered Waller's higher mathematics: "The factor 4 comes in in 'real parts', etc.—phooey!" On occasion, he was convinced of the rigor of his own work, and that Waller was in error: "Waller actually had factor of c wrong in preceding equation." But he also appealed to looser

28. Dirac, "Quantum Theory of the Emission and Absorption of Radiation," 245.
29. Beller, *Quantum Dialogue*, ch. 11.
30. Waller material starts halfway down page "A". Box 8, Folder: Notes on Dirac and Waller, Furry Papers.
31. Waller, "Die Streuung von Strahlung durch Gebundene und freie Elektronen nach der Diracschen relativistischen Mechanik," 841, eq. 10.
32. See Shapin and Schaffer, *Leviathan and the Air-pump*, 65–69.

standards: "I have not been able to transform this [Jacobian] to the form [...]. This should follow from the conservation laws & is probably O. K." [33]

As he worked through the material, Furry performed fewer mathematical simplifications than Waller. Where Waller took it that the electric field was always positive, Furry allowed it to take any value; where Waller simplified his expressions into scalar quantities, Furry tended to retain them as vectors (meaning, roughly, that Furry worked with four times as much information about those quantities). This gave him insight into the physical interpretation of mathematical simplifications and approximations in scattering calculations. He explained what Waller meant by "leaving out" certain mathematical expressions:

The "leaving out" of terms [in the expression for the intensity of scattered radiation] consists in taking $y_1 = y_2 = y$, and assuming that the frequencies involved [...] are not large enough to compass the difference between a positive and a negative energy of the electron.

He analyzed the consequences of Waller's simplification. "Then only the "Zwischenzustände" [intermediate state, table 3.1] can differ from the initial state in the sign of the energy." By keeping mathematical distinctions throughout his calculations, Furry gave himself more work, but he gained insight. Just below his remark about the "Zwischenzustände," he wrote and underlined in red pencil: "This does not remain true when the quanta considered are harder than $12XU_1$" or about 1.4×10^8 electron Volts. Furry calculated that Waller's equations for scattering did not remain valid for very high energy radiation—higher energy than most laboratory experiments, but within the range of cosmic ray observations, and more than enough energy for pair creation. [34]

Coming to Waller's final scattering formulae, Furry found that his approach revealed both trivial and significant differences. Furry wrote out the four spacetime components of an electron's motion in Dirac's relativistic theory both before, $U^{(1)}$, and after, $U^{(2)}$, scattering. Furry wrote that "Waller has listed $U^{(2)}$ wrong, also in his E', $k_r + k_s$ must still be vectors." As he proceeded, Furry found that Waller had "cos θ where I have sin θ," but he thought this was "surely a misprint or unannounced change of convention" in the Zeitschrift für Physik. On the next page, Furry found himself "off by a factor of 2, and also by having $\cos^2 \theta$ instead of $\sin^2 \theta$." In the next line, these difficulties manifest in a more serious form. Waller's results had reproduced the familiar Klein-Nishina formula for Compton scattering. [35] Furry's expression for the intensity of scattered radiation had "$\cos^2$" where Klein and Nishina had "$\sin^2$." Worse, Furry's result was expressed in terms of a different part of the electromagnetic radiation. "Here $\vec{h}_s$ is the magnetic vector, not the electric." He left off mid-page after writing: "This is a serious disagreement." [36] The failure to replicate an accepted formula could indicate that new calculations were in error.

33. Box 8, Folder: Notes on Dirac and Waller, Furry Papers.
34. Box 8, Folder: Notes on Dirac and Waller, Furry Papers.
35. Waller, "Die Streuung von Strahlung durch Gebundene und freie Elektronen nach der Diracschen relativistischen Mechanik," 850.
36. Box 8, Folder: Notes on Dirac and Waller, Furry Papers.

These calculations give a picture of a general method of working in theoretical physics. Furry calculated between the lines of Dirac's landmark paper; then he worked through Waller's detailed calculation in its published form. Then he tried to reproduce the same results by a new method. During his calculations, he refrained from making simplifying assumptions. This resulted in more intricate mathematics, but also in insight into the role and physical interpretation of these assumptions. This activity was pedagogical, verificational, and for research. Furry was learning the results and techniques of theoretical physics; he was checking the validity of published results; and he was asking new questions of the material. All of this was centered around a theoretical description of a particular phenomena: the scattering of light by electrons, specifically the Compton effect. Furry was not alone in these practices.[37] In calculations likely written while he was a doctoral student under Max Born in Göttingen, Victor Weisskopf's archive documents a less intense form of reading. As a student, he copied out the results of three of Dirac's papers, as well as work by Born and Oppenheimer, and Heitler and London, among others.[38] (Before the advent of mimeographs and electronic files, these notes likely served as a personal library.) Furry's calculations illustrate another key feature of radiation theory in the 1930s: it was unruly, unstable, and unreliable. In general, correct calculations were supposed to be reproducible.[39] Furry's remarks about formulae no longer applying for very high energy radiation were emblematic of wider concerns. Oppenheimer wrote to his brother Frank in 1934, to express his dismay over the state of theoretical physics, not only over "the divergence difficulties with the positron," but also over "the utter impossibility of making a rigorous calculation of anything at all."[40]

3.3.2 Furry and Oppenheimer's Nascent State

In 1934, Furry and Oppenheimer published a series of papers "On the Theory of the Electron and Positive" (equivalently, electron and positron, or "pair theory"). Their mood was dark. The first paper contains twenty-three uses of "ambiguous" and cognate words.[41] Some of these ambiguities were attributed to infinite integrals; "difficulties [which] tend to appear in all problems in which extremely small lengths are involved" (260).[42] This was similar to Bohr's reaction to Dirac's "hole" theory in 1929 (see section 2.3.1). Bohr cautioned against extending quantum concepts to phenomena at a smaller scale than the nucleus or electron radius (smaller than 10^{-13} cm).

37. Compare the later work of Shih Tsun Ma: "This work was started in an attempt to check the calculations of Schwinger and Wentzel," Ma, "Vacuum Polarization in the Positron Theory," 1113; for other copying-out and generalizing, see D. Kaiser, *Drawing Theories Apart*, 117–18 Warwick, *Masters of Theory*, esp. §§3.7, 5.3.

38. Folder 7, Box 9.01, Weisskopf Papers.

39. The infinities of radiation theory exacerbated this problem. See Peierls, "Vacuum in Dirac's Theory," 425 and fn; Weisskopf, "Probleme der neueren Quantentheorie des Elektrons," 670; Rueger, "Attitudes towards Infinities"; Mehra and Rechenberg, *HDQT*, vol. 6, 769.

40. Galison, *How Experiments End*, ch. 3, quote: 107.

41. Further references to this paper in this paragraph are given in text. Furry and Oppenheimer, "Theory of the Electron and Positive"; see Pais, *Inward Bound*, 379–80.

42. These anxieties led Heisenberg to consider giving up the continuity of space. Carazza and Kragh, "Heisenberg's Lattice World."

Indeed, some of Furry and Oppenheimer's "ambiguity" was attributed to the limits which Bohr's correspondence principle applied to quantum theories.[43] They worried about even asking

> such questions as "Are there any positives present near a nucleus?"; for such a question can be given an unambiguous meaning only by reference to such an experimental arrangement for answering it. (246)

Here, I want to draw attention to ambiguities of the formalism itself. Furry and Oppenheimer reflected on effects which "cannot, as we have seen, be unambiguously formulated in all cases" (259). In their perturbation analysis, higher-order calculations "cannot in any case be made, because in these higher orders the difficulties of the magnetic proper energy make an insurmountable ambiguity" (258). Basic quantities of the theory, such as the energy of particles and interactions became indistinct:

> In fact the distinction between proper energy [e.g. self energy] and interaction energy [...] cannot be made without ambiguity in a system in which the number of particles is necessarily indeterminate. (254)

Unfortunately, "Because of the gauge dependence of the theory no unambiguous calculation of the polarization energy of the pairs in a magnetostatic field can be made" (261 fn10a).[44] Furry and Oppenheimer tried to proceed in pair theory in analogy to Dirac's radiation theory, by treating the electrostatic (Coulomb) interaction between particles as a perturbation—just the formalism Furry had studied (section 3.3.1). But they found that in this "application one meets only the same difficulties as in the Dirac dispersion theory" (253). However, Furry's archives suggest that their pessimism was due mostly to Oppenheimer's epistemic anxiety.

The positive aspect of Furry and Oppenheimer's proposal was a holistic theory of electrons and positives, each a particle (neither a hole). A standard representation of a system with two types of particles would introduce separate quantities for each type. In contrast, Furry and Oppenheimer's "pair" theory was a theory of both electrons and positrons *together* and described them by a *single* wavefunction. They followed Pascual Jordan's formalism of "second quantization" for Dirac's radiation theory to write the wavefunction for a system of N indistinguishable particles:

$$\psi(0 \cdots 1_{r_1} \cdots 0 \cdots 1_{r_i} \cdots 1_{r_N} \cdots), \tag{3.1}$$

which represented a system with one particle in state r_1, zero particles in states between r_1 and r_i, one particle in the state r_i, etc. The r's referred to both positive- and negative-energy states. To move to a positron theory, Furry and Oppenheimer performed

43. See also, Darrigol, "Quantized Matter Waves," 247–50; Rueger, "Attitudes towards Infinities," 317–22, 327.

44. Lack of gauge invariance was a major flaw of Furry and Oppenheimer's analysis. Peierls, "Vacuum in Dirac's Theory," 431 fn; Furry and Oppenheimer, "Theory of the Electron and Positive II," 434.

Figure 3.4. Wendell Furry at Harvard, 1941. Photograph by Samuel Goudsmit, courtesy AIP Emilio Segrè Visual Archives, Goudsmit Collection.

a simple formal change, which corresponds exactly to regarding the emptiness and not the fullness of a state of negative kinetic energy as equivalent to the presence of a particle, gives us at once the theory of electrons and positives.

They restricted the states, r, to only positive energy and introduced a new label for the negative energy states, ρ. Then, Furry and Oppenheimer wrote an expression for the number of positrons, M_ρ, as a constant minus the number of electrons in negative energy states: $M_\rho = 1 - N_\rho$. From this they wrote a new wavefunction for their theory of N electrons and, separated by a semicolon, M positrons: $\psi(N_r; M_\rho)$,

$$\psi(0 \cdots 1_{r_1} 0 \cdots 1_{r_2} \cdots 1_{r_N} \cdots; 0 \cdots 1_{\rho_1} \cdots 0 \cdots 1_{\rho_M} \cdots). \qquad (3.2)$$

An operator applied to the wavefunction would change the number of particles in a given state by plus or minus one.[45]

45. Furry and Oppenheimer, "Theory of the Electron and Positive," 250–52, eq. 2.4-2.14, 2.32; they separated electron and positron wavefunctions according to: Schrödinger, "Über die kräftefreie Bewegung in der relativistischen Quantenmechanik"; see Oppenheimer to Uhlenbeck, Fall 1933, in Smith and Weiner, *Robert Oppenheimer*, 168.

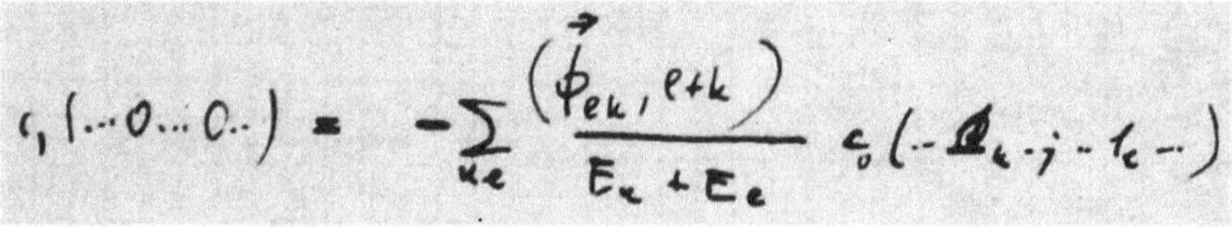

$$c_1 \left(\cdots 0 \cdots 0 \cdots \right) = -\sum_{\ell e} \frac{\left(\vec{\varphi}_{e\ell}, \ell + k \right)}{E_e + E_e} c_0 \left(\cdots \vec{\theta}_e \cdot_j \cdots \ell_e \cdots \right)$$

Figure 3.5. A fragment of Weisskopf's calculation of the vacuum state in his and Pauli's 1934 relativistic scalar field theory. Folder: 2, Box 9.02, Weisskopf Papers, MIT Libraries collection MC-0572.

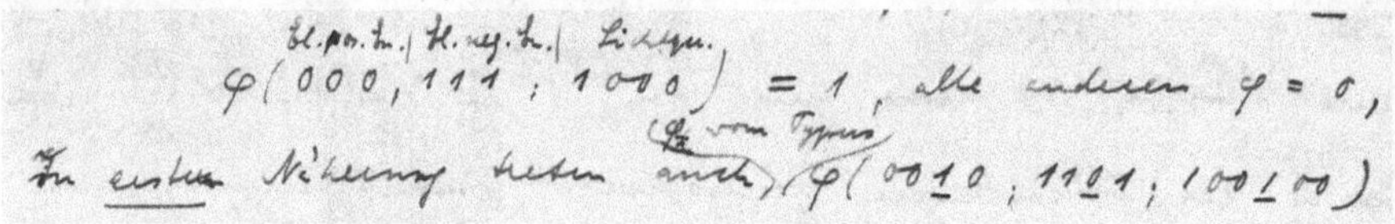

$$\varphi \left(000, 111 ; 1000 \right) = 1, \ldots$$

Figure 3.6. Two lines from Heisenberg's letter to Pauli on February 1, 1935, copyright (2016) Heisenberg-Gesellschaft e. V., München, VR 204617.

Though there were differences in some details, this form of representation was widely shared. Figure 3.5 shows a line of Weisskopf's calculations. On the left-hand side of the equation, he wrote a wavefunction c_1, with zero particles in each state. On the right-hand side, he equated this to a sum of field operators and energy levels, multiplying the wavefunction c_0 with one particle and one anti-particle. Heisenberg used similar notation in a letter to Wolfgang Pauli on February 1, 1935 (figure 3.6).[46] The first expression is the Schrödinger wavefunction, φ, for a light quantum in "hole" theory to zeroth order. Heisenberg labeled the parts of the formalism so that the first set of arguments represented "El.[ectron with] pos.[itive] En.[ergy]," which was empty; the next set represented "El.[ectron with] neg.[ative] En.[ergy]," in which all the states were full; the final set represented light quantum states, with one state occupied. At zeroth order, all other wavefunctions φ of the system were zero. Heisenberg continued to write down the wavefunction in the first-order approximation, underlining the changes from his previous expression. This wavefunction described one positive energy electron, a "hole" (denoted by "0") in the negative energy states, and two light quanta. This formalism was quite capable of representing the vacuum. This was because the wavefunctions represented *every possible state* for particles, regardless of whether or not the state was occupied. The zeros and ellipses of these expressions provided the "well-determined and labelled place" required for Dirac's "hole" theory (section 2.3.3). Heisenberg worked with Dirac's "hole" theory and represented the vacuum as no positive-energy electrons and a full sea. For Weisskopf, in collaboration with Pauli, the vacuum was a state of no particles, $c_1 (\cdots 0 \cdots 0 \cdots)$.

The vacuum state played another crucial formal role in Furry and Oppenheimer's formalism. Rather than representational, this role was algebraic. In Dirac's nonrelativistic radiation theory, he considered statistical ensembles of electrons. Each of

46. Pauli, *Wissenschaftlicher Briefwechsel*, doc. 407.

these electrons could be in a stationary state in an atom, N_r, labeled by a principal quantum number $r = 0, 1, 2, 3, \ldots$ (compare figure 3.2). Say an electron dropped from N_2 to N_0 and emitted a light quantum; then the overall system states would shift $N_2 \to N_2 - 1$ and $N_0 \to N_0 + 1$, where the overall number of electrons remained constant.[47] Instead, Furry and Oppenheimer used the work of Jordan and Eugene Wigner to refashion the number operators, N_r, away from representing the number of electrons in a state, and toward N_r representing the number of electrons that exist at all. Then changing $N_r \to N_r \pm 1$ became creating or annihilating an electron.[48] Furry and Oppenheimer used these number operators—N for electrons, M for positives—on the wavefunction for a system, $\psi_{N,M}$. And they defined a set of operators which created or annihilated an electron $N \to N \pm 1$; created or annihilated a positive $M \to M \pm 1$; and created or annihilated a pair $N, M \to N \pm 1, M \pm 1$.[49] But these operations could not apply to any state of the system: there could not be one fewer particle than the vacuum state. It made no sense to refer to minus one electrons. Algebraically, the vacuum state put a floor under the subtraction of the number of electrons or positives, N and M. States could be counted: two minus one is one; one minus one is zero; zero minus one is zero; or, three minus five is zero; etc.[50] For Furry and Oppenheimer, this algebraic rule took the place of a very physical assumption. It was equivalent to Dirac's interpretation, "according to which 'all negative energy states' should normally be full."[51]

In an October 7, 1933, letter to his brother Frank, Oppenheimer referred to these inventories of Ns and Ms: "The work with the pairs and the MNtory has gone along nicely"; one aspect in particular was

> that the theory gives a consistent interpretation of our inability to localize a particle by experiment within hc/E, and yet does allow us to determine exactly the charge density of a system with unknown number of particles.[52]

How could one know the charge density of a region without knowing how many charged particles were present? (Localization is discussed further in section 4.2.2.) Consider a state with no charge. Was it a vacuum, with no particles? Furry and Oppenheimer wrote the state with no particles as $\psi_{0,0}$. But if they measured the presence of particles by the amount of charge in a given region—say, by a galvanometer—there were a lot more states than $\psi_{0,0}$ that fit that description: $\psi_{1,1}$, $\psi_{2,2}$, etc. all had zero total charge.[53] In pair theory, electric charge was indistinct.

Furry and Oppenheimer explained that when it was possible to absolutely distinguish between electron states and positive states, "there is a unique state (0), for which

47. Dirac, "Quantum Theory of the Emission and Absorption of Radiation," esp. §2; the number of photons was also constant. See Bromberg, "Concept of Particle Creation."

48. Darrigol, "Quantized Matter Waves," 216–25, 250–58; A.I. Miller, *Early Quantum Electrodynamics*, §2.

49. Furry and Oppenheimer, "Theory of the Electron and Positive," eq. 2.4–2.14, 2.32.

50. This was a general feature of Jordan, Klein, and Pauli's field quantization, A.I Miller, *Early Quantum Electrodynamics*, ch. 2.

51. Furry and Oppenheimer, "Theory of the Electron and Positive," algebra: eq. 3.8, 3.13–14, quote: 256.

52. Smith and Weiner, *Robert Oppenheimer*, 164, cf. 166, 171.

53. Furry and Oppenheimer, "Theory of the Electron and Positive," eq. 2.32, 2.33.

the energy is a minimum, and which corresponds to the normal state of the pairs in that field." They labeled the energy of this state $E^{(0)}$ and its wavefunction $\psi_{N,M}^{(0)}$. In all processes of pair creation or destruction, the probability of the process occurring was dependent on $E^{(0)}$. It played a similar role to the absolute zeros of particle numbers, $N = M = 0$. What was the value of this energy $E^{(0)}$? Following Bohr's insistence on considering specific experimental arrangements, Furry and Oppenheimer argued that it depended on the experimental conditions. If there was no electron-positron ψ field at all, "$E^{(0)} \to 0$" as one might expect. Once *any* amount of electron-positive field was present in a region—even if its energy was low and "directly observable pairs are not being created"—$E^{(0)}$ "is the energy of these 'nascent' pairs in the given field." Despite working in a wave-theoretical formalism that did not include Dirac's infinite sea of negative energy electrons, Furry and Oppenheimer nevertheless found "nascent" matter in the vacuum.[54]

The physicality of $E^{(0)}$ and empty space increased when a classical electromagnetic field was introduced. Even if there were no actual electrons or positives, Furry and Oppenheimer argued that "there is an energy $E^{(0)}$ in the pairs formed by the field." This energy E^0 coexisted with the Maxwellian electric field energy E_e. But this Maxwellian energy should not be calculated with the classical expressions for the electric and magnetic field in a vacuum, "for, as we have seen, the field will not be empty." The implication was that one should use Maxwell's equations for fields in matter, in which the dielectric properties and polarization of the medium are included. If the electron-positive field was produced by a particle itself, $E^{(0)}$ would act as a small addition to the electric charge of the particle. Ernest O. Lawrence returned from the 1933 Solvay Conference with news of Dirac and Peierls's polarization of the vacuum calculations, and Furry and Oppenheimer's $E^{(0)}$ resulted in the same shift in the observed charge of the electron.[55]

3.3.3 Crumbs of Comfort

Furry recalled that Oppenheimer chose him as coauthor for this paper because he reacted to Oppenheimer's flashes of insight with further questions, rather than contemplative awe. Working on the "Electron and Positive" paper "was a question largely of spending many hours talking about it at the blackboard in the lab, and probably even more hours pacing the streets of Berkeley, arguing about it."[56] The matter of these discussions can be illuminated by comparing the published paper to an undated draft article on "Theories of the Positron," which closely follows "On the Theory of the Electron and Positive." (Why "positive" vs. "positron"? According to Furry, Oppenheimer disliked Anderson's concatenation of "positive electron" as

54. Furry and Oppenheimer, "Theory of the Electron and Positive," 259.
55. Furry and Oppenheimer, "Theory of the Electron and Positive," 259, 261; Oppenheimer, "[abstract:] The Theory of the Electron and Positive"; cf. Bethe and Heitler's "'twin birth' of a positive and negative electron." Bethe and Heitler, "Stopping of Fast Particles and on the Creation of Positive Electrons," 83; cf. Galison, *How Experiments End*, §3.3.
56. Furry Interview, AIP; cf. Monk, *Inside the Center*, 214–15.

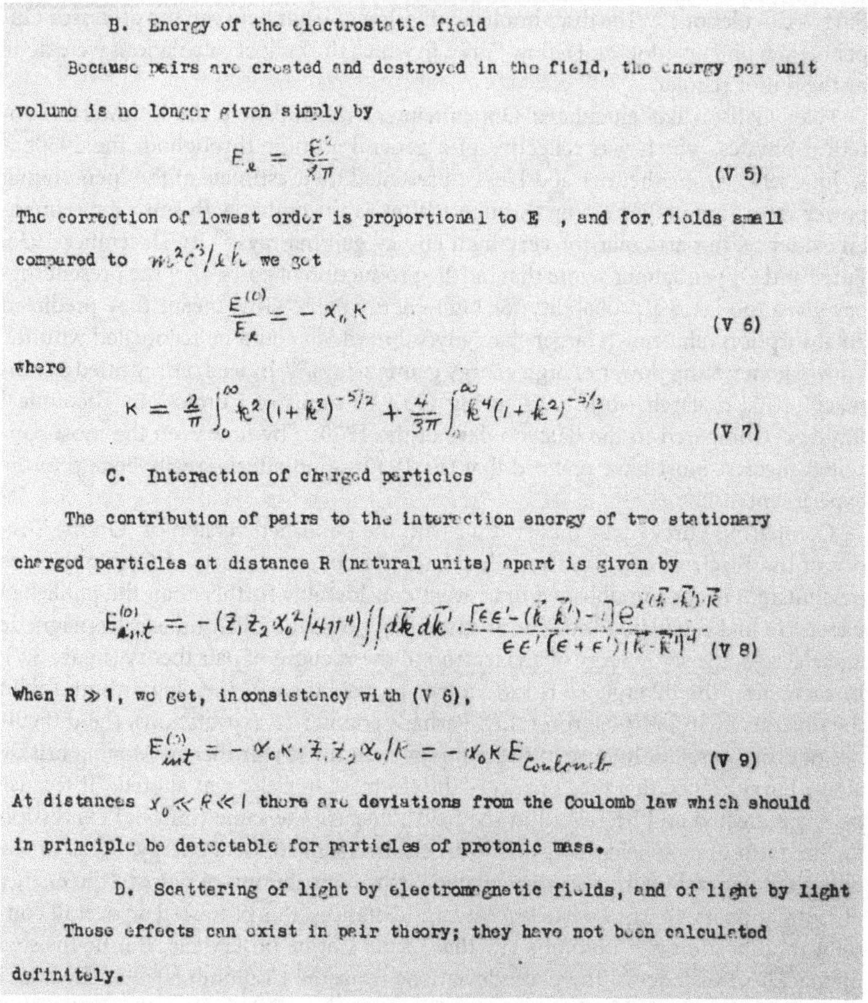

B. Energy of the electrostatic field

Because pairs are created and destroyed in the field, the energy per unit volume is no longer given simply by

$$E_k = \frac{E^2}{4\pi}$$

(V 5)

The correction of lowest order is proportional to E , and for fields small compared to $m^2 c^3 / e \hbar$ we get

$$\frac{E^{(0)}}{E_k} = -x_0 K$$

(V 6)

where

$$K = \frac{2}{\pi} \int_0^\infty k^2 (1+k^2)^{-5/2} + \frac{4}{3\pi} \int_0^\infty k^4 (1+k^2)^{-3/2}$$

(V 7)

C. Interaction of charged particles

The contribution of pairs to the interaction energy of two stationary charged particles at distance R (natural units) apart is given by

$$E_{int}^{(0)} = -(z_1 z_2 x_0^2 / 4\pi^4) \int \int d\vec{k} d\vec{k}' \cdot \frac{[\epsilon \epsilon' - (\vec{k} \cdot \vec{k}') - 1] e^{i(\vec{k}-\vec{k}')\cdot\vec{R}}}{\epsilon \epsilon' (\epsilon + \epsilon') |\vec{k} - \vec{k}'|^4}$$

(V 8)

When R ≫ 1, we get, in consistency with (V 6),

$$E_{int}^{(0)} = -x_0 K \cdot z_1 z_2 x_0 / R = -x_0 K E_{Coulomb}$$

(V 9)

At distances $x_0 \ll R \ll 1$ there are deviations from the Coulomb law which should in principle be detectable for particles of protonic mass.

D. Scattering of light by electromagnetic fields, and of light by light

Those effects can exist in pair theory; they have not been calculated definitely.

Figure 3.7. A page from Furry's report on "New Results Obtained by Pair Theory," Folder: Theories of the positron, Box 6, Furry Papers. Courtesy of the Harvard University Archives.

"positron" because it combined Latin and Greek roots.[57]) Furry's draft included an outline for a full paper that started with a review of "The Experimental Positron" and continued to the "Physical Consequences of the Full-state Hypothesis." "Full-state" referred to Oppenheimer's modification of Dirac's mostly full states of negative energy. The introductory material was only outlined; likely these sections were to be written by Oppenheimer. The later sections of the draft included the results of

57. Furry Interview, AIP

Furry's calculations.[58] The draft implies a division of labor between the professor Oppenheimer and postdoctoral fellow Furry, in which the bulk of calculations were done by the junior scholar.

Peter Galison has elucidated Oppenheimer's pessimistic attitude toward theoretical physics, which was reflective of a general attitude throughout the 1930s.[59] In June 1933, Oppenheimer and Plesset presented their estimate of the "penetrating power" of gamma rays passing through matter as in conflict with some experimental evidence, in particular for very high energy gamma rays.[60] By December 1933, Furry and Oppenheimer wrote that in "the production of pairs, [...] the present theory gives too large a probability for high energy pairs." This meant they predicted an absorption rate "much larger than any value which could be reconciled with the known penetrating power of high energy gamma-rays."[61] Indeed, Furry titled a summary of his research work until spring 1937 "Crumbs of Comfort in Theoretical Physics." Compared to the halcyon days of the 1920s, "by now even the most conceited theorist must have realized that the 1930s—and after?—really belong to the experimentalist."[62]

Comparing Furry's pair theory draft with the published version of "On the Theory of the Electron and Positive," elucidates the concrete effects of Oppenheimer's pessimism. Furry's unpublished draft went considerably further than the published paper. He had calculated the central stumbling points of QED: the electromagnetic and electrostatic self-energy of an electron in the vacuum of pair theory (figure 3.7). In each case "the divergence is less violent on pair theory" (i.e., logarithmic) than the divergence in pre-positron QED. Perhaps counter to expectations, the difficulties of QED were *less* intense in the complex vacuum of pair theory. Moving briskly along, Furry argued that the "Energy of the electrostatic field" was altered, "[b]ecause pairs are created and destroyed in the field"; he gave the same (infinite) expression for the ratio of pair-field energy, $E^{(0)}$, to electromagnetic field energy, E_e, as in the published paper. Finally, Furry calculated "[t]he contribution of pairs to the energy of two stationary charged particles." At large distances, this produced an overall constant modification of Coulomb's law that would not be observable. But he insisted that, at short distances, "there are deviations from the Coulomb law which should in principle be detectable for particles of protonic mass." A final section noted that

58. Box 6, Folder: Theories of the positron, Furry Papers. The draft is undated; the published paper was received at the *Physical Review* on December 1, 1933. Furry referenced experimenters who had been published and discussed in California and at the Solvay meeting in October, which Lawrence attended: Anderson; Blackett and Occhialini; Curie and Joliot; Edwin McMillan; Lauritsen and Crane; Jean Thibaud. For details see Heilbron and Seidel, *Lawrence and His Laboratory*, vol. 1, §IV.1; Institut International de Physique Solvay, *Structure et propriétés des noyaux atomiques*. A second draft follows, which includes Furry's analysis of Peierls, Dirac, and Heisenberg's "hole" theory work. This parallels Furry and Oppenheimer's follow-up publications: Furry and Oppenheimer, "Theory of the Electron and Positive II"; Furry and Oppenheimer, "On the Limitations of the Theory of the Positron."

59. See J. R. Oppenheimer to Frank Oppenheimer, 4.6.1934. Smith and Weiner, *Robert Oppenheimer*, 181; Galison, *How Experiments End*, §3.6; Galison, "Discovery of the Muon"; cf. Pais, *J. Robert Oppenheimer*, 28–29.

60. Oppenheimer and Plesset, "Production of the Positive Electron," 55.

61. Furry and Oppenheimer, "Theory of the Electron and Positive," 260.

62. Folder: Summary of work 1937, Box 7, Furry Papers.

"scattering of light by electromagnetic fields, and of light by light [...] can exist in pair theory; they have not been calculated definitely."[63]

Furry's result for the electromagnetic self energy was omitted entirely from the published paper. In print, Furry and Oppenheimer observed: "For this magnetic proper energy the present theory, of course, gives new results; but these in no way remove the familiar difficulties of the radiation theory." Before introducing a modified version of Coulomb's law of electrostatic attraction, Furry and Oppenheimer undertook a page-long digression on the "limitations [...] met in all applications of the quantum mechanical formalism to wave fields," which were

> apparently not to be overcome merely by modifying the electromagnetic field of an electron within these small distances [below the classical "radius" of the electron], but require here a more profound change in our notions of space and time, on which ultimately the quantum mechanical methods rest.[64]

These concerns were redolent with Bohr's foundational questions for quantum theories. Weisskopf also expressed misgivings about calculating field strengths "at 'the edge of the electron.'"[65] More than just overlaying these discussions on technical results, Furry and Oppenheimer's published paper omitted novel calculations and cast doubt on the wisdom of resolving "the familiar difficulties of the radiation theory" with "new results," at all.[66]

Oppenheimer's pessimism manifested in the published discussion of Furry's result. In lower-energy, longer-scale processes, "the difference between the [modified] charge and the true charge Ze will not be observable." In a footnote, they continued:

> This can be said in another way: Because of the polarizability of the nascent pairs, the dielectric constant of space into which no matter has been introduced differs from that of truly empty space. For fields which are neither too strong nor too rapidly varying the dielectric constant of a vacuum then has the constant value $[\sim (1 - E^{(0)}/E_e)]$. Because it is in practice impossible not to have pairs present, we may redefine all dielectric constants, as is customarily done, by taking that of a vacuum to be unity.[67]

In 1935, Robert Serber called this "renormalization."[68] Furry and Oppenheimer's note connected the vacuum physics of the 1930s to earlier debates among Maxwellian

63. The identical expression for κ in the published version, eq. 4.5, is found here in eq. V.7. The draft version, eq. V.8, expressed $E_{int}^{(0)}$ in terms of the radius between the particles, R; the published version, eq. 4.6, uses the ratio of R and a minimum length λ_0. Folder: Theories of the positron, Box 6, Furry Papers.

64. Furry and Oppenheimer, "Theory of the Electron and Positive," 258, 260.

65. Weisskopf, "Electrodynamics of the Vacuum," 219; see also Pais, *J. Robert Oppenheimer*, 28; Cassidy, "Cosmic Ray Showers," 6; Schweber, "Hacking the Quantum Revolution."

66. Furry and Oppenheimer, "Theory of the Electron and Positive," 258.

67. I performed a substitution using eq. 4.4 to make clear the dependence on $E^{(0)}$. Furry and Oppenheimer, "Theory of the Electron and Positive," 261; Oppenheimer, "[abstract:] The Theory of the Electron and Positive."

68. Serber, "Linear Modifications in the Maxwell Field Equations," 546; Schweber, *QED*, 87; Monk, *Inside the Center*, 217ff.

aether theorists. In 1890, Heinrich Hertz had defined away the electric and magnetic permittivities from Maxwell's equations in a free aether. Oliver Heaviside reproached him for using an aether theory but hiding the density and elasticity of the medium.[69] (The ambivalence of subatomic processes and formal renormalization continued through the 1940s. In 1946, e.g., Hans Bethe and Oppenheimer remarked that the self-energy of the electron "may be regarded as caused by the virtual emission and reabsorption of a quantum" *or* "it may be regarded as a renormalization of the probability" in question. "These are, of course, equivalent descriptions."[70])

Dirac and Peierls had optimistically suggested that vacuum polarization would cause measurable deviations from the Klein-Nishina scattering formula (section 2.4). But even for very-high energy interactions, Furry and Oppenheimer pessimistically suggested that the effects may be "unobservable." The trouble was that at these high energies, the uncertainties of pair theory collided with the uncertainties of quantum electrodynamics "which arise from our ignorance of the reaction of the electron to its own radiation field." Perhaps in a concession to Furry's calculations, they considered heavier particles such as protons, "for which we may perhaps suppose" the critical energy would be lower, and so "the deviations from the Coulomb law should in principle be detectable." This was substantiated in 1935. Edwin Uehling, another student of Oppenheimer, translated Furry and Oppenheimer's Coulomb-interaction conceptual experiment into a robust calculation of the predicted shift in spectral lines of hydrogen "due to the creation and annihilation of electron-positron pairs."[71] Uehling's paper became a landmark, but it was not optimistic. He predicted an increase in the separation of doublet lines; experimentalists had established a decreased separation.

Furry and Oppenheimer's practice of theoretical physics had a dynamics. They began by adapting an existing formalism for relativistic wave mechanics for the case of pair theory. They manipulated this formalism to achieve several desiderata. It must represent electrons and positrons; it must avoid reference to Dirac's "sea"; its MNntory of particles, $\psi_{M,N}$, must have the proper arithmetic. What did this formalism allow Furry and Oppenheimer to do, to calculate, to say about experiments or about the world? To share and to use their formalism, they applied the paper tools of perturbation analysis and "conceptual experiments." For example, Furry and Oppenheimer considered their function for the number of electrons and positives $\psi_{M,N}$ perturbed by an external field, such as an experimentalist's gamma-ray. Because this perturbation would create new pairs, $\psi_{M,N}$ would remain ambiguous, and the link between the formalism of pair theory and experiment remained tenuous. Oppenheimer's epistemological commitments, and his pessimism, often flowed through his

69. Darrigol, *Electrodynamics*, 256–57; units and constants in electromagnetism is a subtle issue. See Jackson, *Classical Electrodynamics*, Appendix on Units and Constants; Bethe and Heitler connected these "effective charge" phenomena to Born's nonlinear electrodynamics; Bethe and Heitler, "Stopping of Fast Particles and on the Creation of Positive Electrons," 105 fn.

70. Bethe and Oppenheimer, "Reaction of Radiation on Electron Scattering," 453.

71. Furry and Oppenheimer, "Theory of the Electron and Positive," 262; Uehling, "Polarization Effects in the Positron Theory," 55; Schweber, *QED*, 86–88.

reasoning with thought experiments. Within these epistemological limits, perturbation analysis provided the tools to connect pair theory to experiment.[72] Then another layer of pessimism descended—another application of critical, philosophical, qualifications. Furry and Oppenheimer were at the same time building a new formalism and expressing its severe limitations; they were deriving new predictions of physical phenomena while casting doubt on their ability to be observed, even in principle.[73] Some of Furry's novel results were left on the cutting room floor. One can almost hear Furry and Oppenheimer walking the streets of Berkeley, arguing over whether this theory just amounts to redefining charge, whether vacuum effects would be observable at all.

3.4 Heisenberg and Euler's Virtual Possibilities

Vacuum phenomena have often been described in "virtual" terms—virtual quanta, virtual photons, virtual electrons, virtual creation or annihilation. Historical discussions of quantum theory often describe historical content in modern "virtual" terminology, perhaps by necessity. This has had the effect of obscuring the origins of "virtual" particles and the role of the "hole" theory vacuum in relativistic quantum theory. It will be useful to briefly clarify the uses of "virtual" in this context.[74]

Generally, in scientific applications, the sense of the adjective "virtual" is of something being taken as what-it-is-not. It signals the use of a metaphor, or "as if." Consider: to the sailor of a small ship, Lake Superior is virtually an ocean; in virtue of its size, Lake Superior should be treated *as if* it was something it is not. "Virtual" can establish a distancing or reservation, as in "virtually certain." In optics, a "virtual image" is an image which does not exist at the location that geometric optics would have it. Your image in a mirror is a virtual image.[75] This basic "as if" sense of "virtual" was in common use in physics from the eighteenth century. Mechanics used "virtual velocities" and "virtual displacements" (motions of a system *as if* it had been pushed out of equilibrium). In the nineteenth century, William Rowan Hamilton built his (Hamiltonian) mechanics on the "virtual work" done under such displacements. "Virtual displacements" could also mean "possible displacements," particularly in contrast to any "actual displacements" a system undergoes. "Virtual" concepts were central to physics pedagogy from the nineteenth century though

72. Furry and Oppenheimer, "Theory of the Electron and Positive," 254, 258.

73. See also J. R. Oppenheimer to F. Oppenheimer 7.1.1934 reporting on new work in pair theory, which "makes the nonobservability of the susceptibility of pairs even more certain." Mehra and Rechenberg, *HDQT*, vol. 6, 909.

74. E.g., Wentzel used "virtual" language to describe the "hole" theory work of Otto Halpern and Max Delbrück, which did not use "virtual." Wentzel, "Quantum Theory of Fields," 390; see also Schweber, *QED*, ch. 2; Halpern and Delbrück each described light scattering off of the Dirac "sea." O. Halpern, "Scattering Processes Produced by Electrons in Negative Energy States"; Meitner and Kösters, "Über die Streuung Kurzwelliger γ-strahlen."

75. Maxwell, *Treatise*, vol. 1, §157.

the twentieth.[76] "Virtual" concepts were deployed in quantum theory. For example, Arnold Sommerfeld distinguished between real and virtual atomic orbits. "[V]irtual not real electronic orbits" were really free paths that were treated *as if* they were atomic orbits.[77] "Virtual" displacements were at the center of angular momentum in (classical and) quantum theory. Heisenberg appealed to virtual changes in magnetic fields to build his 1921 model of atomic cores.[78] Virtual displacements were a pillar of Hermann Weyl's influential *The Theory of Groups and Quantum Mechanics*.[79] This "virtual" language was part of the standard training and practice of modern physicists.

As is well-known, in 1924–1925, Bohr, Kramers, and John Slater (BKS) proposed a new "Quantum Theory of Radiation." Their theory invoked a set of "virtual oscillators" modeled on Planck and Einstein's description of blackbody radiation. At one level, the BKS "virtual oscillators" invoked the familiar "as if" sense of "virtual." On another level, BKS ascribed a quite physical role to this virtual "time-spatial mechanism," such that, for example: "the induced transitions of Einstein's theory occur in consequence of the virtual radiation in the surrounding space due to other atoms."[80] However, the collective views of each author and the reaction to the theory was far more equivocal. The BKS proposal was conceptually and formally daring, and it had wide influence on the development of quantum mechanics. However, it was rapidly refuted by new experiments in 1925.[81]

The BKS theory continued to influence theorists. In the context of Kramers's dispersion theory, John Van Vleck referred to Kramers's dispersion frequencies and their "corresponding fictitious or 'virtual' oscillator."[82] Gregor Wentzel (1898–1978) preferred "*Ersatzoszillators*."[83] In 1933, Felix Bloch invoked Bohr's virtual oscillators in an influential contribution to radiation theory.[84] Dirac provides an enigmatic example. In one series of notes, likely from 1934, Dirac was trying out different approaches to positron theory. One proposal was to treat the negative-energy sea with Jordan's "2nd Q. theory for electrons, with field not quantized." On the same page he wrote: "Try Wave fn. = fn.(particles) × q(field) for a BKS theory." Below this line, the page is blank.[85]

76. A "Typical Outline" of Berkeley's Physics 105A, "Analytic Mechanics," from November 1958 included two hours on "Statics of rigid bodies and virtual work." AIP Manpower and Statistics, Box 9, Folder: Survey of Education in Physics...Appendix 7; Smith and Wise, *Energy and Empire*, 376–94, Garber, *Language of Physics*, 127–28, 240; virtual concepts date from the middle ages, see Jammer, *Concepts of Force*.

77. Sommerfeld, *Atomic Structure*, 190, see also 541–42, 557; Seth, *Crafting the Quantum*, 214–15.

78. Cassidy, "Heisenberg's First Core Model," 206.

79. Weyl, *Theory of Groups and Quantum Mechanics*, 186ff.

80. Bohr, Kramers, and Slater, "Quantum Theory of Radiation," 164–65; virtual oscillators originated in the work of Rudolf Ladenburg and Fritz Reiche. Duncan and Janssen, "Umdeutung in Minnesota. I," 559.

81. Hendry, "Bohr-Kramers-Slater"; Pais, Niels Bohr's Times, §11(d); Beller, Quantum Dialogue, ch. 2; Kragh, "Bohr-Kramers-Slater Theory."

82. Van Vleck, *Theory of Electric and Magnetic Susceptibilities*, 361.

83. Wentzel, "Wellenmechanik der Stoß- und Strahlungsvorgänge," 781.

84. Bloch, "Zur Bremsung rasch bewegter Teilchen beim durchgang durch Materie."

85. Dirac Papers, Series 2, Box 36, Folder 22.

The BKS scheme likely contributed to two of the most important associations for "virtual" in physics: the violation of the conservation of energy, and the representation of *internal* processes that could not be independently observed.[86] For example, in 1934 Peierls described the conservation-breaking transitions of Dirac's radiation theory (section 3.2) as "virtual transitions." These were *internal*, intermediate transitions (table 3.1), in which "an electron jumps from a state of negative energy to one of positive energy" even when the external electromagnetic field was relatively weak.[87]

In Leipzig, Heisenberg and his circle of students and assistants applied "virtual" concepts to the problems of relativistic quantum theory. Perturbation analysis led them in particular to the new process of the scattering of light by light. They developed a particular usage that applied the label "virtual"—not to a particle, or a process, or a state—to *possibilities*. Heisenberg's attention to the power of possibilities points to an early alignment of his relativistic physics with neo-Kantian and Aristotelian philosophies of science

3.4.1 Heisenberg and Dirac

During Heisenberg's early intense engagement with "hole" theory, on April 19, 1934, he excitedly wrote to Pauli and Weisskopf in Zurich, that "the polarization of the vacuum is exactly zero." Or, perhaps considering Bohr's concern for the limits of definitions, he rephrased his thought: "More correctly, one might perhaps say: in ["hole"] theory it is not possible to define what the word[s] 'polarization of the vacuum' might mean."[88] In the published proceedings of the Solvay Congress the previous autumn, Pauli was openly critical of Dirac and Peierls's positron theory.[89] Like Oppenheimer (section 3.3.3), Heisenberg sought to render Dirac and Peierls's vacuum polarization unobservable. But in a landmark paper completed in June 1934, he changed his mind and embraced the consequences of the new vacuum physics. Heisenberg argued that the positron showed that "a light quantum will create a matter field in its neighborhood—similar to an electron creating a Maxwell field in its neighborhood."[90] Just as electrons were sources for the electromagnetic field, photons were sources of the matter field. If electrons had an *electromagnetic* self-energy, so too light quanta had an *electron-positron* self-energy. They were reciprocal, or complementary.[91] Heisenberg hoped that "a unified theory of matter and light fields" was the way forward.[92] Heisenberg, Pauli, Jordan, and others were motivated to explore "unified" field theories, in which the electromagnetic field and fields for matter (pairs, neutrons,

86. Cf. *Glossary of Nuclear Science and Technology*, sv. "virtual quantum."
87. Peierls, "Vacuum in Dirac's Theory," 439.
88. Pauli, *Wissenschaftlicher Briefwechsel*, doc. 370, vol. 2, 318; for Heisenberg's "hole" theory, see A.I. Miller *Early Quantum Electrodynamics*, §§4.4, 4.7.4–7.5, 4.11–12; cf. Mehra and Rechenberg, *HDQT*, vol. 6, 920–21.
89. Dirac, "Théorie," 213–14.
90. Heisenberg, "Remarks on the Dirac Theory," 185; cf. Roqué, "Manufacture of the Positron," 101–12.
91. A.I. Miller, *Early Quantum Electrodynamics*, 48.
92. Heisenberg, "Remarks on the Dirac Theory," 187.

neutrinos) were replaced by a single underlying field.[93] Heisenberg tied the motivation for a unified field theory to the positron: "In the Dirac theory of the positron, a clean separation of the fields which actually occur, into matter fields and electromagnetic fields, is scarcely possible any more."[94] Leading physicists were coming to consensus about some of the radical implications of positron theory.

Heisenberg and Dirac developed their "hole" theory collaboratively (section 2.4). Oppenheimer sent a preprint of his work with Furry to Dirac, though he received no reply.[95] But in the final page of Dirac's lectures at Princeton in 1935, he adopted Furry and Oppenheimer's "nascent" framing of the infinities of radiation theory:

> Just as the self-energy of the electron can be regarded as due to many nascent light quanta surrounding it, so the theory gives around each photon many nascent electrons and positrons which give it a self-energy; i.e. the Hamiltonian contains terms corresponding to such transitions as cause the creation of electrons and positrons.

He provided a succinct sketch of calculations of the infinite self-energy of the photon which had been "pointed out by Heisenberg." H is the Hamiltonian function that described the time-evolution of the system.

Consider the Matrix element

$$(\text{one photon} \,|\, H \,|\, \text{arbitrary [number of particles]}). \qquad (3.3)$$

Among the non-vanishing components there will be some with the right-hand side having one <u>new</u> photon, one electron and one positron. In a stationary state such terms will give a change in energy (by the perturbation method)

$$\sum_{\text{arbitrary states}} \frac{|(\text{one photon} \,|\, H \,|\, \text{arb.})|^2}{W_{\text{one photon}} - W_{\text{arb.}}} \qquad (3.4)$$

and this turns out to be infinite.[96]

By 1934, a developing consensus among theorists held that nascent or unobservable particles filled empty space—at least insofar as theory could be trusted. Positrons were being reliably produced in several laboratories, but detailed and quantitative alignments between theory and experiment were elusive.[97] Despite their different attitudes toward Dirac's sea, nascent particles were present for both particle-based "hole" theory (Dirac, Peierls, Heisenberg) and wave-based pair theory (Oppenheimer, Furry,

93. Jordan consistently held a unified view, Bromberg, "Concept of Particle Creation," §3; cf. Duncan and Janssen, "Pascual Jordan's Resolution," 646–49; on Heisenberg: Pais, "Dirac Theory of the Electron"; Rueger, "Attitudes towards Infinities," esp. 324; compare to J. A. Wheeler in, ch. 5.
94. Heisenberg, "Remarks on the Dirac Theory," 187; cf. Furry and Oppenheimer, "On the Limitations of the Theory of the Positron," 904.
95. Monk, Inside the Center, 214ff.
96. I have added a missing minus sign in the denominator. Notes taken by Boris Podolsky and Nathan Rosen, Dirac Papers, Series 2, Box 27, Folder 1, pages 2–41; cf. Kragh, Dirac, 167–68.
97. Schweber, QED, 77–90; Roqué, "Manufacture of the Positron," §5.

Pauli). What produced this agreement? More than anything else, it was their shared practice of applying "the perturbation method" (see appendix A). For Oppenheimer and Furry more than the others, an epistemological analysis of "conceptual experiments" supported arguments for "nascent pairs" in the vacuum. But they too relied on perturbation theory for quantitative analysis (section 3.3.3). Dirac and Peierls struggled to describe a relativistic quantum vacuum, but calculated the deviations from the Klein-Nishina formula by treating the scattering X-ray as a classical perturbation (section 2.4). Heisenberg's work, in particular, put perturbations at the center of calculation and interpretation of new theoretical phenomena.

3.4.2 "Virtual" Possibilities and Perturbations in Leipzig

Heisenberg and his circle in Leipzig used quite particular formulations of "virtual" concepts. They referred to virtual *processes*, as Peierls did. But in place of "transitions" they wrote of creation and annihilation. They used a different phrase in their most developed work on positron theory: the "virtual possibility" of pair creation and annihilation, or the "infinitely many virtually possible transitions [*virtuell möglichen Übergänge*]" for electrons in QED.[98] Heisenberg's research group applied these virtual concepts broadly. This section focusses on a marquee calculation for the new positron theory, the scattering of light by light. Initially, in 1933, this scattering was investigated with the possibility of quantitative contact with experiment; by 1936, that possibility seemed foreclosed. Theorists came to calculate the scattering of light by light as theory for theory's sake. This research demonstrates the high degree to which perturbation methods determined the nature of vacuum phenomena.

Otto Halpern (1899–1982) was the first to consider "scattering properties of the 'vacuum'" of "hole" theory. Halpern received his PhD in Vienna in 1922 at a moment of noted academic anti-Semitism. For two years he was a fellow with Heisenberg, before taking up a position at NYU in the United States in 1930. In 1933, Halpern was struck by Dirac's "hole" theory—its promise and "the almost insurmountable difficulties which the infinite charge-density without field offers to our physical understanding." In a short letter, he proposed the scattering of light by light as a new test for Dirac's theory. The idea was that light would cause pair creation in an intermediate scattering state (table 3.1), and that the direction and frequency of the light would be altered when the electron returned to its hole. Halpern promised the results of quantitative calculations, but they were not published.[99] (Two years later, Halpern copublished another note on scattering with his sixteen-year-old mentee, Julian Schwinger that is discussed in chapter 4.[100]) Halpern had speculated that light-light scattering would have cosmological importance, perhaps leading to new derivations of Hubble's constant for the expansion of the universe. In 1934, cosmological consideration of light-light scattering (with real pair production) was taken

98. Heisenberg to Pauli, 5.2.1934. This was the first such use in Pauli's correspondence. Pauli, *Wissenschaftlicher Briefwechsel*, 273.
99. O. Halpern, "Scattering Processes Produced by Electrons in Negative Energy States."
100. Bederson and Stroke, "New York University Physics Department," 281–84.

up by Gregory Breit, also at NYU, and John Archibald Wheeler, a postdoctoral fellow at Bohr's Institute in Copenhagen (see chapter 5). Heisenberg's first published contribution to "hole" theory was 1934's "Remarks on the Dirac Theory of the Positron." He integrated treatments of the scattering of light by light by Halpern and by Peter Debye (unpublished) into a sophisticated density-matrix framework. He set the scattering as a problem for two of his doctoral students at the University of Leipzig, Hans Euler (1909–1941) and Bernhard Kockel (1909–1987). In 1935, they published a short note on the subject. They described a process "in which two light quanta virtually create a (positron and electron) pair."[101] They put the "virtual" on the *process*. This was, perhaps, a step removed from a physical interpretation of "virtual pairs." One might ask why did physicists calculate *as if* pairs were created? Several association of "virtual" were at play. Euler and Kockel were analyzing light below the energy threshold for creating real, lasting pairs. Therefore, any pair creation process would have to (momentarily) violate the conservation of energy. At a deeper level, why did Euler and Kockel think to calculate these quantities? Why, for example, a single pair and two light quanta? They followed a structure established by perturbation theory.

Perturbation theory proceeded in two steps: first, idealizing the physical phenomena to make it simpler to represent; and second, calculating approximate quantities to compare to observation. Physicists first modeled a simpler *un*perturbed system, such as an isolated hydrogen atom. Then, they considered a perturbing influence, such as electromagnetic radiation, which produced subtle deviations from the unperturbed state. Finally, physicists deployed mathematical approximation schemes to solve the perturbed-system equations. This is discussed more fully in appendix A. The basic idea of the mathematical approximation can be seen in the Maclaurin series for the exponential function from elementary calculus:

$$e^x = \frac{x^0}{0!} + \frac{x^1}{1!} + \frac{x^2}{2!} + \frac{x^3}{3!} + \frac{x^4}{4!} + \cdots. \tag{3.5}$$

The terms were referred to by their place in the series (zeroth, first, second, etc.). A mathematician graphing an exponential function will draw a curve closer to the true exponential increase if they include more terms in their calculations. In a scattering scheme, the zeroth-order term represented radiation passing through matter without changing frequency or direction, and each further term represented some scattering phenomena.

For Heisenberg, there was a very direct relation between terms in the series expansion and physical phenomena. Each instance of a field-variable—such as the electromagnetic potential A_μ—corresponded with an interaction with the quantum of that field—for example, a photon. Without defining all the symbols, here is Heisenberg's fourth-order matrix element for an electron in "hole" theory, perturbed by an electromagnetic wave, from "Remarks on the Dirac Theory of the Positron":

101. Euler and Kockel, "Über die Streuung von Licht an Licht nach der Diracschen Theorie," 246; D. Hoffmann, "Hans Euler."

Finally H_4 is reduced to the term

$$H_4 = \int d\xi \left[-ic\hbar \frac{1}{24} \left(-\frac{ie}{\hbar c} A^\lambda x_\lambda \right)^4 \frac{\partial S_0}{\partial x_0} \right] \tag{3.6}$$

$$= \frac{1}{48\pi^2} \left(\frac{e^2}{\hbar c} \right)^2 \frac{1}{\hbar c} \int d\xi \frac{(A^\lambda x_\lambda)^4}{(x_\varrho x^\varrho)^2}, \tag{3.7}$$

and gives rise to matrix elements which lead to the scattering of light from light (annihilation and creation of two light quanta each, with the same momentum sum).[102]

Apart from constants, this equation was a contribution to the energy of the vacuum in "hole" theory, which Heisenberg derived from Dirac's density matrix formalism (section 2.4). x_λ was a position coordinate representing the distance between pairs of negative-energy electrons. A^λ represented an external (perturbing) electromagnetic field. Performing the integration would add up the contribution of this perturbation to the energy of the system from every electron. This was the starting point from which Euler and Kockel tried to develop a specific, measurable, prediction. The perturbation-first attitude continued throughout their collaboration. In 1936, Heisenberg and Euler wrote:

> Each single term in the expansion of the [vacuum] energy density in powers of [the electric and magnetic field strengths] can concretely be associated with a scattering process, whose cross section is determined by it. The terms of fourth order, [H_4] for instance, stand for the usual light-light scattering, the terms of the sixth order determine the cross section of the process whereby three light quanta are scattered on each other etc.[103]

This formalism-first attitude was not new. In the 1920s, Heisenberg was dissatisfied with the role of Bohr's "correspondence principle" in casting aside various bits of formalism in a given calculation (keeping only those which fit classical expectations). Compared to Kramers, in 1925 Heisenberg was more willing to follow the quantum formalism where it led.[104] (However, Heisenberg joined his peers in discarding meaningless infinite quantities such as the zero-point energy of radiation and the self-energy of the electron.[105])

Euler, shown on a physicists' ski trip in figure 3.8, completed his doctoral thesis "On the Scattering of Light by Light according to Dirac's Theory" in the summer of 1935 (and published the results in early 1936).[106] Euler distinguished two cases of

102. Heisenberg, "Bemerkungen zur Diracschen Theorie des Positrons," 228; Heisenberg, "Remarks on the Dirac Theory," 184.
103. Heisenberg and Euler, "Consequences of Dirac Theory," 16.
104. Dresden, *H. A. Kramers*, 242; Bromberg, "Impact of the Neutron," 328–29; cf. Serwer, "*Unmechanischer Zwang*."
105. Rueger, "Attitudes towards Infinities," esp. 313–16, fn64.
106. Mehra and Rechenberg, *HDQT*, vol. 6, 920ff; for details of the ski trip, see Heisenberg to Pauli, March 22, 1935, and note, Pauli, *Wissenschaftlicher Briefwechsel*, vol. 2, 381.

Figure 3.8. Werner Heisenberg photographed a ski trip in the Bavarian Alps, where "hole theory" was discussed, in February 1935. Left to right: Carl F. v. Weizsäcker, Christian Bohr, Niels Bohr, Hans Euler. Courtesy of the Archives of the Max Planck Society, Berlin.

scattering. When the energy and momentum of the two light quanta was greater than the rest energy of an electron and positron, "*real* [*wirklichen*] pair creation is allowed." This process would create particles that could be observed with a cloud chamber. The second case was when the energy and momentum of the light quanta was not enough for "real" pair creation. Here, too, Euler argued "there must be a scattering of light by light": the two initial quanta "can merge into [*übergehen*] two other light quanta because of the *virtual* possibility of pair creation."[107] Not virtual pairs, nor the virtual creation of pairs, but the virtual possibility of the creation of pairs. This was the same language Heisenberg used in his February 5, 1934, letter to Pauli.

It was the fourth term in Heisenberg's series expansion, H^4, which led to Euler's paper being written. Indeed, at one point Euler explained that "the [fourth-order] terms considered here [...] come about through the virtual possibility of pair creation." He also explained that "real electrons, in contrast to the *virtual* [case] considered here, would be seen in a Wilson chamber."[108] Euler derived his expression for H^4 from a four-part transition of the system (radiation and matter field) between "almost stationary states" ($i \rightarrow n$). At fourth order, these transitions were broken up

107. Euler, "Über die Streuung von Licht an Licht nach der Diracschen Theorie," 399; cf. D. Kaiser, *Drawing Theories Apart*, 34–36; Wüthrich, *Genesis of Feynman Diagrams*, 25
108. Euler, "Über die Streuung von Licht an Licht nach der Diracschen Theorie," 401–405 fn1.

by several transitions to "'*virtual*' intermediate states," k, l, and m. The transitions did not conserve energy:[109]

$$(i \rightarrow k, k \rightarrow l, l \rightarrow m, m \rightarrow n). \tag{3.8}$$

This matrix formalism expressed only the *overall* transition ($i \rightarrow n$) for which energy was conserved. (Recall Fowler and Dirac's debates over chemical reactions and detailed balancing, section 3.2.) In his conclusion, Euler wrote that the scattering of light by light was an observable effect, in principle. But the actual amount of light he expected to be scattered was minuscule, leaving little hope for an actual measurement.[110]

In their 1936 joint publication, Heisenberg and Euler assessed the situation:

> on the one hand, it will not generally be possible to separate the processes in empty space from material processes, since the fields, if their energy is sufficient, can produce matter; on the other hand, even where the energy is not sufficient for the production of matter, a kind of "polarization of the vacuum" and a change in Maxwell's equations result from its virtual potential.[111]

Heisenberg and Euler took it that all the *possible* processes must be accounted for in a calculation of the characteristics of a physical phenomena. This was an underlying assumption of statistical mechanics and the "selection rules" of quantum theory, which ruled out particular atomic transitions. Dirac appealed quite explicitly to a principle of plenitude in 1931: if "quantum mechanics does not really preclude the existence of isolated magnetic poles," then "one would be surprised if Nature had made no use of it." Peierls, Pauli, and Weisskopf all recognized this form of argumentation.[112] Heisenberg and Euler's consideration of "virtual potential" was an extension of this logic. Not only that, it licensed conclusions about phenomena that had no hint of previous experimental demonstration, such as the scattering of light by light.

Heisenberg and his students preferred to analyze "virtual potential," and the "virtual possibility" as opposed to virtual processes or virtual particles. This language may indicate the influence of the neo-Kantian philosophy of quantum mechanics that was debated in the mid-1930s at Leipzig. Carl F. von Weizsäcker had considered studying philosophy before Heisenberg guided him to physics. During 1934–1935, Heisenberg and von Weisäcker had lengthy discussions with Grete Hermann (1901–1984), who was completing her doctorate in philosophy on a Kantian interpretation of quantum mechanics. A prominent thread of philosophical analyses of quantum

109. Euler, "Über die Streuung von Licht an Licht nach der Diracschen Theorie," 415–17. Euler also used the phrase "virtually create a set of pairs" [*virtuell eine Menge von Paaren ... entsteht*] in his "preliminary and intuitive" treatment of the problem, 400.

110. Euler, "Über die Streuung von Licht an Licht nach der Diracschen Theorie," 443.

111. Heisenberg and Euler, "Folgerungen aus der Diracschen Theorie des Positrons," 714.

112. D. Kaiser, *Drawing Theories Apart*, 32; Borrelli, "Emergence of Selection Rules"; Dirac, "Quantised Singularities in the Electromagnetic Field," 71; Kragh, *Dirac*, 270–77.

theory was that it repudiated the claim that causation was an objective characteristic of nature, or that causation was a necessary part of scientific understanding.[113] Immanuel Kant held the causal view, and physicists such as Victor Weisskopf argued that quantum mechanics repudiated Kant's view of scientific understanding (section 4.2). Hermann acknowledged that quantum mechanics did not allow for deterministic predictions of the results of experiments. She described experimental possibilities as possible "causal chains" and she insisted that in retrospect, after the experiment was performed, a definite causal chain could be associated with any given measurement. This recovered a sense of causality within quantum physics. In von Weizsäcker's words, Hermann described "virtual causal chains."[114] Kant had placed at the center of his philosophy the analysis of the *conditions of possibility* of human knowledge and action. One way to read Heisenberg and Euler's "virtual possibility" and "virtual potential" of pair creation is as a signal that the conditions for the real possibility were absent from the situations they described. They specified these conditions in terms of the energy of radiation needed for real pair creation. Where these conditions were absent, for example, "where the energy is not sufficient for the production of matter," virtual considerations entered.[115]

3.5 Weisskopf and the Consolidation of "Hole" Theory

"Hole" theory began with Dirac's "Theory of Electrons and Protons" in 1930 (section 2.3). After a brief burst of calculations it lay fallow until Carl Anderson's "positives" and Patrick Blackett and Giuseppe Occhialini's "anti-electrons" appeared in photographs of cloud chambers in 1932–1933. After that, several new physical processes involving positrons were introduced. These were conjectured to underlie the mounting experimental evidence for the existence of the positron, and its identification with Dirac's "holes." While images of pairs, such as figure 3.1, were compelling, they provided mostly *qualitative* agreement between experiment and theory. A crucial test for *quantitative* questions was the amount of radiation absorbed by metal plates (recall Compton's X-ray absorption apparatus, figure 2.5). Here theory did not match experimental evidence: experimentalists found that much more radiation penetrated through the absorbers than theory expected. First and foremost, Peierls's friend Hans Bethe worked out the theory of how charged particles slow down in matter, in collisions, by exciting atoms, and by *Bremsstrahlung* ("braking radiation"). In particular, he asked: how much does pair production slow down charged particles in cosmic rays? Several historians have discussed the history of this puzzle.[116] This section will concentrate on a different figure who studied a related process: Victor Weisskopf and

113. Forman, "Weimar Culture, Causality, and Quantum Theory."
114. Mehra and Rechenberg, *HDQT*, vol. 6, 713; for much more analysis and translations of Hermann's work, see Crull and Bacciagaluppi, *Grete Hermann*; on Heisenberg: Camilleri, *Heisenberg and the Interpretation of QM*, 133–51; for neo-Kantian influences on Bohr, see Chevalley, "Bohr's Words and Kantianism."
115. Later, pair creation continued to motivate Heisenberg's philosophy. Heisenberg, *Physics and Philosophy*, 130, 139–40.
116. Cassidy, "Cosmic Ray Showers"; Galison, "Discovery of the Muon," §IV; Roqué, "Manufacture of the Positron," 100–5; Schweber, *Nuclear Forces*, 156, 243–56.

the scattering of light by light. The scattering of light by light played a central part in Weisskopf's consolidation of "hole" theory, which culminated in landmark review articles in 1936 and 1939.[117] While early studies of light-light scattering aimed at contact with experiment, by 1937, the collision of light and light made no such contact. The calculation was valued instead for its theoretical qualities. On related findings, Weisskopf argued for the *irrelevance* of difficult experimental evidence for the new vacuum physics. This makes it particularly revealing about the practice of theoretical physics.

In 1934, Weisskopf (figure 3.9) left Dirac and Peierls in Cambridge, to become Pauli's assistant at ETH Zurich. In addition to Pauli, in this period Weisskopf met with Gregor Wentzel and his student Nicholas Kemmer at the University of Zurich; he corresponded with Heisenberg, Weizsäcker, Euler, and Fritz Sauter in Leipzig, and with Furry and Oppenheimer in California, among others. In the summer of 1936 Weisskopf took up a post at Bohr's Institute in Copenhagen. Anti-Semitism was rising to a crisis in central Europe and Weisskopf was Jewish. Weisskopf had opportunities to move to the Soviet Union. Despite his socialist orientation in so-called Red Vienna, he recoiled at Stalinist terror, and he declined. On September 28, 1937, Weisskopf and his wife Ellen arrived in New York City; Bohr had helped him find a job at Rochester University.[118]

In 1934, Pauli had just completed his landmark review of "The General Principles of Wave Mechanics," which became a standard reference work. Pauli argued that the infinite self-energy of the electron was the central obstacle for constructing a useful *relativistic* wave mechanics of radiation *and* matter. Weisskopf was set to work on the electromagnetic self-energy of an electron in "hole" theory (see section 4.2.3)—and eventually also the Delbrück effect and the scattering of light by light. In the course of this work, Weisskopf noticed that not only Dirac's relativistic equation, but also the Klein-Gordon equation could describe particle creation and annihilation. These particles had mass and positive and negative charge, but they grouped together according to Bose-Einstein statistics. The wavefunction $c_0(1_k; 1_l)$ in figure 3.5 represented hypothetical charged bosons. Pauli and Weisskopf's unrealistic model demonstrated the possibility of constructing relativistic quantum theories *with* particle creation and destruction, but *without* negative-energy states.[119] In this period, Pauli's attitude to "hole" theory changed dramatically. After his initial skepticism, further experimental observation and theoretical analysis from Peierls, Dirac, and Heisenberg brought him around to optimism. In early 1934, Pauli proposed to Heisenberg that they and Weisskopf collaborate on a "hole" theory "Dreimänner-Arbeit" (three-man work), referring to the famous 1925 collaboration between Born, Jordan, and Heisenberg). Weisskopf made pages of calculations on "Pauli's hole theory." However, this

117. Weisskopf, "Über die Elektrodynamik des Vakuums"; Weisskopf, "On the Self-energy and the Electromagnetic Field of the Electron."
118. Weisskopf papers, Box 9.02, Folder 10; Pauli, *Wissenschaftlicher Briefwechsel*, doc. 359; cf. Peierls to Pauli 17.7.1933, doc. 317; Miller, *Early Quantum Electrodynamics*, ch. 4; Enz, *No Time to Be Brief*, ch. 6; Weisskopf, *Joy of Insight*, 106.
119. Pauli and Weisskopf, "Scalar Relativistic Wave Equation."

Figure 3.9. Victor Weisskopf, Maria Göppert-Mayer, and Max Born in Göttingen. Photograph by Samuel Goudsmit, courtesy AIP Emilio Segrè Visual Archives, Goudsmit Collection.

optimism did not last and the planned paper was never written.[120] In this rapidly fluctuating environment, Weisskopf undertook his first attempt to consolidate "hole" theory.

3.5.1 Leapfrogging Experiment

In 1935, Weisskopf published a non-technical review article, "Problems in the New Quantum Theory of Electrons," in *Die Naturwissenschaften*, a rapidly publishing magazine covering all natural sciences, similar to *Nature*. After explaining Dirac's theory, negative-energy states, and positrons as "holes," he turned to the polarization of the vacuum:

> External electromagnetic fields change the charge distribution of the [negative-energy] electrons, which can lead to phenomena which can be called a dielectric

---

120. Folder 10, Box 9.02, Weisskopf Papers. See A.I. Miller, *Early Quantum Electrodynamics*, esp. §§3.2, 4.4, 4.7.5, 4.8; Enz, *No Time to Be Brief*, ch. 6.

polarization of the vacuum. The dielectric phenomena of ordinary matter also arise in a similar manner by a charge shift which is induced within matter by the external fields.

Weisskopf recognized that infinite charges in empty space and infinite polarizations were difficult ideas. But he insisted that the role of the vacuum in electron theory was "not necessarily tied to the perhaps dubious conception of hole theory." Dirac's infinite sea remained difficult to defend.

Weisskopf's review next moved to experiment. But rather than quantitative results, he made an argument for analogical extension from known phenomena to conjecture:

> The experimentally established production of electron pairs from the vacuum by high-frequency light waves is already an interaction between an electromagnetic field and the "vacuum," so, one can assume that this interaction is not limited to high-frequency light waves alone.[121]

He argued that the fact that *high-energy* radiation created pairs from the vacuum indicated that *lower-frequency* light should interact with the vacuum as well. Max Delbrück made this argument in 1933.[122] In his canonical monograph on *The Quantum Theory of Radiation*, Walter Heitler was less optimistic than Weisskopf on the prospects of "hole" theory itself. But he shared the viewpoint that "the fundamental fact of the creation of electron pairs by an electromagnetic field" necessitated a new "*non-linear electro-dynamics for the vacuum.*"[123]

This argument for "hole" theory—or at least for a vacuum with complex physical properties—would have been much stronger if there were measurable effects of the interaction of light and the vacuum below the threshold for "real" pair creation. Here Weisskopf performed some rhetorical gymnastics. One possibility was Uehling's calculation of the shift in the spectral lines of hydrogen due to vacuum polarization (section 3.3.3). In fact, experimentalists *did* observe an unexpected shift in the hydrogen spectrum. Unfortunately, Uehling's shift and the experimental shift were in the opposite direction. While visiting Princeton, Pauli wrote to Weisskopf on June 3, 1936, "It is a pity that the Uehling effects have not been empirically confirmed."[124] Another possibility was the scattering of light by light. However, Euler, Kockel, and Heisenberg had found that the effect was too small to be measured. Euler, Weizsäcker, Heisenberg, and Weisskopf each continued to work on the problem.[125] The most direct connection between subtle vacuum effects and experiment was made by Delbrück. But by 1935, this was not an attractive option.

121. Weisskopf, "Probleme der neueren Quantentheorie des Elektrons," 671; similarly, see Weisskopf, "Electrodynamics of the Vacuum," 206–7.
122. Meitner and Kösters, "Über die Streuung Kurzwelliger γ-strahlen," 144; cf. Brown and Moyer, "Lady or Tiger?," 153.
123. Heitler, *Quantum Theory of Radiation* (1944), 191, 194.
124. Pauli, *Wissenschaftlicher Briefwechsel*, doc. 430a, 791.
125. Weisskopf Papers, Box 9.02 Folder 7.

Delbrück proposed his scattering mechanism to explain a set of observations from Lise Meitner's experiments on gamma-ray scattering with H. H. Hupfeld in 1930 and with H. Kösters in 1933. In addition to positrons and Compton-frequency-shifted radiation, they reported on an "unshifted," coherent, portion of the radiation. Delbrück proposed that this "unshifted" line was the result of light scattering *directly off the Dirac sea* nearby to an atomic nucleus (with not even virtual pair creation).[126] This instigated an experimental controversy between Meitner in Berlin and Louis Harold Gray and G. T. P. Tarrant at Cambridge University. The stages of the controversy were recorded in Heisenberg's three-part article on nuclear theory (1932–1933): part two offered an explanation for the Meitner-Hupfeld effect but part three renounced it.[127] In November 1933, Gray and Tarrant critiqued Meitner's work and provided new evidence against the existence of any such scattering.[128] By June 1934, Oppenheimer and the experimentalist Charles Lauritsen argued that "[u]nless existing theory be inapplicable to the calculation of this effect, [Delbrück] scattering can hardly be of importance."[129] Physicists *did* agree on the photoelectric-like production of positrons, which caused deviations from the Klein-Nishina scattering formula (section 2.2.3). But Oppenheimer and Lauritsen argued that these two processes *alone* could account for the experimental data, *without* Delbrück scattering. According to historian Xavier Roqué, the experimentalists' controversy closed in 1937, against the Berlin laboratory. Delbrück himself moved to research at the intersection of physics and biology (for which he became famous).[130] What radiation there might have been was not taken to be the product of coherent scattering off of Dirac's sea. As early as 1936, Heitler's monograph on radiation theory discussed these experiments but made no mention of Delbrück.[131]

For Weisskopf in 1936, "The calculation of pair creation and annihilation which starts from [the "hole" theory] picture shows good agreement with experiment." It was

> In the calculation of most other effects which follow from the positron theory, [that] one always encounters the problem to what extent the behavior of the vacuum electrons actually is to be regarded as the behavior of the vacuum.[132]

These other effects were the scattering of light by light and vacuum polarization. No reference was made to Delbrück scattering.

126. Meitner and Kösters, "Über die Streuung Kurzwelliger γ-strahlen," 144; Brown and Moyer, "Lady or Tiger?"; cf. Meitner, "Die Streuung Harter γ-strahlen."
127. Brown and Moyer, "Lady or Tiger?," 174.
128. Gray and Tarrant, "Anomalous Absorption of High Energy Gamma Radiation. II," 703–5.
129. Lauritsen and Oppenheimer, "On the Scattering of the Th C" γ-rays," 80; cf. Bethe, "On the Annihilation Radiation of Positrons."
130. Roqué, "Manufacture of the Positron," §§3–4; Brown and Moyer, "Lady or Tiger?"; Brown, "Compton Effect," 238; Aaserud, *Redirecting Science*, 94–97, 223, 234–35; this account differs from Mehra and Rechenberg, *HDQT*, vol. 6: 804–5, 821, 906–7.
131. Any coherent radiation was attributed to "secondary" electron-positron annihilation: Heitler, *Quantum Theory of Radiation* (1936), 203, 209.
132. Weisskopf, "Electrodynamics of the Vacuum," 207.

Delbrück scattering itself remained a *possible* process, and became a stringent test for theorists' skills. Weisskopf and Kemmer worked through the scattering of light by light, and the scattering of light by a constant potential. (The constant potential was an idealization of Delbrück's nuclear field.) On January 20, 1936, Pauli wrote to Weisskopf from Princeton with the news that one of Oppenheimer's doctoral students was working on the scattering of light by light, but had found discrepancies with Euler and Heisenberg's result.[133] By that spring, Weisskopf had found *two* new approaches to the scattering phenomena, each of which confirmed Euler and Heisenberg. The first was in collaboration with Kemmer. In an article in *Nature* they employed an (unspecified) method to directly calculate the cross-section (likelihood) of scattering from a constant field. They wrote that the extension of their method to scattering from a Coulomb potential (Delbrück scattering) was in progress. But they were unable to follow through with these more realistic calculations. Kemmer continued to work on the problem. In 1937 he wrote that the scattering of light by light and by nuclear fields were "of less interest for investigating experimental tests than for [investigating] the principles of the theory."[134]

3.5.2 A Textbook of Subtraction Physics

The scattering of light by light was the centerpeice of a second pedagogical article Weisskopf published in the Danish Academy of Sciences' *Letters*, "The Electrodynamics of the Vacuum Based on the Quantum Theory of the Electron."[135] This became a standard reference. He maintained the dielectric-polarization picture of the vacuum from his review in *Die Naturwissenschaften* (section 3.5.1). But rather than simply commenting on published results, in 1936 Weisskopf gave *new derivations* of the effects of vacuum polarization. What was the value in another derivation of Euler, Kockel, and Heisenberg's modification to Maxwell's equations? Chapter 2 discussed how re-derivations of empirically verified results had been an important part of the acceptance of Dirac's theories. But Weisskopf was not offering a derivation of Euler's result *from a new theory*. Rather, this work was appreciated for its *aesthetic*, or formal, qualities.[136]

The value of Weisskopf's "Electrodynamics of the Vacuum" can be assessed in the reactions of his peers. In an undated letter to Weisskopf, Euler praised his work for its simplicity and unification of multiple phenomena. He expressed optimism that this picture could provide a unified treatment of the applications of "positron theory," and

133. The student was Arnold Nordsieck. Pauli, *Wissenschaftlicher Briefwechsel*, doc. 425c.
134. Kemmer and Weisskopf, "Deviations from the Maxwell Equations"; Kemmer, "Über die Licht-streuung an elektrischen Feldern nach der Theorie des Positrons," 112; cf. Achieser and Pomeranchuk, "Kohärente Streuung von γ-strahlen an Kernen"; experiment and theory of the Delbrück effect only converged in the late 1960s: Papatzacos and Mork, "Delbrück Scattering."
135. Weisskopf, "Electrodynamics of the Vacuum"; A.I Miller, *Early Quantum Electrodynamics*, §4.13, a fine point: Miller attributes to Weisskopf the message "to avoid second quantization," 77. However, in the quoted passage, Weisskopf was not referring to second quantization of the electron-positron field, he was referring to the electromagnetic field, which Weisskopf did not quantize even a first time. See 209 and 226.
136. Netz, "Aesthetics of Mathematics"; on aesthetic judgement in physics, see McAllister, "Dirac"; Chevalley, "Physics as an Art."

he slyly referred to Weisskopf's project in Pauli's derogatory terms, as a "'Textbook of Subtraction Physics.'" Similarly, Oppenheimer sent Weisskopf comments on a draft of the paper on June 3, 1936:

I was happy to have this simple and elegant treatment of the problem, where one can really follow the analogies and see where the numbers come from.[137]

On 30 October, 1936, Dirac wrote to Peter Kapitza (who was looking to hire a physicist to the Soviet Union):

when I was in Copenhagen I saw [Weisskopf's] latest work on the polarization effects which occur in the theory of the positron. This is a very good piece of work—it provides a more physical way of dealing with the question than Heisenberg and I have done.[138]

Unity, simplicity, elegance, clarity, physicality. These were the aesthetic virtues that theorists' valued in Weisskopf's work.

Weisskopf's "Electrodynamics of the Vacuum" focused on the interaction of the quantum vacuum of positron theory perturbed by classical electromagnetic waves.[139] He located the source of new phenomena in the "material energy density" of the vacuum.[140] Weisskopf made the physical assumptions underlying positron theory explicit and showed how these assumptions could be *used* within a calculation. In an echo of Oppenheimer and Furry's worries about positron theory, Weisskopf set out his goal as to show that Dirac and Heisenberg's positron theory was "essentially free from any arbitrariness" by showing that it was consistent within the bounds of its physical assumptions. It was an almost axiomatic approach. The theory took

the following properties of the vacuum electrons as being physically meaningless:

I.1 The energy of the vacuum electrons in field-free space;
I.2 the charge- and current-densities of the vacuum electrons in field-free space;
I.3 a field-independent electric and magnetic polarizability of the vacuum, which is constant in both space and time.

These quantities refer only to the field-free vacuum. It may be regarded as obvious that these quantities cannot have a physical meaning. All three quantities prove to be divergent sums after summing the contributions of all the vacuum electrons.[141]

137. Oppenheimer even expressed some optimism about quantum theory: "The shower & burst phenomena, taken at face value, seem to me a very strong argument for the validity of the theoretical formulae up to very high energies." Weisskopf Papers, Box 9.02, Euler: Folder 7; Oppenheimer: Folder 8.
138. Weisskopf, *Joy of Insight*, 100; Rubinin, "Kapitza and Dirac," 21; Kojevnikov, *Stalin's Great Science*, chs. 4–5.
139. A.I. Miller, *Early Quantum Electrodynamics*, 58, 66, §4.13; Schweber, *QED*, 86–89, 117–19.
140. Cf. Pauli's letter to Heisenberg 24.10.1934, on the "*materielle Selbstenergie der Lichtquanten*," Pauli, *Wissenschaftlicher Briefwechsel*, doc. 386.
141. Weisskopf, "Electrodynamics of the Vacuum," 207.

"Field free" space meant Dirac's sea with no perturbing external electromagnetic field. Assumption I.1 was, perhaps, aimed at objections to "hole" theory such as Vladimir Fock's: that the negative-energy electrons should have gravitational effects (section 2.3.3). More directly, it eliminated the "Coulomb energy of the 'vacuum,'" from Weisskopf's self-energy calculation.[142] Assumption I.2 was familiar from Dirac's "hole" theory—the redefinition of electric charge (section 2.2.3). The third assumption was newer, stemming from vacuum polarization. Like Furry, Oppenheimer, and Serber, Weisskopf argued that such a "polarizability [I.3] could in no way be observed, but would only multiply all charges and field strengths by a constant factor."[143] These clearly stated principles would provide some order in positron theory.

These were more than framing assumptions. Weisskopf integrated these principles in his calculations. As an "illustration of our physical interpretation of the subtraction terms," Weisskopf calculated the charge density of the vacuum, ρ, induced by an external electric potential, V:[144] He expanded the density in a series, and kept only the zeroth- and first-order terms, $\rho = \rho^{(0)} + \rho^{(1)}$. The zeroth-order term, $\rho^{(0)}$, was a divergent integral adding up the charges of the vacuum sea. The first-order term, $\rho^{(1)}$, was calculated with an expression similar to the self-energy of the electron, equation (3.4):

$$\rho^{(1)} = e \sum_i \sum_k \frac{H_{ki}(\psi_k^* \psi_i + \psi_i^* \psi_k)}{W_i - W_k}, \tag{3.9}$$

where H_{ki} was the interaction Hamiltonian, ψ_i was the electron wavefunction for the state, i, and W_i was the energy of the state. Weisskopf "introduce[d] the potential [V] as a perturbation" in equation (3.9). To attack the resulting equation, Weisskopf made a second series expansion, this time in the electric potential strength $\vec{g}$. He obtained the following complicated expression:

$$\rho^{(1)} = \frac{e^2 V_0}{8\pi^3 h^3} \int d\vec{p} \frac{c^2}{W^3(p)} \exp\left(\frac{i\vec{g} \cdot \vec{r}}{h}\right)$$
$$\times \left\{ \frac{g^2}{2} - \frac{c^2(pg)^2}{W^2(p)} - \frac{c^2 g^4}{4W^2(p)} + \frac{25}{16} \frac{c^4(pg)^2 g^2}{W^4(p)} - \frac{21}{8} \frac{c^6(pg)^4}{W^6(p)} + \cdots \right\}. \tag{3.10}$$

Here is where he applied his assumptions. "Now which part of this charge density has physical meaning? Because of the assumption [I.2], $\rho^{(0)}$ must be deleted." This subtraction was not novel. But in the face of such a tangle as equation (3.10), Weisskopf applied his third assumption, that a constant polarization of the vacuum was unobservable. And he did so *term by term* inside the brackets { } in equation (3.10):

142. Weisskopf, "Self-energy of the Electron," 162.
143. Weisskopf, "Electrodynamics of the Vacuum," 207; this argument was broadly accepted, i.e., Heitler, *Quantum Theory of Radiation* (1944), 191; cf. Pauli's dismissal of the unobservable electromagnetic zero-point energy, Pauli, *General Principles of Quantum Mechanics*, 180.
144. In the following I suppress complex conjugates.

the terms with g^2 [are not proportional to the external field strengths] and must be deleted because of [I.3]. One should note that it is also only these terms which lead to divergences. The remainder yields finite integrals.[145]

This was what Dirac might have appreciated as a *physical* reason to delete certain terms from a calculation: those terms resulted in an unobservable effect.[146] And just those same terms removed the divergent integrals, leaving a finite result—the terms with g^4—for the expected scattering of light by light.[147] In this way, Weisskopf clearly showed how the foundational assumptions of the theory could be applied *within* calculations to produce finite results from a theory struggling to cope with meaningless infinities. The physics of principles could be synthesized with the physics of problems.[148]

3.5.3 The Vacuum as Resource

In an undated handwritten letter (likely from 1936) Euler gave Weisskopf a sense of the field from Leipzig. Weisskopf had sent Euler a draft of "Electrodynamics of the Vacuum," which Heisenberg and von Weizsäcker then discussed. Heisenberg tasked Euler and Kockel with investigating the vacuum solutions to Maxwell's equations in Dirac's theory. They were working on a non-linear electrodynamics to describe *Bremsstrahlung*, the braking radiation produced by decelerating charges. They had made some progress toward finding the form of the addition to the classical Lagrangian from vacuum polarization. Euler explained:

> How such a formula, calculated from light-fields, should be given for nuclear fields is our question. For example: It appears that one cannot—from the Hamiltonian function of formula I—construct the matrix elements which stand for [*bedeuten*] Delbrück scattering of a light quantum on a Coulomb field (coherent Meitner-Hupfield nuclear scattering). On the other hand, it seems one can do something right by building the matrix elements which stand for Williams splitting of a light quantum into two light quanta in a Coulomb field. But where the moral of these permits and prohibitions lies is not clear to us.[149]

Euler told Weisskopf that Delbrück scattering was impossible to describe with their existing formalism.[150] In 1935, Evan Williams had proposed another new scattering

---

145. Weisskopf, "Electrodynamics of the Vacuum," 222; cf. Heisenberg, "Remarks on the Dirac Theory," 179.
146. I might identify it as a philosophical or epistemological reason: the application of the operationalist principle that whatever is unobservable should be discarded.
147. These results were considered correct in renormalized QFTs. Schwinger, "Gauge Invariance and Vacuum Polarization," fn2; but see Silva and Freire, "Concept of the Photon."
148. See Seth, *Crafting the Quantum*.
149. Weisskopf Papers, Box 9.02, Folder 7.
150. Namely: $1/E_0^2 \sqrt{1 + (\hbar/mc)^2 \square} [\alpha(H^2 - E^2)^2 + \beta(HE)^2]$.

process: that an incoming photon might "split" into two photons in the field of a nucleus.[151] This too was a difficult calculation.

Euler reported more severe trouble. He was exploring a new formula, but it was incompatible with Weizsäcker's "equivalent photon" method. (Here, "nuclear field" referred to the electromagnetic field emitted by a nucleus.)

> If, however, the [new formula] is also applicable to nuclear fields, the corresponding braking formula can not be calculated by the Weiszäcker method, since the non-linearity of the field equations makes the Fourier analysis of the nuclear field [the method] uses questionable.

Weisskopf was acutely aware of this issue. The Weizsäcker-Williams method required analyzing the Coulomb field of a nucleus with a Fourier decomposition into frequencies. Then the nuclear field could be represented as a collection of photons.[152] In order for the method to work, this had to be mathematically possible, which required that the strength of each frequency component to be independent of the others.[153] This was precisely the basis of Oppenheimer's criticism in January 1935, when he asked, "Are the Formulae for the Absorption of High Energy Radiations Valid?" When electromagnetic fields are nonlinear, their components could not be simply added together like notes into a chord.[154]

On February 9, 1935, Weisskopf wrote with some objections to Oppenheimer, in defense of the Weizsäcker-Williams method. He argued that the physical scenario Oppenheimer considered implied the violation of conservation of momentum.[155] Oppenheimer replied to Weisskopf on February 27, 1935, thanking Weisskopf for his "long and careful letter." Oppenheimer explained some features of his calculations, but concluded that he did not understand what Weisskopf was arguing. Weisskopf kept on the issue, and circulated a draft publication among colleagues. In California, Felix Bloch showed it to Oppenheimer. On January 25, 1936, Oppenheimer wrote a four-page letter to Weisskopf that began:

> Over the Christmas holidays Pauli visited me here [in Pasadena], & we came to talk of your objections to my arguments about the validity of the high-energy formulae. [...] Pauli & I went into the question very carefully; Bloch & I have discussed it at some length; & both are now completely convinced that your

---

151. Williams, "Correlation"; Thorsen, "Introduction [General Theory of Penetration]," 212ff; Carson, "Exchange Forces—II," 122, fn72.

152. They built on previous work by Bohr and Gaunt. For a high-energy collision, the passing charge/Coulomb field could be approximated as a pulse of electricity, or a photon. Weizsäcker, "Ausstrahlung bei Stößen sehr schneller Elektronen"; Williams, "Correlation"; Cassidy, "Cosmic Ray Showers," 15–16; Galison, How Experiments End, 101–10; Meyenn, "Von Weizsäcker's Semi-empirical Mass Formula," §8.

153. Weizsäcker, "Ausstrahlung bei Stößen sehr schneller Elektronen"; Bhabha, "The Creation of Electron Pairs by Fast Charged Particles."

154. Oppenheimer, "Are the Formulae for the Absorption of High Energy Radiations Valid?," 48; Williams, "High Energy Formulae"; Cassidy, "Cosmic Ray Showers," 17; Galison, "Discovery of the Muon," 132–33.

155. Pauli, Wissenschaftlicher Briefwechsel, vol. 2, 380.

objections are unfounded. But it seems to me that I ought to try to convince you too.[156]

On February 14, Pauli wrote to Weisskopf and displayed his acerbic bite. Upon re-reading Weisskopf's draft, Pauli considered the possibility that "[the author] is so confused or I am so stupid that I cannot understand him at all." Pauli wrote to Isidor Rabi, an editor at the *Physical Review*, on March 5. With Weisskopf's authorization, Pauli retracted Weisskopf's submission.[157] Working through a collegial network, Oppenheimer's objection to the applicability of the Weizsäcker-Williams method was effective.

According to Euler, when the same point was made to Heisenberg and Weizsäcker, they took it seriously. With one method known to fail, where did Heisenberg's group look to the future? To Weisskopf's dielectric vacuum. Euler's letter to Weisskopf continued:

> Therefore, Heisenberg and Weizsäcker's view is that your general reasoning is correct: In hole theory the vacuum has dielectric properties which can attenuate the interaction of fast electrons and nuclei and thus the braking [radiation].[158]

This letter cannot be taken to represent Heisenberg and Weizsäcker's lasting views. It was a chaotic period in every sense.[159] What the letter shows is that the vacuum—with its dielectric properties—had become a *resource* for physicists who were looking to build new theories and solve theoretical problems.[160] Formally, this manifested in non-linear additions to the equations of classical electromagnetism. These considerations were strong enough to unseat Weizsäcker's own theory. And they offered a frame for a new picture of *Bremsstrahlung* in terms of the material energy of the vacuum.[161] Rather than trying to eliminate the physical qualities of the vacuum from quantum theory, physicists were increasingly putting the physicality of empty space at the center of their formalism and their physical worldview. Even pragmatic Heitler was convinced that "the idea of *an electromagnetic field in vacuo* has to be *abandoned*."[162]

---

156. Weisskopf Papers, Box 9.02, Folder 8.

157. Pauli, *Wissenschaftlicher Briefwechsel*, doc. 427b, 425d.

158. "*Deshalb sind Heisenberg und Weizsäcker der Ansicht, das Ihre allgemeine Überlegung richtig ist: Das Vakuum habe in der Löchertheorie dielektrische Eigenschaften, könne also die Wechselwirkung zwischen Kerne und schnellen Elektronen und damit dei Bremsung abschwächen.*" Weisskopf Papers, Box 9.02, Folder 8.

159. Both Euler and Kockel suffered because they refused to participate in the Nazi Party (E) or had socialist history (K). Cassidy, *Uncertainty*, 340–41; D. Hoffmann, "Hans Euler"; Cassidy, "Cosmic Ray Showers"; Galison, "Discovery of the Muon"; Monk, *Inside the Center*, 234–37.

160. This goes somewhat farther than Roqué's sound argument about "hole" theory as a heuristic. Roqué, "Manufacture of the Positron."

161. Weisskopf made calculations of *Bremsstrahlung* by pair creation, starting from Møller, Bethe, and Heitler's radiation theories, and applying the "Weizsacker Method." He wrote an (unsent?) letter to Christian Møller dated 18.1.1936. Weisskopf Papers, Box 9.02, Folder 13.

162. Heitler, *Quantum Theory of Radiation* (1944), 192; cf. in wave mechanics: "Intuitively, this [non-linear electrodynamics] can be understood as: the vacuum polarization generated by a field reacts back onto the field, and this interaction has a non-linear character (similar to a polarizable medium with field-strength-dependent dielectric constant)." Wentzel, *Einführung in die Quantentheorie der Wellenfelder*, 189; on pragmatic "quantum engineers," see Kragh, "Relativistic Collisions."

3.6 Conclusion

This chapter showed how physicists learned new material, how they built up formalisms, and how they worked out the consequences of positron theory. One focus was on the inevitability or persistence of rich physical properties being ascribed to the vacuum. Werner Heisenberg initially argued that, in "hole" theory, vacuum polarization was not real, or even definable. But he soon reversed course, and like Paul Dirac and Rudolf Peierls (section 2.4), he investigated its consequences. Wendell Furry and Robert Oppenheimer did not apply "hole" theory's many-electron formalism. But even with a wave-field description of electrons and positrons—without Dirac's sea—they discovered "nascent pairs" which were practically impossible to eliminate (section 3.3.2). Heisenberg's attitudes to vacuum polarization changed in time. Furry's and (particularly) Oppenheimer's multiple attitudes were *synchronic*, which resulted in their deeply ambiguous analysis.[163] They put in question whether they could make any conclusions at all. They argued some vacuum effects would be in-principle unobservable. And they made calculations for the observations of other vacuum effects.

These vacuum effects were related to a suite of new physical processes in the vacuum. Probably the most important of these was the photoelectric production of pairs from Dirac's sea. This was given as the underlying mechanism for the production of electron-positron pairs in cloud chamber photographs (figure 3.1) and it underlay the calculation of relativistic radiation absorption (*Bremsstrahlung*). This chapter focused on an adjacent group of phenomena, which were not so directly relevant to experiment.[164] Dirac and Peierls argued that vacuum polarization would appear as deviations in the Klein-Nishina scattering formula, for X-rays with unheard-of energy. Furry and Oppenheimer remarked that vacuum polarization would produce a shift in atomic spectral lines, and Edwin Uehling calculated a measurable effect (section 3.3.3). Unfortunately, Uehling's result did not agree with experiment. The scattering of light by light (with intermediate pair creation and annihilation) was another such process. Heisenberg's group at Leipzig produced the most concerted efforts to understand it. After mammoth effort, Heisenberg and Hans Euler produced a prediction for the amount of scattering, but it was so minuscule there was no serious effort to measure it. Finally, Max Delbrück suggested that an anomalous experimental result was the result of light scattering directly off Dirac's sea. Then came a double calamity: the theoretical details proved impossible to calculate, and the experimental results were rejected as an error. These processes were nevertheless important to theorists. Increasingly, they were acknowledged as valuable to theory, as such. By 1936, the new materiality of empty space was stable enough to count as a *resource* for the construction of new theories. In light of this situation, physicists such as Victor Weisskopf and Walter Heitler made a leapfrog argument for the reality of these vacuum phenomena. From the impressive photographs of real pair production,

---

163. Of course, they saw quantum theory as ambiguous. As Suman Seth might say, Furry and Oppenheimer were ambiguous about the ambiguity.

164. This is not a complete list. See Cassidy, "Cosmic Ray Showers"; Mehra and Rechenberg, *HDQT*, vol. 6, §IV.5.

they argued for the reality of subtler effects, without relying on the (lack of) relevant measurements.

Furry summarized the scene in 1937:

An interesting feature of the Dirac theory of the positron is the possibility of processes in which electrons and positrons, though they are not actually created so as to be experimentally observable, are able to play the part of catalytic agents producing effects which otherwise could not appear.

He mentioned the scattering of light by light and by a Coulomb field.

[B]ut the transitory or virtual presence of charged particles provides a means of mediating indirect interactions. [... This] transitory presence of the pair is possible even when there is not enough energy available to produce an actual observable pair.[165]

Like Heisenberg's group at Leipzig (section 3.4.2), Furry took into account the possibility of electron-positron processes, even when doing so would violate the conservation of energy. These vacuum phenomena were transitory, indirect, and unobservable—strange attributes for a physical theory. But these processes were important. Furry's chemical metaphor neatly dovetails with the chemical origins of Henrik Kramers's and Dirac's intermediate states (section 3.2). Elusive vacuum processes were seen as catalysts for observable phenomena.

Where did physicists get the idea for these processes? They were unobservable in the sense that an experimentalist could never capture one in a photograph of a cloud chamber. But they were also unobserved, in the sense that the secondary phenomena they catalyzed into existence had not been measured, either. Consider the scattering of light by light. Euler explained that the electrons and positrons in intermediate states of the scattering process would not be observed in a cloud chamber. But also: there were no hints of evidence that light scattered off of itself.[166] I do not know how Otto Halpern or Heisenberg or Euler first came to explore the possibility of light-by-light scattering. In print, Heisenberg and Euler explained that the scattering was indicated by certain mathematical terms, and that these terms were produced by perturbation theory. They repeatedly explained that the energy of the vacuum came from "matrix elements of the perturbation theory," and "perturbation terms." The algorithm of perturbation theory generated their results. (See appendix A.) But they were not only led by mathematics. In an alternating fashion, they explained physical effects in terms of these matrix elements and terms, and they also tried to make these terms physically "plausible" or rooted in "virtual possibility."[167] Historians David Kaiser and Adrian Wüthrich have shown how Heisenberg and Euler's coordination of perturbation terms with atomic processes fits into the lineage of Richard Feynman's

165. Furry, "A Symmetry Theorem in the Positron Theory," 125.
166. As of this writing, hints have arrived. Ellis, Mavromatos, and You, "Light-by-light Scattering."
167. Heisenberg and Euler, "Consequences of Dirac Theory," 15.

diagrammatic approach to quantum theory (see also section 4.4).[168] Feynman's early work did not follow Heisenberg and Euler's strict terms-to-processes approach, but Freeman Dyson re-established this relationship between perturbation terms and the (physical) processes that were represented by Feynman graphs.[169]

The centrality of perturbation theory was hardly specific to the Leipzig group. According to Heitler's influential 1936 monograph *The Quantum Theory of Radiation*: "The creation and annihilation of pairs can then be considered as the result of a *perturbation*."[170] But there was a characteristic ambiguity in physicists' discussion of perturbations. Was Heitler referring to the algorithm of perturbation analysis, to a mathematical term it produced? Or was he referring to the electromagnetic radiation that the mathematical terms represented? The best answer, I think, is both of the above. This fusion of a mathematical method with an ontological picture was characteristic of theoretical physics. This ambiguity was particularly intense in virtual, transient, unobservable, material, physical, vacuum physics.

168. D. Kaiser, *Drawing Theories Apart*, 34–36; Wüthrich, *Genesis of Feynman Diagrams*, 25.
169. D. Kaiser, *Drawing Theories Apart*, esp. 175–76; Schweber, *QED*, §§9.9 and 9.11; on realistic representation, see D. Kaiser, "Stick-Figure Realism"; Wüthrich, *Genesis of Feynman Diagrams*, §1.6.
170. Heitler, *Quantum Theory of Radiation* (1936), 193.

4

Weisskopf, Schwinger, Feynman: Vacuum Loops and Fluctations

4.1 Introduction

Chapter 3 showed how Paul Dirac's "hole" theory reverberated among theoretical physicists. Rudolf Peierls and Werner Heisenberg each developed the formalism of "hole" theory toward greater conceptual consistency and calculability. Continuing skepticism about the reality of Dirac's sea of electrons in negative-energy states motivated J. Robert Oppenheimer, Wendell Furry, Wolfgang Pauli, and Victor Weisskopf to develop alternative formulations of a relativistic quantum theory of electrons and positrons. Agreement on this material vacuum, I argue, was produced by theorists' practices. In particular, the common application of perturbation methods to the relativistic vacuum produced the quantitative agreement among theorists. Once the principles of relativistic vacuum calculation were elucidated, the vacuum became a new resource for theorists in their research. Whether or not theorists started out on Dirac's sea, they converged on the startling terrain of "nascent pairs" or a "material energy-density of empty space." This vacuum was *dynamic*: it changed in response to fields and charges; it affected physical measurements; and it created new phenomena such as the scattering of light by light.

In the light of contemporary discussions of the quantum vacuum, there is a surprising absence in the story told in Chapter 3. The agreement about the presence of infinite matter in the vacuum—the conclusion about negative-energy, or nascent, or virtual, electrons—was established independently of *fluctuations*. Chapter 4 traces the origin of "vacuum fluctuations" as they were taken up in the development of modern "renormalized" Quantum Field Theory (QFT). The central questions I try to answer include: how did fluctuations enter QFT, both in physicists practices and in the constitution of the vacuum as an object of theory? What is the relationship between fluctuations in QFT and fluctuations in measurement?[1]

Section 4.2 continues Victor Weisskopf's story. Pursuing theoretical unification, in 1934, he connected electromagnetic field fluctuations, the self-energy of the electron, the zero-point energy of quantum electrodynamics, and Dirac's radiation theory.[2] At

---

1. The idea that space was filled with fluctuating zero-point energy was first proposed by the chemist Walther Nernst in 1916. However, it is not clear how much influence this proposal had on subsequent developments. Nernst took this to be the zero-point energy of molecules of aether, leaving some ambiguity as to whether this should be considered a zero-point energy of radiation itself. Kragh, "Preludes to Dark Energy," §4.

2. This is slightly earlier in Weisskopf's career than the attribution in Milonni, "Why Spontaneous Emission?," Sam Schweber once attributed the connection between the zero-point energy and spontaneous

More than Nothing. Aaron Sidney Wright, Oxford University Press. © Oxford University Press 2024.
DOI: 10.1093/oso/9780190062804.003.0004

this time, physicists considered two types of fluctuations: fluctuations that were inherent in the formalism of quantum theory, and fluctuations that were a consequence of physical measurements. The confusing literature was exemplified by a famous 1933 paper by Niels Bohr and Léon Rosenfeld, which Weisskopf helped see to publication. In the event, however, Weisskopf went his own way. He was convinced of the reality of randomness and fluctuations, and in several influential analyses he simply asserted their existence. By 1939, he elevated vacuum fluctuations to the status, in quantum theory, of a principle. I describe these as *primitive* fluctuations, because they do not rely for their existence on any other entity or action, such as a measurement.

Section 4.3 picks up the new "renormalized" QFT in the post–World War II work of Julian Schwinger. The standard view, as expressed by historian Sam Schweber, is that Schwinger followed Weisskopf's view of primitive vacuum fluctuations.[3] However, this is not quite right. Schwinger was much closer to Bohr, and, later in his career, Schwinger explicitly rejected primitive vacuum fluctuations. He saw vacuum polarization and the self-energy of the electron as reciprocal—or complementary—processes, in which the quanta of one field interacted with the vacuum fluctuations of the other. He also made the vacuum central to QFT calculations. Schwinger's "vacuum expectation values" became perhaps *the* central object of calculations in QFT.[4] For example, physicists used Feynman diagrams to evaluate vacuum expectation values.[5] Schwinger constructed a formalism for the vacuum and used it to calculate not only the physical properties of empty space, but the physical properties of everything else in QFT. This work set the course for modern quantum theory. In Schweber's words, "in the post–World War II period Schwinger set the standards and priorities of theoretical physics."[6]

During the 1940s, the problems of QED that vexed both Furry and Weisskopf were addressed—if not fully solved—by Tomonaga Sin-Itiro, Schwinger, and Richard Feynman; in 1965 they shared the Nobel Prize in physics. Both physicists and historians have seen this success as a victory of conservative, incremental, and pragmatic approaches to QED; none of them succeeded by re-founding electrodynamics or electron theory.[7] In Section 4.4, I present the limits of this conservative attitude. In the 1940s, Feynman had difficulty with the incorporation of loops in his diagrammatic calculations. These difficulties were eventually tamed, and he was able to use loops to represent processes such as vacuum polarization. In the 1960s, Feynman attempted to extend his success in QED to a diagrammatic quantum theory of gravity. There, the vacuum loops were too much to overcome. The tried-and-true techniques had found their limit.

emission to Fermi: Schweber, "Weimar Physics," 291. However, I do not find this in E. Fermi, "La théorie du rayonnement"; or E. Fermi, "Quantum Theory of Radiation," esp. 95. I regret that Sam passed away before I could ask him about this.

3. Schweber, *QED*, 336.

4. E.g., by Dyson: Dyson, "Radiation Theories," 494; Schweber, *QED*, §§9.9, 9.11.

5. Wüthrich, *Genesis of Feynman Diagrams*, 20, 31, §6.2.2.

6. Schweber, "Schwinger's Green's Functions," 7783; though, as he did to Dirac, Pauli ridiculed Schwinger as a "prophet" of mathematics, Schweber, *QED*, §7.3.

7. Schweber, "Empiricist Temper"; Schweber, *QED*; cf. Wüthrich, *Genesis of Feynman Diagrams*, 21.

4.2 Weisskopf's Fluctuations

In his early academic career, Weisskopf persistently argued for the existence of the un-caused, the spontaneous, the finite, and the unpredictable. In August 1930, Weisskopf published a popular article on "Modern Physics and the Problem of Causality" in *Der Kampf*, the monthly magazine of Vienna's Social Democratic Party. He argued that the "principle of strong causality must be broken in the new physics. There are, therefore, processes which it is in-principle impossible to predict."[8] Weisskopf saw randomness thoughout the subjects of his early career, beginning with the width of spectral lines

4.2.1 Spontaneous Emission and Line Widths

One of the elementary processes in Albert Einstein's 1916 theory of radiation (section 3.2) was *spontaneous* emission (*Ausstrahlung*). He supposed that molecules in a black-body chamber could spontaneously jump from a higher-energy state to a lower-energy state and emit radiation "without external causes." He modeled this on Ernest Rutherford's 1900 description of radioactive decay.[9] Einstein's "A" coefficient was the rate of occurrence of spontaneous emission. Dirac's 1927 radiation theory gave the first "proper [quantum mechanical] account of spontaneous emission," and Dirac wrote that his work "must presumably give the effect of radiation reaction on the emitting system, and enable one to calculate the natural breadths of spectral lines."[10] In a follow-up paper on dispersion, Dirac explained that calculating line widths involved divergent results in second order perturbation theory. As for Rutherford and Einstein, Dirac simply assumed that excited atomic states had a fixed "life time."[11]

In 1928, after two years at the University of Vienna, twenty-year-old Weisskopf moved to Göttingen University to study with Max Born. His doctoral dissertation aimed to understand the widths of spectral lines. In the tradition of spectroscopy, the first goal of an observation was to distinguish the number and frequency (color) of lines in the spectrum of the radiant light emitted by the system under study, such as a vial of hot hydrogen gas. But there were other features of spectrographic data that were taken to give information about the experimental systems, including the width and intensity (brightness) of the lines.[12] Spectral line width was also at the center of foundational debates in quantum theory. For example, Erwin Schrödinger's critique of discrete quantum "jumps" argued that if Bohr's picture of near-instantaneous jumps was correct, then spectral lines should be infinitesimally sharp—in contrast to

8. Weisskopf Papers, Box 2.01, "Folder: My Life Werke 1930–35," *Der Kampf* 22(8) 1930: 344. Weisskopf, *Joy of Insight*, 40; for causality see Carson, Kozhevnikov, and Trischler, *Weimar Culture and Quantum Mechanics*; Stöltzner, "Causality Debates."

9. Einstein, "Quantum Theory of Radiation," 223.

10. Dirac, "Quantum Theory of the Emission and Absorption of Radiation," 265.

11. Dirac, "Quantum Theory of Dispersion," 713, 711.

12. See Hentschel, *Mapping the Spectrum*.

their observed width.[13] Bohr's defense came from Pauli in 1926. He argued that "the natural width of the spectral lines is therefore a statistical effect which has nothing to do with the temporal course of the emission process." But he deferred a confident treatment until he possessed "a complete knowledge of the laws governing the temporal occurrence of the transitional processes."[14] That knowledge was provided by Dirac's radiation theory. The problem and its proposed solution were connected by Weisskopf and his co-advisor in Berlin, Eugene Wigner. In 1930 they co-wrote a landmark paper on the "Calculation of Natural Line Widths on the Basis of Dirac's Light Theory." They explicitly assumed that atoms in stationary states—contrary the original meaning of "stationary"—had some likelihood of spontaneous transitions to lower-energy states, which they modeled on the "radioactive law" of nuclear decay.[15] Weisskopf and Wigner's calculations were very *close* to the observed phenomena of line widths. But they did not find a result that could be unambiguously compared with an experimental measurement. Their "natural" line width was overshadowed by larger effects, such as those due to the motion of hot radiating atoms.[16]

He continued to study the problem of line widths during his rapid series of post-doctoral positions from 1931 to 1933 in Leipzig (with Heisenberg), Berlin (with Schrödinger), Copenhagen (with Bohr), and Cambridge (with Dirac). In a review of line widths for astronomers, Weisskopf argued that "Heisenberg's Uncertainly Principle" required there to be another quantum source of line widths. "The existence of a natural width of line leads us to the conclusion that the energy of the terms is not quite definite." If Weisskopf was being consistent within this article, he was referring here to the "existence of a natural line width" within quantum *theory*, because "[t]he observed widths of lines are considerably greater than the values calculated above." That is, observation was incapable of determining whether Weisskopf's natural width occurred.[17] This was dangerously close to circular reasoning. The initial assumption that stationary states were not stationary led Weisskopf to the reality of (unobserved) natural line widths, which in turn led him back to a conclusion about the states (that their energy levels were not definite).

Strikingly, Weisskopf argued that even the *ground* state of an atom was impermanent: "Speaking precisely, the normal state also has a finite life; its life is ended by absorption processes."[18] This was more than a departure from Bohr's stationary states.

13. Tanona, "Bohr's Response to the Heisenberg Microscope," §5; Beller, "Schrödinger's Interpretation of QM," §2.

14. Pauli, "Quantentheorie," 69.

15. Weisskopf and Wigner, "Berechnung der natürlichen Linienbreite auf Grund der Diracschen Licht-theorie," 62, 71; Dirac had found their central result three years earlier, but did not publish: Kragh, *Dirac*, 125.

16. Weisskopf, *Joy of Insight*, 41–44; The situation had improved by 1936, but still: "To classify the various causes contributing to the width and shift of spectral lines is not an easy matter, nor is it entirely unambiguous," Margenau and Watson, "Pressure Effects on Spectral Lines," 23, cf. 41–42; see also Mehra and Rechenberg, *HDQT*, vol. 2, §III.4.

17. Weisskopf, "Intensity and Structure of Spectral Lines," 301, 308; Weisskopf suggested that at the edges of a line, the contributions from the natural shape and other broadening could be distinguished. Korff and Breit, "Optical Dispersion," 487–89; Margenau and Watson, "Pressure Effects on Spectral Lines," 24–25; Weisskopf first gave these arguments in his dissertation: Weisskopf, "Zur Theorie der Resonanzfluoreszenz," 24.

18. Weisskopf, "Intensity and Structure of Spectral Lines," 303.

It challenged the very stability of atoms. In 1932, Weisskopf visited the Physical Institute in Charkow (now Kharkiv), Ukraine. There, he had long conversations with the young firebrand, Lev Landau. At Charkow, Weisskopf did not shy from the radical consequences of his un-stationary states. In the *Physikalische Zeitschrift der Sowietunion* he argued for an increasing role for the "natural" line breadth within the complex of processes that produced the observed widths.[19] Across these works, Weisskopf argued that the existing framework of quantum mechanics was inadequate to address the width of spectral lines; to explain this phenomenon an additional hypothesis was necessary: even stationary states and ground states were unstable and could decay, like radioactive nuclei. In 1934, he applied an idiosyncratic picture of fluctuations to QED.

4.2.2 Uncertainty and Fluctuations

In the canon of theoretical physics, fluctuations were most often discussed in this period in terms of measurements. Here is the simplest scenario: you wish to know how far a balcony extends over a lawn. You mark out a coordinate system on the grass, climb to the balcony, and carefully drop ten golf balls from the ledge. The ten golf balls will fall in ten different positions. You calculate the average position of the balls. From this you can calculate the amount of fluctuation around the mean value by finding the average distance of each ball from the mean position. This helps you evaluate the quality and amount of uncertainty of your measurement. In 1905, Einstein transformed fluctuation analysis into one of the premier paper tools of theoretical physics. A fuller discussion is provided in appendix B. This section picks up the analysis of fluctuations, uncertainty, and measurement of quantum fields.

In his 1929 lectures at the University of Chicago, Heisenberg extended his analysis of quantum measurements and uncertainties from his famous gamma-ray microscope thought experiment to quantum fields. His task was "to trace the origin of the uncertainty in a measurement of the electromagnetic field to its experimental source." Note, his task was not oriented to primitive uncertainties that might inhere in things in themselves. To examine measurement uncertainty, Heisenberg considered a small cube of space with volume $\delta v = l^3$. To simultaneously measure the electric, E, and magnetic, H, field strengths in the cube, parallel cathode rays (i.e., electrons) streamed in opposite directions through δv, as in figure 4.1. The deflection, up or down, of the cathode rays would be proportional to (both) the electric and magnetic field strengths. Heisenberg emphasized the role of the physical characteristics of this thought experiment: this arrangement measured only the *average* values of the field strengths throughout δv; and "the field must not vary appreciably across the width of the rays," marked d in figure 4.1 (i.e., measurement was possible only for electromagnetic radiation with wavelength greater than d). Heisenberg then argued that there were two sources of limitations in the accuracy of the measurement, which illustrated his view on wave-particle duality. One arose "[b]ecause of the natural spreading of the matter rays," considered as continua.[20] The other "essential factor" viewed the cathode rays as streams of individual electrons that would deflect one another inside the cube δv. "The amount of this modification is uncertain to some extent, since it is not known at which point in the cathode ray the electron is to be found." To find the total uncertainty, $\Delta E \Delta H \geq hc/(\delta l)^4$, Heisenberg multiplied the two contributions and emphasized that "the simultaneous consideration of both the corpuscular and wave picture of the process taking place is again fundamental."[21]

In this passage, it is evident that Heisenberg followed Bohr toward grounding quantum theory in *more detailed* thought experiments, and crucially in locating the uncertainties *in these details* of the measurement process. For example, Heisenberg did not attribute an uncertainty to the interaction of two cathode-ray electrons; there *was* an interaction, but the *source* of the uncertainty was the impossibility of *locating* each electron within the physical area of measurement δv.[22] This was consistent with his general discussion of the interpretation of the quantum formalism. After introducing the Dirac-Jordan transformation theory, Heisenberg wrote that "we must add the express condition that the experiment under consideration actually affords a determination of [the quantity] α."[23] This agreement between Heisenberg and Bohr was juxtaposed by their understanding of wave-particle duality. As philosopher Kristian Camilleri has shown, where Bohr saw wave and particle as two necessary-but-incompatible aspects of atomic phenomena, Heisenberg argued that "both the corpuscular and wave picture of the process" must be applied *simultaneously*.[24] (I will argue in section 4.3 that Schwinger's QFT shared many of these characteristics.)

Heisenberg also saw fluctuations as built into the formalism of quantum mechanics. He was perhaps reflecting on the fact that Dirac's general transformation theory was a generalization of his own "exchange" analysis of fluctuations of helium (section 2.3.3). And he reproduced Pascual Jordan's calculation of the fluctuations in blackbody radiation.[25] But Heisenberg made no mention of fluctuations in his analysis of field measurements. The only sources of uncertainty in his analysis were the natural broadening of cathode rays and the unknown position of electrons within the rays. He did not appeal to any uncertainty or fluctuation inherent in the electromagnetic field that was being measured. As Heisenberg returned to the question of

20. This spreading of "rays" was a consequence of classical optics and was a key part of Bohr's response to Heisenberg's 1927 uncertainty principle; Bohr, "The Quantum Postulate and the Recent Development of Atomic Theory," 582; Tanona, "Bohr's Response to the Heisenberg Microscope."

21. Heisenberg, *Physical Principles*, 52–54.

22. Compare Bohr's discussion of the uncertainty in the Compton effect of the position of the electron after the scattering: the uncertainty arises "on account of the impossibility of attributing a definite instant to the recoil"; Bohr, "The Quantum Postulate and the Recent Development of Atomic Theory," 583.

23. Heisenberg, *Physical Principles*, 147.

24. In agreement with Camilleri, *Heisenberg and the Interpretation of QM*, §4.4; Pauli also followed Bohr here: Pauli, *General Principles of Quantum Mechanics*, 7.

25. Heisenberg, *Physical Principles*, §V.7, 145–48; shortly thereafter, Heisenberg revisited this calculation with increased attention to the details of the idealized fluctuation measurement. Heisenberg, "Über Energieschwankungen in einem Strahlungsfeld," 122; on Jordan, see Duncan and Janssen, "Pascual Jordan's Resolution"; Kojevnikov, "Einstein's Fluctuation Formula," 202, 212; Kragh, "Preludes to Dark Energy," 224; Bacciagaluppi, Crull, and Maroney, "Jordan's Derivation of Blackbody Fluctuations"; note, however: Plato, *Creating Modern Probability*, 95–107.

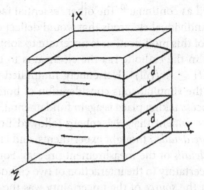

Figure 4.1. Heisenberg's figure for the "Discussion of an Actual Measurement of the Electromagnetic Field."

fluctuations in the mid-1930s, he continued to argue that fluctuations and uncertainty were created by acts of measurement themselves.[26]

Bohr and his assistant Léon Rosenfeld (BR) came closer to a universal view of fluctuations and uncertainty but still fell short of Weisskopf's proposal. Their paper, "On the Question of the Measurability of Electromagnetic Field Quantities" was published in December 1933 after more than a dozen drafts and proofs.[27] Its central result was to show that Heisenberg and Pauli's 1929–1930 QED produced the same uncertainties as were derived from an analysis of ideal measurements. BR sought to align the mathematical formalism with the conceptual analysis of experiments.[28] The paper was prompted by a critical argument from Landau and Peierls: that measurements in quantum field theory were impossible.[29] Here, the relevant aspect of this complex paper is the role BR gave to fluctuations and the zero-point energy.

Following a line of thought about field measurements from James Clerk Maxwell, BR argued that the measurement of quantities such as field strengths in a region were really measurements of the motion of charged test bodies. To measure the field in a region, put a charged cube in it and observe its acceleration. Translated into quantum theory, this procedure rooted the origin of uncertainties of field quantities in the uncertainty of the position and momentum of the charged test bodies. BR accounted for the finite size of these test bodies, much as Heisenberg had added the width of cathode rays to his field measurements, or the width of the walls of his blackbody cavity. In order to make these measurements conform to special relativity, BR also added a "width" in the duration of field measurements.[30]

26. A.I Miller, *Early Quantum Electrodynamics*, 57.

27. For fuller treatments, see Pais, *Niels Bohr's Times*, 16(d); Darrigol, "Cohérence et complétude"; Jacobsen, *Léon Rosenfeld*, 61–81.

28. Heitler, *Quantum Theory of Radiation* (1936), 81; Wentzel, *Einführung in die Quantentheorie der Wellenfelder*, 26, 111; Pais, *Development of the Theory of the Electron*, 13–14.

29. Landau and Peierls, "Extension of the Uncertainty Principle," 49; Hall, "Purely Practical Revolutionaries," chapter 7; Kojevnikov, *Stalin's Great Science*, chapter 4.

30. Bohr and Rosenfeld, "On the Question of the Measurability of Electromagnetic Field Quantities," 360–68; Maxwell, *Treatise*, vol. 1, §44.

According to BR, whenever physicists specified the number of light quanta in a region—as in Heisenberg's one light quantum state, $\phi(000; 111; 0100)$, figure 3.6— the magnitude of the electromagnetic field strengths would fluctuate. These were "the peculiar fluctuation phenomena which derive from the basically statistical character of the [quantum] formalism."[31] To make this quantitative, BR calculated $\mathfrak{E}_x^{(G)}$, the average of the x-direction field strength, analogously to the average Heisenberg considered in figure 4.1, except also considering average in time.[32] They calculated the mean-squared fluctuation (see appendix B) of $\mathfrak{E}_x^{(G)}$, in terms of the density of light quanta, ω.

From [that formula] we see that for a given light quantum composition the fluctuations in question can never vanish, since even when $\omega = 0$, i.e., in the complete absence of light quanta, they assume a finite positive value $[S_0]$.[33]

Roughly, in a series of field-strength measurements, the fluctuations would contribute to the uncertainty as, $\mathfrak{E}_x^{(G)} \pm S_0$. According to BR, these fluctuations were present *only* when a phenomenon was described in terms of the number of light quanta. But the fluctuations were present in *all* such phenomena, even when no light quanta were present. Likely the most important example of such a phenomenon was the focus of Weisskopf's research—spontaneous emission. In order for emission to be "spontaneous" and not "stimulated," no light quanta can be present.

BR used the magnitude of zero-quanta fluctuations, S_0, as a reference point for all light-quanta fluctuations.

The square root of $[S_0]$ may be regarded as a critical field strength, $\mathfrak{S}$, in the sense that only when considering field averages essentially larger than $\mathfrak{S}$ are we allowed to neglect the corresponding fluctuations.

BR considered two geometries in which $\mathfrak{S}$ would be relevant, in comparison to a measure of the uncertainties in the fields (in their macroscopic measuring bodies), $\mathfrak{A}$. They found that $\mathfrak{A}$ was dominant; "therefore, in testing the characteristic consequences of the formalism we can to a large extent disregard the field fluctuations" $\mathfrak{S}$.[34] In the style of nineteenth-century fluctuation analyses, BR went to great lengths to calculate a fluctuation, which was important because it could be ignored.

31. I read this "statistical character" as referring, as Heisenberg did, to the roots of the Dirac-Jordan transformation theory in Heisenberg's fluctuation analysis of helium. This is independent of Born's probabilistic interpretation of the wavefunction.

32. Explicitly, they averaged the x-component of the field, $\mathfrak{E}_x$, over a spacetime region, G, of volume, V, and duration, T: $\mathfrak{E}_x^{(G)} = \frac{1}{VT} \int_T dt \int_V \mathfrak{E}_x dv$.

33. I have suppressed BR's polarization index. Bohr and Rosenfeld, "On the Question of the Measurability of Electromagnetic Field Quantities," 365–66; it is not perfectly clear *why* BR thought the quantum formalism resulted in these fluctuations. Similar views were expressed by, e.g., Pauli: "The measurement of the field strength is connected with a finite but indefinite change in the number of photons." Pauli, *General Principles of Quantum Mechanics*, 188, however, BR did not use Pauli's argument, which conflicted with BR's larger field measurement scheme.

34. Bohr and Rosenfeld, "On the Question of the Measurability of Electromagnetic Field Quantities," 365.

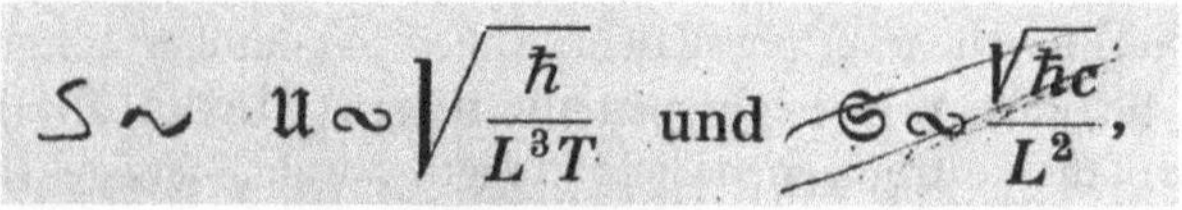

$$S \sim \; \mathfrak{u} \varpropto \sqrt{\frac{\hbar}{L^3 T}} \quad \text{und} \quad \mathfrak{S} \varpropto \frac{\sqrt{\hbar c}}{L^2},$$

Figure 4.2. Marginalia at equation (14), page 18, of an offprint of Bohr and Rosenfeld (1933) dedicated to Oppenheimer. Reprinted by permission of the Royal Danish Academy of Science and Letters.

BR insisted that zero-quanta fluctuations limited only the testability of QED *as a theory*. They did not treat them as a universal physical phenomenon. Their analysis "in no way implies that these fluctuations set an absolute limit for any utilization of field measurements."[35] (This ambivalence about the role of fluctuations in *idealized* measurements was also found among engineering and experimental studies of "noise."[36])

For many physicists, the BR paper was taken to comfortably close the controversy with Landau and Peierls over the future of QED. However, positive reactions were not universal. Bohr inscribed an offprint of the BR paper to "Professor R. Oppenheimer[,] with kindest regards and best wishes for the new year." The offprint includes marginalia that I take to be Oppenheimer's. The most dramatic marks were applied to BR's discussion of the magnitude of electromagnetic vacuum fluctuations, S_0, and its associated "critical field strength," $\mathfrak{S}$ shown in figure 4.2.[37] Oppenheimer carried the S from BR's equation (11) across six pages to equation (14); with two lines Oppenheimer crossed out the critical field symbol, and the numerator of BR's expression. Oppenheimer and Furry had heard a version of the BR analysis from Bohr himself in Pasadena, and they accepted it as an analysis of *electromagnetic field* measurements (chapter 3). But they argued that if *charges* were to be measured in pair theory, Bohr's discussion was inadequate.[38]

In 1935, Oppenheimer rejoined the discussion with two arguments: charge measurements would fluctuate if measured by BR's highly charged ideal devices, because these devices would create electron-positron pairs; and the effect of these fluctuations should be *combined* with BR's light-quantum vacuum fluctuations, what Oppenheimer called "the fluctuations arising from the corpuscular character of electromagnetic radiation." His fluctuation of charge, $(\Delta \epsilon_r)^2$, was the square of BR's critical field strength $\mathfrak{S}$. Oppenheimer gave BR's value (figure 4.2) as

$$(\Delta \epsilon_r)^2 \sim \frac{hc}{L^4}. \tag{4.1}$$

35. Bohr and Rosenfeld, "On the Question of the Measurability of Electromagnetic Field Quantities," 387; Darrigol, "Cohérence et complétude," 171–73.

36. Yeang, "Tubes, Randomness, and Brownian Motions"; many scientists contributed to both lines of research: Wheaton, *Tiger and the Shark*, 142–47; Fürth, "Über Strahlungsschwankungen nach der Lichtquantenstatistik."

37. Barchas Collection, Department of Special Collections and University Archives, Stanford Libraries, Stanford CA, 94305 USA. The dedication is unsigned. I thank Finn Aaserud and Christian Joas of the Niels Bohr Archive in Copenhagen for assistance in confirming Bohr's handwriting.

38. Furry and Oppenheimer, "Theory of the Electron and Positive," 247–48.

With the addition of electron-positron fluctuations, the total fluctuations became

$$(\Delta\epsilon_r)^2 \sim \frac{\hbar c}{L^4}\left(1 + \frac{e^2}{\hbar c}\right),\tag{4.2}$$

where he combined BR's length and time averages L^3T into a single four-dimensional interval L^4. "From the form of these expressions" Oppenheimer concluded that

> it seems quite doubtful whether the physical significance of the modifications introduced by the pairs can be legitimately evaluated without taking into account the atomicity of the electric charge with which the field measurements are necessarily made.[39]

Even though Oppenheimer agreed with BR in almost every respect, still Oppenheimer thought that pair theory undermined the foundations of BR's work. It was just "taking into account the atomicity" of measuring devices that propelled Landau and Peierls's original critique of QED.[40] Oppenheimer challenged the foundational assumption of Bohr's complementarity: that measuring instruments be considered as classical objects.

On May 30, 1935, Heisenberg wrote to Bohr to defend Oppenheimer's 1935 calculation of charge fluctuations. In this letter, he presented a revised picture of the ground state of wave theory. The lowest-energy state became relative to the scale of each measurement. It *could* be a single electron, when the wave field was being measured over a region larger than e^2/mc^2. But: "If one averages over much smaller areas, the state of lowest energy looks more like: an electron and a large number of pairs."[41] This picture of an electron and a cloud of pairs echoes Furry and Oppenheimer's "nascent pairs" (chapter 3) but was newly connected to measurement fluctuations.

4.2.3 Weisskopf's Perturbative Vacuum

Weisskopf was engaged with research on spontaneous emission and the "natural" width of spectral lines, on uncertainty and fluctuations in field measurements. These were connected in his picture of vacuum fluctuations. In 1932–1933, before moving to Cambridge, Weisskopf held his Rockefeller fellowship at Bohr's Institute in Copenhagen. There he served as one of Bohr's scribes who took dictation for draft upon draft of the measurability paper. At ETH Zurich, he gave a seminar on Bohr and Rosenfeld's work which led to a correspondence with the authors on the details of ideal experiments in December 1933.[42] There were, I think, two important nodes for connections between these ideas for Weisskopf.

---

39. I have inferred the second equation from the first line of Oppenheimer's table 4, Oppenheimer, "Note on Charge and Field Fluctuations," 145.

40. Oppenheimer's student Peter Morrison worked on fluctuations for his PhD, and added a consideration of emission of multiple-quanta at low energies; P. Morrison, "Energy Fluctuations in the Electromagnetic Field."

41. Kalckar, "Correspondence Included (1996)," 456; A.I Miller, *Early Quantum Electrodynamics*, 61–62; for later developments, see Blum and Joas, "Dressed Electrons," 3.

42. Cf. Weisskopf to Rosenfeld 2.12.1933, Bohr to Weisskopf 5.12.1933, Kalckar, "Correspondence Included (1996)," 512–13.

The first node was ground states. To calculate the self-energy of the electron in "hole" theory, Weisskopf did not try a standard approach: evaluate an expression for the electromagnetic field energy created by a single electron. Following Ivar Waller and Oppenheimer, he instead took "the difference between the state with a normal electron of positive energy when the negative energy levels are occupied and the self-energy of the 'vacuum' ": $E_{Vac+1} - E_{Vac}$. Both terms were infinite; the goal of this line of research was to balance the infinities in an unambiguous way, leaving a finite result.[43] In Dirac's presentation of his "hole" theory (of protons) in 1930, he made an analogy between the lowest-energy state of his electron theory and the lowest-energy state of the hydrogen atom, which was familiar from Bohr's atomic theory (section 2.3).[44] Even though Wigner and Weisskopf challenged the idea that "stationary" atomic states were truly stationary, in their radiation theory atoms did stop transitioning to lower states. This happened when the atom reached an "absolutely stationary state—the energy of which we can make the zero point [Nullpunkt] of the energy scale."[45] Earlier, in 1927, Landau took the "zero state" of the line width problem to be the ground state of the atoms.[46] In 1934, Weisskopf had to consider one system which was formally similar to his approach to line widths, E_{Vac+1}, and one without a (positive-energy) system, E_{Vac}. Each needed a ground state. But in "hole" theory, Oppenheimer argued in 1930 that "there is no normal state [i.e., ground state] for the matter, because states of infinite negative energy are possible." This paper included Oppenheimer's calculation of the self-energy of the electron, and also the width of spectral lines breadths.[47] (Section 3.3.3 discussed Furry and Oppenheimer's vacuum arithmetic.) In effect, Weisskopf extended the unstable ground state of his atomic theory to an unstable vacuum state.[48]

The second node was zero-light quanta states. In the first order of perturbation theory, Weisskopf's E_{Vac+1} was a state with one positive-energy electron, a full Dirac sea, and no light quanta. In his work with Wigner on line breadths, Weisskopf considered a state with one atom in an excited state and zero light quanta; when the atom transitioned to a lower-energy state, it emitted light.[49] (The overall situation they analyzed was atoms being bombarded with light quanta—but Weisskopf and Wigner isolated spontaneous emission as a special source of line breadths. Spontaneous emission was, by definition, independent of the incident "stimulating" radiation.) The initial state of these atoms was precisely the "stationary" state that they assumed had a finite lifetime. These zero-quanta states were the cases in which Bohr and Rosenfeld gave fluctuations the most importance.

43. Weisskopf, "Self-energy of the Electron," 164; cf. Oppenheimer, "Note on the Theory of the Interaction of Field and Matter," 463, 466; Peierls, "Vacuum in Dirac's Theory," 426.

44. Dirac, "Theory of Electrons and Protons," 361; Kragh, Niels Bohr and the Quantum Atom.

45. Weisskopf and Wigner, "Berechnung der natürlichen Linienbreite auf Grund der Diracschen Lichttheorie," 73.

46. Landau, "Damping Problem in Wave Mechanics," 17; cf. the matrix elements for the ground state, $(0|V|0)$ in Dirac and Harding, "Photo-electric Absorption in Hydrogen-like Atoms," 211.

47. Oppenheimer, "Note on the Theory of the Interaction of Field and Matter," 469, 475–76.

48. Compare Heisenberg to Bohr, 12.3.1934: "it seems to me that in the quantum theory of wave fields the problem of a single point charge is analogous to the treatment of a quantum mechanical system in the lowest quantum state [...]." Kalckar, "Correspondence Included (1996)," 449.

49. Weisskopf and Wigner, "Berechnung der natürlichen Linienbreite auf Grund der Diracschen Lichttheorie," 61–62; cf. Dirac, "Quantum Theory of Dispersion," 718–19.

Unlike previous treatments, in 1934 Weisskopf calculated "The Self-Energy of the Electron" in terms of "the zero-point fluctuations of the field strengths which exist in empty space." But he did not follow Heisenberg's analysis of uncertainty in measuring devices. Neither did he apply the standard "mean-squared" fluctuation analysis that had been applied by Bohr and Rosenfeld (appendix B). Instead, Weisskopf applied a perturbation analysis to the self-energy problem and simply *labeled* one term in a series expansion as the fluctuations. Specifically, he wrote down an expansion of the transverse part of the electric potential, $\vec{A}_{tr} = \vec{A}_{tr}^{\,0} + \vec{A}_{tr}^{\,1} + \cdots$. The zeroth-order term would simply be zero in classical electromagnetism. But not so in "hole" theory.

In fact, however, $[\vec{A}_{tr}^{\,0}]$ represents the zero-point fluctuations of the field strengths which exist in empty space. In general, one can take into account the effect of this zero-point field strength, which causes spontaneous emission, by writing down the light field existing in space in the following fashion in a Fourier decomposition.

This made $\vec{A}_{tr}^{\,0}$ a periodic function of time.[50] These zero-point fluctuations caused an "additional current density," $\vec{i}_{tr}^{\,1}$, which Weisskopf calculated by a further perturbation method. He found that the electromagnetic contribution to the self-energy depended directly on $\vec{i}_{tr}^{\,1}$ and $\vec{A}_{tr}^{\,0}$. In this way, "the zero-point fluctuations of the field strengths" determined Weisskopf's result.[51]

In his 1936 monograph, Heitler joined Weisskopf in first using perturbation theory to calculate an infinite self-energy of the electron, and then attributing it to "zero-point fluctuations of the electric field strength." Following Bohr and Rosenfeld, Heitler added: "The zero point fluctuations of E have no direct connection with the zero point energy [...] which is of purely formal character."[52] Unlike Weisskopf's self-energy paper, Heitler *did* make a connection to the uncertainty principle. He wrote an uncertainty relation between the number of light quanta, N and the field strength: $\Delta N \Delta E \sim E$. He explained that,

if N has a given value, the electric field strength has no definite magnitude but will fluctuate about a certain average value. This is the case even if no light quanta at all are present ($N = 0$).[53]

---

50. Weisskopf, "Self-energy of the Electron," 164; for the perturbation terms, see Pais, *Development of the Theory of the Electron*, 28, eq. 17; Schweber, *QED*, eq. 2.5.42; compare to the formalisms in Yeang, "Two Mathematical Approaches."

51. Weisskopf made a sign error in his calculation of the electrodynamic self energy which was pointed out by Wendell Furry, section 3.3. However, the error was in the evaluation of the integrals further along in the work, the use of A_{tr}^{0} was not questioned. Weisskopf, "Self-energy of the Electron," 165, beween eqs. 22 and 23; A.I. Miller, *Early Quantum Electrodynamics*, 61; Schweber, *QED*, 122–25.

52. Heitler, *Quantum Theory of Radiation* (1936), 67 fn, 193; see also Wentzel, *Einführung in die Quantentheorie der Wellenfelder*, §8.

53. Heitler, *Quantum Theory of Radiation* (1936), 67. A difficulty that was not mentioned—so far as I know—is that a positive definite magnitude cannot "fluctuate" *about zero*. The average of any set of positive numbers will always be greater than zero.

The perspicuous feature of Heitler's discussion was that his appeal to an uncertainty principle was *post hoc*, after the calculation of the self-energy effect itself. This was paralleled in his discussion of Weisskopf and Wigner's work on spontaneous emission and line widths (section 4.2.1). Heitler introduced Weisskopf and Wigner's radioactive decay as an extra assumption in perturbation theory and proceeded to calculate the distribution of the intensity of emitted radiation. The shape of the distribution depended on the lifetime of the atomic states $1/\gamma$. Heitler then appealed to "Heisenberg's *uncertainty relation* for energy and time: $\Delta E \Delta t = \hbar$," so his readers could "understand" the distribution. He argued *from* the assumption of the radioactive decay *to* an uncertainty in the energy of the state, $\Delta E = \hbar \gamma$. (He did not argue from this uncertainty relation to a finite lifetime.) Conversely, Vladimir Fock developed a similar uncertainty relation without reference to fluctuations. He wrote "a relation expressing expressing the 'complementarity' (in the spirit of Bohr)" between light-quantum theory and classical field amplitudes, $\Delta E \geq 1/2\hbar\omega$. Fock presented this as a useful interpretive device, but he did not rely on it for detailed calculations.[54]

In the mid-1930s, vacuum fluctuations, uncertainty relations, and the self-energy of the electron were connected in a complicated web. They were unnecessary for Fock; ambiguous for Bohr and Rosenfeld; a heuristic for Heitler; and central for Weisskopf. By the late 1930s, Weisskopf's picture of electromagnetic vacuum fluctuations became more widely accepted. In 1937, Bohr came to adopt Weisskopf's picture of vacuum fluctuations causing spontaneous emission.[55] And in a conference presentation in 1946, Bohr cited Weisskopf and applied his picture to the self-energy of the electron: "we must in quantum electrodynamics, owing to the fluctuations of the field intensities even in photon-free space, take into account a new contribution to the self-energy."[56] In a well-known review of the theory of the electron, Abraham Pais commented that Weisskopf's result for the self-energy of the electron in "hole" theory

> recalls the great caution which is needed in judging any approach to the problems of elementary particle physics in which the many-body character of the problem is not taken into account from the outset.[57]

This positions Weisskopf's "hole" theory as a capstone to the argument of chapter 2, that Dirac's relativistic theory was also a many-particle theory.

Weisskopf further developed this picture in a 1939 review article "On the Self-Energy and the Electromagnetic Field of the Electron." His central claim was that

54. This paper established the modern mathematical space for QED, "Fock space." Fock, "On Quantum Electrodynamics," 347; Fock's analysis was close to: Pauli, *General Principles of Quantum Mechanics*, 188; cf. Pechenkin, "Early Statistical Interpretations," §4; and Jammer, *Conceptual Development*, 355.

55. Bohr, "Field and Charge Measurements in Quantum Theory," 201.

56. Bohr, "Problems of Elementary-particle Physics," quote: 220; cf. Kalckar, "Introduction"; Wheeler, "Field and Charge Measurements."

57. Pais, *Development of the Theory of the Electron*, 29 but see 28 fn31 for reservations about fluctuations.

the self-energy was only logarithmically divergent *in all orders* of perturbation theory.[58] Fluctuations played a much more prominent role in 1939 than their earlier identification with $\vec{A}_{tr}^{0}$. Now, Weisskopf presented the existence of vacuum fluctuations as an *independent postulate* of QED. "The quantum theory of the electromagnetic field postulates the existence of field strength fluctuations in empty space."[59] These fluctuations were not produced by anything else, whether an inherent uncertainty or the act of measurement. Weisskopf presented a similar picture of the electron's charge to the one Heisenberg described to Bohr in 1935 (section 4.2.2). He argued that

> we also find around the electron a cloud of higher charge density coming from the displaced [vacuum] electrons. The total effect is a broadening of the charge of the electron over a region of the order h/mc as it is indicated schematically in figure [4.3]."[60]

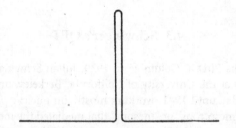

FIG. 1a. Schematic charge distribution of the electron.

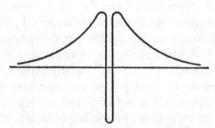

FIG. 1b. Schematic charge distribution of the vacuum electrons in the neighborhood of an electron.

Figure 4.3. Weisskopf's two schematics for the charge distribution of an electron (top) and the charge distribution of Dirac's "sea" surrounding it. Reprinted figure with permission from Victor Weisskopf, "On the Self-Energy and the Electromagnetic Field of the Electron," *Physical Review* 56 (1939): 72–85. Copyright (1939) by the American Physical Society.

---

58. Pauli wrote to Dirac on 12.5.1939, generally endorsing Weisskopf's work. But Pauli doubted the claim about all orders of perturbation theory, Pauli, *Wissenschaftlicher Briefwechsel*, vol. 2, doc. 564; a decade later, Freeman Dyson would realize this result more fully; Schweber, *QED*, chapter 9.

59. Weisskopf, "On the Self-energy and the Electromagnetic Field of the Electron," 73, his $\vec{A}_{tr}^{0}$ reappears as A_0 on 80; cf. Weisskopf, "On the Self-energy of the Electron."

60. Weisskopf, "On the Self-energy and the Electromagnetic Field of the Electron," 73–74.

After World War II, Weisskopf's 1939 paper was cited as the received view on the self-energy by Feynman, Schwinger, and Koba Zirô and Tomonaga.[61]

Weisskopf's paper was published on July 1, 1939, from his fresh position at the University of Rochester in New York. On August 2, prompted by Leo Szilard, Einstein wrote to President Franklin Roosevelt explaining some of the possibilities for nuclear power and bombs. He recommended that the US military begin coordinating with nuclear scientists. Weisskopf became an American citizen and, like many of his peers, joined the war effort. As of May 10, 1945, the Theoretical Division of the Manhattan Project at Los Alamos was lead by Hans Bethe, with Weisskopf as deputy and group leader; other leaders included Robert Serber; Rudolph Peierls led a group that included Klaus Fuchs (who admitted to spying for the USSR in 1950); Richard Feynman led a group that included his friend Theodore Welton.[62] There is not space the space to even sketch the enormous impact World War II had on physics and physicists, not to mention the wider world.[63]

4.3 Schwinger's QFT

After completing his PhD at Columbia in 1939, Julian Schwinger took up a postdoctoral fellowship at the University of California, Berkeley, under Oppenheimer. He stayed at Berkeley until 1941, working mostly on nuclear physics and Yukawa Hideki's theory of "mesotrons" or "mesons" that mediated the interaction of protons and neutrons in atomic nuclei.[64] Oppenheimer and Schwinger's working relationship closely paralleled Oppenheimer and Furry's five years earlier (section 3.3). In 1932–1933, Oppenheimer rapidly produced theoretical estimates of the positron phenomena Carl Anderson observed; in 1939, Oppenheimer produced a theoretical estimate for William A. Fowler and Charles Crane's studies of pair creation by fluorine that had been bombarded by protons.[65] In both periods, Oppenheimer turned to his postdoctoral fellows to develop these estimates into more careful analyses. On October 29, 1939, Oppenheimer and Schwinger sent just such a letter to the *Physical Review*: "On Pair Emission in the Proton Bombardment of Fluorine." According to Schwinger's recollections, this work convinced him of the reality of vacuum polarization. In chapter 3, I showed how Oppenheimer's pessimism about theoretical physics led to qualifications and omissions of Furry's results in pair theory. Similarly, Schwinger recalls that he argued from Fowler and Crane's results to—in the letter's words—the "radical suggestion" that "there is a nonelectromagnetic coupling between nuclear particles and the pair field." When Schwinger saw these results in

---

61. Feynman, "Relativistic Cut-off for Quantum Electrodynamics," 1433; Schwinger, "QED1," 1440; Koba and Tomonaga, "Radiation Reactions in Collision Processes," 292; cf. Pauli, "Dirac's New Method of Field Quantization," 201–4, 207.

62. Weisskopf papers Box 2.01, Folder 6. For a fuller account see Rhodes, *Making of the Atomic Bomb*.

63. For an overview, see Roland, "Science, Technology, and War."

64. Mehra and Milton, *Climbing the Mountain*, chapter 3.

65. N.B. not Ralph Fowler. Fowler and Lauritsen, "Pair Emission from Fluorine," 841.

print, he discovered that Oppenheimer had roughly doubled the length of the letter with qualifications and alternatives to his radical proposal.[66] Oppenheimer was again reining in his fellow's ambitions. After Berkeley, Schwinger was hired by Purdue University in Indiana and joined the Manhattan Project in 1943—first at the Metallurgical Laboratory at the University of Chicago and then the Radiation Laboratory at MIT. After the war he joined Furry at Harvard, where he made his path-setting contributions to QED.[67]

In June 1947, Schwinger and Weisskopf—by then at MIT—discussed QED prior to attending a landmark conference organized by the National Academy of Science at Shelter Island, NY. On the train from Boston and during the conference, they discussed recent experimental results from Willis Lamb and Robert Retherford that showed a deviation from Dirac's formula for the fine structure of the spectral lines of hydrogen.[68] According to Hans Bethe's report of the conference, Schwinger and Weisskopf, and also Oppenheimer, proposed that this discrepancy could be accounted for by the self-energy of the electron. Bethe gave his own non-relativistic treatment, in which the line shift was caused by the reaction of the electron to its own electromagnetic field.[69] Schwinger's early relativistic calculation attributed the shift to "the logarithmically divergent self-energy of a free electron, which arises from the virtual emission and absorption of light quanta."[70] He mentioned vacuum polarization, but only to remark that "it has long been recognized that such a term is equivalent to altering the value of the electron charge by a constant factor," e.g. it was unobservable (chapter 3). Similarly, Tomonaga applied his "self-consistent subtraction method" with Koba to the Lamb-shift problem.[71] None of these approaches highlighted fluctuations or the vacuum.[72]

Weisskopf pursued a calculation of the spectral line shift on the model of his earlier work in Dirac's "hole" theory: calculate the energy difference between Dirac's sea and the sea with one positive-energy electron $\Delta E = E_{Vac+1} - E_{Vac}$. Before the news from Lamb and Retherford, Weisskopf had given the question of hydrogen's spectral lines in "hole" theory to an MIT graduate student, Bruce French. Once Weisskopf returned from Shelter Island he continued his discussions with Schwinger—who aimed at a properly relativistic calculation—and he began combining his non-relativistic self-energy calculation with French's non-relativistic work. By the spring of 1948, French and Weisskopf had one result, while Schwinger and Feynman each had his own. None of them agreed on the underlying reason for the discrepancy, though Schwinger's and

66. Oppenheimer and Schwinger, "Pair Emission in the Proton Bombardment of Fluorine," 1066; Mehra and Milton, *Climbing the Mountain*, 178.

67. Schweber, *QED*, chapter 7; Mehra and Milton, *Climbing the Mountain*, chs. 7–8.

68. Lamb and Retherford, "Fine Structure of the Hydrogen Atom"; Schweber, *QED*, 303; Mehra and Milton, *Climbing the Mountain*, chapter 7; Schweber, "Shelter Island Conferences Revisited," 78–84.

69. Bethe, "Electromagnetic Shift of Energy Levels," Schweber, *QED*, §5.6

70. Schwinger, "Quantum-electrodynamics and the Magnetic Moment of the Electron," 416.

71. Koba and Tomonaga, "'Self-consistent' Subtraction Method"; Schweber, *QED*, §5.7.

72. Similarly, Pais mentioned fluctuations explicitly in his relativistic discussion only in reference to Weisskopf's perturbation calculation. Pais, *Development of the Theory of the Electron*, 26; e.g., fluctuations were absent from the self-energy discussions in: Dancoff, "Radiative Corrections for Electron Scattering"; Pauli, "Relativistic Field Theories of Elementary Particles"; Dirac, "Quantum Electrodynamics"; Wentzel, "Recent Research in Meson Theory."

Feynman's *relativistic* calculations eventually converged to the same value. Facing this opposition, French and Weisskopf delayed publication of their result until French discovered the source of the discrepancy: it was Feynman and Schwinger who were in error, as they then admitted. The younger physicists' confidence in their "highbrow" relativistic methods was misplaced.[73]

Weisskopf's views on fluctuations were cautiously increasing in scope. His physical postulate of *electromagnetic* vacuum fluctuations appeared in the Lamb shift work of his postdoctoral fellow Theodore Welton (appendix B).[74] But Weisskopf had made no mention of vacuum fluctuations *of electrons and positrons* until just this period of intense discussion of the Lamb shift with his student French, Welton, Schwinger, Bethe, and others. In a 1949 review article, "Recent Developments in the Theory of the Electron," Weisskopf argued that in "hole" theory,

> The new aspect of the vacuum has a decisive effect upon the problem of the self-energy of the electron. The properties of the vacuum with respect to the electrons are now, in some aspects, analogous to its properties in respect to the electromagnetic field. There exist also zero-point fluctuations of the electric charge and the electric current in the vacuum.

He was not as confident about these phenomena as he was in the electromagnetic fluctuations. These latter were responsible for the Lamb shift, which was "a direct effect of the zero-point oscillation"; however "[f]uture experiments" were needed to determine whether polarization of the vacuum effects "can be considered as real."[75]

Weisskopf and Schwinger collaborated on Lamb shift calculations, but fluctuations did not play a role in Schwinger's early presentations of QED. The coauthored abstract of Schwinger and Weisskopf's paper, read at the annual meeting of the American Physical Society in January 1949, did not mention fluctuations.[76] Neither did Schwinger mention fluctuations at the Pocono Manor conference, held March 30– April 1, 1948.[77] But Niels Bohr interpreted Schwinger's work in terms of fluctuations. He began a comment on Schwinger's and Feynman's presentations with a departure from his position in the 1930s (section 4.2.2):

> It was a mistake in the old days to be discontented with field and charge fluctuations. They are necessary for the physical interpretation. Schwinger's treatment is an advance in bringing the theory into a regular order where effects can be interpreted.[78]

---

73. French and Weisskopf, "Electromagnetic Shift of Energy Levels," "hole" theory: 1240; For "highbrow" see Oppenheimer to Peierls 30.12.1948, and in inverted commas, Dyson to Peierls 16.1.1949, Peierls, *Correspondence*, vol. 2: 159, 161; on this episode: Schweber, *QED*, 92, 243–45, §5.8.

74. Welton, "Quantum-mechanical Fluctuations"; Schweber, *QED*, 406–8.

75. Weisskopf, "Theory of the Electron," 311, 314; compare Heitler's comment: Institut Solvay, "Electron Theory," 286.

76. Schwinger and Weisskopf, "Electromagnetic Shift of Energy Levels."

77. John Wheeler took notes that were mimeographed and distributed to participants. I have made minor grammatical and typesetting changes. Schwinger's talk is on pages 9–45, Box 19, Folder 28, Schwinger Papers, UCLA; cf. Schweber, *QED*, chapter 4, table 5.8.1; D. Kaiser, *Drawing Theories Apart*, chapter 2.

78. Schweber, "Shelter Island," 298.

After the Pocono conference, Schwinger lectured at the University of Michigan summer school in theoretical physics, July 19–August 7, 1948 (Freeman Dyson was a student). Beginning in his lecture notes, and in a series of publications (QED1–3), Schwinger made vacuum fluctuations central to his physical picture.[79] But he did not go so far as Weisskopf—he did not postulate *primitive* fluctuations, as independent postulates.[80]

4.3.1 Complementary Fluctuations

On July 21, 1948, Schwinger arrived at Ann Arbor to deliver a series of lectures at the University of Michigan summer school. His plan was not to merely demonstrate his techniques for calculating the Lamb shift. On July 29, the *Physical Review* received the first paper in a series that put forward his relativistically "Covariant Formulation" of the core of QED.[81] Like Tomonaga and his Japanese coauthors, Schwinger developed a generalization of Dirac, Vladimir Fock, and Boris Podolsky's 1932 QED that was consistent with special relativity. The relativistic generalization of the Schrödinger equation is now called the Tomonaga-Schwinger equation,[82]

$$i\hbar c \frac{\delta \Psi[\sigma]}{\delta \sigma(x)} = H\Psi[\sigma], \tag{4.3}$$

where the states, $\Psi[\sigma]$, were specified on a space-like surface, σ, in Minkowski spacetime (figure 4.4). In a letter to Peierls, Oppenheimer described Schwinger and Tomonaga as "two [physicists] who have carried out essentially parallel and very beautiful developments in quantum electrodynamics."[83] After the Pocono conference, Oppenheimer received a letter from Tomonaga, which he circulated among the American physicists.[84] Like their peers, in 1948 Tomonaga and Koba responded to "the problem lying close at our hand," Bethe's calculation of the Lamb shift.[85] Tomonaga and Koba treated the self-energy as an electron transitioning to virtual states and the polarization of the vacuum as virtual pair production "*viz.* a polarization charge produced out of vacuum." An advantage of their theory was that it "reduced [the divergences] into two quantities: the self-energy and the vacuum polarization."[86] But in Tomonaga's reformulation of QED, he did not offer new physical interpretations of these phenomena. This was, perhaps, due to his disinclination toward revolutionary hypotheses.[87] However, Tomonaga had specific concerns that the mathematical

79. Recent Developments in Quantum Electrodynamics, Summer Symposium, University of Michigan, page 55, Box 24, Folder 13, Schwinger Papers.

80. Schweber describes Schwinger's 1948 publications as delivering Weisskopf's 1949 outlook, but Schweber does not distinguish the full range of their views, Schweber, *QED*, 336.

81. Schwinger, "QED1"; D. Kaiser, *Drawing Theories Apart*, 73 fn121; Mehra and Milton, *Climbing the Mountain*, chapter 8.

82. Tomonaga, "Relativistically Invariant Formulation. I"; Schweber, *QED*, chapter 6.

83. Oppenheimer to Peierls and Mark Oliphant 8.4.1948, Peierls, *Correspondence*, vol. 2, 136.

84. Pais, *J. Robert Oppenheimer*, 116.

85. Tomonaga and Oppenheimer, "On Infinite Field Reactions in Quantum Field Theory," 224.

86. Koba and Tomonaga, "Radiation Reactions in Collision Processes," 297, 301.

87. Darrigol, "Yukawa et Tomonaga"; Schweber, *QED*, §6.4.

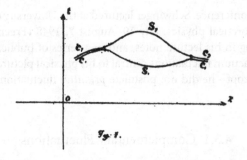

Figure 4.4. Tomonaga's figure showing how a transformation matrix, $T[C_2; C_1]$, could be defined between states on two arbitrary space-like surfaces, C_1 and C_2, which only differed between the finite portions, S_1 and S_2. Time runs on the vertical axis, and one spatial dimension runs on the horizontal axis. Tomonaga, Sin-Itiro. "On a Relativistically Invariant Formulation of the Quantum Theory of Wave Fields." *Progress of Theoretical Physics*, 1, no. 2 (1946): 27–42, by permission of the Physical Society of Japan.

treatment of the photon in renormalized QED was illegitimate. (As Schwinger would later admit, Tomonaga's concerns were well-founded.) Tomonaga was likely reticent to place a physical foundation under a still-shaky structure. This view was shared by other senior physicists who had the skills to master the new QED, such as Pauli.[88]

Schwinger similarly positioned the self-energy of the electron and vacuum polarization as the seat of QED's divergences;[89] but he did not share Tomonaga's reservations about physical interpretation. In his relativistically covariant formulation of QED there were *two* vacua: one of the electromagnetic field and one of the electron-positron field. "External" (classical) fields of one type interacted with the quantum fluctuations of the other.

> The phrase "polarization of the vacuum" describes the modification of the properties of an electromagnetic field produced by its interaction with the charge fluctuations of the vacuum. In the language of perturbation theory, the phenomenon considered is the generation of charge and current in the vacuum through the virtual creation and annihilation of electron-positron pairs by the electromagnetic field.

Schwinger identified the same phenomena physicists understood by "vacuum polarization" in the 1930s: the modification of electromagnetism, as calculated, for example, by Heisenberg and Hans Euler (section 3.4.2). He also found the same logarithmic form of the divergent currents in the vacuum, and applied Weisskopf's rule (section 3.5.2) for discounting "a uniform renormalization that has no observable consequences." In a complementary way, Schwinger described the self-energy of the electron as

---

88. Schwinger, "Tomonaga Sin-Itiro," 41; Pickering, *Constructing Quarks*, 174; on technical concerns, see Itzykson and Zuber, *QFT*, 217–24; on "cultural" explanations in this context, see Ito, "Cultural Difference."
89. Schwinger, "QED1," 1439.

The interaction between the electromagnetic field vacuum fluctuations and an electron, or more exactly, the electron-positron matter field, modifies the properties of the matter field and produces the self-energy of an electron.

According to Schwinger, this "mechanism [...] is commonly described as the virtual emission and absorption of a light quantum by an otherwise free electron."[90] Schwinger grounded the formalism of perturbation theory in the physical hypothesis of complementary vacuum fluctuations. As will become clear from Schwinger's technical discussion, the *interaction* played a crucial role in these fluctuations.

4.3.2 Calculating with Schwinger's QED (1948–1949)

In addition to the centrality of vacuum fluctuations in the physical interpretation of Schwinger's QED, the vacuum played *the* central role in its formalism. Schwinger worked in the tradition of wave theory, or field theory (like Furry and Oppenheimer, section 3.3). Formally, Schwinger required that the vacuum state be "an absolute minimum" of energy, and that it should be identical for observers in any inertial frame of reference. Earlier work invoked a *single* vacuum state—for example, a full Dirac sea and no positive-energy electrons, or the state with no electrons or positrons. By contrast, Schwinger defined *two* vacuum states, one each for "the vacuum of the electromagnetic fields" and "the matter field vacuum."[91] These vacua were the reference-states by which all other physical states were built. Defining them was no trivial matter. In the second paper in his reformulation of QED, section 4.1 "Definition of the Vacuum" was spread over seven pages.[92]

Once his vacua were defined, Schwinger used the vacuum states to construct "vacuum expectation values" (his coinage), which were the central expression he used to calculate measurement outcomes.[93] Expectation, or expected, values were inherited from statistical mechanics—informally, they meant the average value of a quantity, such as temperature.[94] In order to construct an average, many measurements were necessary. What appeared on the page as a representation of a single incidence of a physical process—or an observable "having a single value" for a single state—was really a representation of the *average* result of many measurements.[95] Quantum theorists were recasting their phenomena of inquiry from individual events to wholes: entire experiments, statistical ensembles, or "species" of particles. In this environment, vacuum expectation values became the central formal object of QED. Fluctuations might be expected to fit into Schwinger's

90. All quotes from Schwinger, "QED1," 1440.

91. Schwinger, "QED2," 652, 656.

92. Most of this was taken up with defining the commutation relationships between field components.

93. As a matter of formalism, expressions such as $(0|V^2|0)$ were introduced in 1932, where the zero represented the ground state of an atom: Dirac and Harding, "Photo-electric Absorption in Hydrogen-like Atoms," 211.

94. Neumann, *Mathematische Gundlagen der Quantenmechanik*, §III.3; Pauli, *General Principles of Quantum Mechanics*, 17, 128, 132; Plato, *Creating Modern Probability*, §3.2, 215; Pechenkin, "Early Statistical Interpretations."

95. London and Bauer, "Theory of Observation"; Dirac, *Principles* (1967), §12.

formalism just here. In the standard formalism for fluctuations, they were *defined* as the deviations from an expectation value (appendix B). However, Schwinger did not calculate fluctuations in the standard form.[96] Instead, like Weisskopf (section 4.2.3), he *referred* to fluctuations, but relied on perturbation theory to characterize them.

Some notation:[97] An expectation value of an observable, such as electric charge, Q, was written $\langle Q \rangle$. To calculate it, physicists specified the state of the system for which the observable was being measured, Ψ, then following the Dirac-Jordan transformation theory, this was translated into an inner-product, which was evaluated with linear algebra. For the vacuum state, Ψ_0, this was written,

$$\langle Q \rangle_0 = (\Psi_0, Q\Psi_0). \tag{4.4}$$

The electron-positron field was represented by ψ. The adjoint of ψ was written $\bar{\psi}$. Schwinger defined the vacuum of electron-positrons as the state Ψ_0, which could not be made lower by an annihilation operator $\psi^{(+)}$,[98]

$$\psi^{(+)}(x)\Psi_0 = 0. \tag{4.5}$$

This was Schwinger's version of the vacuum arithmetic Furry and Oppenheimer deployed (section 3.3.2). The vacuum was required to be the lowest possible state, and there could be no negative number of particles. Equipped with this vacuum state, Schwinger constructed vacuum expectation values, equation (4.4), for observables which were composed of field components such as $\psi^{(+)}(x)$. The most important of these was the current density four-vector, j_μ, which represented the electric charge and current in a region.[99] At the Pocono conference, he explained that his definition of j_μ was equivalent to the current in Heisenberg's "hole" theory, in which the contribution "for all neg[ative] energy states full all positive empty" is subtracted from j_μ.[100]

Schwinger insisted that there was no current in the vacuum. He calculated that the vacuum expectation value of the current was zero:

$$\langle j_\mu \rangle_0 = 0. \tag{4.6}$$

Later in the paper he returned to this quantity: "We must require that, in the absence of a real electromagnetic field, no current be induced in the vacuum."[101] He did not

96. Appendix B discusses a possible reason for this—the standard fluctuation calculation could not be made relativistic.

97. I have removed some extraneous symbols from quotations in Schwinger's published works, where they are not necessary for the discussion. Schwinger also constructed a host of "auxiliary fields" which are not pertinent here.

98. Schwinger, "QED2," 657.

99. Schwinger defined the electron-positron current as $j_\mu = -\dfrac{iec}{2} Tr \left[\psi(x), \bar{\psi}(x) \right] \gamma_\mu$, where Tr means the current was proportional to the trace (diagonal sum) of the matrices, and γ_μ were the 4×4 Dirac matrices.

100. Pocono conference, page 17, Schwinger Papers, Box 19, Folder 28; cf. Schweber, "Shelter Island."

101. Schwinger also evaluated the stress-energy tensor, $\Theta_{\mu\nu}$, which was not as obliging to his demands. Its vacuum expectation value diverged to infinity. In response, Schwinger *re-defined* $\Theta_{\mu\nu}$ so that its vacuum expectation value *would* be zero. Schwinger, "QED2," 657–58, quote: 659.

calculate the mean-squared fluctuations of $\langle j_\mu \rangle_0$ (see appendix B). The same was true of the analogous electromagnetic field quantities. In Schwinger's first QED, the vacuum of isolated quantum fields did not fluctuate.

Later, Dyson described the integral in Schwinger's equation (4.6) as "a highly divergent integral which is mathematically meaningless." "However," Dyson had "no doubt that correct physical results are obtained by putting $\langle j_\mu \rangle_0 = 0$ simply." He gave two reasons for asserting equation (4.6) as correct. Then he explained Schwinger's method:

> When a calculation leads to a divergent integral or a mathematically indeterminate expression, we use physical arguments or arguments of Lorentz invariance to find a definite value for the quantity we are unable to calculate. This is the reason for the great success of the covariant formulation of electrodynamics introduced by Schwinger.[102]

In chapter 3, I discussed how Weisskopf applied this method of applying physical principles to remove infinite expressions in "hole" theory. Dyson would likely observe that Schwinger's method was an advance on earlier physicists who simply discarded infinities *because they were infinite*. Here Dyson saw "physical arguments" at work.[103]

Instead of arising from primitive fluctuations, Schwinger described the polarization of the vacuum as arising from classical perturbations. Schwinger turned to "the induction of a current in the matter field vacuum by an electromagnetic field—the polarization of the vacuum." Schwinger adapted the familiar description of vacuum polarization (e.g., from Weisskopf's 1936 review, section 4.2.2). "It is supposed that the matter field, initially in its vacuum state, is perturbed by the establishment of an externally generated electromagnetic field, described by the potential $A_\mu(x)$." The "externally generated" field was "external" to the quantum system; a classical perturbation. But Schwinger's relativistic description needed to include the time in which polarization occurred—his descriptions were *narratives*. "It will be convenient to assume that the potential vanishes prior to the creation of the field, as well as after the eventual removal of the field." This has a familiar beginning, middle, and end structure. Formally, this eliminated the possibility that the initial vacuum state involved the potential $A_\mu(x)$ and allowed Schwinger to use the vacuum state of the isolated electron-positron field, equation (4.5).[104] We might flesh out this scenario by imagining an experimenter with an isolated vacuum chamber who turns on a nearby electromagnet, which sends electromagnetic waves through the vacuum, and then turns it off. Tomonaga et al. did not make Schwinger's turning "on" and "off" idealization; as a consequence, they applied perturbation theory directly to the initial and final states of scattering problems.[105] By contrast, Schwinger could only apply his technique to initial and final states that were "at" infinity, resulting in such expressions as the initial condition, in the infinite past $\Psi[\sigma] \rightarrow \Psi_0, \sigma \rightarrow -\infty$,

102. Dyson, *Advanced Quantum Mechanics*, II–112.

103. Assent to the validity of such argument was not universal. For Pauli's objections see Schweber, *QED*, §8.11 and chapter 10.

104. Schwinger, "QED2," 658; and Schwinger, "QED3," Appendix.

105. Generally: Tomonaga, "Relativistically Invariant Formulation. I," 37, figure: 40; applied to scattering: Itô, Koba, and Tomonaga, "Elastic Scattering of an Electron," 280.

meaning that an arbitrary state $\Psi[\sigma]$ became the vacuum state Ψ_0 as the space-like surface retreated infinitely far away in the infinite past.[106] Imagine the surfaces in figure 4.4 receding past the top and bottom edges of the page. (Chapter 6 discusses Roger Penrose's improvement of this representation.)

Schwinger's formalism was a method for pulling the vacuum state forward in time while accounting for the action of the perturbing external field. In this case, the perturbation, $A_\mu(x)$, was included in the interaction part of the Hamiltonian. This changed the Tomonaga-Schwinger equation (4.3) to

$$i\hbar c \frac{\delta\Psi[\sigma]}{\delta\sigma(x)} = H_I \Psi[\sigma] = -\frac{1}{c} j_\mu A_\mu(x)\Psi[\sigma]. \tag{4.7}$$

Following the algorithms of perturbation theory (appendix A), Schwinger plugged into equation (4.7) a series expansion for the *state* $\Psi[\sigma]$ and approximated by cutting the series at the first term. The result was

$$\Psi[\sigma] = \left(1 + \int_{-\infty}^{\sigma} j_\mu(x')A_\mu(x')d\omega'\right)\Psi_0, \tag{4.8}$$

$$= U[\sigma, -\infty]\Psi_0,$$

where the action of the terms between the brackets $(\cdot)$ was collected into a time translation operator, $U[\sigma, -\infty]$, that brought the vacuum $\Psi_0[-\infty]$ through an infinite duration, to $\Psi[\sigma]$.[107] Without the perturbation A_μ, the operator $U[\sigma, -\infty]$ would just be 1, and leave the vacuum unchanged.

With this preparation, we can now see exactly how the vacuum featured in Schwinger's calculation of vacuum polarization. The key quantity was familiar from the case of no external field, the vacuum expectation value of the current, but now with reference to the perturbed state $\Psi[\sigma]$,

$$\langle j_\mu(x)\rangle_\sigma = \left(\Psi[\sigma], j_\mu(x)\Psi[\sigma]\right). \tag{4.9}$$

Schwinger used the operator $U[\sigma, -\infty]$ to rewrite the physically complicated state, $\Psi[\sigma]$, as the vacuum Ψ_0 and the transformations it had undergone from the infinite past until σ. This changed the expectation value:

$$\langle j_\mu(x)\rangle_\sigma = \left(\Psi_0 U^{-1}[\sigma, -\infty], j_\mu(x)\, U[\sigma, -\infty]\Psi_0\right)$$

which was rewritten:

$$= \left\langle U^{-1}[\sigma, -\infty] j_\mu(x) U[\sigma, -\infty]\right\rangle_0. \tag{4.10}$$

Schwinger started with an expectation value for an arbitrary state, $\langle\cdot\rangle_\sigma$, and transformed it into a vacuum expectation value, $\langle\cdot\rangle_0$.[108] It is difficult to overstate the

---

106. Schwinger, "QED2," 658; Schweber, *QED*, 325; Dirac had used the infinite past for initial states, Dirac, "Developments in Quantum Electrodynamics," 12.

107. Schwinger, "QED2," 658.

108. I have added the σ to the expectation value. Schwinger, "QED2," 658.

importance of this progression. Not only could states of matter be built up from the vacuum with creation operators, complicated states of matter could be *decomposed* back to simple vacuum states. This movement from equation (4.9) to equation (4.10) was reproduced in Weisskopf's calculation of vacuum polarization, figure 4.5.

In lectures given in 1954 at Harvard, Schwinger emphasized the centrality of the vacuum objects:

we deal only with the transformation function $\langle 0\sigma_{+\infty}||0\sigma_{-\infty}\rangle$, the vacuum-to-vacuum amplitude from $t = -\infty$ to $t = +\infty$. As we will show later, this is usually the

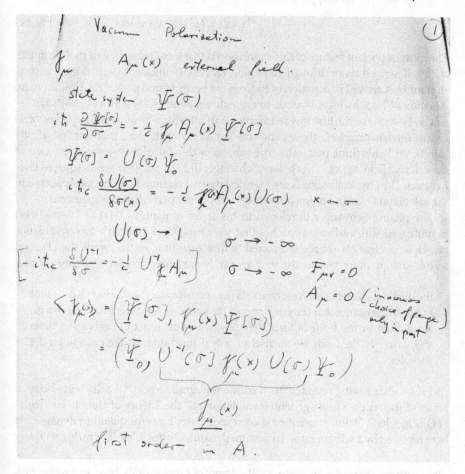

Figure 4.5. The first lines of Weisskopf's seven-page calculation of vacuum polarization (undated). Weisskopf followed Schwinger's calculation (see text). The text at bottom right reads: "(innocuous choice of gauge only in past)"; at bottom: "first order in *A*." At the end of the calculation, he identified an expression which "leads to the 1/15," which was likely a reference to (either or both of) French and Weisskopf (1949) equation (17), and Schwinger's QED1 equation (2.47). The values of such numerical factors were the central site of disagreement among calculations of the Lamb shift (Schweber 1994, table 5.8.1). Folder 12, Box 9.02, Weisskopf Papers, MIT Libraries collection MC-0572.

quantity of physical interest. Naturally, we can always start from the vacuum state as a reference, and create the desired state with appropriate sources.[109]

In Schwinger's QFT, the vacuum was the measure of all things.

The burden of evaluating vacuum polarization had been shifted from the states Ψ to the operator $j_\mu(x)$, and how *it* transformed under the transformations U. Schwinger applied the perturbation expansion, equation (4.8), to find

$$\langle j_\mu(x) \rangle_\sigma = \langle U^{-1}[\sigma, -\infty] j_\mu(x) U[\sigma, -\infty] \rangle_0$$

$$= \frac{i}{\hbar c} \int_{-\infty}^{\sigma} \langle [j_\mu(x), j_\nu(x')] \rangle_0 A_\nu(x') d\omega'. \qquad (4.11)$$

The most important feature of this expression is that it contains the external potential, $A_\nu(x)$. It arrived there through equation (4.8), the (truncated) perturbation expansion for U; it arrived in equation (4.8) from the perturbation part of the Hamiltonian, equation (4.7). Schwinger went on to evaluate equation (4.11) with virtuosity. His result had two parts: the first was an infinite constant which "produces an unobservable charge renormalization"; the second was "a finite contribution" that was "physically significant." This finite part—the true success of this calculation—was a function of the external field $A_\mu(x)$.[110] As before, Schwinger did not calculate mean-squared fluctuations. The physically significant manifestation of vacuum polarization appeared in the calculation because of the *interaction* term in the perturbation expansion.

The picture Schwinger developed in his series of papers QED1-3 (1948-1949) fit quite well with earlier approaches that were motivated by Bohr's correspondence principle: a classical source interacted with a quantum system, with the classical source and quantum system always different. For example, in 1946 Bohr argued that

> the dualistic character of these concepts [i.e., particle and field], which has its root in the circumstance that the properties of the particles, like their mass and charge, are defined by the fields of force they produce or the effect of fields upon them and, inversely, the fields are themselves only defined through their action on the particles.[111]

In 1951—after having revisited his wartime waveguide work[112]—Schwinger began a series of papers on a more general formulation of "The Theory of Quantized Fields" (TQF) in which "[t]he conventional correspondence basis for quantum dynamics is here replaced by a self-contained quantum dynamical principle."[113] Schwinger clearly

---

109. Schwinger papers, Box 16, Folder 9, "The Theory of Coupled Fields, Notes on Lectures by J. Schwinger, Harvard, 1954," notetakers unknown, 1.
110. Actually, Schwinger showed that the result was dependent on the field strengths, which he further wrote in terms of an external current. Gauge invariance motivated the shift away from A_μ. Schwinger, "QED2," 662, eq. 2.44.
111. Bohr, "Problems of Elementary-particle Physics," 222.
112. Lippmann and Schwinger, "Variational Principles for Scattering Processes. I"; Schweber, *QED*, §3.3; Galison, *Image and Logic*, chapter 4.
113. Schwinger, "TQF1," 914; this work is largely discussed for this "dynamical principle" and Schwinger's further development of Green's functions. Schweber, *QED*, §7.9; Mehra and Milton, *Climbing the Mountain*, chapter 9; Schweber, "Schwinger's Green's Functions."

aimed for a more general theory, and one that was "self-contained." Duality was all well and good when there were only two options. What of muons or neutrinos? Schwinger was trained as a *nuclear* physicist, and his 1941 paper with Oppenheimer was about precisely this subject: a non-electromagnetic interaction. In QED1–3 there was only one type of external source; in The TQF series (1951–1954) Schwinger introduced both Bose-Einstein and Fermi-Dirac classical sources.[114] This allowed him to specify the characteristics of each vacuum *on its own terms*, not in terms of the interaction perturbation, H_I, with a *different* field.[115] In my view, this switch—to perturbing electrons with an electron source—recovers something of Heisenberg's interpretation of wave-particle duality: "both/and" not "either/or" (section 4.2.2). In his later work, Schwinger investigated the particle-aspect of electrons with the wave-aspect of electrons.

In the 1960s and 1970s, Schwinger abandoned his search for a *fundamental* theory. He replaced it with an idiosyncratic "source theory," which had more direct contact with (idealized) experiments. Commenting on a student at Harvard in 1970, Sidney Coleman wrote to a colleague that "I have seen him less frequently than I used to since he disappeared into the wilds of source theory to work with Julian Schwinger."[116] Whatever its reception, source theory was a recommitment from Schwinger to the importance of "external" sources in QFT as representations of experimentation. These sources were not modeled on the details of actual experiments. Rather, "[t]he source concept is an *idealization* of the realistic processes."[117] I see this as a QFT development of Bohr and Rosenfeld's ideal measuring devices (section 4.2.2). Schwinger's source theory was an attempt to account for the quantum nature of even "external" fields—addressing a dissatisfaction with classical fields in QED that was expressed by Landau, Peierls, Oppenheimer and others in the 1930s.[118] Schwinger's radical departure was making his sources obey quantum commutation relations, albeit non-relativistic ones. One aspect of the "conservative" success of renormalization was the persistence of central concerns of the Copenhagen interpretation(s) of quantum theory.[119]

4.3.3 Conservatism

At the 1948 Solvay Congress in Brussels, Oppenheimer gave a report on "Electron Theory," in which he served as an elder statesman of the new QED. In contrast to his pessimism about pair theory and theoretical physics in the mid-1930s (chapter 3), Oppenheimer's postwar views were moderate—as judged by the reactions of boosters

---

114. Bose-Einstein: Schwinger, "TQF4," eq. 120; Fermi-Dirac: Schwinger, "TQF5," eq. 14.

115. This was still a perturbation analysis. Consider the abstracts to the TQF papers, or Schwinger, "Green's Functions II," 458.

116. Coleman to R. G. Moorhouse, 11.3.1970, Coleman Papers.

117. Emphasis added. Quoted in Mehra and Milton, *Climbing the Mountain*, 453, generally: chs. 12–13.

118. E.g., "[Dirac's] *manière de tenir compte de l'existence d'un champ extérieur n'est nullement satisfaisante; toute la conception de «champ extérieur» ne l'est pas.* [...] *En vérité, les «champs extérieurs» proviennent d'autres électrons et de protons,*" Schrödinger, "Théorie relativiste de l'électron," 278.

119. Schwinger's viewpoint was "in fact a pragmatic variant of the standard Bohr interpretation," Schweber, *QED*, 356, conservative: xxiv, similarly for Dyson: 548–49.

and cynics. Dyson was a partisan for Tomonaga, Schwinger, and Feynman's work. After seeing Schwinger lecture at Michigan, on July 20, 1948, Dyson wrote home to his parents that "I think in a few months we shall have forgotten what pre-Schwinger physics was like." When Dyson read Oppenheimer's Solvay report, he was moved to write Oppenheimer a memorandum attacking his treatment of the new QED.[120] In contrast, critical Pauli wrote to Bohr on February 18, 1948:

> At present I feel both inspiration and uneasiness about [the new QED calculations]. But Oppenheimer had also this time a positive and stimulating effect on me, namely as an antidote against my old sickness overpessimism.[121]

To pessimistic Pauli, Oppenheimer was an optimist. To optimistic Dyson, Oppenheimer was a pessimist. Oppenheimer sat in between his divided colleagues.

Though they approached the topic from opposite directions, both Dyson and Pauli saw new physical ideas at work. For example, Dyson argued that the unambiguous calculation of Schwinger's time-evolution operator $U(\infty)$ "is achieved by introducing into the theory what is really a new physical hypothesis."[122] Pauli's chief complaint was that Schwinger hid new physical ideas. Pauli wrote to Bohr on December 31, 1948, after studying a draft of QED2, that "[Schwinger] is a master in concealing physical difficulties with mathematical elegance." On January 29, 1949, Pauli wrote to Bethe:

> That is why I find Schwinger's illusion that he can deduce his final results [...] without new assumptions, not only logically confusing, but also an obstacle to further progress and an inappropriate reduction of Schwinger's own (good) physical instinct in favor of an untenable pseudo-conservatism!

And in a letter to Oppenheimer on February 22, 1949, Pauli wrote that "[Schwinger] must have strong psychological reasons for the very conservative appearance of his theory."[123] In Brussels, Oppenheimer trod a middle road between boosterism and cynicism. For him, "the earliest [of the new] methods must be characterized as encouraging but inadequate." He acknowledged the successes of the Lamb shift calculations, but ended his report with a series of unanswered questions, particularly about the extension of relativistic methods outside electrodynamics to nuclear physics.[124]

Against Dyson's and Pauli's novelty, Oppenheimer emphasized the continuity in prewar and postwar electrodynamics. He was conservative. Oppenheimer's report began with a historical discussion of electron theory, in which he established this continuity. "The problems with which we are concerned go back to the very beginnings

120. Schweber, *QED*, quote: 335, memo: 521–26; D. Kaiser, *Drawing Theories Apart*, 87–97.

121. Pauli, *Wissenschaftlicher Briefwechsel*, vol. 3, 508.

122. "[N]amely that the electron-positron field always acts as a unit and not as a combination of two separate fields." Dyson, "S Matrix in Quantum Electrodynamics," 1754.

123. My reading of this last letter differs from Schweber's, who sees it in terms of Pauli and Schwinger's personal relationship, which took a sour turn when Schwinger did not reply to a letter from Pauli sent in January 1949. However, Pauli's letter to Bohr expressed similar sentiments *before* that dispute began. Schweber, *QED*, 352; Pauli, *Wissenschaftlicher Briefwechsel*, vol. 3, 589, 623.

124. Institut Solvay, "Electron Theory," 273, §III; cf. Pais, *J. Robert Oppenheimer*, 114–16.

of the quantum electrodynamics of Dirac, of Heisenberg and Pauli."[125] Like Bohr's Pocono comments, Oppenheimer's report emphasized "the decisive, if classically unfamiliar, role of vacuum fluctuations" in QED. He asserted that this role

> was perhaps first shown—albeit in a highly academic situation—by Rosenfeld's calculation of the (infinite) gravitational energy of the light quantum, and came prominently into view with the discovery of the problem of the self-energy of the photon due tot [sic] the current fluctuations of the electron-positron field,

citing Rosenfeld's 1930 paper.[126] In fact, Rosenfeld's paper was written years before the positron appeared, and made no mention of fluctuations. Oppenheimer, consciously or not, was projecting his contemporary understanding backward, and thereby rhetorically establishing a continuity between past and his present.

Oppenheimer framed the development of electrodynamics as a search for "a consistent and reasonable modification" of the theory. In an implicit rejection of revisions of classical theory by Born, Infeld, Kramers, Dirac, and others, Oppenheimer wrote that

> it has proved decisive that it is *not* sufficient to develop a satisfactory classical analogue; rather one must cope directly with the specific quantum phenomena of fluctuation and pair production.[[127]] (272)

But despite renormalization, Oppenheimer thought that "the singularities [of QED] indicate that it is not a completed consistent theory." In parallel to his 1935 fluctuation analysis (section 4.2.2), he thought that in future developments "electron theory and electrodynamics will no longer be so clearly separable from other atomic phenomena" (273).[128]

Oppenheimer's report put (vacuum) fluctuations at the center of QED—its contemporary state, its relationship to competing proposals, and its history. Insofar as it addressed fluctuations, Oppenheimer's narrative was not accurate. Rhetorically, inventing continuity with the past is the inverse of invoking radical discontinuity and revolution. It was the opposite of the interwar rhetoric of a "crisis" in physics, and the bifurcation of "classical" and "modern" or "quantum" physics.[129] Oppenheimer's history was consistent with his moderate, conservative, interpretation of renormalization theory. Unlike the Weimar era, the late 1940s was a boom-time for theoretical

125. Subsequent citations to this paper will be given in text, Institut Solvay, "Electron Theory," 269–70.
126. Rosenfeld, "Über die Gravitationswirkungen des Lichtes."
127. For these alternatives, see Pais, *Development of the Theory of the Electron.*
128. He expressed this in Heisenberg's terminology as QED being an "almost closed, almost complete system. [...W]e look precisely to the absence of complete closure to brings [sic] us away from the paradoxes that still inhere in it." Institut Solvay, "Electron Theory," 273, cf. 280, 281; Bokulich, "Open or Closed?"
129. This was similar to E. T. Whittaker's well-known misconstrual of the history of relativity; Hunt, "Whittaker. A History of the Theories of Aether"; Gingras, "Collective Construction"; see Seth, "Crisis and Construction"; Gooday and Mitchell, "Rethinking 'Classical Physics'"; Staley, "Trajectories."

physicists in Allied nations. Who would want a revolution?[130] In 1953, Oppenheimer gave the BBC Reith lectures. His topic was *Science and the Common Understanding*; in his words, "science, for all its revolutions, is conservative."[131]

The discussion in Brussels after Oppenheimer's report was revealing. Dirac was the first to speak, defending the possibility of finding *exact*, not perturbative, solutions for QED equations.[132]

> Mr. Dirac.—[...] All the infinities that are continually bothering us arise when we use a perturbation method, when we try to expand the solution of the wave equation as a power series in the electron charge. Suppose we look at the equations without using a perturbation method, then there is no reason to believe that infinities would occur. (282)

Dirac and Walter Heitler had made similar remarks at the first postwar conference on physics on the east side of the Atlantic, held at the Cavendish Laboratory, Cambridge, in 1946.[133] In Brussels, Dirac was answered by

> Mr. Bohr.—[...] I must deeply sympathize with the approach by Oppenheimer. [...] There is hardly any practical way, and that has been through ages, to attack a problem, except by approximations. (283)

Bohr agreed with Oppenheimer that the next developments would be led by applications outside of electrodynamics. Dirac's mathematical claims were challenged by Pauli; but the challenge was to the possibility of non-perturbative approaches, not to the perturbation diagnosis of the infinities.[134] This discussion provides further evidence that the practice of perturbation methods drove relativistic quantum theory.

Schwinger took up many aspects of this picture in his relativistic QED, but he did not accept vacuum fluctuations as an additional postulate. He was more conservative than Weisskopf about fluctuations. For Schwinger—formally and conceptually—fluctuations were brought about in the interaction between a quantum field and classical sources. For Schwinger, "vacuum fluctuations" did not necessarily change

130. On the politics-physics connection, see Carson, Kozhevnikov, and Trischler, *Weimar Culture and Quantum Mechanics*; on boom-time: Kevles, *Physicists*, chs. XXI–XXIII; D. Kaiser, "Cold War Requisitions"; Krige, *American Hegemony*.

131. Oppenheimer, *Science and the Common Understanding*, 50; cf. Schweber, *Einstein and Oppenheimer*, chapter 4, 31, 234-35.

132. Kragh, *Dirac*, 183.

133. "We can conclude quite generally that the wave equation never has a solution of the form of a power series." Dirac, "Difficulties in Quantum Electrodynamics," 12; cf. Dirac, "Developments in Quantum Electrodynamics"; Heitler: "If one tries to expand the wave function in a power series of the coupling constants the individual terms are all infinite, except the term of lowest non-vanishing order." Heitler, "Quantum Theory of Damping," 189.

134. In Cambridge, Pauli diagnosed the infinities as arising from two sources: the point-like radius of the electron, which was present in classical relativistic electron theory; and "the fluctuation singularities, which are essentially quantum-theoretical." Specifically, infinite fluctuations occurred in all attempts to quantize *waves* or *fields*: systems with an infinite number of degrees of freedom. Pauli, "Difficulties of Field Theories and of Field Quantization," 5; he expressed this view to Dirac on 18.7.1939. Pauli, *Wissenschaftlicher Briefwechsel*, vol. 2, doc. 575.

the meaning of "the vacuum." His most cited paper, "On Gauge Invariance and Vacuum Polarization" (1951), was directed at the calculation of vacuum expectation values for vacuum polarization effects. It made no mention of fluctuations.[135] This was consistent with his later views. In 1972, Schwinger spoke at a conference on "the physicist's conception of nature." Earlier, John Archibald Wheeler had given a free-ranging lecture that included a discussion of the quantum fluctuations of the geometry of spacetime (chapter 5). Schwinger dryly observed:

> Unfortunately, this episode ["hole" theory], and discussions of vacuum fluctuations, seem to have left people with the impression that the vacuum, the physical state of nothingness (under controlled physical circumstances), is actually the scene of wild activity.

He insisted that "the vacuum is not only the state of minimum energy, it is the state of zero energy, zero momentum, zero angular momentum, zero charge, zero whatever."[136] Unlike Weisskopf, Schwinger did not see the need for introducing a new fluctuation postulate into quantum theory. Schwinger's view was continuous with Heisenberg and Oppenheimer's pre–World War II fluctuation studies: fluctuations were created by measurement interactions, or their theoretical representations.

4.4 Feynman's Vacuum Loops

Despite Pauli and Dyson's characterization of the novelties of renormalized QED, there is some truth to the conservative picture. In the 1940s, Tomonaga, Schwinger, and Feynman did not enact the sort of conceptual change that Einstein had propelled in 1905. Such change had been possible. Feynman was John Archibald Wheeler's graduate student at Princeton University, and together they worked on a revolutionary particle scheme that eliminated *fields* from radiation theory. Had Wheeler and Feynman's action-at-a-distance scattering theory succeeded, physics may have been jerked from Einstein back to Newton (see section 5.2). This section examines Richard Phillips Feynman's analysis of vacuum polarization—electron-positron loops in his eponymous diagrams.[137] While Schwinger was publishing his initial QED trilogy in *Physical Review* during 1947–1949 and establishing vacuum polarization as a paradigmatic calculation, Feynman was struggling to include vacuum polarization in his developing framework. Later, in the 1950s and 1960s, Feynman's diagrammatic quantum theory of gravity was pulled to a standstill by vacuum loops. Even for one of the most inventive and disinhibited physicists, theory was not a free invention. It may

135. Schwinger, "Gauge Invariance and Vacuum Polarization"; citation analysis: Mehra and Milton, *Climbing the Mountain*, 308.
136. Schwinger, "Report on Quantum Electrodynamics," 413, 416.
137. For biographical details, see Gleick, *Genius*; Schweber, *QED*, chapter 8; Mehra, *Beat of a Different Drum*.

be that scientific failures are at least as revealing as successes for understanding the constraints of scientific practice.[138]

4.4.1 The "Theory of Holes"

Feynman's thinking about the vacuum came initially as a rethinking and reformulation of Dirac's "hole" theory. After the war, Feynman followed Hans Bethe to Cornell, where he was a professor until moving to Caltech, in 1950. At Cornell he taught Physics course 491 "ADV[anced] Q.M., Q.E.D." A set of lecture notes in Feynman's hand is preserved in his archives at Caltech.[139] Though the notes are undated, textual clues allow an approximate dating to 1947–1948. We should pause to note the degree to which Feynman integrated the newest results from his theory into his pedagogy, much as Roger Penrose did in the 1960s (see chapter 6). Neither physicist shied away from integrating the bleeding edge of research into their advanced teaching. In this Feynman may have taken up the practices of his doctoral advisor, Wheeler, who was well known for introducing quantum theory to freshmen (see chapter 5).

In his course 491, Feynman introduced Dirac's theory of the electron and addressed the "Klein paradox" where a reflective potential barrier becomes transparent at certain energies on Dirac's theory.[140] He had already introduced the Dirac equation and the existence of negative and positive energy states in the theory. He then turned to the "Theory of Holes" in which "all negative energy states are occupied." But to Feynman this was "Meaningless. Leads to difficulties."[141] On the verso of this page Feynman continued that under Dirac's theory "Pair creation → sends neg. energy to + energy. ∴ electron + hole. Anhil. [sic] electron goes into hole."[142] Under a line across the page Feynman presented two contrasting pictures of scattering processes, in drawings and text (figure 4.6). At the top of figure 4.6 he showed a picture in Dirac's theory. The image is not quite a spacetime diagram and not quite a graph. It has axes like a graph, but traces the paths of particles like a diagram. Time is implied to be the vertical axis, space the horizontal. A particle, + (perhaps a proton), approaches a potential, A, from the bottom at 1. It passes through A and is deflected to the left toward 2. A positive energy electron, el, appears above A, along with the rightmost, lighter pencil, arrows labelled "electron with neg E," going to 2'. Parallel with the x-axis he noted the equivalence of this physical picture with a mathematical sum of simpler pictures. The text accompanying this topmost diagram reads

convention:

(1) Electron can make transition to various states of + Energy, and ampl[itude]

---

138. On constraint and freedom in scientific practice, see Pickering, *Mangle of Practice*, §1.4; Galison, *Image and Logic*, 800–3.

139. Feynman papers, Box 7, Folder 4, Advanced Quantum Mechanics at Cornell. These notes have been discussed by several historians. See D. Kaiser, *Drawing Theories Apart*; Wüthrich, *Genesis of Feynman Diagrams*; Gross, "Pictures and Pedagogy."

140. Kragh, *Dirac*, 89.

141. Feynman papers, Box 7, Folder 4, Advanced Quantum Mechanics at Cornell, p. 13.

142. Feynman appears to have had trouble spelling "annihilation," and often wrote "anhillation" and abbreviated versions of such. No further "*sic*" will be included in my text when he made this error.

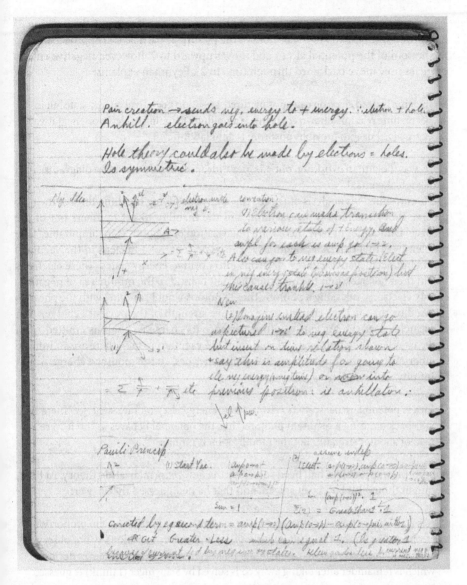

Figure 4.6. Feynman's drawing of two views of scattering processes, just below "My Idea," from his Cornell Course 491 lecture notes. Feynman papers, Box 7, Folder 4, Advanced Quantum Mechanics at Cornell, p. 13 verso. Courtesy of the Archives, California Institute of Technology.

for each is amp go 1 → 2. Also can go to neg[ative] energy states. (Elect[ron] in neg[ative] energy state behaves as positron) but this causes trouble 1 → 2'.[143]

143. Feynman papers, Box 7, Folder 4, Advanced Quantum Mechanics at Cornell, p. 13 verso.

Then came Feynman's new picture. In the bottom image of figure 4.6 the position 2′ has moved from the top to the bottom of the picture. A particle still approaches from the bottom of the potential at (1) and moves upward to 2, however, negative energy particles now move backward through time to 2′. Feynman explains:

> Imagine instead electron can go as pictured 1 → 2′ to neg[ative] energy state but insist on time relations shown & say this is amplitude for going to ele[ctron] neg[ative] energy [...] or into previous positron: is anhillation.[144]

Here was Feynman working out his new idea, alongside his new diagrams, in a pedagogical context.

He continued his analysis of Dirac's theory by specifying the difficulties of the hole theory. They centered on the vacuum, and the inability to get numbers out of the theory: "Difficulties of hole theory are polarization of vacuum is incomputable." The problem was that external potentials (say, the presence of a nucleus in the vacuum) would bounce negative-energy electrons into positive-energy states. These electrons would then fall back into the hole they had left behind in the infinite sea of negative-energy electrons, releasing a photon. These photons would combine with the photons released by the nucleus, changing the effective strength of the nucleus's charge. That is, particles near the nucleus would experience its charge as photons emitted from itself, as well as the photons it induced from the vacuum. In fact, there are an infinite number of these photons with up-to-infinite energies, so the modified charge would be infinity. In Feynman's words:

> Idea is potential ~~make~~ scatters (virtual) neg. energy electron into +energy. There it falls back emitting a ~~quantum~~ photon. Thus the potential behaves as tho it were a field of another strength. But the modification is infinite.[145]

But this prewar problem had been solved by the new renormalization theory. At least, Feynman thought it had been solved, and that he had guessed the answer:

> It is now known however what to do. The ∞ term makes potential act in proportion stronger ∴ equivalent to changing charge (charge renormalization). ∴ not observable, by hypothesis removed. Remainder makes physical sense. Problem of deciding which is remainder after taking away ∞ is solved by Schwinger (I think) & guessed by me.

What of the more general case, where more than one photon was emitted by a cascading negative energy electron? "If two or more photons are emitted before re-anhillation (including the anhill one) everything is definite finite & O.K. (Discovered last weekend, Saturday)." Here was closeness between research and pedagogy![146]

144. Feynman papers, Box 7, Folder 4, Advanced Quantum Mechanics at Cornell, p. 13 verso.
145. Feynman papers, Box 7, Folder 4, Advanced Quantum Mechanics at Cornell, p. 14.
146. The question of multiple light-quantum emission in light-light scattering was considered by Halpern and by Breit and Wheeler in 1933–1934; see section 3.4.2. These statements help date these notes to before

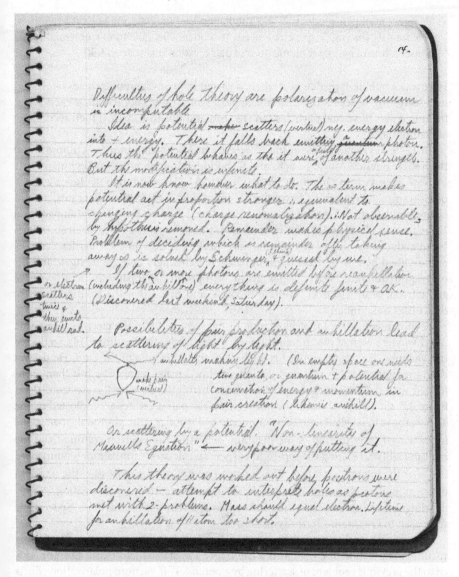

Figure 4.7. Feynman's diagram of the scattering of light by light. Feynman papers, Box 7, Folder 4, Advanced Quantum Mechanics at Cornell, p. 14. Courtesy of the Archives, California Institute of Technology.

Feynman continued to consider the scattering of light by light (chapter 3).

Feynman's 1949 series of publications in *Physical Review*, and probably before Schwinger's (1948) QED1. Feynman Papers, Box 15, Folder 6.

Possibilities of pair production and anhillation lead to scattering of light by light. [Figure 4.7] (In empty space one needs two quanta, or quantum + potential for conservation of energy & momentum in pair creation (likewise anihill).

Here was a bubble-type diagram where a closed loop of vacuum particles intercedes in a scattering process. The diagram depicts two photons approaching each other, making a pair of virtual particles (an electron and positron) which then recombine and annihilate.[147]

In 1949, Feynman spoke at the University of Michigan Summer Symposium on theoretical physics. Morton Fuchs and R. J. Riddell Jr. took notes on Feynman's talk, "A New Approach to Quantum Electrodynamics." Here Feynman deployed a range of diagrammatic reasoning, beyond what count now as "Feynman diagrams," to elucidate his theory.[148] Feynman began his lecture by going back to Heisenberg and Dirac's work "about 1930" and referenced Hendrik Lorentz's older electron theory. He introduced the apparently infinite mass of the electron on these earlier theories, and outlined how to address them in terms of Schwinger and Tomonaga's, and his own "quite different," work. In parallel he introduced the apparently infinite charge and current of an electron on these theories, and noted that the solution to these difficulties was similar.

Feynman considered the calculation of the self-energy of the electron—one side of Schwinger's coin of renormalization processes. Under the heading "Polarization of the Vacuum" Feynman set out his analysis of the other side of the coin. He drew a diagram, figure 4.8. "In this case the potential [$\underline{a}$] creates a pair which is then annihilated with the emission of a photon which then interacts with the electron."[149] In figure 4.8, the potential $\underline{a}$ is at the bottom left. The pair it creates are indicated by the arrows above it, labelled with their momenta $\underline{p}_{-a} + \underline{q}$ and $\underline{p}_{-a}$. The photon the pair emits is the wavy line at the top of the diagram, deflecting the electron coming from the bottom right and exiting the top-right corner. The electron's initial momentum is $\underline{p}_1$, and after the interaction it is $\underline{p}_2$. This is a more complicated picture than the standard interaction of an electron with a potential, with no intervening pair creation. A similar diagram and treatment occurs in Feynman's 1949 published article, "Space-time Approach to Quantum Electrodynamics."[150]

Compared to Schwinger, Feynman was less certain of the necessity of including vacuum polarization in QED. "This term [figure 4.8], if it actually exists, must be added to the previous problem of scattering by a potential." If vacuum polarization effects were real, Feynman supposed he must consider them. The effect of the virtual pair "essentially yields a change in strength of the potential $\underline{a}$." However, this modification was infinite, because there was no limit to the loop momentum q. "We might want to throw out the closed loops as they lead to divergence difficulties, but there are certain reasons for including them (such as conservation of probability)."[151] Feynman did not

147. Feynman papers, Box 7, Folder 4, Advanced Quantum Mechanics at Cornell, p. 14.
148. Wüthrich, *Genesis of Feynman Diagrams*, §5.5; Gross, "Pictures and Pedagogy."
149. Feynman papers, Box 15, Folder 7, A New Approach to Quantum Electrodynamics, p. 45.
150. Feynman, "Space-time Approach to Quantum Electrodynamics," 779.
151. Feynman papers, Box 15, Folder 7, A New Approach to Quantum Electrodynamics, p. 45.

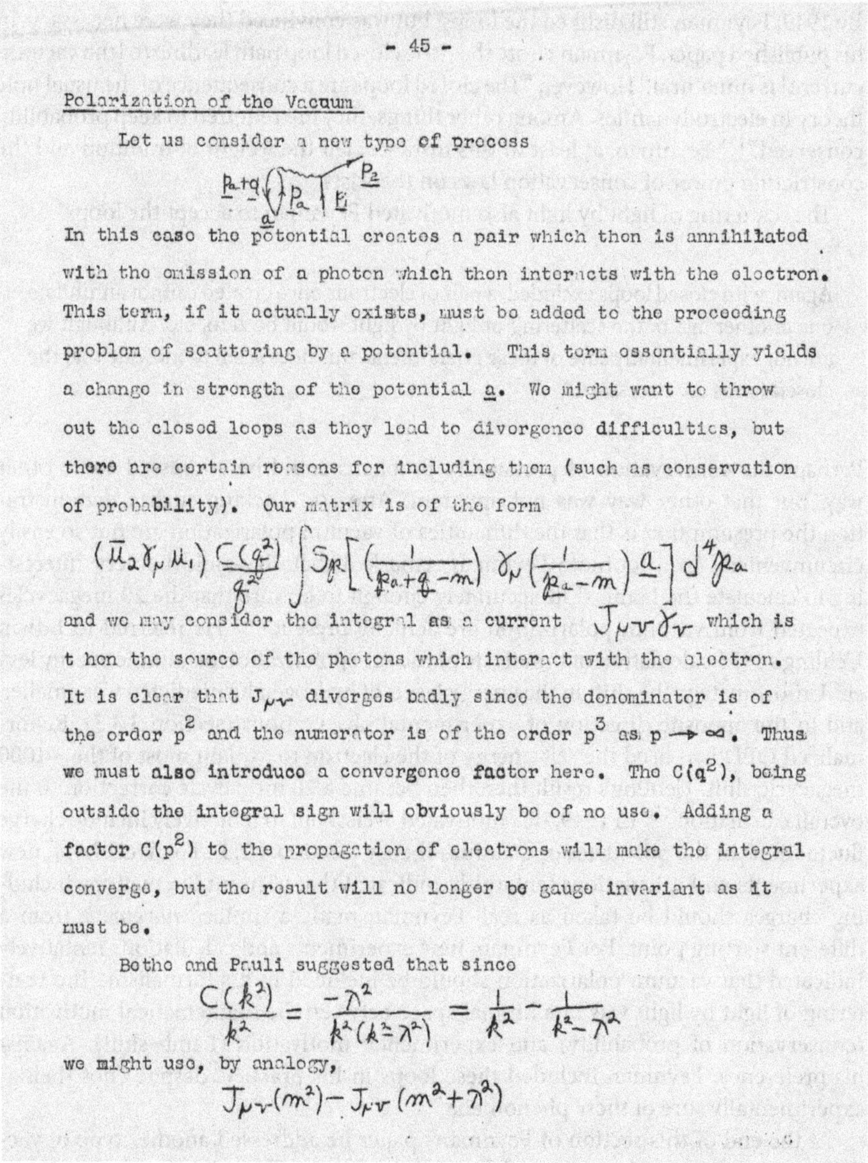

- 45 -

Polarization of the Vacuum

Let us consider a new type of process

$$\frac{p_1+q}{a}\Big/\Big\langle\frac{}{p_a}\Big|\frac{}{E}\Big\rangle^{p_2}$$

In this case the potential creates a pair which then is annihilated
with the omission of a photon which then interacts with the electron.
This term, if it actually exists, must be added to the proceeding
problem of scattering by a potential.   This term essentially yields
a change in strength of the potential a.   We might want to throw
out the closed loops as they lead to divergence difficulties, but
there are certain reasons for including them (such as conservation
of probability).   Our matrix is of the form

$$(\bar{u}_2\gamma_\mu u_1)\,\frac{C(q^2)}{q^2}\int S_{\mu}\left[\left(\frac{1}{p_a+q-m}\right)\gamma_\nu\left(\frac{1}{p_a-m}\right)\frac{a}{}\right]d^4p_a$$

and we may consider the integral as a current $J_{\mu\nu}\gamma_\nu$ which is
then the source of the photons which interact with the electron.
It is clear that $J_{\mu\nu}$ diverges badly since the denominator is of
the order p^2 and the numerator is of the order p^3 as $p \to \infty$.   Thus
we must also introduce a convergence factor here.   The $C(q^2)$, being
outside the integral sign will obviously be of no use.   Adding a
factor $C(p^2)$ to the propagation of electrons will make the integral
converge, but the result will no longer be gauge invariant as it
must be.

Bethe and Pauli suggested that since

$$\frac{C(k^2)}{k^2} = \frac{-\lambda^2}{k^2(k^2-\lambda^2)} = \frac{1}{k^2} - \frac{1}{k^2-\lambda^2}$$

we might use, by analogy,

$$J_{\mu\nu}(m^2) - J_{\mu\nu}(m^2+\lambda^2)$$

Figure 4.8. Vacuum polarization from Feynman's "A New Approach" lectures. Courtesy
of the Archives, California Institute of Technology.

neatly integrate vacuum polarization into QED, as Schwinger had. Historian Adrian
Wüthrich has shown that as early as 1947, Feynman tried to show that loops canceled
themselves out, and could be rejected.[152] Loops could be confusing to Feynman.[153]

152. Wüthrich, *Genesis of Feynman Diagrams*, §4.6.
153. D. Kaiser, *Drawing Theories Apart*, 194 fn43.

By 1949, Feynman still disliked the loops, but was convinced they were necessary. In his published paper, Feynman wrote that "the closed loop path leading to [the vacuum current] is unnatural." However, "The closed loops are a consequence of the usual hole theory in electrodynamics. Among other things, they are required to keep probability conserved."[154] Feynman, at least in this instance, felt the weight of tradition and the constricting power of conservation laws on theorists.

The scattering of light by light also motivated Feynman to accept the loops.

> Again, with closed loops excluded, a pair of electrons once created cannot annihilate one another again, the scattering of light by light would be zero, etc. Although we are not experimentally sure of these phenomena, this does seem to indicate that the closed loops are necessary.

Perhaps the conservation of probability problems could be addressed some other way. But that other way was not apparent. And so, "Lacking such a demonstration the presumption is that the difficulties of vacuum polarization are not so easily circumvented." In a footnote Feynman remarked that "It would be very interesting to calculate the Lamb shift accurately enough to be sure that the 20 megacycles expected from vacuum polarization are actually present."[155] He referred to Edwin Uehling's 1935 calculation of the effects of vacuum polarization on atomic energy levels. Unfortunately, the shift in the energy levels of hydrogen he predicted was smaller, and in the opposite direction of experimental observations (section 3.3.3). Renormalized QED had used the self-energy of the electron to explain most of the ~1000 megacycle shift. Uehling's result then then became a 20 megacycle correction to the overall calculation.[156] In 1949, this motivated Weisskopf to tentatively include charge fluctuations in the postulates of quantum theory (section 4.2.3). For Weisskopf, new experiments and calculations tentatively indicated that vacuum fluctuations including charges should be taken as real. Feynman made a similar *movement* from a different starting point. For Feynman, new experiments and calculations tentatively indicated that vacuum polarization should be included in his formalism. The scattering of light by light was in a liminal space between the mathematical motivation (conservation of probability) and experimental motivation (Lamb shift). Against his preference, Feynman included these loops in his practice, despite "not [being] experimentally sure of these phenomena."

At the end of this section of Feynman's paper he addressed another type of vacuum process. Using a logic of unobservability—familiar from Eddington and Dirac (section 2.3.3)—Feynman declared that these loops could be ignored.

> There are loops completely without external interactions. For example, a pair is created virtually along with a photon. Next they annihilate, absorbing this photon.

154. Feynman, "Space-time Approach to Quantum Electrodynamics," 779.
155. Feynman, "Space-time Approach to Quantum Electrodynamics," 779, 780.
156. Schweber, *QED*, 232–37, 243–45.

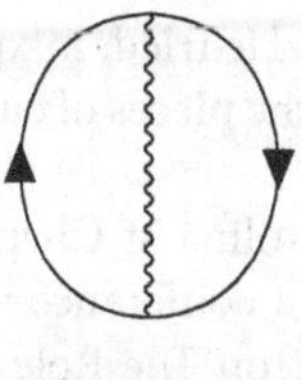

Figure 4.9. A bubble Feynman diagram. Credit: author

Such loops are disregarded on the grounds that they do not interact with anything and are thereby completely unobservable.[157]

A diagram picturing the process Feynman described in his footnote is shown in figure 4.9. These processes were unobservable in a different way than the "renormalized" mass or charge of the electron. For Feynman, they were unobservable not because of formal manipulations, but because they never connected with a source or measuring instrument. Two notions of observability were present in Feynman's thinking: a formal, mathematical understanding of unobservability used in renormalization, and a more physical understanding of observability, as defined by measurement interactions. Dyson agreed that diagrams such as figure 4.9 should be "eliminated" because they arose from "totally irrelevant fluctuations of fields in the vacuum."[158]

4.4.2 Cutting Rings into Trees

In the late 1950s and 1960s, Feynman turned to the problems of constructing a quantum theory of gravity, in which the vacuum and loop divergences were of primary importance. Little published work came of Feynman's engagement with gravity, though records survive in conference proceedings, lecture notes, and private calculations.[159] Feynman found that quantizing gravity was more of a challenge than quantizing other fields such as electromagnetism and the electron-positron field. The particles of the gravitational field, gravitons, are bosons, but act differently than the particles of the electromagnetic field (photons). The main problem is that gravitons couple to themselves. Photons do not. Photons couple with (interact with) charged particles, but they do not have charge themselves. Gravitons, on the other hand, interact with all particles, including themselves. Moreover, the mathematical framework of the theory must be subject to stricter laws of invariance than special relativistic theories. The Lorentz invariance that was so successfully integrated into QED would not be enough: invariance under general coordinate transformations would be required. In quantum gravity, Feynman tried to apply his familiar methods: a diagrammatic perturbation analysis. But he found more loops, and new kinds of loops, littering his

157. Feynman, "Space-time Approach to Quantum Electrodynamics," 780; cf. Wüthrich, *Genesis of Feynman Diagrams*, §5.5.2.
158. Dyson, "Radiation Theories," 496.
159. Feynman, "Role of Gravitation in Physics"; Feynman, "Quantum Theory of Gravitation."

pages of unpublished calculations. He tried, mixing mathematical metaphors, to cut his loops into "trees" and to order the pieces of cut loops in "baskets." Feynman could not square his circles.

At the University of North Carolina at Chapel Hill in 1957, Bryce DeWitt and Cécile DeWitt-Morette organized a conference under the aegis of the US Navy and a newly founded physics institute, on The Role of Gravitation in Physics.[160] There were forty-five participants from nine countries.[161] Wheeler and his students were a force at this conference (see section 5.4). In the introduction to the first section of the conference dedicated to quantized general relativity, Peter "BERGMANN first asked the question,

> "Why quantize?" His reply was that physical nature is an organic whole, and that various parts of physical theory must not be expected to endure in "peaceful coexistence." An attempt should be made to force separate branches of theory together to see if they can be made to merge, and if they cannot be united, to try to understand why they clash.

It was thought that quantizing gravity would "have a profound effect on our notions of space and time."[162] Though by the early 1960s, Feynman would explore the opposite of this view, and it dominated the conference.

Throughout the conference Feynman participated by interjecting and asking sharp questions of participants, which were recorded. And he gave remarks of his own, proposing thought experiments, for example a gravitational double-slit experiment and a Stern-Gerlach experiment with gravitational apparatus. He argued for either the necessity of quantizing gravity, or the invention of a physical principle: that macroscopic objects aren't quantum. He thought that either way new physics would result; a quantum theory of gravity, or a new physical principle. At the closing session he produced "Critical comments" on the state of the field. He expressed reasons for studying the quantum theory of gravity, but noted that the difficulty was the lack of experiments. He was disdainful of overly mathematical approaches to the theory. His view was different from that of André Mercier or Roger Penrose (chapter 6). Feynman counseled: "Don't be so rigorous or you will not succeed."[163] The conference was left in an uneasy state. Bergmann's summary,[164] concluded with quantization: "We have been given several word pictures and we must build a framework to hang the pictures on."[165]

The next year, the motivating question of the Chapel Hill conference was still on Feynman's mind. In his research papers, under the heading "Problems" on October 15, 1958, he wrote:

> What is the connection of gravity to the rest of physics? From one view gravity determines the geometry of space & time in which other phenomenon lie. Yet

160. DeWitt-Morette, *Role of Gravitation in Physics*; Kennefick, *Speed of Thought*, 117–18, 231, chapter 12.
161. USA, Japan, France, UK, West Germany, Israel, Sweden, Turkey, and Denmark.
162. DeWitt-Morette, *Role of Gravitation in Physics*, 74.
163. DeWitt-Morette, *Role of Gravitation in Physics*, 150.
164. Bergmann, "Summary of the Chapel Hill Conference."
165. DeWitt-Morette, *Role of Gravitation in Physics*, 155.

the "other phenomenon" are not so distinct from gravity that the latter might not simply be "just another particle," this time of spin 2, mass 0 to be added to the 30 already known. So gravity would be part of the larger problem below—what particles exist?[166]

Like his advisor Wheeler, he addressed his sheets of paper as if they were an interlocutor, asking questions, proffering partial answers which led to more questions (chapter 5). (Unlike Wheeler, once he was outside the military strictures of classified research Feynman stopped using large hardbound notebooks). This note is emblematic of Feynman's approach to quantizing relativity. As he would express in correspondence with Weisskopf in 1961 (see below), he treated gravity as just another field, to be treated with the same perturbative framework as all other particles.[167]

At La Jolla, California, in 1961, Feynman gave a talk on the quantum theory of gravity. This was on his mind enough that he gave this talk at a conference on the Theory of Weak and Strong Interactions.[168] Gravity was on Feynman's mind, and he thought there was a formal connection to weak and strong interaction theories.[169] He developed a perturbation analysis in which the spacetime metric $g_{\mu\nu}$ was approximated linearly:

$$g_{\mu\nu} = \delta_{\mu\nu} + h_{\mu\nu}, \tag{4.12}$$

where h was taken to be a small change from the δ function.[170] Feynman continued his talk by contrasting three approaches to quantization. First, searching for general, nonperturbative, quantum operators, was "beset with difficulties—maybe too general." The second was "Diagrams & they [their] calc[ulation] to various orders." This was "what [Feynman] chose—concentrated on problem of guage [sic] invariance & renormalization." The third way was something he called "Real to imaginary formulation," on which his graduate student Harold Yura was working.[171]

Feynman began his analysis by writing down an approximate Lagrangian and displaying a gravitational version of a paradigmatic calculation, the Compton effect. He then wrote down the equations of motion for the problem, and asked, "How to solve? Not original" and proceeded to set up a solution using complex analysis. Then he began to draw his diagrams. He started with a theory with no closed loops at all. Feynman could then calculate the scattering amplitude of the Compton effect in quantum gravity by considering diagrams such as figure 4.10. "With no closed loops all is well." He went on to construct gauge-invariant expressions for the propagator and drew

166. Feynman papers, Box 66, Folder 3.

167. D. Kaiser, "A ψ is just a ψ," §3.

168. This was the conference at which G. F. Chew gave a famous talk on the S-matrix. Cushing, *Theory Construction and Selection*, §6.4; D. Kaiser, *Drawing Theories Apart*, 306–14.

169. He made a formal analogy to Chen-Ning Yang and Robert Mills's field theory for the weak interaction. Pickering, *Constructing Quarks*, §6.1.

170. Feynman papers, Box 52, Folder 20: Renormalization of the Gravitational Field International Conference on the Theory of Weak and Strong Interactions, La Jolla, 1.

171. Feynman papers, Box 52, Folder 20: Renormalization of the Gravitational Field International Conference on the Theory of Weak and Strong Interactions, La Jolla, 1.

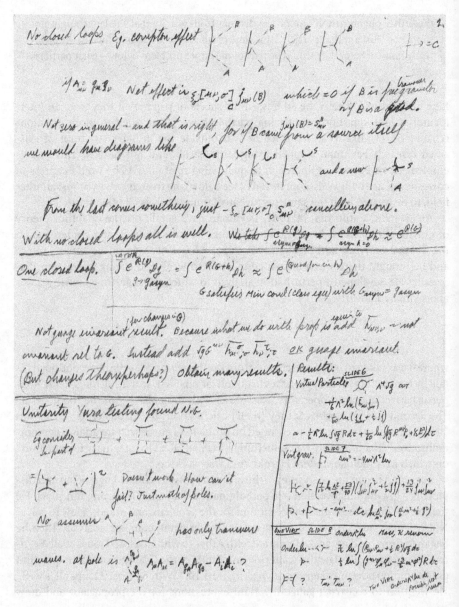

Figure 4.10. Feynman's 1961 notes for his presentation to the La Jolla conference. At the top of the page: drawings of "No closed loops Eg. Compton effect." At the middle of the page "One closed loop," at right, "SLIDE 6 Results: Virtual Particle" and "SLIDE 7 Virt[ual] grav[iton]". Bottom left: Harold Yura's results on "Unitarity." The solid lines represent particles such as electrons, the dashed lines represent gravitons. Courtesy of the Archives, California Institute of Technology.

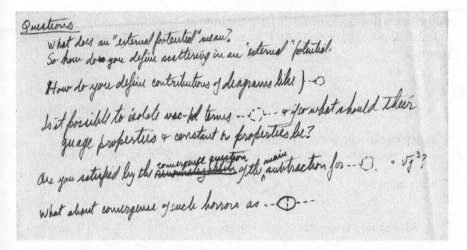

Figure 4.11. Feynman's undated Questions on quantum gravity. Courtesy of the Archives, California Institute of Technology.

some diagrams, noting that Yura was working on the unitarity (conservation of probability) of the theory.[172] In the simplest cases, for the lower orders of perturbation theory, everything fit into place.

What about loops, such as that depicted on his notes for a slide, figure 4.10? Feynman calculated that the virtual-particle loop depicted in figure 4.10, "slide 6," would depend on the relativistic cutoff momentum as Λ^4, far far more divergent than the logarithmic divergences encountered in QED. Feynman presented his results for the divergence of different processes in quantum gravity. He saw three "ways out" of the difficulties he encountered.

a) Find a new prop[agator]; i.e. method for integrating closed loops
b) Try admitting an extra scalar field with virtual waves, just to cancel these. Or believe consequences so scalar waves are real.
c) Modern approach. (ie [Yura])[173]

(Feynman's suggestion that he admit an extra scalar field is a root of the modern theory of "ghosts" in particle theory. They are not interpreted realistically.[174]) Closed loops, the vacuum polarization of the graviton field, were the sticking point in Feynman's theory.

In more than 100 pages of notes, in different series, some of which are dated to the winter and fall of 1961, Feynman searched for a way to remove loop diagrams from his quantum gravity. He tried to cut open loops, removing vacuum polarization from

172. Feynman papers, Box 52, Folder 20: Renormalization of the Gravitational Field International Conference on the Theory of Weak and Strong Interactions, La Jolla, 2.
173. Feynman papers, Box 52, Folder 20: Renormalization of the Gravitational Field International Conference on the Theory of Weak and Strong Interactions, La Jolla, 2.
174. Weinberg, *QTF* vol. 2, Section 15.6; Cao, *Conceptual Developments*, §10.3.

the quantum theory of gravity.[175] He asked himself "Questions" such as "What does an 'external potential' mean?" And "So how do you define scattering in an 'external' potential."

Is it possible to isolate vac-pol terms $----\big(\ \ \big)----$ & if so what should their guage [sic] properties & constant h properties be?

Are you satisfied by the ~~renormalizability~~ convergence question of the naive subtraction for $-----\big(\ \ \big) = \sqrt{g}^3$?

What about convergence of such horrors as $-----\big(\ \ \big)----$ [?]

See figure 4.11.[176] Feynman proceeded in dialogue with himself. Asking questions with a mix of words, formalism, and diagrams, he interrogated himself alongside his subject.

In a three-page "Summary of situation (maybe)" of this work, dated January 15, 1961, Feynman expressed that he wanted "what we can write as $\int e^{\int R(g)d\tau} K(g)\mathscr{D}g$ where $K(g)$ is a form from matter." That is to say, he was looking to fit gravity in the formalism of his path integral approach to the quantum.[177] "The central problem is: Precisely how do you do the g integral in view of the fact that $R(g)$ is singular? The answer is still not known to me."[178] Here Feynman expressed his difficulty in mathematical terms: how to integrate a given function. $R(g)$ was the Ricci curvature tensor, which Feynman wrote as a function of the metric, g. He continued to study the case in which there is no matter in the world. In this case he could integrate his function using integration by parts (though he wrote that he was "a bit confused here—but correct in idea").[179]

The next step was to consider graviton loops in this theory. Again he turned to his integral "$\int e^a \mathscr{D}h$. a is a quad[ratic] form in h, but it is singular. How to integrate it? One way is to add something to it to make it non singular. How can this be done without changing the result in an arbitrary fashion? I don't know." He played with different forms of $h_{\mu\nu}$ to add to a that would make it non-singular. (Recall that $h_{\mu\nu}$ was the small deviation from a flat metric, equation (4.12).) His second plan offered an encouraging result. By adding a term, b, to a above, he found a gauge-invariant and finite result.[180] "We must make sure it is legitimate."[181]

175. Compare Takeda Gyō's QED "opened loops," D. Kaiser, *Drawing Theories Apart*, 139–41.
176. Feynman papers, Box 66, Folder 3.
177. Feynman, "Space-time Approach to Non-relativistic Quantum Mechanics."
178. Feynman papers, Box 66, Folder 2: Gravity, p. "TS1."
179. Feynman papers, Box 66, Folder 2: Gravity, p. "TS1."
180. The formula for b was $\sqrt{G}h^\nu_{\mu;\nu}h^{\mu\sigma}_{;\sigma}$.
181. Feynman papers, Box 66, Folder 2: Gravity, p. "TS2."

Checking the legitimacy of his new result meant considering a situation that included matter. To first order, everything could be worked out in terms of tree diagrams (no loops). He plugged his linear approximation, equation (4.12), into expressions for the curvature and propagator.

Next if we expand $R(G + h)$ and $K(G + h)$ up to 2nd order each we have

; but we have left out one which is of the same order, namely

G G

G

$R(g + h)$ to <u>third</u> order and $K(G + h)$ to first: $\times$-- $\bigcirc$ [...] This is very

annoying—and needs study.

Feynman was discovering the necessity of analyzing loops in the theory.

G G

In short we must return to $+$ because a new

problem is here, namely a closed loop connecting to matter. This is where we <u>begin</u> and <u>end</u>.

Closed loops—vacuum polarization—would interact with matter fields, creating untamable divergences.[182] (The vacuum effects of similarly tadpole-shaped diagrams are discussed in chapter 7.)

In a letter to Weisskopf at MIT drafted January and February 1961, Feynman justified his approximate approach to quantizing GR:

> As you know, I am studying the problem of quantization of Einstein's General Relativity. I am still working out the details of handling divergent integrals which arise in problems in which some virtual momentum must be integrated over. But for cases of radiation, without radiation corrections, there is no difficulty (i.e., it is about where electrodynamics was in 1946).
>
> I won't be concerned with fields of arbitrary strength, but will be "practical" and expand in powers of $GE^2/\hbar c$, where E is the energy of a <u>particle</u> (not an extended object) and G is the gravity constant. Since $GE^2/\hbar c \sim 10^{-38}$ for a proton the most "practical" thing to do is to go to zero order and forget the whole problem all

182. Feynman papers, Box 66, Folder 2: Gravity, p. "TS3."

together. I am surprised that I find people objecting to an expansion of this kind, because it would be hard to find a situation where numerically a perturbation series is more justified.[183]

Here Feynman joked to his friend that the most practical thing to do would be to forget the gravitational problem altogether. More seriously, he found that objections to his perturbation approach were unfounded. The smaller the parameter in a perturbation series, the more mathematically accurate were the results. QED had wide success with a coupling constant of roughly $1/137$ or $\sim 10^{-2}$. Compared to this, who could object to $\sim 10^{-38}$?

In the spring of 1962, Feynman began preparations to give a lecture at a conference on General Relativity that was organized by Einstein's collaborator Leopold Infeld in Poland. (The same conference at which Penrose introduced his conformal methods; see section 6.4.1). In a letter from Feynman to Infeld sent May 8, 1962, he announced that he would

talk at the Relativity Conference on "The Quantum Theory of the Gravitational Field." I shall try to send you a short summary in a few days—but it will, I hope, not be too complete or accurate, as I expect to do more research before my talk [in July]. Thank you for reserving 45 minutes, as my subject is quite involved with problems requiring discussion and I could not do as well with a shorter time.

In fact Feynman would write to Infeld and Andrzej Trautman as late as July 10, 1962—fifteen days before the conference was to begin—with his abstract. He was "still doing the research and was not sure of what [he] could say until [he] worked it out."[184]

Feynman's presentation was published from a tape recording in *Acta Physica Polonica* in 1963.[185] He maintained his philosophy of not approaching the implications of quantizing geometry, but rather working from a perturbation expansion (equation (4.12)). Even this he found ridiculous. As he had noted in his letter to Weisskopf, he disparaged the entire project, with humor:

My subject is the quantum theory of gravitation. My interest in it is primarily in the relation of one part of nature to another. There's a certain irrationality to any work in gravitation, so it's hard to explain why you do any of it; for example, as far as quantum effects are concerned let us consider the effect of the gravitational attraction between an electron and a proton in a hydrogen atom; it changes the energy a little bit. Changing the energy of a quantum system means that the phase of the wave function is slowly shifted relative to what it would have been were no perturbation present. The effect of gravitation on the hydrogen atom is to shift the phase by 43 seconds of phase in every hundred times the lifetime of the universe! [...] This irrationality is shown also in the strange gadgets of Prof. Weber, in the absurd creations of Prof. Wheeler [...]. It is therefore clear that the problem we are

183. Feynman papers, Box 66, Folder 6: Gravity Waves.
184. Feynman papers, Box 5 folder 3: Warsaw conference.
185. Feynman, "Quantum Theory of Gravitation."

working on is not the correct problem; the correct problem is what determines the size of gravitation? But since I am among equally irrational men I won't be criticized I hope for the fact that there is no possible, practical reason for making these calculations.[186]

Here we can see how Feynman kept up some of the spirit of the Chapel Hill conference five years prior. Like Peter Bergmann, he expressed a concern for the relation of one part of nature to the other. The unity of nature motivated him to explore how gravity must fit into the framework of the quantum. However, he would stick with his perturbative approach, not delving into the changing notions of space and time that Bergmann envisioned.

As he had at Chapel Hill, Feynman bemoaned the lack of experiments. And again he imagined new ones:

I am investigating this subject despite the real difficulty that there are no experiments. Therefore there is so [sic, no] real challenge to compute true, physical situations. And so I made believe that there were experiments; I imagined that there were a lot of experiments and that the gravitational constant was more like the electrical constant and that they were coming up with data on the various gravitating atoms, and so forth; and that it was a challenge to calculate whether the theory agreed with the data.[187]

Feynman sought to recreate the situation in which he had succeeded in QED. He would pursue his approximate methods in an imagined world of developing experimental evidence. He imagined a gravitational equivalent of Willis Lamb's measurement of the shift in spectral lines of hydrogen.

Feynman was at first optimistic:

Now, I thought that this would be very easy and I'd just go ahead and do it, and here's what I planned. I started with the Lagrangian of Einstein for the interacting field of gravity and I had to make some definition for the matter since I'm dealing with real bodies.

As he had at La Jolla, he wrote the Einstein Lagrangian in terms a small perturbation $h_{\mu\nu}$. He considered gravity interacting with a scalar field. Then Feynman looked at what he would have done if gravity were no different from any other field. He derived equations of motion for the curvature, $h_{\mu\nu}$, in terms of its interaction, $\bar{S}_{\mu\nu}$, with the scalar field, ϕ:[188]

$$h_{\mu\nu,\sigma\sigma} - \bar{h}_{\sigma\nu,\sigma\mu} - \bar{h}_{\sigma\mu,\sigma\nu} = \bar{S}_{\mu\nu}(h,\phi). \qquad (4.13)$$

186. Feynman, "Quantum Theory of Gravitation," 697.
187. Feynman, "Quantum Theory of Gravitation," 697–98.
188. Feynman's bar notation is read as $\bar{x}_{\mu\nu} = \frac{1}{2}(x_{\mu\nu} + x_{\nu\mu} - \delta_{\mu\nu}x_{\sigma\sigma})$. The dynamics of the scalar field was given by the Klein-Gordon equation, $\phi_{\sigma\sigma} - m^2\phi = \chi(\phi,h)$.

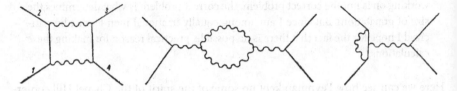

Figure 4.12. Feynman's (708, 1963) loop diagrams for gravitons. Courtesy of the Polish Academy of Sciences.

Feynman proceeded conservatively, in strict analogy to QED:

> The problem is to solve those equations in succession, and to use the usual methods of calculation of the quantum theory. Inasmuch as I wanted to get into the minimum of difficulties, I just took a guess that I use the same plan as I do in electricity; and the plan in electricity leads to the following suggestion here: that if you have a source, you divide by the operator of the left side of [equation (4.13)] in momentum space to get the propagator field.[189]

But there was a problem his audience knew well: equation (4.13) was singular. Feynman worked around this difficulty, exploiting the coordinate invariance of general relativity to come up with a description of two bodies interacting by the exchange of a virtual graviton. This was fine, but did not include higher-order radiative corrections. Feynman had succeeded in replicating Einsteinian gravity to first order, but that was not good enough.

A propagator alone was not enough technical machinery to support Feynman's diagrammatic techniques. He needed to understand what happened when lines met at vertices. This, it turned out, was straightforward. Drawing out a diagram for two scalar particles interacting with a graviton, he proceeded in his usual way to write out the vertex rules for calculation. He then took up a calculational example: the gravitational analogue of the Compton effect. "However the next step is to take situations in which we have what we call closed loops, or rings, or circuits, in which not all momenta of the problem are defined."[190] These vacuum polarization and self-energy processes required a new approach (figure 4.12).

What was the relationship between these problematic loop, or ring, or circuit diagrams and the straightforward no-loop or "tree" diagrams? "I proved that if you have a diagram with rings in it there are enough theorems altogether, so that you can express any diagram with circuits completely in terms of diagrams with trees and with all momenta for tree diagrams in physically attainable regions and on the mass shell."[191] Feynman could break open rings, such that it seemed that virtual-particle-loops consisted of real, free particles (figure 4.13). But there was another subtlety. In order to maintain gauge invariance, only the complete set of tree diagrams that accorded

189. Feynman, "Quantum Theory of Gravitation," 698–99.
190. Feynman, "Quantum Theory of Gravitation," 703.
191. Feynman, "Quantum Theory of Gravitation," 705.

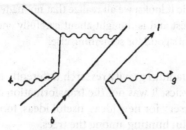

Figure 4.13. Feynman's (709, 1963) diagram of a cut-open loop. Courtesy of the Polish Academy of Sciences.

to a given loop diagram was gauge invariant. For each diagram all the trees must be summed together to maintain gauge invariance. Feynman called these baskets or barrels.

> The question is: Will any odd tree diagrams be left out or can they all be gathered into processes? The question is: Can we express the closed ring diagrams for some process into a sum over various other processes of tree diagrams for these processes?
>
> Well, in the case with one ring only, I am sure it can be done, I proved it can be done and I have done it and it's all fine. And therefore the problem with one ring is fundamentally solved; because we say, you express it in terms of open parts, you find the processes that they correspond to, compute each process and add them together.[192]

So far, so many successes. "Now, the next question is, what happens when there are two or more loops? Since I only got this completely straightened out a week before I came here, I haven't had time to investigate the case of 2 or more loops to my own satisfaction." Here trouble brewed. "The preliminary investigations that I have made do not indicate that it's going to be possible [to] so easily gather the things into the right barrels. It's surprising, I can't understand it; when you gather the trees into processes, there seems to be some loose trees, extra trees."[193] Feynman didn't understand what was going wrong. His perturbation series directed him to draw diagrams that did not fit his expectations. In QED, his diagrammatic method was famed for its ability to organize complex calculations. In quantum gravity, this ability was failing.

However, he was not entirely dissatisfied with his work. After all, the gravitational corrections to scattering processes are absurdly small. And he had found interesting theorems that would be relevant in Weak and Strong Interaction theories. But it was precisely this broader view that confounded him. He could not let these calculations alone:

> And so I'm stuck to have to continue this investigation, and of course you all appreciate that this is the secret reason for doing any work, no matter how absurd and

192. Feynman, "Quantum Theory of Gravitation," 708–9.
193. Feynman, "Quantum Theory of Gravitation," 710.

irrational and academic it looks; we all realize that no matter how small a thing is, if it has physical interest and is thought about carefully enough, you're bound to think of something that's good for something else.[194]

Feynman justified his quantum gravity research differently than most of his peers. It was not the unity of physics. It was not the transformation of space and time. It was instead justified as a nursery for new ideas, useful ideas, tools to be used. Feynman's view of the forest kept him hunting among the trees.

After his presentation, Bryce DeWitt pressed Feynman on points of style and substance.[195] DeWitt (1923–2004) studied under Schwinger and received his doctorate from Harvard in 1950. His focus was quantum gravity (often in collaboration with his wife, the mathematician Cécile DeWitt-Morette).[196] By the time of the Warsaw conference, DeWitt was publishing a series of papers on a general theory of quantization of fields with infinite-dimensional invariance groups (the gravitational field being the chief example).[197] These papers were characterized by their abstract and mathematical approach. DeWitt was searching for the type of general solutions Feynman thought were *too* general.

At Warsaw, DeWitt pushed Feynman for more mathematical details and rigor in his approach. This exchange reflects how Feynman's conservative approach was a radical one within the society of relativists. Feynman's attempt to simply extend— non-rigorously—his approach to QED for gravity cut against the grain of existing work in quantum gravity. DeWitt began his questions:

I should like to ask Prof. Feynman the following questions. First, to give us a careful statement of the tree theorem; and then outline, if he can to a brief extent, the nature of the proof of the theorem for the one-loop case, which I understand does work. And then, to also show in a little bit more detail the structure and nature of the fictitious particle needed if you want to renormalize everything directly with the loops. And if you like, do it for the Yang-Mills [i.e., Weak Interaction], if things are prettier that way.

Feynman demurred:

I usually don't find that to go into the mathematical details of proofs in a large company is a very effective way to do anything; so, although that's the question that you asked me—I'd be glad to do it—I could instead of that give a more physical explanation of why there is such a theorem; how I thought of the theorem in the first place, and things of this nature; although I do have a proof—I'm not trying to cover up.

194. Feynman, "Quantum Theory of Gravitation," 710.

195. Other questioners included Christian Møller, Nathan Rosen, Dmitry Ivanenko, James Anderson, Joseph Weber, and Rainer Sachs.

196. DeWitt, "Quantum Perturbation Theory"; DeWitt and DeWitt, "The Quantum Theory of Interacting Gravitational and Spinor Fields"; DeWitt, "Dynamical Theory in Curved Spaces. I."; Blum and Hartz, "1957 Quantum Gravity Meeting."

197. Morette-DeWitt, *Pursuit of Quantum Gravity*, §III.2.3.

But DeWitt pressed: "May we have a statement of the theorem first?"

> Feynman: That I do not have. I only have it for one loop, and for one loop the careful statement of the theorem is …—look, let me do it my way. First—let me tell you how I thought of this crazy thing. I was invited to Brussels to give a talk on electrodynamics.[198]

Whence followed a long, almost three-page, description of how Feynman came to his idea. In a metonymy of his sum-over-histories approach to quantum mechanics, rather than a rigorous presentation of his theory, Feynman provided his own research history.

Feynman ended his story with his current predicament—a theorem to cut open rings that would work for only one loop. He was cautiously optimistic. "But I've only had a week, gentlemen; I've only been able to straighten the difficulty of a single ring a week ago when I got everything cleaned up." He felt a little besieged, being asked only about the things he hadn't yet been able to tackle properly. "I'm sorry […] I hoped that I would be able to get it. I still have a few irons to try; I'm not completely stuck—maybe."[199]

Despite this deluge of detail, DeWitt pressed on, asking for more technical information.

> Because of the interest of the tricky extra particle that you mentioned at the end […] have you got far enough on that so that you could repeat it with just a little more detail? The structure of it and what sort of equation it satisfies, and what is its propagator.
>
> Feynman: Give me ten minutes. And let me show how the analysis of these tree diagrams, loop diagrams and all this other stuff is done [in a] mathematical way. Now I will show that I too can write equations that nobody can understand.[200]

He then proceeded to produce five pages of details for DeWitt, unhappily.

Feynman's struggle with the vacuum showed both a conservatism and a radicalism. Historian Sam Schweber has argued for Feynman's essential conservatism in his approach to the quantum.[201] What he meant was this: rather than looking for a new scientific revolution, like that of the development of the quantum theory in the first place, Feynman proceeded incrementally, within the framework of the existing theory, not looking for revolutionary new ideas. Despite the novelties Pauli and Dyson saw in renormalized QED, there is an undeniable germ of truth in Schweber's conservative picture. However, the label "conservative" does not sit well over Feynman's work, or his exuberant personality. Consider that his doctoral thesis, published after the war, was a reformulation of quantum mechanics itself. To Heisenberg's matrix

---

198. Feynman, "Quantum Theory of Gravitation," 714.
199. Feynman, "Quantum Theory of Gravitation," 717.
200. Feynman, "Quantum Theory of Gravitation," 717.
201. Schweber, QED, xii, xxvi, 92, 202, 595.

174 MORE THAN NOTHING

mechanics, Schrödinger's wave mechanics, and Dirac's general transformation theory, we may add Feynman's path integral approach to the quantum.[202] Peter Galison has described this work as "radical in its way."[203] It is true that Feynman did not take Dirac's route back to classical physics to search for an end to the infinities of QED. But neither did Feynman stick closely to the received formulation of quantum mechanics (as Tomonaga and Schwinger did). In order to characterize his work, "conservative" needs to be understood in more detail. Feynman was clearly not a *social* conservative, in the sense of someone who adheres to convention and propriety. (As is well known, he self-fashioned his iconoclastic persona.) This applied in aspects of his personal life, and in his willingness to transgress the conventions of relativity. In the material discussed in this chapter, Feynman showed a *methodological* conservatism. He preferred to keep his time-tested methods. But on the other hand, his path integral formalism was an enormous methodological innovation. In terms of ontological commitment, his perturbative approach to gravitation—had it succeeded—would have been a major revision of Einsteinian curved spacetime. Perhaps, then, it is best to see Feynman's work as a radicalism within a generalized conservatism. Radical conservatism may be a paradox. This paradox was closely associated with Feynman's doctoral advisor, and the subject of chapter 5, John Archibald Wheeler.

4.5 Conclusion

This chapter further developed the history of the active picture of the vacuum in quantum theory. Chapters 2 and 3 emphasized quantum theories of electrons and positrons that usually represented electromagnetism as a classical phenomenon (despite aspirational "light quantum" or "photon" language). Chapter 4 followed these developments as they were integrated with the quantum theory of electromagnetism. While Dirac established the first theory of quantum matter interacting with quantum electromagnetism in 1927, Tomonaga, Schwinger, and Feynman were the first to establish this fuller quantum picture in a relativistic formalism (see table 4.1). I emphasize the role of classical perturbations for two reasons: it is an underappreciated aspect of these theories, and because of the important role of perturbation theory in this history. As for the vacuum phenomena discussed in chapter 3, perturbation methods were the theoretical tool by which abstract formalism was related to experimental phenomena—and, I argue, it led physicists to propose the existence of never-before-seen phenomena, such as the scattering of light by light. The centrality of perturbation theory was a major continuity between pre- and post-World War II relativistic theory.[204] Weisskopf applied the same form of second-order perturbation theory in his work on spontaneous emission, line breadths, and the self-energy of the electron. The new relativistic renormalization techniques allowed new, more precise, calculations of the effects of vacuum phenomena that were aligned with new,

202. Feynman, *Feynman's Thesis*.
203. Galison, "Feynman's War," 395.
204. Schweber, *QED*, §9.17; D. Kaiser, *Drawing Theories Apart*, 195–220; Wüthrich, *Genesis of Feynman Diagrams*, §6.2.2; for light-by-light scattering, see Jost and Luttinger, "Vacuumpolarisation und e^4-ladungsrenormalisation für Elektronen"; Karplus and Neuman, "Non-linear Interactions."

Table 4.1. Quantum and classical parts of dynamical systems

	Dynamical System	
	Quantum parts	Classical perturbation
Dirac's radiation th.:	assembly of electrons, assembly of light quanta	electromagnetism
"hole" theory:	distribution of electrons	electromagnetism
pair theory:	electron-positron field	electromagnetism
Schwinger's QED:	electron-positron field, electromagnetic field	electromagnetism
Schwinger's TQF:	electron-positron field, electromagnetic field	electromagnetism, or electron-positron field

more precise, experiments. Vacuum phenomena were conceptualized in the multiple vocabularies of relativistic quantum theory: radiation transitions and scattering; semi-classical and more fully quantum; "holes" and matter waves; vacuum currents and virtual particles. These pairings—and interactions between them—coexisted from the 1930s through the 1950s.

In a series of influential papers within the "hole" theory framework, Weisskopf argued for

the fact that the vacuum has physical properties, which enables it to absorb light and to produce electrons. Hence, the physical description of the vacuum is bound to be more complicated than hitherto and must contain the latent electron pairs which can be created.[205]

This quotation, from 1949, would not be out of place in any other presentation of the "hole" theory. But this same paper also claimed that there were "zero-point fluctuations" of the electromagnetic and electric charge and current (section 4.2.3). This was particular to Weisskopf. As this chapter has shown, Weisskopf expressed his conviction in an underlying disorder or randomness from the beginning of his scientific career. While Weisskopf identified zeroth-order perturbation terms as representing vacuum fluctuations, he did not undertake what was usually meant by a fluctuation analysis (or a measurement analysis). Schwinger incorporated Weisskopf's perturbation-fluctuations into renormalized QED.

This chapter used the opinions of Pauli, Dyson, and Oppenheimer to historicize the question of whether quantum electrodynamics was conservative, or whether it developed fundamentally new ideas. There is no question that renormalized QED introduced new formal techniques. This chapter examined Schwinger's calculation of vacuum polarization. The calculation exemplified his use of vacuum concepts, and clarified his view of fluctuations. Schwinger ascribed the origin of quantum fluctuations to the interaction of a field with a classical source. Formally, this was similar to Weisskopf, but Schwinger did not think fluctuations were *inherent* in fields, or

205. Weisskopf, "Theory of the Electron," 310.

deserving of an independent postulate in quantum theory. The calculation also ex-emplified the use of vacuum expectation values. Schwinger provided a method for building up states with particles from the vacuum and them breaking down to the vacuum. Crucially, Schwinger's method translated the states *in time*. Vacuum expec-tation values represented a transition from vacuum state to vacuum state and from infinite past to present (or, with Dyson, the infinite future). This was so useful because the vacuum state was the simplest possible state. This simplicity would be spoiled if fluctuations were a property of isolated fields in themselves. The formal utility of Schwinger's methods and his underlying physical picture were mutually reinforcing.

The most famous of the new formal tools were, of course, Feynman diagrams. This chapter examined the "loop" and "bubble" diagrams for vacuum polarization. When Feynman turned to the quantum theory of gravity in the late 1950s and early 1960s, these loops caused insuperable problems. In one sense, Feynman's approach to quan-tum gravity was conservative: he took gravity as just another force to quantize, like the electric Coulomb force. And he applied the quantization methods he developed in the 1940s to gravity. But in another sense, Feynman's approach to quantum gravity was iconoclastic. Few relativists shared his outlook, and Feynman resisted his peers' demands for mathematical rigor or respect for the foundations of general relativity (such as general covariance). If Sam Schweber is right that renormalized QED was a conservative victory, this conservatism had its limits.

5

John Archibald Wheeler: Everything from Nothing

5.1 Introduction

"6 May 1952 5.55 pm. Learned from Shenstone 1/2 hour ago the great news that I can teach relativity next year. I wish to give the best possible course. To make the most of the opportunity, would be good to plan for a book on the subject."[1] So began John Archibald Wheeler's first volume in his long series of "Relativity Notebooks": with the "great news" from Princeton University's Physics Department chair Allen Shenstone that Wheeler would be given permission to teach a graduate seminar in relativity the next year. At this point in his career Wheeler (1911–2008) was an established physicist. He was born in Jacksonville, Florida, to Joseph and Mabel Wheeler, both public librarians. John began his university studies at Johns Hopkins University in Baltimore, Maryland, at age sixteen. He graduated with a combined BA and PhD in physics in 1933, at age twenty-one. He took up formative postdoctoral fellowships with Gregory Breit at New York University and at Niels Bohr's Institute in Copenhagen. He collaborated with Breit, and at Copenhagen, with J. Robert Oppenheimer's former postdoc Milton Plesset, on the scattering of light quanta (section 3.3). At Copenhagen, Wheeler formed a lasting relationship with Bohr. In the late 1930s, they developed George Gamow's "liquid drop" model of the nucleus and in 1939 they used the model to predict that uranium-235 and plutonium would be fissile. Also in the late 1930s, at the intersection of nuclear models and scattering theory, Wheeler developed what he called the "resonating group method." The idea was to use group theory to classify all the possible intermediate states a nucleus could inhabit in a scattering process. The work used high-powered mathematical techniques (drawing on mathematicians for assistance) but was at same time "phenomenologically" oriented toward nuclear experiments. It was the founding of scattering-matrix (S-matrix) theory.[2] His first long-term academic position was at the University of North Carolina, Chapel Hill, but he soon moved to Princeton University, where he remained throughout the period discussed in this chapter. Though not an early joiner, Wheeler eventually played an important role in the American war effort. He worked at the University of Chicago Metallurgical Laboratory, at the E. I. duPont deNemours company in Delaware, and at the Hanford Engineering Works in Washington State. Government service played a strong role in his postwar life, and he continued to work

---

1. JAW Series V, Relativity Notebook I, p. 1. See also D. Kaiser, "A ψ is just a ψ," 322.
2. Breit and Wheeler, "Collision of Two Light Quanta"; Plesset and Wheeler, "Ineslastic Scattering"; Wheeler, "Mathematical Description of Light Nuclei"; Wheeler, "Molecular Viewpoints"; Stuewer, "Liquid-drop Model"; Cushing, *Theory Construction and Selection*; Wheeler, "Men and Moments."

More than Nothing. Aaron Sidney Wright, Oxford University Press. © Oxford University Press 2024.
DOI: 10.1093/oso/9780190062804.003.0005

on military projects, particularly on hydrogen bomb designs. When Wheeler received word about teaching relativity, he was intensely engaged with hydrogen bomb design in the run-up to the "Mike" shot that proved the feasibility of igniting fusion reactions on Earth on November 1, 1952. He established a hydrogen bomb design laboratory at Princeton, called Project Matterhorn.[3]

Wheeler's attention to relativity would dominate the second phase of his academic career and lead him to make daring ontological proposals that put gravitation at the heart of contemporary physics. This chapter will explore a central facet of Wheeler's work on relativity, his engagement with the vacuum.[4] Reflecting on this facet of his work illuminates my primary object of investigation in tracing the histories of the vacuum in modern physics: the relationship between physicists' epistemologies and their ontologies. In later chapters the primary epistemic strategies physicists employed will be formal and mathematical. But for Wheeler, I argue that his primary impulse was philosophical—a philosophical searching for the simplest and most basic structures that explain empirical phenomena. As Gerald Holton has argued, scientists are often moved by more-or-less acknowledged *themata*, presuppositions that cannot be justified by induction from data nor by deduction from formal principles. "Simplicity" is a prime example of a scientific thema.[5] In chapter 2, I discussed Paul Dirac's conviction that physicists should search for beautiful mathematics with which to build new theories. Dirac's aesthetics was not merely a standard by which theories could be judged, after they had been developed. Dirac's aesthetics formed a *methodological* dictum, that was meant to guide theorists' practices. Similarly, this chapter demonstrates that Wheeler's commitment to simplicity determined his epistemic practices—his way of knowing and building theories. These practices, in turn, set the path of his changing ontologies of the vacuum, and of everything else. Compared to the earlier chapters of this book, this narrative is less concerned with mathematical tools. In a broad sense, Wheeler's decision to work in relativity forced him to make use of differential geometry, but the practical procedures of calculation mattered less in this work than it did in Dirac's or Roger Penrose's (chapter 6). The most important reason for this was the difficulty of making any calculations in quantum gravity, at all. Over time, Wheeler constructed layer upon layer of abstractions in his attempts to tunnel around this impasse.

Wheeler's changing ontologies are sketched in figure 5.1. At the time he started teaching relativity, in 1952, Wheeler sought to reduce the diversity of "fundamental" particles to a base of four "zero-rest-mass" fields (sections 5.2 and 5.3). By the mid-1950s, a formal result found by Wheeler's graduate student Charles Misner led Wheeler to further reduce the number of fundamental entities in his ontology. Misner's work exploited the fact that in general relativity distributions of mass-energy change the shape of spacetime. He showed that it was possible invert the arrow of

3. Galison and Bernstein, "In any Light," esp. 321–22; for biographical details: Wheeler and Ford, *A Life in Physics*, esp. 20, 65, 103, 219; see also Christensen, "John Archibald Wheeler"; Stachel, "Andover Conference"; Galison, "Re-reading the Past"; Galison, "Structure of Crystal"; Misner, "John Archibald Wheeler."

4. Recently, some of this terrain has been mapped by others, including Rickles, "Geon Wheeler"; Brill and Blum, "Revival of General Relativity."

5. Holton, "Role of Themata."

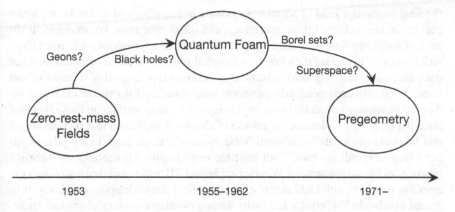

Figure 5.1. Wheeler's ontological progression.

inference: not only could he reason from distributions of mass-energy to curves of spacetimes, but he could reason from curves in spacetime to distributions of mass-energy. Simply by reading the ripples in spacetime—by examining the equations that describe curved empty space—Misner and Wheeler could deduce the form of, for example, the electromagnetic field. Rather than believe in the fundamentality of *multiple* zero-rest-mass fields, then, Wheeler could further simplify his world picture to take only curved empty space—the vacuum—as fundamental, as a "magic building material" that constituted the universe. He re-christened General Relativity as "geometrodynamics," the dynamics of the geometry of empty space, which was to be as real as the electromagnetic field of electrodynamics (section 5.4).[6] Not satisfied with progress in geometrodynamics, Wheeler increased his level of abstraction—his object shifted from from spacetime to a space of possible spacetimes—and searched for a simpler, more basic "pregeometry" (sections 5.5 and 5.6). By 1971 he had come to the idea that the simplest structure that could possibly build a geometry was the yes-no structure of the calculus of logical propositions. Much as he felt that a theory that started with particles could never explain the existence of particles, he felt that a theory that started with geometry could not explain geometry. This work had roots in his explorations of the foundations of quantum theory, and the work of his PhD student Hugh Everett. As he began a second phase of his career, after retiring from Princeton at age sixty and taking a position at the University of Texas, Austin, Wheeler moved from the foundations of relativity to the foundations of quantum mechanics.[7] He was never to realize his dream of unifying geometry and the quantum.

This chapter is based on a study of Wheeler's series of research notebooks. These are large, hardbound in black, with letter-size sewn pages that are ruled and numbered. Relativity Notebook I is a Shaw's brand account book, number 149, with 152 pages. Wheeler left space at the beginning to construct a table of contents and used

6. See also Rickles, "Geon Wheeler."
7. Wheeler and Ford, *A Life in Physics*, 318.

the final pages as a kind of scrapbook, pasting in the titles and abstracts of physics papers, cartoons clipped from magazines, and other ephemera. For example, at the back of Relativity Notebook III, on page 295, he pasted a typewritten line from Sherlock Holmes: "I have no data yet. It is a capital mistake to theorize before one has data. Insensibly one begins to twist facts to suit theories, instead of theories to suit facts." While rigorous notebook protocols were standard for experimental and industrial physicists, not all theorists kept notebooks. Dirac worked on loose sheets of plain paper. Sidney Coleman, the subject of chapter 7, explained to a correspondent that he could not locate his original "false vacuum" calculations "in the piles of paper I laughably call my files."[8] But wartime work imposed notebooks on theorists' practice. In his autobiography, Wheeler explained: "[I]nspired by Enrico Fermi's example as well as by rules laid down at Los Alamos, I started keeping my thoughts in bound notebooks."[9] (Fermi was a rarity among twentieth-century physicists: an accomplished theorist *and* experimentalist.) Such research notebooks offer historians the opportunity to follow science in the making; the steps, missteps, and reorientations that are so often excluded from the published record of scientific research.[10] Recently, historians such as Hans-Jörg Rheinberger and Christoph Hoffmann have advanced the study of research notebooks by asking a further question, beyond what such notebooks can tell historians in retrospect. They ask, What is the *epistemic value* of research note-taking practices? Compared to a pile of loose sheets, What does a notebook add to scientific knowledge making?[11] Wheeler's notebooks demonstrate that Rheinberger's influential description of the laboratory life sciences—in terms of shifting research questions and novelty-generating "epistemic things"—maps well onto theoretical practices and objects of theory.[12] Wheeler's notebooks ordered his research program and provided material continuity through his shifting questions and objects. They allowed him to respond to his own previous thoughts, either by glancing at a single page, or surveying months or years of research. They were often a sounding board and sometimes diaristic, but they were not strictly personal and private. They recorded the thought and words of others, and Wheeler sometimes invited his students to peruse their pages or check his calculations. They were invited into more than a world on paper, a universe, a place for thinking "great darings."

5.2 Philosophy of Approach to Elementary Particle Problem

Wheeler's career was motivated by philosophical concerns from the beginning. In the 1940s, he collaborated with his PhD student Richard Feynman on a critique of the concept of "field" in physics. They daringly reinvigorated the pre-Maxwellian action-at-a-distance approach to electromagnetism. Did quantum theory make it possible to eliminate fields from physics? Wheeler sent a draft of their coauthored paper to Feynman in 1942 and wrote to Albert Einstein about field-free electrodynamics in

8. Sidney Coleman to Kenneth Young, 13.9.1978, SRC.
9. Wheeler and Ford, *A Life in Physics*, 439.
10. Holmes, Renn, and Rheinberger, *Reworking the Bench*.
11. Rheinberger, "Scrips and Scribbles"; C. Hoffmann, "Processes on Paper."
12. Rheinberger, *History of Epistemic Things*; Rheinberger, "Cultures of Experimentation."

1943.[13] Immediately upon the end of the war, Wheeler returned to this topic. In 1950, with his student John Toll, he studied the scattering of light by light, centered on calculating the "*refractive index* of a polarized vacuum" (see section 3.4.2).[14] Wheeler described this work in a November 10, 1949, letter to Feynman:

> I am having interesting times trying to develop the theory of world lines, about which we once talked a little; a description of nature which makes no use of the concepts of space and time. But that is something else!

Feynman replied on December 8, "I would be very anxious to hear more details about your results." Somewhat later, on May 4, 1951, Feynman wrote to Wheeler at Los Alamos (five days before the "George" hydrogen bomb test). "I wanted to know what your opinion was about our old theory of action at a distance." New experiments at MIT had shown that "positronium"—like a hydrogen atom, but with a positron in place of a proton—existed, and that it rapidly decayed. This work was a brilliant confirmation of Feynman's QED, but was counter to Wheeler and Feynman's action-at-a-distance expectations. Feynman wrote: "So I think we guessed wrong in 1941. Do you agree?"[15] If Wheeler replied, it has been lost. Before Wheeler invested in Einstein's field theory of gravity, he made exhaustive attempts to eliminate fields altogether. But *experimental evidence* ruled out a particle-based, action-at-a-distance theory as a viable possibility.[16] He threw himself into fields, whole hog.

Writing at Los Alamos on July 18, 1953, Wheeler jotted notes on the "Philosophy of approach to elementary particle problem." He exhibited a "radical conservatism": a willingness to take existing concepts as far as possible before introducing new ideas.[17] He began by noting that "Humor = humbleness, willingness to find self wrong, to learn." Wheeler had a non-dogmatic attitude toward his work. He found joy in it. Wheeler continued to acknowledge that physicists hold *some* philosophical attitude at work in physics: "In addition to whatever written down in learn[ed] journal papers, there is a conscious or unconscious way of doing things—Einstein and Infeld."[18] He was likely referring to a passage from Albert Einstein and Leopold Infeld's popular book, *The Evolution of Physics*. It discussed a philosophy very much in keeping with Wheeler's views: "the belief that it *is* possible to describe all natural

13. Wheeler to Feynman 26.3.1942. Folder 3.10, Box 3, Feynman Papers. Wheeler to Einstein 3.11.1943, Einstein to Wheeler 6.11.1943, and Wheeler to Einstein 3.12.1943. Einstein Archive, docs. 23-442 to 445. My thanks to Diana Kormos-Buchwald for access to material from the Einstein Archives. Wheeler and Feynman, "Interaction with the Absorber"; Wheeler and Feynman, "Classical Electrodynamics"; Plass and Wheeler, "Blackbody Radiation"; Gleick, *Genius*, 92–120; Schweber, *QED*, 384–87.

14. Toll and Wheeler, "Pair-Theoretic Applications"; see also Rickles's discussion of the work with Toll and Wheeler to Bohr 3.9.1949 in Rickles, "Geon Wheeler," 247–50, figure 1.

15. Folder 3.10, Box 3, Feynman Papers. Deutsch, "Formation of Positronium"; Schweber, *QED*, xxviii, 461–62.

16. Wheeler expressed two more objections in a 1953 letter to Alfred Schild: electrons were a poor building block of a particle theory: (1) they had too few degrees of freedom to account for other particles; and (2) "dressed" electrons in renormalized QED were no longer simple. Rickles, "Geon Wheeler," 250.

17. On radical conservatism, see Thorne, "John Archibald Wheeler (1911–2008)"; Misner, "John Archibald Wheeler," 19–20; in 1962, Feynamn referred to "Wheeler's principle of 'radical conservatism'" in Institut Solvay, "Present Status of QED," 89.

18. JAW, Relativity Notebook II, pp. 35–37.

phenomena in terms of simple forces between unalterable objects. Throughout the two centuries following Galileo's time such an endeavor, conscious or unconscious, is apparent in nearly all scientific creation." Earlier, Hermann Helmholtz connected this to the assumption that nature was *comprehensible*. For Einstein and Infeld, even if physicists abandoned simple forces and unalterable objects, the assumption of comprehensibility remained.[19]

At this time, Wheeler was preparing to give a series of talks in Tokyo.[20] He set out a

> Proposed Tokyo program: Be as conservative as possible about introducing new elements into description. Make basics as clean & simple as possible. Is only the consequences that are complicated: ice; elem[ementary] particles; meteorology; geology.[21]

Here Wheeler was gesturing to the immense power of physics to describe complex phenomenon with only simple-seeming sets of equations and principles. Though the principles of thermodynamics and hydrodynamics may be written down compactly, they in a sense command the climate of the Earth. In 1951, Enrico Fermi had given a list of "fundamental particles" discovered to date. It had twenty-one entries, including the electron, positron, proton, antiproton, neutron, antineutron, photon, neutrino, graviton, and pions, muons, and V-particles.[22] This was a far cry from Paul Dirac's claim that his electron-proton theory made real "the dream of philosophers" by reducing all matter to a single substance.[23] What was responsible for the length of Fermi's list? Could there really be twenty-one "fundamental" kinds of being? This was Wheeler's "elementary particle problem."

In this drive to simplicity, Wheeler saw Einstein as an ally. Since 1933, Einstein also held to a principle that the laws of physics should be simple (perhaps in a logical or mathematical sense).[24] On May 13, 1953, from 4:10 to 5:10 pm, Wheeler brought his GR class to visit their Princeton neighbor. A passage from Wheeler's notes reads:

> On meson theory, if Lord created 5 kinds of mesons fields, it may be so, but in that case Ein[stein] not interested in the universe.[25]

This surely resonated with Wheeler.

In order to evoke the principles of his conservative attitude, Wheeler's Tokyo Program went on to imagine a band of physicists cast upon Wake Island, a US atoll that had been captured by the Japanese in 1941, and was perhaps the world's most isolated island.

19. Einstein and Infeld, *Evolution of Physics*, 57–58, 312–13.
20. Wheeler, "Origin of Cosmic Rays."
21. JAW, Relativity Notebook II, p. 36.
22. E. Fermi "Fundamental Particles"; Brown, Dresden, and Hoddeson, "Pions to Quarks."
23. Kragh, *Dirac*, chapter 5.
24. Dongen, *Einstein's Unification*, 48–49, 58; Norton, "Einstein and Mathematical Simplicity"; Renn and Sauer, "Pathways out of Classical Physics"; Ryckman, *Einstein*, chapter 10.
25. Relativity Notebook II, p. 11, Wheeler Papers.

Desert island philosophy: imagine selves cast up on wake island [*sic*] with library of all theory and exp[eriment] up to now, to solve elem[entary] particle problem.— What to use as starting points?—Others not ambitious enough? Go whole hog now!

Then followed a list of sixteen "elements" of the description of nature. For "(1) Models of a theory," Wheeler would look to "gravitation and electrodynamics." Referencing his work with Feynman on an action-at-a-distance formulation of electrodynamics, he asserted: "(2) Idea of field theory and action at a distance as equivalent." But more than a quick allusion to his earlier work, he continued to ask for a "Deeper basis for feeling that this is an important idea? Geometrization of nature—field implies a 4-dimensionality that may well be too restrictive. Geometrization needed as basis for making all sources [of fields] alike?"[26] Wheeler's "Philosophy of approach to elementary particle problem" note demonstrates his open, pedagogical attitude toward his work, and a radical conservatism that sought to limit the complexity of the basic elements of a picture of nature. At this stage in his research, the vacuum of GR was not an object of research in its own right; it entered as part of the "geometrization of nature" to explain the diversity of particle physics. The possible elements that might explain the existence of elementary particles included an imaginary library of all of physical theory and experiment, narrowed to a list of 16 with such broad topics as spin 1/2, statistics, and the correspondence principle. After his return from Tokyo in September this list would be cut down considerably.[27] Characteristically, he strove for *deeper* theories which described *simple* objects.

5.3 Zero-rest-mass Fields

Wheeler's initial gesture toward solving the elementary particle problem—how to explain the origin and form of elementary particles—already led him toward considering fields as the fundamental elements at play. By the summer of 1953 he was considering the "geometrization of nature" as a key for solving problems of particle physics.[28] Wheeler was a novice to the relativist literature.[29] In a letter to Behram Kursunoglu (copied to Einstein), Wheeler described himself as "only an innocent student of this subject trying to get some perspective on how things really stand."[30] Beginning in 1952, Wheeler visited colleagues at Princeton and the Institute for Advanced Study. He visited Einstein, Hermann Weyl, Wolfgang Pauli, Freeman Dyson, and others to learn the lay of the land, and try out new proposals.[31]

After his return from his Japanese visit he would elaborate on this gesture with a more concrete proposal for the fundamentality of only zero-rest-mass fields.

26. Paul Dirac (chapter 2) made an influential argument to explain why the charges of all electrons are alike in 1931.

27. JAW, Relativity Notebook II, pp. 35–37.

28. JAW, Relativity Notebook II, p. 35.

29. This had real consequences. He assigned a PhD problem to Charles Misner that had been solved twenty-five years earlier, see section 5.4.

30. Wheeler to Kursunoglu, 3.11.1953, Einstein Archive doc. 14-360.

31. JAW Papers, Relativity Notebook I, p. 111.

This work was never published, though it was circulated in pre-print form among colleagues. It forms an important prelude to Wheeler's later development of his "quantum foam" picture and helps show the development of his thought, especially his radical conservatism.

Wheeler's views of the vacuum came in three phases. The first, briefest, phase relied on the notion of a "zero-rest-mass" field. In modern notation, the Dirac equation (section 2.2.3) for a relativistic electron-positron field, ψ is written,[32]

$$i\hbar\,\gamma^\mu\partial_\mu\psi = mc\psi. \tag{5.1}$$

In mathematical terms, because of the mass term on the right-hand side, this is an *inhomogeneous* differential equation. It is vastly easier to manipulate, solve, and make use of *homogeneous* differential equations. Some of these equations appeared in physicists' theories naturally (as they saw it). For example, the wave equation for a relativistic electromagnetic field, A_μ, was written:[33]

$$\partial_\mu\partial^\mu A_\mu = 0. \tag{5.2}$$

Some of the most influential papers in twentieth-century physics used an approximation in which an electron's mass was taken to be zero—making equation (5.1) more like equation (5.2)—in order to employ powerful mathematics.[34] On November 3, 1953, Wheeler wrote that "electron, as example, seems to have practically no true mass. Mass seems to originate in large measure from interaction. The more we undress the electron, the less we see in it."[35] Other physicists worked with existing zero-rest-mass fields in hopes of building more realistic theories of massive particles, including Roger Penrose (chapter 6). As discussed in section 1.3, there is a sense in which a space that is filled with zero-rest-mass fields is still a vacuum. The region between the Sun and the Earth is the vacuum of space, even though light is constantly shining through it. Wheeler regularly referred to spaces filled with zero-rest-mass fields as "empty." In this period, physicists considered the following to be zero-rest-mass: electromagnetic fields (photons), gravitational field (gravitons), and the neutrino field (neutrinos).

Wheeler started his reflections on this matter in a notebook devoted to Quantum Electrodynamics on October 11, 1953, in Princeton. Under the heading "Attempts at unified theory," he wrote that his "Japanese visit emphasized difficulties of view that each particle has a field associated with it, and suggested the alternative view that only the fields of zero rest mass should be regarded as fundamental." As an example he proffered the electron, which was to be regarded as "no more fundamental than any other particle," given the picture of the electron as a zero-rest-mass "neutrino

---

32. In which i is the imaginary number; $\hbar$ is Planck's constant divided by 2π; the index μ stands for each of the four spacetime directions $\mu = t, x, y, z$; γ^μ are the four four-by-four Dirac matrices; ∂_μ is a partial derivative operator; m is the electron rest mass; and c the speed of light.

33. In Lorentz gauge.

34. E.g., Gell-Mann and Low, "Quantum Electrodynamics."

35. JAW, Relativity Notebook II, p. 72.

with a charge loaded on its back."[36] But from where would any mass come from if only massless fields were fundamental? Wheeler appealed to QED, noting that "The mass, and charge, due to interaction with the field is very great, as seen in quantum electrodynamics, in comparison with the initial or undressed mass and charge of the electron, if indeed these initial values are not zero." The point is that this greatness of the charge can be seen as arising not from a lone particle itself, but rather as arising from the interaction between particles. Referring to Victor Weisskopf's work on the self-energy of the electron in the 1930s (section 4.2) and earlier work by Hendrik Lorentz, Wheeler wrote that "Weisskopf treatment of 'self'-interaction energy suggests feasibility of reverting to program of Lorentz, à la mode moderne, & regarding all mass as due to field interaction." He followed this remark with a short order-of-magnitude computation of the ratio of the dressed and undressed masses of the electron (meaning the mass of the electron with surrounding virtual particles, or without).[37] Wheeler had already been thinking along these lines, including in his July list of sixteen elements "(8) Electromagnetic origin of mass and the self energy story."[38] As he had months earlier, Wheeler insisted that he "Be conservative," and here that meant "tak[ing] e[lectro]m[agnitism] and grav[itatio]n seriously down to very smallest distances."[39] This was indeed a *radical* conservatism, compared to J. Robert Oppenheimer's refusal to publish his and Wendell Furry's calculations in small-distance "pair" theory (section 3.3).

But at this stage for Wheeler "The biggest issue [was] the neutrino," or v. This was probably because it was both massless and had the electron's spin-1/2 character that Wheeler knew he needed to have any hope of making an "attempt at unified theory." Wheeler asked: "Is it <u>additional</u> to e[lectro]m[agnitism] & grav[ity]; or is it in some sense already <u>contained</u> in grav[itatio]n? The <u>conservative</u> approach is to assume that the v is already a part of grav[itatio]n, until this view is proven wrong."[40] How could the neutrino be somehow already included in gravitation? Earlier in the spring of 1953 in Washington, Wheeler had made notes on the "Nature of a Unified Theory" commenting on a paper by Einstein and Nathan Rosen from 1935 called "The Particle Problem in the General Theory of Relativity."[41] Einstein and Rosen proposed that particles could be included in relativity if they were seen not as mathematical point objects, but rather as the span of a "bridge" between two sheets of spacetime, the metric g. Wheeler would later adopt this idea into a picture of "wormholes" between these sheets (see section 5.4). Wheeler's comments included the note that "(2) Their bridge idea is most intriguing—two sheets of g [i.e., spacetime] meeting at each singularity, get neutrino?" More than this, Wheeler connected the Einstein-Rosen bridge to further relativistic and atomic-physics concepts: "(3) Note the duality all over the place—two sheets of g; two directions to describe world lines; two signs of

36. Until the late 1990s it was thought that neutrinos were massless. Physicists now believe they are massive particles.
37. JAW, Notebook of Quantum Electrodynamics, p. 19. See section 4.2.3 and Blum and Joas, "Dressed Electrons."
38. JAW, Relativity Notebook II, p. 36.
39. JAW, Notebook of Quantum Electrodynamics, p. 19.
40. JAW, Notebook of Quantum Electrodynamics, pp. 19–20.
41. Einstein and Rosen, "Particle Problem."

e [i.e., charge]; two light cones; spin 1/2 for neutrino and for electron. The opportunity is at hand to tie all these together in a way that ought, if found, to appear to be inevitable."[42] Wheeler pitched his spin-sheet picture to Einstein during his relativity class's visit in May 1953. But Einstein did not like the idea.[43] In his October notes, Wheeler's view of the Einstein-Rosen picture had changed: it was "too primitive"; with "no spinor character."[44]

Wheeler and Einstein shared a belief in the fundamental importance of simplicity; they differed in their engagement quantum theory and with experiment. Einstein acknowledged that an eventual unified theory would have to account for quantum phenomena, but he saw this as a secondary goal.[45] By ignoring all but two of Fermi's "fundamental" particles (keeping only the electron and proton), Einstein cut himself off from central advances in the 1930s, 1940s, and 1950s on nuclear decays, cosmic rays, and quantum electrodynamics.[46] Compounding this narrow subject matter, over the course of his career Einstein retreated from performing, inventing, and engaging with experiment. In a measured analysis, Jeroen van Dongen "observes a gradual decoupling of the pragmatic and empirical side of the field of physics from the core of [Einstein's] professional science: his efforts in theory construction."[47] Wheeler was certainly a theorist in his academic work. But at this stage of his career he embraced experiment and observation whenever he could, including his speculations on neutrinos and gravitation.

On April 21, 1954, Wheeler wrote to his Princeton colleague Howard Percy Robertson, an expert on GR and cosmology:

> Dear Bob:
> We were talking some weeks ago about the relation, if any, between neutrinos and the expanding universe.

In Wheeler's early relativistic picture, he estimated that the density of matter in the universe was about 1,000 times too small compared to astronomical data. Would he abandon his theory? No. "Instead of saying that this discrepancy is a difficulty, suppose we say it is a situation. Then we conclude that there is something else present [beyond dust and electromagnetic radiation] that gives a density of the required order."[48] Wheeler was attempting a transmogrification—as he might say—of conflict with empirical evidence into new theoretical possibilities. Neutrinos were the catalyst for this reaction.

> A uniform density of neutrinos is however something that one has to expect. The question indeed is not whether, but how much. Now we come to the heart of this note: Unpublished work of Reines and Cowan and collaborators at Los Alamos leaves open the possibility that all of the desired [matter] is in the form of neutrinos.

42. For light cones, see chapter 6. JAW, Relativity Notebook I, p. 151.
43. JAW, Relativity Notebook II, p. 11.
44. JAW, Notebook of Quantum Electrodynamics, p. 20.
45. Sauer, "Einstein's Program," 286–87.
46. On his isolation, see Dongen, Einstein's Unification, 186–89.
47. Dongen, Einstein's Unification, 80; cf. Galison, "Theoretical Predispositions."
48. Folder 6.18, Robertson Papers.

Fred Reines and Clyde Cowan Jr. were studying the immense flux of neutrinos produced by a weapons-plutonium-producing nuclear reactor at Hanford, Washington.[49]
In his letter to Robertson, Wheeler was less interested in Reines and Cowan's *results*
than in their measurement of the background signal in their detector.[50] In an effort to
reduce this background, Wheeler explained, "they took their detector underground
[in a] cave," but still measured some background of neutrinos, presumably coming
from outer space. Wheeler estimated the "corresponding density of neutrino energy
in space," and it filled the gap between his neutrino-less estimate and astronomical
data.

> In other words, there is no present means, either theoretical or experimental,
> to exclude the possibility that neutrinos constitute by far the largest part of the mass
> of the universe; and in fact, taking seriously the general principles [of GR] men-
> tioned earlier seems to indicate that there is no evident escape from the conclusion
> that the neutrinos are indeed this important.

> Regards,
> John A. Wheeler[51]

This attempt to reverse the fortunes of a theory at odds with experiment was characteristic of Wheeler's thought, but far from unique to him.[52] While Wheeler was not directly performing these experiments, he was integrating the latest (still-unpublished)
results from particle detectors to compare with astronomical data in his relativistic
research. All that was needed was a unified theory of gravity and neutrinos.

Instead of the Einstein-Rosen picture, Wheeler turned in his Notebook of Quantum Electrodynamics to an analogy from an equation that describes the curvature
of empty space, $R_{ik} = 0$, to Paul Dirac's relativistic theory of the electron. "We look
to see if there is a square root of the eq[uatio]n $R_{ik} = 0$ in some sense analogous to
the square root that Dirac took of the wave eq[uatio]n."[53] A square-root of sorts had
allowed Dirac to construct a linear theory of quantum mechanics that was compati-
ble with the theory of relativity (section 2.2.3). Wheeler sought to modify one of the
fundamental equations of general relativity, the formula for the line element, $(ds)^2$,
that represents how distances are to be calculated in a curved space. "To take square
root suggests that we have to take root of g's. One way to do this: replace

$$\sqrt{(ds)^2} = \sqrt{-g_{\mu\nu}dx^\mu dx^\nu} \text{ by}$$
$$ds = \gamma_\alpha dx^\alpha.$$

Here the γ's are to be regarded as f[unction]s of position, just as the g's were," where
Wheeler's notation recalls the γs in equation (5.1). Wheeler continued with his "hope

49. Wheeler cc'd the nuclear physicists. In 1956 they discovered anti-neutrinos, Reines and Cowan,
"Reines-Cowan Experiments"; Galison, *Image and Logic*, §6.3.
50. On experimental background, see Galison, *How Experiments End*, 2–3, 71, 88, §4.7.
51. Folder 6.18, Robertson Papers.
52. See Pickering, *Constructing Quarks*, 262, 387; Collins, "Lead into Gold."
53. A Dirac operator in quantum mechanics is an operator that is the square-root of the Laplacian
operator. Dirac's gamma matrices are in a sense square roots of 1, or the identity matrix.

that the geometry built up from these spinors [i.e., γs] contains: (1) general relativity (2) the ν (3) <u>room</u>, at least, for electromagnetism (4) quantum mechanics: by $\psi \sim \sum_{paths} e^{i\gamma_\alpha dx^\alpha}$?" In this last questioning note, he proposed using Feynman's sum-over-histories version of quantum mechanics. This was one guise of the role of Wheeler's "quantum principle," here not fundamental but something to be derived from the basics. We see here a grand vision for how simple reworkings of theory—taking a square root—might refigure all of physics. Wheeler pondered a possible path for calculation along these lines, establishing a geometry. "Let us look at two orders of combining dx^1 and dx^2 [figure 5.2]." He held himself back. "But don't plunge yet into these trivial math. points. General philosophy first. (1) γ's describe, we hope, ν-grav[itatio]n field. (2) Some c [sic]."[54] Here the October note ends. But Wheeler wrote enough to give a picture of his methodology: he was not interested in quick paths to calculation, but rather put "general philosophy first."

On November 4, 1953, Wheeler set out "points for [a] proposed article" that never reached publication, with the title "Elementary Particles from Massless Fields—An Assessment."[55] He again demonstrated his questioning after the basic constituents of things, his commitment to a basic philosophical attitude, and a moral sense of what avenues are correct to pursue in physics. Wheeler knew he was thinking against the grain of modern field theory, and his radical conservatism was a strategy for justifying this.[56] Wheeler began his note by asking, "Suppose the basic idea of meson field theory is wrong? What then do we have for a broad plan to attack the elementary particle problem?" Note that this discussion, though not explicitly about unified field theory, takes place in one of Wheeler's "Relativity" notebooks. Wheeler continued to emphasize that what he was after was a general philosophy of how to do this sort of physics—a general philosophy of how to approach the problem of the existence of so many differing types of particles. "By plan we mean, not theory, but the general idea of the scheme of things that precedes a theory." Wheeler first analyzed the current state of meson-field theory: "What is this general scheme in the case of the meson field idea? That the π meson is represented by a field of its own, over and above the electromagnetic field; that the electron is represented by such a field; and that the nucleons are represented by such fields." However, if each of the known particles requires its own field, and a set of "n fields, [requires] $\frac{n(n-1)}{2}$ coupling constants," physics would be far from ideally simple. Wheeler was worried about "Convergence," though it is not clear exactly what he meant. More definitely, he was concerned with the origin of the coupling constants, stating that they must have arrived from "a deeper theory—otherwise why a particular numerical value." These difficulties motivated Wheeler to set out some of his philosophy of searching for alternate basic structures to describe the world: "Alternative important to consider, if 50–50 chance idea of meson theory wrong, irresponsible not to see what else." This "what else" was the zero-rest-mass field hypothesis, though not expressed in so many words here. Wheeler used his notebook to work though his thoughts before preparing another document in

54. JAW, Notebook of Quantum Electrodynamics, pp. 19–20.
55. JAW, Relativity Notebook II, p. 72.
56. In his notes from an April 2, 1953, discussion with Hermann Weyl, Wheeler wrote: "Why don't I get into modern field theory? Because one can't see the forest for the trees." JAW, Relativity Notebook I, p. 111.

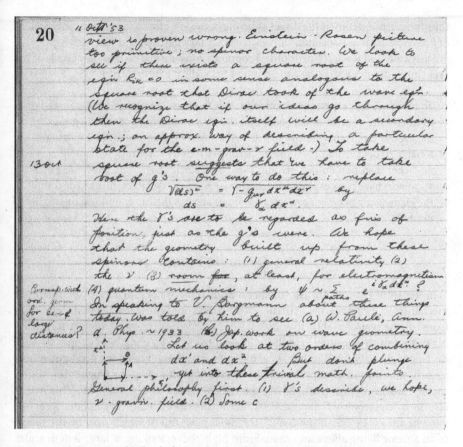

Figure 5.2. Page 20 of Wheeler's Notebook of Quantum Electrodynamics showing his "two orders of combining x^1 and x^2," at bottom left. Path A first travels dx^1 and then travels dx^2. Path B first travels dx^2 and then dx^1. Establishing whether A and B arrive at the same spot, in general, was not a given in spinor geometries. Above, Wheeler records speaking about this topic with Valentine Bargmann, who contributed to the theory of relativistic spinor fields. Courtesy of the American Philosophical Society.

his efforts to transform his unease at contemporary particle physics into a research program.

At the turn of the new year, January 19, 1954, Wheeler made ten copies of a working paper at Los Alamos to circulate among close colleagues. This preprint encapsulates many of the ideas detailed above, including the belief in the fundamentality of zero-rest-mass fields, and Wheeler's radical conservatism. It was entitled "The Zero Rest Mass Fundamental Field Hypothesis." According to the chronology established here, this is the final example of Wheeler's zero-rest-mass hypothesis. However, a glimpse of what was to come appears in this preprint. He began: "An attempt is made to assess the hypothesis that each elementary particle is a disturbance in a common fundamental field, or system of fields, of zero rest mass: the electromagnetic field, the gravitational field, and the neutrino field." Wheeler here was again swimming against the grain. "This picture is the direct opposite of the view that a new characteristic field of finite rest mass is associated with each elementary particle, and that with n particles are associated $n(n-1)/2$ primordial coupling constants." The first paragraph of his preprint ends with a striking expression of his radical conservatism.

> To renounce free experimentation of ideas is the ideal in mind. Instead, the goal is to extend the domain of application of the already existing and solidly grounded theories of electromagnetism, of gravitation, and of quantization. It appears reasonable to believe that from these ideas there follows a network of consequences far richer than past limited investigations have so far sought out. Until these consequences have been explored, it seems premature to invent new ideas to deal with the elementary particle problem, or even to assume that invention is going to be needed.[57]

5.4 Quantum Foam

Wheeler would eventually ask whether a unified theory of gravitation and quantum mechanics that explains the existence of particles could come out of a basic "quantum foam" at the seat of reality. I write "ask" because Wheeler would literally formulate his proposals as questions. He did not see himself as pushing bold proposals onto his fellow physicists, but rather as asking questions that revealed the presuppositions already extant in their experimentally verified theories. (Or, one might view it as a rhetorical strategy for downplaying the radical nature of his proposals. This would also fit a "conservative" approach.) Wheeler developed many metaphors to evoke what he meant by "quantum foam," the most powerful of which may be the image of an aviator flying above an ocean. From high above, looking at large length scales the aviator sees a smooth surface—analogous to the smooth manifold of spacetime observed when studying macroscopic distances. If the aviator swoops down closer to the surface, she will observe waves propagating, and interfering with one another. These waves represent gravitational waves and also the ripples in spacetime caused by the existence of distributions of matter and energy. Closer still, an aviator equipped with a telescope may zoom down to the millimeter scale and observe foam at the tip of breaking waves. At these shortest distance scales the regular surface of the ocean is disrupted. If the ocean surface can be thought of as a flat two-dimensional sheet, the foam contains structures of other dimensionalities: three-dimensional structures of bubbles. Still at the micro-scale, bubbles in the foam are constantly being created and destroyed, and connections between neighboring parts of the water are being opened and closed. This smallest scale represents Wheeler's picture of the quantum nature of space and time, frothing with quantum fluctuations as in Weisskopf's quantum vacuum.

Two developments drew Wheeler to this picture. First, his investigation of "geons" and second, a result found by his graduate student Charles Misner. Misner's result was the rediscovery that because distributions of mass energy create ripples in spacetime,

---

57. JAW, Relativity Notebook II, pasted into p. 100, page [1] of preprint.

one could "read off" distributions of mass energy from studying the gravitational field alone. A geon was an object in the world of classical general relativity—what Wheeler would call "geometrodynamics" to signify that relativity was the study of a dynamic geometry. Geons first appeared in Wheeler's notebook on January 25, 1954, while he was in Rochester, New York, for a conference on High Energy Nuclear Physics.[58] Geons were originally christened "Kugelblitz" (Ball-lightning), apparently at the suggestion of his student Arthur Wightman: "Ball of light held together by gravitational forces as classical model for an elementary particle = fireball = (Wightman name) Kugelblitz."[59] The idea was to investigate concentrations of electromagnetic radiation—light, a zero-rest-mass field—that were so dense they would hold together under their own gravity. He would come to call this "mass without mass" because from far away a geon looked like any other massive object, but it consists only of fields of zero rest mass. Misner's work pointed the way for Wheeler to reduce the number of fundamental zero-rest-mass fields to one, the gravitational field of empty spacetime.

An obvious model for a geon was a regular star—*massive* particles held together by their own gravity. It is also possible that he was inspired by the work of Project Matterhorn. This was the military research project he established at Princeton. It had two divisions. Wheeler's "Matterhorn B" (for bomb) designed thermonuclear weapons. Lyman Spitzer's "A" division worked on controlled fusion reactions. Spitzer's "stellarator" reactor design was a lemniscate of hot plasma. The two worlds were far from separate. Wheeler's doctoral students could work for Project Matterhorn.[60] Looking forward to April 27, 1968, Misner and Kip S. Thorne each worked through notes on problem #58 for Princeton University course 570—Relativity, "Gravitational Radiation from H-Bomb."[61]

Wheeler's geon work maintained his philosophy of radical conservatism. In an October 12, 1954, letter to John von Neumann, Wheeler wrote:

> Dear Johnny, [...] I am still trying to understand elementary particles in terms of existing concepts. The general philosophy is this, how does one know what to invent, or even that invention is necessary, until one has explored further the rich consequences of what one already has. For example, you will see from a quick glance at the enclosed MS [manuscript] that out of gravitation and electromagnetism (or neutrinos, or both) one can construct very interesting models for elementary particles ("geons").[62]

Geons were almost preposterous objects, enormous on a scale hard to imagine and eventually found to be unstable. However, they were allowed by the theory. This was

58. Noyes et al., *Fourth Annual Rochester Conference.*
59. JAW, Relativity Notebook II, p. 96.
60. Lyman Spitzer, *Propsed Stellarator*; Wheeler, *Project Matterhorn*; Bromberg, *Fusion*, 14–17; Ford, *Building the H Bomb*, chapter 11.
61. Likely they were considering this for inclusion in *Gravitation*. Charles W. Misner Papers, Special Collections, University of Maryland Libraries, Box 1, Folder: J A Wheeler, course handouts, Fall 1968, Princeton.
62. JAW, Relativity Notebook III, pasted onto p. 79.

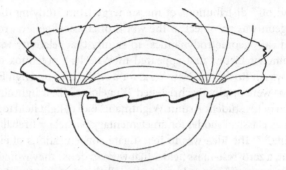

Fig. 7. Schematic representation of lines of force in a doubly-connected space. In the upper continuum the lines of force behave much as if the tunnel mouths were the seats of equal and opposite charges.

Figure 5.3. Wheeler's drawing of electromagnetic field lines trapped in a multiply connected space. Reprinted figure with permission from John Archibald Wheeler. "Geons." *Physical Review* 97, no. 2 (1955): 511–36. https://doi.org/10.1103/PhysRev.97.511, copyright (1955) by the American Physical Society.

consistent with the principle of plenitude expressed in the 1930s by Dirac and Werner Heisenberg (section 3.4.2).[63]

Wheeler's project with geons had an explicitly philosophical aim. His major paper on the subject (probably the manuscript referred to in his letter to von Neumann) was received by the *Physical Review* on September 8, 1954. In it he asserted that "geons furnish for the first time time a completely classical, divergence free, self-consistent picture of the Newtonian concept of body" over an enormous range of (large) masses.[64] He argued that in a purely classical picture, the discontinuity and quantum-nature of point charges should be ignored. How, then, to account for free electricity in the theory, not electricity bound into geons? Wheeler noted that he, and most physicists, had tacitly assumed that the spacetime manifold of relativity was simply connected. Simply connected manifolds are those without holes or loops. The surface of a sphere is a simply connected manifold; a punctured sphere is not. The possibility of non-simply connected spacetime became one of the foundations of Wheeler's view on the vacuum as quantum foam. Wheeler's letter to von Neumann continued:

> If this line of reasoning has any value, it suggests (last section of MS) that one might understand the existence of electric charge, not through the presence of real carriers, but through the existence of multiple connected topologies in the spacetime continuum, that permit the electric lines of force to continue without end.[65]

63. Cf. Wheeler and Ford, *A Life in Physics*, 232, 241.
64. Wheeler, "Geons," 535.
65. JAW, Relativity Notebook III, pasted onto p. 79.

At this classical stage in Wheeler's paper, the violent chaos of a foaming vacuum was not yet in place, and he considered only a spacetime "which on the whole is nearly flat except except in two widely separated regions, where a double-connectedness comes into evidence as symbolized in [figure 5.3]."[66] Here each end of the "tunnel" behaves as an isolated charge when viewed from far away, but yet no electromagnetic field lines ever actually end. They are constantly looping though the "tunnel mouths."

However, never one to shy away from seeking the fuller picture, Wheeler even in this paper about the perfection of the classical Newtonian concept of body introduced the quantum principle. "Let one pass from the classical theory [...] to the corresponding quantum theory, either by the prescription of Feynman,

$$\psi \sim \sum_{\text{all conceivable relevant field histories}} \exp[(i/\hbar)(\text{classical action for each field history})]$$

(5.3)

or by any other standard method." He asserted that once one has crossed into the quantum world "one is forced to recognize the existence of fluctuation in all fields." (Again, as for Weisskopf, section 4.2.[67]) While at large distances, fluctuations of the metric were considered irrelevant, the mathematics of point charges requires integrating over entire regions, "including regions of the very smallest spatial extension." Wheeler at this point was appealing to a formal requirement about integration to motivate following physics down to the smallest scales. (Later he would use black holes as a more physical motivation.)

> Here the inevitable fluctuations force on space time a most complicated structure. Because it is the essence of quantum mechanics that *all* field histories contribute to the probability amplitude, the sum [equation 5.3] not only may contain doubly and multiply connected metrics; it must do so. General relativity quantized, leaves no escape from topological complexities of which figure [(5.3)] is only an oversimplified symbol.

By May 6, 1955, Wheeler "S[aw] no reasonable alternative to idea that lines of force loop about in space time."[68] Wheeler was trying to extend the force of physicists' reasoning about all possible physical processes to all possible geometries.

In 1956, Wheeler took up a position as a visiting scholar at the Instituut Lorentz in Leiden, the Netherlands, for several months. He brought three others with him—graduate students Charles Misner, Peter Putnam, and Joe Weber.[69] The fourth in his series of Relativity Notebooks has a "reward for return" page that lists addresses in Princeton, the Instituut Lorentz, Los Alamos, Copenhagen (where he went to visit Bohr), and his cottage in South Bristol, Maine. On March 10, 1956,

---

66. Wheeler, "Geons," 535.

67. Later, Wheeler cited Welton's analysis of the Lamb shift, Dyson's presentation of Welton's results, and Hendrik Casimir. Wheeler, "Superspace," 255, 301 fn37. See appendix B.

68. JAW, Relativity Notebook III, p. 152.

69. For details of Weber's career see Collins, *Gravity's Shadow*, pt. 1; Kennefick, *Speed of Thought*, chapter 10.

he returned to the "Elementary particle problem," now equipped with his insight about the centrality of quantum fluctuations. "Following late afternoon's talk with Weber, thought more about problem." Wheeler began to play with the fundamental constants that characterize the quantum ($\hbar$), gravitation (G), and relativity (c). He constructed expressions for quantum fluctuations he likely learned while a postdoctoral fellow of Bohr (see section 4.2.2).[70] "Granted only GEO [i.e., massless] field, have $\left(\frac{\hbar G}{c^3}\right)^{1/2}$ = 1.6×10^{-33}cm and $\left(\frac{\hbar c}{G}\right)^{1/2}$ = 2.2×10^{-5}gm as natural units. Elementary fluctuations characterized by that mass and that energy and size." Even in the tiny scales of atomic physics, 1.6×10^{-33}cm was considered a fantastically small distance; however, 2.2×10^{-5}gm would be considered enormous. The mass of the proton was only 1.7×10^{-24}gm. If fluctuations in a vacuum with zero-rest-mass fields were characterized by energies of 2.2×10^{-5}gm, they would overwhelm conventional particles.

> Hence particle represents only a trivially small disturbance in the vacuum. [Consider cane seat, local concentration of caning pattern; it alone appears when viewed from large distance, but is quite negligible on small scale][*sic*]. Therefore, to understand particle have first to understand vacuum.[71]

The vacuum of GR was attractive because of its simplicity and its ability to give rise to other forces than gravitation. Quantum fluctuations provided another motivation for Wheeler to promote the vacuum from a tool for explaining particles to an object of theory in its own right. Quantum fluctuations also provided a new set of (productive) contradictions.

One obvious feature of the vacuum was that it has no mass. Wheeler wondered "How [to] have mass compensate to zero?" *Charge* fluctuations—as in Oppenheimer and Furry's pair theory—easily balanced-out to zero overall charge, because charge came in equal positive and negative quanta (positron and electron). There was no similar negative mass. That is, given the $\left(\frac{\hbar c}{G}\right)^{1/2}$ mass scale of fluctuations, was it possible to find *some* negative energy contribution that would compensate for the positive mass and balance this out to zero? (Earlier, Wheeler had considered the negative and positive energy balance in his letter to Robertson; section 5.3.) In his notebook, Wheeler considered the gravitational interaction of two fluctuating elementary particles. "At first sight, gravitation negligible. But let's see. What is grav[itatio]n energy of two masses $\left(\frac{\hbar c}{G}\right)^{1/2}$ at distance $\left(\frac{\hbar G}{c^3}\right)^{1/2}$?

$$\delta m = \frac{E}{c^2} = -\frac{Gm_1m_2}{c^2 r_{12}} = -\frac{G\left(\frac{\hbar c}{G}\right)}{c^2\left(\frac{\hbar G}{c^3}\right)^{1/2}} = -\left(\frac{\hbar c}{G}\right)^{1/2}.$$

So we <u>do</u> have all the qualitative requirements for compensation. Is compensation a theorem or a definition? Don't know. Is a particular value of the undressed charge

70. See, Wheeler, "Geometrodynamics and the Issue of the Final State," 507–9.
71. JAW, Relativity Notebook IV, p. 9.

needed for compensation? Will have to study later the worm hole characteristics." Here Wheeler combined Einstein's famous formula $E = mc^2$ with the simplest calculation of gravitational energy from Newtonian gravitational theory, $E = Gm_1m_2/r_{12}$, and found that the "mass," δm, of the energy of gravitational attraction compensated the mass of fluctuations perfectly. This was only a "rough and ready analysis [...]. To go further, need a way to do calculation of energy for vacuum state and really show energy is zero." What he was looking for was a quantum theory of gravity, along the lines he presented in his "Geons" paper where a sum-over-histories approach to the quantum would be applied to geometry. "Have to sum over all contortions of grav[itationa]l field & electromagnetic field but take only zero energy part." Here vacuum fluctuations were seen as more basic than the electromagnetic field itself. "Vacuum polarization a fundamental property of vacuum, not merely of electron field."[72] Three days later at the Instituut Lorentz while attempting to write up his ideas for a High Energy Nuclear Physics conference in Rochester, Wheeler asked, "How come mass of vacuum vanishes so exactly. Seems fantastic this should come out! Is it defined as the answer? Or calculable. Would be an enormous triumph to get a simple proof. THE VACUUM HAS NO MASS!"[73]

Wheeler continued his discussions with Weber on the role of the vacuum in explaining the elementary particles. In notes from March 22, 1956, Wheeler shared his conviction that a theory of particles must have as its root something other than particles themselves.

—But first let me state my point about electron problem—we tried our best to get along without facing the problem of electron structure. Of course, only way was to assume the electrons already there to begin with, but to "call them unobservable." Worked pretty well, but Lord caught up with us on vac[uum] polz[arizatio]n, etc. Only way really to avoid is to face electron structure problem, talk about something out of which electrons can be created without electrons actually being there in the first place. We thought we could get by without being detected in our cheating but the Lord has found us out.

Wheeler continued to cast about for metaphors to aid his thinking. Rather than appeal to the "cane seat" of a few days before, particles were now seen as fluctuations in a primordial porridge. "I say, is electron pair created out of porridge; percentage wise a fantastically small disturbance in the porridge:

$$\frac{\text{extra density of em energy in the electron}}{\text{density of e.m. energy in the vacuum}} \sim [...] \sim 10^{-88},"$$

where "em" and "e.m." stand for electromagnetic. But if the origin of particles was to be found in enormous statistical fluctuations of some basic "porridge," why was it that experimentally observed electrons are incredibly stable objects? That is, given the fact that electrons do not spontaneously decay into other particles, how was it possible

72. JAW, Relativity Notebook IV, pp. 9–10.
73. JAW, Relativity Notebook IV, p. 10.

to understand them as consisting of fluctuations? "Hard to understand problem of electron stability. Start, not with 1st approx[imation] (in development in powers of 10^{-88}!) but with 0th approx[imation]—the vacuum. WHY DOESN'T VACUUM HAVE WEIGHT?" In marginalia next to the word "VACUUM" Wheeler wrote "Great darings— / Archimedes weigh earth / Sound=mechanics / Heat=atomic motion."[74]

In the midst of his 1956 struggles with the mass of the vacuum, Wheeler's student and guest at Leiden, Charles Misner, was working on a project that was supposed to become his dissertation. Misner was gifted at mathematics and was searching for a unified picture of gravitation and electromagnetism. And he succeeded, in words borrowed from Hugh Everett, in creating an "already unified field theory" in which electromagnetism and gravitation were unified into one geometrical picture.[75] This was only possible within the classical geometrodynamic framework, GR with massless fields. However, this work did not turn out to be Misner's doctoral thesis. Presumably after the circulation of a letter or preprint of this material to Peter Bergmann and his relativity group at Syracuse University, Misner learned that his results had been found already in 1925 by George Y. Rainich. Wheeler's notebook retains a letter from Misner to Bergmann sent June 27, 1956: "Many thanks for calling my attention to the work of Rainich and Williams."[76] Misner's predicament was likely caused by his supervisor's relatively recent entry into relativity. Instead Misner's thesis work applied Feynman's prescription for field quantization to GR.[77]

Wheeler circulated an "Informal Mid-Summer Letter" from his summer house on High Island, Maine, in 1956. In it he expressed the central concerns of his paper with Misner:

> The center of interest [over last year] possessed a remarkably sharp focus, in the issue: Is curved space only an <u>arena</u> for physics, or is it an appropriately extended sense the building material for <u>all</u> of physics? Thanks especially to Charles Misner, it became clear during the year that all of classical physics can be described in terms of curved empty space and nothing more.[78]

Note the caution surrounding the radicalism. The center of interest was an "issue" framed as a question. Classical physics could be "described" in terms of curved empty space, but that was only a description. The progression of Wheeler's thought was not one of a man moving from certainty to certainty, but one continually searching for answers to deep questions. Wheeler continued to describe his and Misner's description of nature in terms of paradoxical formulations: "gravitation without gravitation," "in Einstein's standard way" of eliminating the force of gravity and replacing it with the curvature of space; "electromagnetism without electromagnetism," in the work of Rainich, "one can find all that needs to be known about the electromagnetic

---

74. JAW, Relativity Notebook IV, p. 23.
75. Misner and Wheeler, "Classical Physics as Geometry," 527 fn3.
76. JAW, Relativity notebook IV, p. 130. See also Misner to Wheeler, August 1956, reproduced in Rickles, "Geon Wheeler," figure 8; on Rainich, see Rainich, "Electrodynamics"; Lehmkuhl, "General Relativity as a Hybrid Theory."
77. Misner, "Feynman Quantization."
78. JAW, B:W546 Series V, Relativity Notebook V, pasted on p. 157, p. 1.

field" from purely geometrical quantities; "charge without charge (lines of force in a multiply connected space); mass without mass (geons)."[79] Misner and Wheeler published these leading questions in a joint publication in 1957. It became a central reference for others who published on geometrodynamics. Radical conservatism was on full display. "In describing classical physics [...] as geometry, we invent no new ideas." Rather they simply accepted "Maxwell's 1864 electrodynamics of empty space, his formulation of the stress-momentum-energy tensor of the electromagnetic field, and Einstein's forty-one-year old description of gravitation in terms of curved space."[80] Wheeler's fluctuation calculations resulted in a somewhat pessimistic companion article, which concluded that *"pure quantum geometrodynamics must be judged deficient as a basis for elementary particle physics."*[81]

It was in this context of ambiguity and ambivalence—forcefully promulgating a metaphysical proposal in a seventy-nine-page paper only to call it into question in an eleven-page postscript—that evidence of Wheeler's willingness to shake foundations arises. At the same time as he developed his picture of the vacuum as a sort of froth, he appears to renounce Bohr's correspondence principle:

> On an atomic scale the metric appears flat, as does the ocean to an aviator far above. The closer the approach, the greater the degree of irregularity. Finally, at distances of the order [1.6×10^{-33} cm], the fluctuations in the typical metric component, $g_{\mu\nu}$, become of the same order as the $g_{\mu\nu}$ themselves. Then the character of the space undergoes an essential change, as indicated schematically by figure [5.4]. Multiple connectedness develops, as it does on the surface of an ocean where waves are breaking.[82]

Wheeler developed his metaphor of the aviator flying over spacetime and finally reaching a scale where vacuum fluctuations reach the same magnitude as the magnitude of the background metric components. However, immediately following

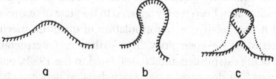

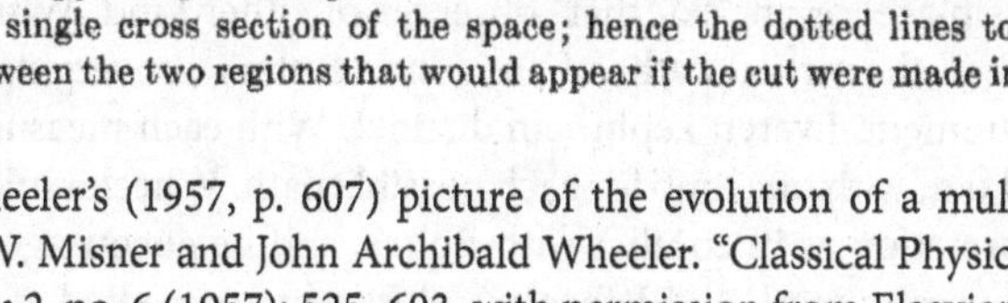

a   b   c

FIG. 1. Effect of a local fluctuation in the metric in introducing a multiple connectedness. In (c) the upper region of space is not disconnected from the lower one, as might appear from a single cross section of the space; hence the dotted lines to symbolize the connection between the two regions that would appear if the cut were made in another place.

Figure 5.4. Wheeler's (1957, p. 607) picture of the evolution of a multiply connected space. Charles W. Misner and John Archibald Wheeler. "Classical Physics as Geometry." *Annals of Physics* 2, no. 6 (1957): 525–603, with permission from Elsevier.

79. JAW, B:W546 Series V, Relativity Notebook V, pasted on p. 157, p. 1.
80. Misner and Wheeler, "Classical Physics as Geometry," 526.
81. Wheeler, "Quantum Geometrodynamics," 613.
82. Wheeler, "Quantum Geometrodynamics," 607.

this he asserted that "Of course it is not necessary to propose a method to observe these fluctuations in order to note how directly and inescapably they follow from the quantum theory of the metric. One does not have to use the word 'observation' at all." Rather, one can simply appeal to formal properties of the quantum-mechanical propagator functional. This assertion runs directly counter to the thread of fluctuation analyses identified with Heisenberg, Bohr, Rosenfeld, and Oppenheimer, but was in accord with Weisskopf (section 4.2). Going forward, Wheeler would make further breaks with Bohr, even if he did not recognize them as such.

On March 10, 1957, Misner sent a long letter to Wheeler, in anticipation of the latter's planed publications in the *Reviews of Modern Physics*. Misner set out his feelings on "the hopes and prospects for gravity in the future," with certain passages marked as available for quotation.

What are the prospects that "Physics as Geometry" will become a valid physical theory prepared to meet all experiments confidently, instead of just a playground universe where physics is simpler than we know it to be? Very good I think. You should, I imagine, want to go rather wide open in stating your wild ideas about wormholes, foaming space, spin without spin, etc. in order to establish your property rights along these roads. Not every one will be the right road; but they at least cross the right roads, which might never be found by a more cautious adventuring.[83]

The publication to which Misner referred was likely Wheeler's contributions to the conference on "The Role of Gravitation in Physics" at the University of North Carolina, Chapel Hill (at which Feynman spoke; see section 4.4.2).[84]

In struggling to integrate quantum theory and QR, Wheeler was influenced by another of his students. In 1956, Hugh Everett III was completing his doctorate under Wheeler at Princeton. He proposed an ambitious new interpretation of quantum theory that treated the measuring instruments of an experiment and the atomic system it measured on an equivalent basis. Rather than referring to "classical" apparatus and "quantum" phenomenon, Everett only referred to the state of one relative to the other. Wheeler called it the "relative state formulation of quantum mechanics." This was particularly interesting for cosmology. The "orthodox" interpretation of quantum mechanics was being established and consolidated in the 1950s, out of a cacophony of earlier positions.[85] It described a necessary division between classical and quantum descriptions. If the system in question was the entire universe, however, there was no obvious place for an "external" observer of either kind. While the orthodox view described all the virtual paths of an atomic system converging into one at the point of measurement, Everett kept them distinct. With each measurement, his universe branched out, with one limb for each possible path. Wheeler initially did not see Everett's interpretation as in conflict with Bohr's, and he encouraged them to come to an accord. Everett's work and Wheeler's gloss were presented at the Chapel Hill

83. JAW, Series I, Box: Mel-My, Folder: Misner, Charles W.
84. Wheeler, "Everett's 'Relative State' Formulation"; Rickles and DeWitt, *Role of Gravitation*.
85. Howard, "Who Invented the 'Copenhagen Interpretation'?"; Camilleri, "Myth of the Copenhagen Interpretation."

conference in 1957.[86] Subsequently, Everett left academia for weapons research, and his interpretation lay dormant until Bryce DeWitt revived it in the late 1960s.[87] It also recurred in Wheeler's thinking.

In July, 1959, Wheeler spoke at the "Enrico Fermi" International School of Physics at Varenna, Lake Como, Italy, on "Neutrinos, Gravitation and Geometry." In the 130-page published version of his lectures, he developed the centrality of the neutrino to elementary particle physics and pondered ways to incorporate it into his geometrodynamics. This tied together his earlier exchange with Robertson (section 5.3) and further developed his break with Bohr's correspondence principle.

> The neutrino is the only Fermi field which has no mass and no charge. It transmits energy from place to place with the speed of light. It is impossible to ever catch up with a neutrino and transform it to rest. In this sense the neutrino has the character of a field, rather than the character of a particle, like the electron. [...] It is the only field endowed with spin 1/2 and antisymmetric statistics which can be supposed to be as fundamental in character as the electromagnetic field and the gravitational field.[88]

But he admitted that the present state of geometrodynamics did not have a place for the neutrino. He repeated his questions from 1956: "Are fields and particles foreign objects that move about in the arena of space and time? Or are they in some way constructed of space? Is the metric continuum a magic medium [...]? In other words, is physics at bottom a matter of pure geometry? Is geometry only an *arena*, or is it everything?"[89] After a review of geometrodynamics, Wheeler promised another look at this problem.

Rhetorically, Wheeler's rather leading questions about the fundamentality and fucindity of his geometrodynamical outlook were motivated by its power to explain "gravitation without gravitation," "mass without mass," "electromagnetism without electromagnetism," "charge without charge," and yet other paradoxical formulations. These slogans referred to what might be called *reductionist* successes: the force of gravity could be described without a gravitational field (instead, by curved empty space); electromagnetism could be described without an electromagnetic field (instead, by curved empty space); etc. These successes depended on fiercely policing the boundary around what counted as classical—no point masses or charges. How would geometrodynamics build particles? Wheeler dismissed his earlier attempts. Geons were "a concrete quantum geometrodynamical model for an elementary particle," but their instability caused the project to collapse. Worse:

> Of course regions of spacetime in which the curvature is growing ever sharper can be imagined to break up into particles much as the crest of an ocean wave break

86. I.e. Everett and Wheeler's 10.3.1957 letter to colleagues, asking for "comments or criticisms," Everett Papers, Box 1, Folder 8, Everett, "'Relative State' Formulation"; Freire, *Quantum Dissidents*, §3.

87. DeWitt, "Everett-Wheeler Interpretation."

88. Wheeler, "Neutrinos," 1-2.

89. Wheeler, "Neutrinos," 8. In 1958, he repeated this at King's College London; JAW, B: W564 Series V, Relativity Notebook V, p. 177.

up into foam and droplets. But such a picture is far too fuzzy and speculative in character to supply anything with the sharpness of a model.

In order to make progress on the connection to elementary particle physics, Wheeler expounded the "*Necessity of seeking guidance from elementary particle physics and quantum electrodynamics*."[90] Even in 1962, Wheeler was distinguished among (quantum) gravity researchers in his engagement with the non-gravitational (majority of) physics research.

This threw Wheeler into a discussion of fluctuations in quantum electrodynamics and quantum geometrodynamics. He emphasized that "Quantum geometrodynamics has thus this much point of contact with [...] quantum electrodynamics, that all space in both accounts is the seat of violent fluctuations."[91] He embraced Dirac's "hole" picture of the vacuum.

> A world containing one electron [...] does not make sense. The electron would drop to a negative energy state in a fantastically short time, as Dirac long ago showed. The observed stability of electrons against this type of radiative catastrophe demands that all negative energy states be occupied. Out of this infinite but unobservable sea it is possible anywhere to raise up an electron into tangible being.

Wheeler argued that "familiar experiments on pair production" such as the demonstration of the Lamb shift in the fine structure of hydrogen "stand witness" to Dirac's picture.[92] For Wheeler, geometrodynamics and QED were tied together by the mutual violence of their vacua.

"But what of the other characteristic feature of electrodynamics—the 'undressed electron?'" Wheeler asked, setting up the frame for his most explicit identification of wormhole-mouths with particles:

> *If there is to be any correspondence between pure quantum geometrodynamics*—with no extra fields, no coupling constants, and no arbitrary inventive elements—*and quantum electrodynamics, then the "undressed electron" must be identified with the elementary wormhole.* [...] Only this identification can provide in straight quantized general relativity a natural picture for the virtual pairs throughout all space which are demanded in the most compelling way by quantum electrodynamics.[93]

Here again, Wheeler demonstrated his questioning pose ("If there is to be...") and the centrality of his conservatism ("no arbitrary inventive elements") to his thought. However, while a place for the undressed electron was necessary, it was not the fundamental starting point. Wheeler appealed to his earlier order-of-magnitude analysis of the relative density of nuclear matter, $\sim 10^{14} \text{g/cm}^3$, to the unheard-of density of vacuum fluctuations $\sim 10^{94} \text{g/cm}^3$.

90. Wheeler, "Neutrinos," 68–69.
91. Wheeler, "Neutrinos," 77–78.
92. Wheeler, "Neutrinos," 69.
93. Wheeler, "Neutrinos," 78.

In other words elementary particles do not form a really basic starting point for the description of nature. Instead they represent a first order correction to vacuum physics. That vacuum, that zero order state of affairs, with its enormous densities of virtual photons and virtual positive-negative pairs and virtual wormholes, has to be described properly before one has a fundamental starting point for a proper perturbation theoretic analysis.[94]

Even though Wheeler was searching for a simple, fundamental building block of nature, he was at the same time trying to establish a basis for applying perturbation theory (appendix A). As for Dirac's appreciation of beauty and generality (chapter 2), Wheeler did not strive for perfect mathematical expressions. He wanted a perfect ontology, not a perfect formalism.

In the next sections I will lay out how Wheeler's dissatisfaction with the prospects for quantum geometrodynamics led him to reevaluate the status of geometry—curved empty space—as fundamental. Already in 1959 he can be seen shuffling through his library of concepts and testing their centrality. Here the casualty was Bohr's correspondence principle. This is somewhat surprising because Wheeler was a man steeped in the Copenhagen spirit of Bohr's philosophy.[95] In the second of his 1959 Lake Como lectures, Wheeler wrote under the heading "Correspondence principle inapplicable to neutrino field; 'non-classical two-valuedness'":

It is not possible to find any classical description of the neutrino field along correspondence principle lines. In this respect this third field of zero rest mass is quite different in character from gravitation and electromagnetism. For them it is possible to go from the quantum theory to the classical limit, for example in the sense that the vibrational quantum numbers or "occupation numbers" of the various modes of oscillation of the electromagnetic field are allowed to become very large. But the neutrino satisfies Fermi statistics and cannot have occupation numbers greater than unity. Thus there is no obvious classical starting point for an analysis of the neutrino field. A wider framework of ideas seems necessary.

Wheeler specified what he meant, here, by Bohr's correspondence principle: that quantum systems exhibit classical behavior in the "classical limit." Photons would behave as electromagnetic waves. Gravitons would behave as gravitation. But there seemed to be no classical analog of the neutrino for a quantum system to behave like in a "classical limit." There were no known classical, massless, spin-1/2 entities. This lack of a classical limit was relatively well accepted; Julian Schwinger made similar remarks in 1953, as did Sidney Coleman in the early 1970s (section 7.5.1).[96] For Wheeler, a central pillar of Bohr's approach to quantum mechanics, the correspondence principle, could not apply. "So far as one can judge, quantum considerations must be accepted from the beginning in dealing with the neutrino and spin."[97]

94. Wheeler, "Neutrinos," 129.
95. Such as it was; see Howard, "Who Invented the 'Copenhagen Interpretation'?"
96. Schwinger, "TQF2," 716.
97. Wheeler, "Neutrinos," 88.

Wheeler's foundations were shifting—was gravitation primary, or the quantum principle? At the same time, what Wheeler *meant* by "quantum" was shifting away from Bohr's correspondence with classical ideas.

5.5 Natural Philosophy

In his 1959 Lake Como lectures, Wheeler set up an analogy between the singularities of gravitation theory and the singularities of atomic theory. In so doing he explained one sense of what he meant by the "quantum principle": the impossibility of localizing quantum objects too precisely. Wheeler presented a capsule history of quantum theory that served a particular rhetorical purpose. Wheeler gave a rational reconstruction of the development of quantum theory. It eliminated actual historical events, such as Bohr's 1913 stationary state hypothesis—in which the laws of electromagnetic radiation did not apply to stationary electrons in an atom.[98] It replaced that picture with uncertainty-based reasoning. Similarly to Oppenheimer's history of QED (section 4.3.3), Wheeler's reconstruction served to create an imagined continuity between his past and present.

According to Wheeler, the 1910s were characterized by a striving to solve the problem of atomic stability. It was well known that the Coulomb force law predicted that two oppositely charged objects should attract one another. What stops the negatively charged electrons from smashing into the positively charged nucleus of an atom? One answer was that the electron progressed in orbits around the nucleus and the Coulomb force provided the centripetal force that kept electrons in their orbits. But this interpretation ran afoul of the laws of electromagnetic radiation. For a constantly accelerating electron orbiting around the nucleus should radiate energy, and spiral into the nuclear center. Even in the orbiting picture, then, the stability of atoms was a mystery. Proposals were made to modify the Coulomb force law, or perhaps to modify the laws of electromagnetism to explain atomic stability. Wheeler characterized the eventual collision of electron and nucleus as a "singular state."[99]

> The singular state for a pair of charged particles is avoided, not by changing the law of interaction between electron and the nucleus, not by giving up the laws of electromagnetic radiation, but by adding the quantum principle. The electron cannot be confined to too small a localization distance L without making the associated spread of momentum, $\Delta p \sim \hbar/L$, so great, and the corresponding kinetic energy, $\hbar^2/2mL^2$, so large, as to overcompensate the Coulomb binding, $-e^2/L$. The electron stops radiating, not because the electromagnetic field is incapable of absorbing further energy, but because there is no lower state to which the electron can drop.

This argument was not, in fact, made by Bohr or Arnold Sommerfeld in their atomic theories. Nor was it made in Pauli's or G. Charles Darwin's quantum mechanical

---

98. Jammer, *Conceptual Development*, §2.2.

99. Perhaps because the Coulomb force between two charges, $F = k_e q_1 q_2 / r_{12}^2$, depends inversely on the distance between the charges. When an electron hit a nucleus, the distance would be zero, and the force singular.

description of the hydrogen atom (section 2.2). Rather than atomic theory, it had more in common with descriptions of field measurements and uncertainty (section 4.2.2). Wheeler continued: "In loose terms, the quantum of action [i.e., $\hbar$] forces on physical processes a smoother dependence on space and time than would have been expected from the unquantized version of the laws of motion."[100] It was unusual to see the quantum principle in terms of "a smoother dependence on space and time," especially considering Heisenberg and Bohr's rejection of classical trajectories.[101] But it served to align Wheeler's history with the structure of his developing theory: for both, the uncertainty principle was at the foundation.

In 1962, Wheeler won a Fulbright fellowship to be held at the University of Kyoto. At 2 p.m., May 18, 1962, in Hiraicho, Kyoto, Wheeler marked in his notebook a newfound centrality of the quantum principle to his thought. Whereas earlier the quantum was something to emerge out of geometry—for example, as wormhole throats—it now was placed at the center of his ontology. With frequent underlining and set out from the page in red outline he wrote:

The underline{interval ratio} $\frac{CD}{AB}$ independent of the comparison route—geometrodynamics says yes—but how is this result achieved?—that is physics—to paraphrase the words Newton used in another connection in the opening of his Principia. Quantum principle does everything else so it underline{must do this}, I say. How exciting to see how! ~~Extra~~ Supra dimensional world?

Unfortunately the only diagram accompanying the text may not be related (figure 5.5). It lies outside the black and red outlining and does not contain labels ABCD. Regardless of the relevance of the nearby diagram, we can apprehend that Wheeler was thinking of the ratio found between two spacetime intervals. In order to find the ratio between two separated intervals, one would have to "slide" one interval through the intervening spacetime, bringing the two intervals into contact. In flat Euclidean space this is trivial, but in the curved spacetime arena of general relativity—or geometrodynamics—sliding an interval around the space affects it in more complicated ways. Wheeler here asserted that the simple result that the ratio of two spacetime intervals should remain the same regardless of the route of comparison needed an explanation. Wheeler asserted that the "Quantum principle does everything else so it underline{must do this}." Wheeler tried to *invert* the position of geometry and the quantum. "See—or try to see—the geodesic equations of motion as a special consequence of this quantum principle."[102] The geodesic equations of motion were a central piece of general relativity that described how unaccelerated bodies move—for example, the Earth's motion around the Sun is geodesic motion.[103] What of Wheeler's reference to Isaac Newton? In his author's preface to the *Principia*, Newton wrote that "*geometry* is founded on mechanical practice and

100. Wheeler, "Neutrinos," 67.
101. Beller, *Quantum Dialogue*, esp. chapter 4.
102. JAW, Relativity Notebook VIII, p. 254.
103. Later, Wheeler described Einstein's original spacetime as "nothing if it was not a home for geodesics," but that later work by Einstein, Grommer, Infeld, and Hoffman shifted the emphasis to the field equations. Wheeler, "Superspace," 305–6 fn69; see Lehmkuhl, "General Relativity as a Hybrid Theory."

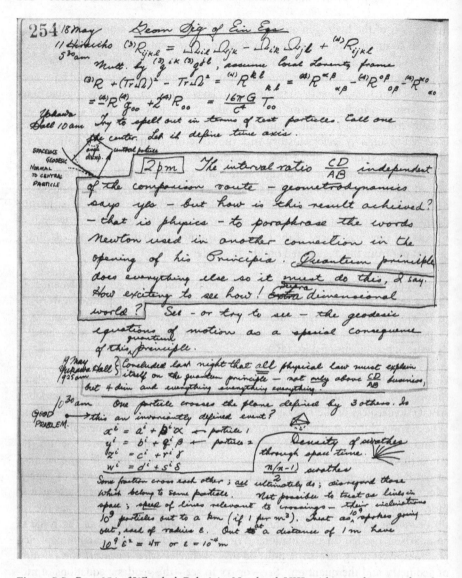

Figure 5.5. Page 254 of Wheeler's Relativity Notebook VIII, with some boxes and underlining in red. Courtesy of the American Philosophical Society.

is nothing other than a part of *universal mechanics* which reduces the art of measuring to exact propositions and demonstrations.[104] (Wheeler's "interval ratio $\frac{CD}{AB}$" may have referred to any number of Newton's diagrams of parallelograms.) The next morning at 9:25 a.m. Wheeler continued (figure 5.6): "Concluded last night that <u>all</u> physical law must explain itself on the quantum principle—not <u>only</u> above

104. Newton, *Principia*, 382.

$\frac{CD}{AB}$ business, but 4 dim[imensions] and everything everything everything."[105] Part of Wheeler's struggle to understand the universe was to understand why we should be living in a four-dimensional universe. Wheeler worked by considering what elements of theory could serve as the *most basic*, and drawing out the consequences.

Early the next month, Thursday, June 7, 1962, Wheeler returned to his Newtonian ruminations on a train from Kyoto to Osaka. He had come to the end of his previous hardbound, 300-page notebook and bought a Japanese book to replace it. Under the heading "Relativity and topology" he reflected: "Such a beautiful little book! Use for relativity notes? Yes—because I don't think anything I have ever done in my life is as important as the questions of relativity and topology which I am getting into now. Let's summarize—

(1) Newton, Principia, p.1, geometry asks one to draw a line, only physics supplies the means. $\sum e^{iJ_H/\hbar}$, constructive interference. Everything can happen. Law of combination of probability amplitudes alone, <u>no other laws</u>.[106]

Here the quantum principle appears in the formal guise of Feynman's sum-over-histories prescription, $\sum$, for forming a quantized theory from the exponential of a classical action.[107] The quantum-mechanical notion that all classical histories of a particle's path are to be considered (summed over) has been transmuted into the metaphysical principle of plenitude: "everything can happen." Again, Wheeler was searching for a singular principle that would manifest the totality of physics. With the quantum combination of probability amplitudes for each possible path of a particle alone, perhaps Wheeler would need "<u>no other laws</u>."

Wheeler continued his summarizing list with a reference to his previous notebook, and a connection to the changing topologies of his quantum foam picture:

(2) Apparent contradiction in fact interval $\frac{CD}{AB}$ independent of route of comparison—so this too must arise from principle of constructive interference. Hence <u>one</u> argument for relation between apparently distant points. Another:

(3) Differential equations relate what goes on at one point to what goes on at a nearby points. But when topology changes two points formerly near are no longer near. The cut off in influence is to be compared with the cut off of light passing from A to B as the two prisms in the drawing [figure 5.6] are separated.

Describing a function (via differential equations) on a space with a changing topology was a critical and unsolved mathematical difficulty for geometrodynamics. Wheeler found an analogy in Newton's geometrical optics. A beam of white light entering the the connected prisms at A in figure 5.6, when separated, would produce a beam split into the colors of the rainbow at B. Wheeler continued in his third point to grapple with how to integrate his picture of foaming topologies with the basic structures of geometrodynamics, Einstein's field equations. On a 4 p.m. train back to Kyoto, Wheeler continued his thoughts about the origin of the dimensionality of the world.

105. JAW, Relativity Notebook VIII, p. 254.
106. JAW, Relativity Notebook IX, p.3.
107. On "interference," see Wheeler, "Superspace," 243–45.

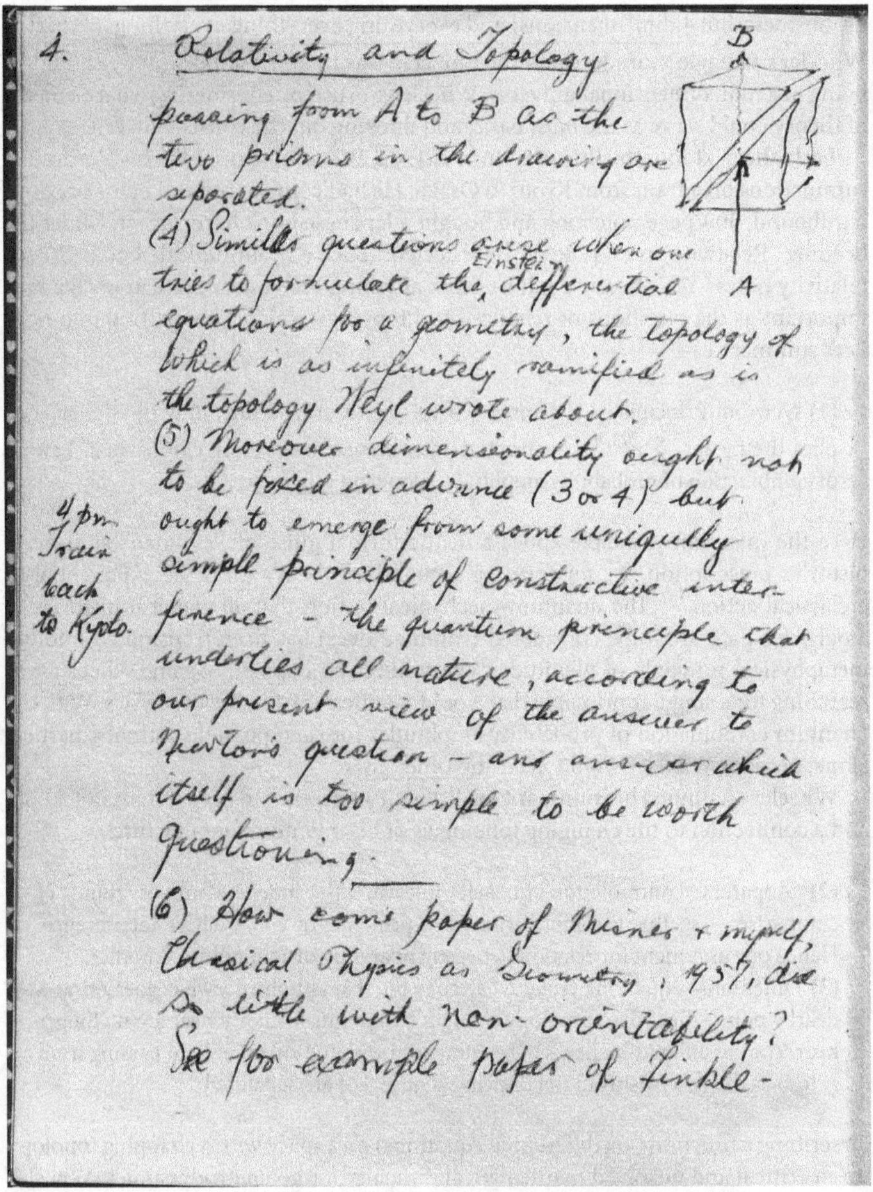

4.   Relativity and Topology

passing from A to B as the
two prisms in the drawing are
separated.
(4) Similar questions arise when one
tries to formulate the differential
                     Einstein
equations for a geometry, the topology of
which is as infinitely ramified as is
the topology Weyl wrote about.
(5) Moreover dimensionality ought not
to be fixed in advance (3 or 4) but
ought to emerge from some uniquely
simple principle of constructive interference — the quantum principle that
underlies all nature, according to
our present view of the answer to
Newton's question — and answer which
itself is too simple to be worth
questioning.

(6) How come paper of Misner + myself,
Classical Physics as Geometry, 1957, did
so little with non-orientability?
See too example paper of Finkel-

*4 pm
Train
back
to Kyoto.*

B

A

Figure 5.6. Page 4 of Wheeler's Japanese-paper Relativity Notebook IX, with two prisms at top right. Courtesy of the American Philosophical Society.

(5) Moreover dimensionality ought not to be fixed in advance (3 or 4) but ought to emerge from some uniquely simple principle of constructive interference—the quantum principle that underlies all nature, according to our present view of the answer to Newton's question—an answer which itself is too simple to be worth questioning.

Here Wheeler's aesthetics of physics shines through, with simplicity serving to separate what it was possible to question.

Jumping two years to May 22, 1964, Wheeler recorded a conversation with the British relativist Dennis Sciama (Penrose's mentor). It underlined the troubles the field faced in finding a quantum theory of general relativity.

> I mentioned to Dennis Sciama that I (in contrast to him) think there's hope in strug-
> gling through the quantum issues. He discouraged because Dirac, Schwinger and
> many lesser lights have dealt with q[uanti]z[atio]n, don't seem to get anywhere. I
> argued that the real problem is what questions to ask—and for that have real hope
> that just straight common sense will lead in—"I'm perhaps a fool who goes in where
> you [sic] the angel fears to tread."[108]

Sciama was skeptical because the leading lights of quantum theory had floundered. Calculations were nearly impossible. Wheeler responded to Sciama's skepticism by redirecting the conversation to foundations and fundamentals. For Wheeler the prob-lem was not that the questions that had so far been asked had no satisfying answers. The problem was that the prior step had not been correct. The right questions needed to be formulated.

Wheeler's record of his conversation with Sciama continued with a peculiar metaphor that he thought wrapped up the central issues. He returned to the founda-tions of quantum mechanics, and Everett's branching interpretation of interactions:

> I spoke of big double decker red bus going down to War Memorial [on Hills
> Road, Cambridge] splitting wave f[unctio]n into 2 parts, one of which goes to
> R[ail]R[oad], other down hill road [sic]—What does a "half" a bus look like—have
> such issues in this—grand mixture of all the most important: q[uantum] theory
> elem[entary] part[icles] cosmology causality baryon conservation—how can one
> stay away from?

What would it mean for a bus to split into two paths, quantum mechanically? What would half a wavefunction look like? Half a bus in each or two buses progress-ing in different "worlds"? Did such a split create a new cosmos? Wheeler had been uncomfortable with Everett's "splits" from the beginning. He made marginal notes on a preparatory essay for Everett's thesis, titled "Probability in Wave Mechanics" (1955). Where Everett wrote of an observer "splitting each time a measurement is made," Wheeler commented: "Split? Better words needed."[109] But how could Wheeler stay away from this complexion of the most interesting problems? Later that day, Wheeler's notes allowed him to record further reactions to the conversation, con-necting this richness of problems in subfields of physics to the existence and fertility of paradoxes. "Later thought reaction of mine—elem[entary] particle physics no

108. JAW, Relativity Notebook XII, p. 188.
109. Page 7, Folder 6, Box 1, Everett papers.

paradoxes that I know about, whereas collapse <u>does</u> have such paradoxes—therefore <u>latter</u> is the field for hope."[110]

Wheeler did receive pushback on his philosophical questioning. On Tuesday evening, September 14, 1965, at Princeton, Wheeler recorded some friction with one of his doctoral students, Kip Thorne. Under the heading "The worry about Geometry as All," Wheeler "Discussed with [Alfonso] Campolattaro and [Kip] Thorne my concern about coming through in my life with Einstein's dream.

> (1) How to get <u>any</u> clue at all to particle-quark machinery from foamy universe picture? (2) How <u>get</u> spin 1/2? [...] On former spoke of need to construct a model. Thorne asked, what reason to think of this "everything-as-geometry" view as being any more final than 19th century view that all was mechanics. I spoke of all physics then being imbedded in geom[etry], not identical <u>with</u> geometry. He said, only a diff[erence] in degree. I said, how wonderful to understand e[lectro]m[agnetism], charge, mass as aspects <u>of</u> geom[etry], not imbedded in geom[etry]. He said Einstein dream as little foundation to him as Mormon theology. Best arg[ument] in his eyes for Ein[stein] dream not its beauty but productive leads it have [*sic*] given people like me.[111]

Thorne's comments were similar to an anti-realist philosophy of science that views current successes to be only as reliable as the convictions of earlier scientists, now discarded. Except that Thorne, in Wheeler's telling, saw the possibility of a *pragmatic* argument from the productivity of geometrodynamics. Wheeler was still struggling with how to move from his evocative-but-fuzzy quantum foam idea to the specifics of modern particle physics. In response to skepticism, Wheeler responded with wonder and beauty.

The next Spring, on April 9, 1966, Wheeler was preparing for a talk in Berlin on "Quantum Geometrodynamics."[112] He was fifty-five years old and he was struggling to formulate new ideas. He began with contrasting the idea of resonance and the idea of propagation. "One is transverse to direction of propagation, [the] other along that direction. What mixture of these two effects occurs?" He pictured resonance as a "system passing back & forth between one micro 3-geometry and another." Conversely, "At macro-level wave packet concept applies." These two dynamics would couple together in the phenomenon of gravitational collapse. But a quantum geometrodynamics reconfigured the basic concepts of physics, and had to be struggled with anew.

> The new feature in the present ~~picture~~ problem is the role of time—"before" and "after" being lost. Reincarnation? Even that is a misleading comparison because fourth incarnation "after" third incarnation. We think more of <u>parallel</u> existences. How to speak of this? Everett's "relative state" formulation of quantum mechanics. London's discussion of measurement problem. Lee & Wick, also Wigner, on superselection rules. Bohr's question, what does a wave function mean in a closed space.

110. JAW, Relativity Notebook XII, p. 188.
111. JAW, Relativity Notebook XVIII, pp. 221–22; cf. Wheeler, *Einsteins Vision*.
112. This probably became Wheeler, *Einsteins Vision*.

Without a definite conception of before and after, quantum geometrodynamics was lost at sea. Wheeler struggled to find paths into his ideas, from the metaphor of reincarnation to probing the foundations of quantum mechanics. His first thought here was not to Feynman's formal sum over histories approach but Everett's relative state formulation. It in a sense offered an answer to "Bohr's question" about the meaning of a wavefunction in a closed space by providing a way to formulate quantum mechanics without reference to an external observer. This also explains the relevance of the "measurement problem."[113] The measurement problem can be stated as the problem of explaining the interaction between macroscopic measuring apparatuses (or observers) and quantum particles. It deals with the relation between the quantum and an external, macro-sized reality. Superselection rules were mathematical rules applied in the formalism of quantum mechanics to create more general pairs of orthogonal quantum states.[114] Wheeler continued:

> My feeling that we have a time bomb in our midst—that (q[uanti]z[e]d) general relativity may provide a model to discuss, not merely everything, but more than everything. Want to deny any mystification. How talk about things that we can never see? Not a new idea. ψ formalism not itself seeable; only the consequences seeable. See why position seeable, or velocity, but not both. Model displays all this. Further step in this direction to speak of parallel existences. Not simultaneously realizable—yet interacting.
>
> What words to use now in speaking of all these nebulous ideas? Surely want to be conservative, Yet want to indicate scope of issues. What is the thesis of this part? That we have a time bomb. That the[115]

And there the note ends.

Wheeler returned to his Berlin preparations three weeks later on April 30, 1966. After working on scratch paper (which he discarded), he turned to his notebook to record that he felt "that the 'more than everything' idea is the truly new and impressive feature of the thinking at this stage." In what sense more than everything? Perhaps in the sense that a quantum geometrodynamics would posit a multiplicity of universes, a multiplicity of "everything." This quantum idea was still tied carefully to Wheeler's conviction that curved empty space—pure geometry—was the correct description of the world. Under the title "More than everything; & Principle of purification" Wheeler continued to develop his ideas, searching for a way to put them into words, culminating in an almost religious conviction in the power and centrality of geometry.

> Try for word to indicate idea of purifying geom[etry] of particles and of spinor fields. Could it be that electromagnetism has to do with slipover from one topology to another [...?] Great thing about this idea is that it gives the appearance of

113. London and Bauer, "Theory of Observation"; for a modern introduction, see Myrvold, "Philosophical Issues in Quantum Theory," §4.
114. Wightman, "Superselection Rules."
115. JAW, Relativity Notebook XIV, pp. 18–19.

being _testable_ with a reasonable amount of work. [...] If one _could_ show that e[lectro]m[agnetism] can be understood in this way, it would give marvellous feeling that one is on the right track in whole program.

Here, Wheeler's notebook records a clear example of concepts shifting in the course of an investigation.[116] Not only an example, but a reflexive comment on the changes in progress.

If just trying to _formulate_ principle of purity leads to idea like above, let's work harder at set of words to be used for it [...]. Of course, since I started job (at top of page), idea of what principle _is_ has changed. Now more than ever idea of nothing but geometry [...]. Moreover, it isn't at all clear that there really _is_ an idea common to geodesic eq[atio]n of motion, spin, & electromagnetism [...]. Much more solid unit along the line "nothing but geometry." [There is no God but God, and Mohammed is his prophet—reword—There is nothing but Geometry, and Everything is Geometry, and more than Everything is Geometry!] [_sic_][117]

Wheeler's religious references were informed by his commitment to the Unitarian Church. He attended throughout his life, and he and his wife Janette helped found a Unitarian church in Princeton, NJ.[118] To Wheeler, geometrodynamics was in a _worse_ situation than electrodynamics had been in, before Tomonaga, Schwinger, and Feynman showed the way to do relativistic calculations in the 1940s (chapter 4). Then, physicists at least roughly agreed on the correct quantities to _try_ to calculate (Compton scattering, the self-energy of the electron, etc.). In New York on January 27, 1967, Wheeler wrote that he "must explain why others so slow. Less because anyone disbelieves in any positive sense. More because so few see what to do next. Not so much difficulty of cal'n as _what_ to calc[ulate]. Conceptual difficulties."[119]

5.6 Increasing Abstraction: Superspace

Geometrodynamics had reached something of an impasse. Several early results and Wheeler's charismatic presentations, pedagogy, and organizational powers had found their place in the post-1955 "renaissance" of general relativity. But he had been unable to make his metaphors quantitative. Thorne and Sciama were worried about the lack of progress (section 5.5). In Feynman's 1962–1963 Lectures on Gravitation at Caltech, he recalled his doctoral studies at Princeton and reflected on making Wheeler's images quantitative:

On another occasion I received a phone call from [Wheeler] in the middle of the night, when he said to me "I know why all electrons and positrons have the same charge!" Then he explained further, "They are all the same electron!" His idea was

116. Compare the description of laboratory bench work in. Rheinberger, "Cultures of Experimentation."
117. JAW, Relativity Notebook XIV, pp. 19–20.
118. Wheeler and Ford, _A Life in Physics_, 73, 102, 153.
119. JAW, Relativity Notebook XIV, p. 141.

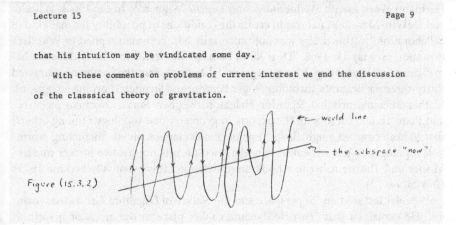

Lecture 15   Page 9

that his intuition may be vindicated some day.

   With these comments on problems of current interest we end the discussion

of the classsical theory of gravitation.

Figure 5.7. Feynman's 1963 sketch of an electron moving forward and backward in time. Courtesy of the Archives, California Institute of Technology.

that is the same object has a world line which is extremely complicated, when we look at it in the subspace "now", we see it at many different places. (Figure [5.7]) Later on, I was able to make this kind of idea quantitative, by interpreting a positron as being an electron whose phase is going backwards in time—and developing simplified methods for calculating matrix elements involving annihilation and creation of pairs.

Feynman framed his path-integral QED as a quantification of Wheeler's qualitative image. To his class, Feynman was optimistic about geometrodynamics. "It would truly be very beautiful if the idea of wormholes, and geometrodynamics, could be perfected to improve our understanding of nature—and knowing Wheeler, it does not seem to me unlikely that his intuition may be vindicated some day."[120] In the late 1960s, Wheeler made three attempts to make quantum geometrodynamics quantitative. First, as a physicist in mathematical trouble, he tried to convince some mathematicians that geometrodynamics had interesting problems. Second, he compromised on the strictures of general relativity by breaking its four-dimensional spacetime manifold into three dimensions of space and one of time. This was in service of his third, step. He abstracted away from individual spacetimes to "superspace."

   From July 16 to August 31, 1967 at the Battelle-Seattle Center near the University of Washington, Wheeler and French-American mathematician Cécile DeWitt-Morette organized a conference bringing together mathematicians and physicists. Their introductory essay likened the estrangement of mathematicians and physicists to the Tower of Babel. Earlier, in the late 1930s, Wheeler's S-Matrix theory relied not only on mathematical publications, but also on personal interactions with mathematicians

120. Feynman Papers, Box 9, Folder 5: Lectures on Gravitation, lecture 15, pp. 8-9; Feynman, *Feynman Lectures on Gravitation*, 205; Schweber, *QED*, 418, 449, 577.

Hermann Weyl, Joseph Wedderburn, and Eugene Wigner.[121] In the 1960s, Wheeler and DeWitt-Morette's goal was to create the conditions of possibility for that kind of collaboration.[122] This desire was not universally felt. Feynman replied to Wheeler's invitation on May 19, 1966: "Dear John, I am not interested in what today's mathematicians find interesting. Kindest regards."[123] Nevertheless, the conference attracted thirty-three participants, including Roger Penrose and Stephen Hawking (chapter 6). Mathematicians included Sigurdur Helgason, Stephen Smale, Norman Steenrod, and René Thom. These were the experts in geometry and topology (among others) that Wheeler hoped might help him make calculations about fluctuating wormhole topologies. (Also in Seattle, Wheeler tried to convince his two former students Misner and Thorne to write a book on gravitation. This eventually became 1973's *Gravitation*.[124])

Wheeler lectured on "Superspace and the Nature of Quantum Geometrodynamics." He explained that "Particle dynamics takes place in the arena of spacetime. Geometrodynamics takes place in the arena of superspace."[125] This was an increase, or an iteration, of abstraction. First, in standard GR an abstraction of physical spacetime to a mathematical pseudo-Riemannian space; second, seeing these pseudo-Riemannian spaces as merely one element in a "super" space of spaces. Wheeler began by discussing the nature of motion for quantum particles, illuminating his earlier remarks in his notebook about how the quantum principle and superposition could form the basis for geometry.

The concept of superspace required that the geometry of spacetime *not* be considered four-dimensional, but rather $(3 + 1)$-dimensional; that is, spacetime is to be understood as a sheet of three-dimensional space evolving in time, denoted $^{(3)}\mathcal{G}$.[126] Splitting the spatial dimensions from the time dimensions was a strategy from the efforts to build a Hamiltonian form of GR, to which Misner made major contributions.[127] Recall Dirac's and Hermann Minkowski's commitment to four-dimensionality (section 2.2.2). A $(3 + 1)$-dimensional spacetime picture was a conceptual break from the more general spacetime of orthodox relativity, which Wheeler denoted $^{(4)}\mathcal{G}$.[128] Wheeler illustrated the manifold of superspace as being composed of points that each represent a $^{(3)}\mathcal{G}$ geometry with figure 5.8. He continued to make an analogous argument for the dynamics of superspace to the earlier argument about the dynamics of a particle in spacetime: that the dynamics only makes sense when considering a superposition of probability amplitudes. "In summary, (1) the $^{(3)}\mathcal{G}$'s allowed by [constructive interference] are the basic building

121. Wheeler, "Mathematical Description of Light Nuclei," 1118fn20, 1119fn21a.

122. DeWitt-Morette and Wheeler, "Battelle and Babel." Recall the worries about physics and mathematics discussed in section 1.2.1.

123. Feynman Papers, Box 3, Folder 3.10: Wheeler.

124. Notes copied from Misner's bound research Notebook, August 31, 1967, Nordita. Folder: CW Misner's earliest notes re development of MTW book, Box 1, Misner Papers; Misner, Thorne, and Wheeler, *Gravitation*.

125. Wheeler, "Superspace," 242.

126. Wheeler, "Superspace," 246.

127. Arnowitt, Deser, and Misner, "Dynamical Structure and Definition of Energy"; see Blum and Hartz, "1957 Quantum Gravity Meeting."

128. Bergmann, "Sandwich"; cf. Stachel, "Rise and Fall of Geometrodynamics," 44.

blocks; (2) their inter connections give $^{(4)}\mathscr{G}$ its existence, its dimensionality, and its 'magic structure'; and (3) in this structure every $^{(3)}\mathscr{G}$ has a rigidly fixed location of its own."[129] The further abstraction of superspace provided enough structure to give each possible geometry a *mathematical* location (recall that Dirac's many-body quantum theory did similar work). But superspace was an arena in which quantum wavefunctions of geometry could combine and interfere. That was at least the hope.

Wheeler's aim was a quantum theory of the evolution *of* spacetime *in* superspace. Referencing Misner's doctoral thesis, Wheeler spoke of a Feynman "sum over histories" approach to quantum geometrodynamics.[130] Rather than $^{(3)}\mathscr{G}$'s orderly progressing along the time axis of a $^{(4)}\mathscr{G}$—the $^{(4)}\mathscr{G}$'s "magic structure"—in quantum geometrodynamics the configuration of $^{(3)}\mathscr{G}$'s must arise from summation over an infinite variety of histories. This reconfigured the concepts of the classical world.

> These considerations reveal that the concepts of spacetime and time itself are not primary but secondary ideas in the structure of physical theory. These concepts are valid in the classical approximation. However, they have neither meaning nor application under circumstances when quantum-geometrodynamical effects become important.

More than a break from 4 dimensions to 3 + 1, Wheeler abstracted away from spacetime itself.

> Then one has to forgo that view of nature in which every event, past, present, or future, occupies its preordained position in a grand catalog called "spacetime." There is no spacetime, there is no time, there is no before, there is no after. The question what happens "next" is without meaning.[131]

At the quantum level, fluctuations demanded that "Geometry in the small fluctuates not only from one microscopic pattern of curvature to another, but much more, from one microscopic topology to another." Wheeler's concept of the foam-like structure of space and time was given a new arena: fluctuations in superspace. "In other words, space 'resonates' between one foamlike structure and another."[132] This was another level of mathematical difficulty. Bryce DeWitt had made headway describing transitions in superspace between one curvature and another—such as the evolution in time of a space rippling with gravitational radiation.[133] But Wheeler aspired to describe spacetime fluctuating between, for example, spherical and toroidal topologies (*B* to *C* to *E* in figure 5.8).

A few weeks before the Battelle Center conference, Wheeler gave a similar lecture at the College de France, June 19–23, 1967.[134] On the morning of Sunday, June 25, at

129. Wheeler, "Superspace," 252.
130. Misner, "Feynman Quantization."
131. Wheeler, "Superspace," 253.
132. Wheeler, "Superspace," 264.
133. DeWitt, "Quantum Theory of Gravity. I."
134. Lichnerowicz and Tonnelat, *Fluides et champ gravitationnel.*

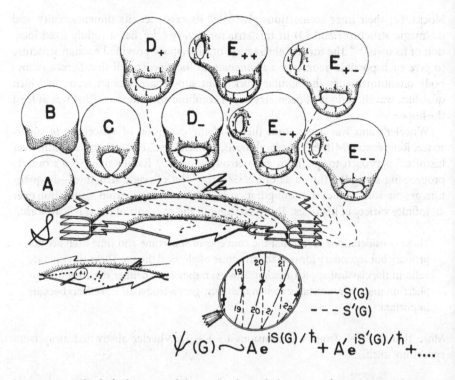

Figure 5.8. Wheeler's diagram of the multi-sheeted character of superspace when spin is considered. Republished with permission of Springer, from John Archibald Wheeler, "Superspace and the Nature of Quantum Geometrodynamics." In *Topics in Nonlinear Physics: Proceedings of the Physics Session*, 615–724. New York: Springer, 1968; permission conveyed through Copyright Clearance Center, Inc.

the Chateau Frontenac in Paris, Wheeler reflected on the future of his program for physics, and the role of superspace. He again expressed his distaste for symmetries (section 5.5).

> As [Hans-Peter] Dürr remarked yesterday, it would be terrible if sym[metry] analysis of elem[entary] particles gave everything, for then there would be no hope of going further. Vice versa, it seems to me here that if this business is to be right, it ought to go "all the way," because there isn't going to be any easy way ever to see any later clue to going further.

In the notes to his Battelle address he cited Dürr and the latter's mentor Heisenberg's distaste for symmetries in particle physics. Wheeler relegated quarks to mere bookkeepers.[135] His note continued:

> But then the story of superspace ought to carry the whole show. The existence of 3-dimensionality and the form of the Einstein-Schroedinger equation (DeWitt

135. Wheeler, "Superspace," 300 fn32.

eq[uatio]n.) ought both to follow from the structure of superspace. This would be magical and beautiful, and would make one feel that he perhaps had the whole business in hand.[136]

The "Einstein-Schroedinger equation" was first published by DeWitt in 1967.[137] De-Witt credits the work of Asher Peres and Dirac with leading him to the equation in a conversation with Wheeler around 1965 or 1966.[138] For definiteness, what is now known as the Wheeler-DeWitt equation for the quantum-mechanical evolution of a universe can be written

$$H\psi\left(^{(3)}\mathscr{G}\right) = 0, \tag{5.4}$$

where H is a Hamiltonian constraint functional and $\psi(^{(3)}\mathscr{G})$ is the wavefunction of 3-geometries in a closed universe.[139] If Wheeler's superspace could determine the form of equation (5.4), it "would be magical and beautiful."

At the top of these notebook pages Wheeler had marked: "Law without Law or Law out of Being" and "Law by Being." He was exploring another (productive) para-doxical formulation, now centered on the self-referential idea of law emerging out of the very existence of things. Wheeler's thoughts again became theological: "Su-perspace would be the great arena, and its very existence would give it the law of its life—it could have no other law. Out of nothing would come everything. 'I am that I am.'"[140] Wheeler was searching for another way to turn an arena into everything, an-other way to make things out of nothingness. He was still thinking in terms of curved empty space, "nothing," but also of a common translation of the response Moses re-ceived upon asking God's name (Exodus 3:14). Wheeler returned again and again to paradoxical formulations, and strove to explain ever more basic facets of reality.

Later that evening at the Trianon Palace, hotel Versailles, Wheeler continued his ruminations on superspace, the possibilities of paradoxes, and the search for the simplest constituents of reality. Was the metric, $g_{\mu\nu}$, a basic concept for relativity?

What a condition! But is there anything simpler? Why should there be a metric in superspace? And built on a metric in 3-space! If can get eq[uatio]ns of motion from H.J. eqn [Hamilton-Jacobi equation], can one not get metric also? And wouldn't that be still more exciting? But how could one possibly expect to get that without idea of prob[ability] amp[litude] being on top? Say one does idealize each sector of superspace with a given topology as a single point to play a [surface?]. Then what kind of structure does one have? n wormholes, 2^n layers. [figure 5.9] One gets impression each is fantastically cut off from others. However, that appearance is a defect of the model. Yet can move around point to being back.

136. JAW, Relativity Notebook XIV, pp. 190–91.
137. DeWitt, "Quantum Theory of Gravity. I," 1123.
138. Interview with Drs. Bryce DeWitt & Cecile DeWitt-Morette by Kenneth W. Ford at the University of Texas at Austin February 28, 1995, AIP http://www.aip.org/history/ohilist/23199.html.
139. See DeWitt's eq. 5.15 for the expression for H. DeWitt, "Quantum Theory of Gravity. I," 1123.
140. JAW, Relativity Notebook XIV, pp. 191–92.

These sketches were probably the prototype for Wheeler's published "problems" at the end of his long Battelle lecture, figure 5.8. Recall that the metric in relativity determined how distances were measured locally. It was an instruction for how to determine how far apart one point is from another.[141] This functioned well in a stable topology. But in Wheeler's wormhole picture of the vacuum wormholes forced the topology to be dynamic, with neighborhoods of particles changing at each instant. Rather than a problem, however, Wheeler reconfigured this worry into a possibility (as he had with cosmological neutrinos, section 5.3). Wheeler saw the opportunity to explain the origin of the basic concept of the metric (distance).

> Maybe that's how idea of distance arises—just how much moving around a worm-
> hole must undergo to undo something of this sort. Would seem very natural for 3
> reasons. First, one cannot expect distance to be a primordial notion. Second loads
> of wormholes offer among all topological ideas I have ever imagined the most nat-
> ural building material for idea of distance. Third, there would seem hope along this
> line to avoid ever expanding distances. Idea of infinite superspace terrifies me as
> much as idea of open 3-space antagonizes me.

Wheeler reflected on his thinking, his place and time, and formulated a new paradox with which to think:

> [Recall Bohr saying "I'm a dilettante" by which he meant "I'm an amateur" by which
> he meant "I'm not a professional worker"][sic]. This to me is a new idea tonight,
> dreamed up in the Salle a Entrer of the Trianon Palace Hotel at Versailles as I lie in
> wait for O. Klein, (coming with L. Rosenfeld for the Conference on Life), as I lis-
> tened to beautiful music playing in the halls where Woodrow Wilson, Clemenceau,
> and others did the Treaty of Versailles. What an inspiring idea! How stupid not to
> have thought of it before! What a challenge to develop it now! And how quickly
> it came out as I started to get quantitative—an argument for "L'audace, toujours
> l'audace"! Absolutely wonderful to think of distance without distance.[142]

The next morning, Monday, June 26, 1967, at the Chateau Frontenac, Wheeler directed himself to "See notes at right made yesterday evening at Versailles—idea of
'geometry without geometry' as step toward 'anything without anything.'" By evening
he had decided to settle on an idea worked out the day before: "for time being still
think of the geometry as being 3 dimensional so as to have a stomping ground for
wormholes and their spatial arrangements."[143] Then his picture was of

> 3 dim[ensionality], wormholes all over the place, and connected with one another in
> all sorts of ways. But if break connection in one place, remake somewhere else, then
> can in principle connect any point with any point. This I have recognized previously
> but what good has it done!

141. According to Wheeler. Galison, "Re-reading the Past."
142. JAW, Relativity Notebook XIV, p. 193.
143. JAW, Relativity Notebook XIV, p. 193.

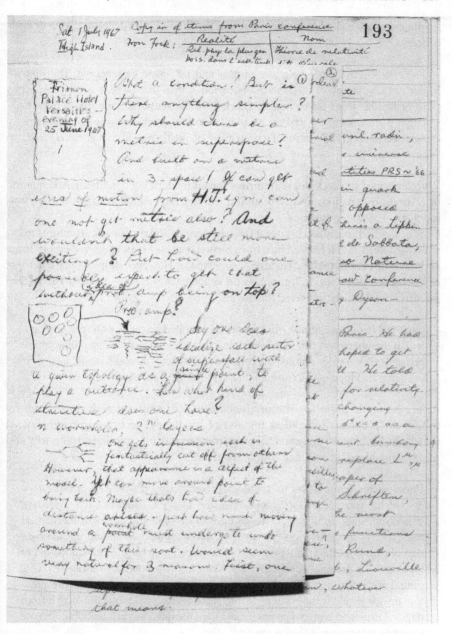

Figure 5.9. Page glued to page 193 of Wheeler's Relativity Notebook XIV. Courtesy of the American Philosophical Society.

Again, calculations had been elusive. Again the solution was deconstruction of basic concepts.

> But now feel that the whole idea of near and far has to be refigured. The wonderful thing about all this line of discussion is this, that there is some prospect of being able to go far enough to check it, in the sense that one would hope to <u>derive</u> 3-dimensionality and the Einstein-Hamilton-Jacobi eq[uatio]n or Einstein-Schroedinger eq[uatio]n from 1st principles—if not, then must change either tactics or strategy.[144]

Wheeler hoped to check this with formal means by deriving a fundamental equation of relativistic dynamics and of quantum geometrodynamics from his ontological-topological considerations.

Wheeler had continued his quest for ever simpler and more basic structures to explain the world, and it had led him from zero-rest-mass fields, to curved empty space and quantum foam, to the structure of superspace. However, despite the excitement of his French notes in 1967, progress on superspace was slow. Back in Paris seven months later, Wheeler reflected that he

> Can't feel happy at unfinished character and limited scope of work presently going on on superspace. Not sufficiently illuminated by any convincing philosophical principle. Takes as for granted the concept of distance, whereas it would be my expectation that topological considerations would stand ahead of metric concepts.[145]

Rather he would move to yet more fundamental structures, foreshadowed in a selection of "problems" offered at the end of his Battelle lectures: pregeometry. At Battelle he brought up pregeometry in an ironic mood, first quoting an authority and introducing his ideas, but then dismissing them.

> Under such circumstances [of topological complexity] it would seem difficult to uphold the concept of dimensionality at the smallest distances. [...] Moreover, if electromagnetism and other fields have to do with the quantum-mechanical resonance of space between one topology and another, why should not the concept of metric itself be likewise a derived concept, going back for its foundation to topological or pretopological—and at any rate to pregeometrical—ideas [...]. One cannot even mention these topics without recalling the universal sway of the quantum principle, and without stressing the "order of creation" as one thinks of it from physical evidence: Not first geometry and then the quantum principle, but first the quantum principle and then geometry!

Wheeler was looking for foundations out of which to build a theory of nature, and he maintained his view that quantum mechanics would hold of that foundation. But he pulled himself back.

---

144. JAW, Relativity Notebook XIV, p. 192.
145. JAW, Relativity Notebook XIV, p. 238.

It is enough to raise these issues, with all their depth, to see into what difficulties one can get with quantum geometrodynamics if one tries to think of it as an "ultimate" theory. However, physics has never depended for its progress on having an ultimate theory. There is no reason to think that the situation is different today. While one can raise ultimate issues of all kinds, there is no reason to believe that they all have to [be] settled now! Nor that one *can* resolve them now![146]

Perhaps here he was remembering Thorne's September 14, 1965, criticisms of final theories (section 5.5).

5.7 Pregeometry

Despite his knowledge of the limitations and prospects for "ultimate" theories, Wheeler continued his search for foundations that would undergird his physics and philosophy: a pregeometry. The meaning of pregeometry changed over time, starting with the vague thoughts of his published Battelle lectures; to a connection to Leibnizian monads and self-reference; moving toward an admittedly "too literal-minded" physical picture of spacetime as a sort of chainmail; and ending with the thought of geometry emerging from a pregeometry "isomorphic" to the calculus of logical propositions. Here again, Wheeler's notebooks not only record his changing concepts, but by providing an archive, a sounding board, and a scene for new combinations of ideas they shaped the concepts' development.[147]

On August 11, 1968, out on his deck on "a glittering day" at his house at High Island, Maine, Wheeler reflected on his "Strategy on Superspace." He expressed his dissatisfaction in martial metaphors.

Last year has been one of "casing the joint," looking over the issues of gravitational collapse and the structure of superspace first from one side, then from another, without really coming to a pitched battle yet with anything. [...] Not one step nearer did we seem to come to a philosophy of superspace—why superspace has to be.

In terms of the basic constituents of the world, Wheeler had gone from his zero-rest-mass field hypothesis, to seeing the world as a quantum foam, and now to the primacy of superspace.

I have to get through my head the idea that superspace has primacy—from it are to be derived such concepts as 3-dim[ensional] space, "points" in 3-dim[ensional] space, changes in topology, and even metric itself. Anyone who wants to be in where the big machinery is has to get into superspace.

However, Wheeler's conviction of the primacy of superspace combined with his dissatisfaction with the progress on it as of 1968 led him in a familiar direction: to search for further abstraction.

146. Wheeler, "Superspace," 294.
147. See C. Hoffmann "Processes on Paper."

So one should <u>start</u> with superspace and work <u>back</u> from it to all these "derived ideas" rather than proceeding, as we do now, the other way. That there <u>are</u> two ways to proceed is an indication that we have in superspace a handle on physics of real world, fruitful to operate! Evidently we have to visualize an approach to superspace much more abstract than any that we have today, in which 3-dimensionality will not even be mentioned, but be derived (as an approximation).

Here Wheeler returned to the Einstein-Schrödinger equation to ask "Today how far are we along this road? In the quantum mechanical operator

$$\frac{\nabla^2 \psi}{(\delta^{(3)}\mathcal{G})^2} + {}^{(3)}R\psi = 0 \tag{5.5}$$

we haven't any explicit way of writing presently known to be an explicit reference to 3-dimensions—<u>this</u> one wants first of all to kill. Second, is there hidden an assumption about 3-dimensionality in the <u>topology</u> of superspace; and, if so, how? These are really exciting issues!"[148] This equation is a more detailed expression of equation (5.4). The first term indicates that a second derivative of the wavefunction of a universe, ψ, was to be taken with respect to the three-geometries of spacetime, ${}^{(3)}\mathcal{G}$. The second term indicates that a three-dimensional version of the Ricci tensor, ${}^{(3)}R$, was to be multiplied by the wavefunction. Together the "(3)s" on the derivative and the Ricci tensor contained the "explicit reference to 3-dimensions" that Wheeler sought to explain away with something deeper.

Almost one year later, early in June 1969, back at High Island, Wheeler returned to these questions. He had just left a large conference on relativity in Cincinnati. Superspace dominated the proceedings. It was discussed in presentations by Peter Bergmann, Bryce DeWitt, Arthur Fischer, Charles Misner, and Wheeler. "By general demand a special informal session was organized and held on Tuesday night to discuss 'Superspace.' The Superspace Session started at 8:00 P.M. and faded away beginning at 2:00 A.M."[149] The conference gave Wheeler a lot to think about. In his notes at High Island, superspace was twinned with pregeometry; Wheeler vacillated between seeing one or the other as primary. Under the heading "Pregeometry" he sought simplicity. "Have only domain of 3-geometries, superspace. This may be much simpler, I can believe, than it now looks. Recall how few points one needs in [??]'s way of doing Fourier analysis. Recall how few lines a Picasso needs to reveal an individual and his mood." This search for simplicity led Wheeler to consider whether superspace really was the ultimate reality. "Why not likewise a much more elementary object than superspace as presently conceived to reproduce what one today asks superspace to reproduce?"[150] This was abstraction and simplicity at work.

On the next page of his notebook, marked Friday, June 13, 1969, Wheeler continued these thoughts under a new banner "Superspace above Battles of Matter &

148. JAW, Relativity Notebook XV, pp. 106–7.
149. Carmeli, Fickler, and Witten, "Preface," v; Wheeler, "Particles and Geometry."
150. JAW, Relativity Notebook XVI, p. 83

Energy." Here he combined theological and political thinking with an understanding that he must make his ontology respect his epistemology. If superspace was to be primary, Wheeler would have to demonstrate its simplicity:

in the outlook of today [...] the structure of superspace is in effect God given: an arena that tells probability amplitudes how to develop with time, but an arena that does not in turn respond to this probability amplitude. And how can it respond anyway without doing violence to the quantum mechanical principle of superposition? But to have this "autocracy" in contrast to the "democratic give-and-take of the geometry-particle interaction" demands a really high toned aristocracy, one in touch with God himself, one might say. I have spoken so often against the view of a geometry "high above the battles of matter and energy" that I feel uneasy now to be speaking for a superspace that stands unmoved by what goes on within it— yet I see no alternative to this primacy of superspace in the scheme of things. But this judgement carries with it the responsibility to exhibit the unique simplicity of superspace." [151]

Here, Wheeler's overarching commitments shaped a physical proposal. In order to maintain his ontological view of superspace as primary building block of nature, the view must pass muster with his epistemological criterion of simplicity. As we will see, simplicity remained Wheeler's guide even as he abandoned the primacy of superspace in a search for yet more basic structures.

A transition seems to have occurred between Wheeler's summer thoughts and his conversations that fall. On October 14, 1969, Wheeler noted that the previous Saturday he had "discussed with [Robert] Geroch my feeling that we have to move above superspace as presently conceived to get a conceptual framework where changes in topology occur easily and naturally." Geroch had attended the Battelle conference, and in 1967 published a proof that the equations of superspace could not account for changes in topology—at least without a quantum theory. [152] Now, rather than turning to the Einstein-Schrödinger/Wheeler-DeWitt equation for insight into the path ahead, Wheeler asserted that he was willing to move beyond the formalism. "Don't occur in present formalism—so much the worse for the formalism, I said." However, all was not conceptual anarchy. He held fast to the quantum principle. "Also I stressed again my feeling that the quantum principle comes first, everything else is dependent on it—great 'building power.'" [153]

On a Sunday, January 4, 1970, train to Washington, Wheeler meditated on "Pregeometry and the necessity of the universe." Again his proposals increased in abstraction. This took him from superspace, through a childhood memory, and to Gottfried Wilhelm Leibniz's eighteenth-century monads.

Pregeometry and the superspace in which the pregeometry executes its dynamics are two faces of the same machinery. But why should there have been any machinery

151. JAW, Relativity Notebook XVI, pp. 84–85.
152. Geroch, "Topology in General Relativity"; Wheeler, "Superspace," 302 fn51.
153. JAW, Relativity Notebook XVI, p. 255.

there in the first place? At 4, living on Wick Ave in Youngstown, freshly arrived from Glendale, Calif., I used to go around the circular dining room table, in the central supporting columns of which sort I could drop unwanted bread crusts and ask what is at the end of space and have my mother tell me it goes on and on. Now I am dealing with superspace and asking what is the idea that lies beyond it. Whence does this machinery derive its necessity? "I am that I am" are the words of the Lord in the Bible. But with Thomas Browne we have today the debt of our reason to pay to all this beauty. Few men more to my taste than Leibniz can have reflected on this issue. Constricted though he was in not having the quantum principle, he still surely had behind his monad concept some of the concepts of necessity that we seek for here.[154]

Why is there anything at all? God? Since the seventeenth century and the Christian-Baconian writing of Thomas Browne, Wheeler suggested, even if God is the ultimate, we are obliged to pursue rational explanations of the world. And this led him to Leibniz. Wheeler did not explicitly cite a work of Leibniz's. However, we can briefly reconstruct some of Leibniz's metaphysics to give insight to Wheeler's thinking. Monads were simple substances, the ultimate atomic constituents of nature. For Leibniz, each monad was an individual perfection, a sort of inanimate soul.[155] Further, each soul bore the marks of its past, present, future, "and traces even of everything which occurs in the universe." In this way "every substance is like an entire world and like a mirror of God, or indeed of the whole world which it portrays, each one in its own fashion."[156] Wheeler, like Leibniz, was searching for the ultimate constituents of reality, and in that search each had found reasons to imagine individual entities that exist in time in a different way from the ordinary.[157] Monads included the past, present, and future of the universe; fluctuating $^{(3)}\mathcal{G}$ geometries are spacetime and matter but exist in a way such that "the question what happens 'next' is without meaning."[158]

Wheeler continued to reflect on the relationship between laws of nature as written down by physicists on pieces of paper and the actual structures of existence:

We think of no creation; it simply is. What kind of insubstantial nothingness is it, however? Do the words on a piece of paper jump off in the night hours and execute devices? Do the terms in an equation spring to life as midnight trolls? Are we such pale imaginings, too? [...] "We are such stuff as dreams are made on." Matter out of space; pure curved empty space, and nothing more; and space out of pregeometry; and pregeometry out of one or another equation that has taken wing?[159]

Wheeler was considering a path to rationalism, or perhaps a Pythagorean view of the world as mathematics.

154. JAW, Relativity Notebook XVII, 5. See also p. 20.
155. Monadology §§1, 3, 18, 19, Leibniz, Discourse on Metaphysics, 251, 255.
156. Discourse §§VIII, IX, Leibniz, Discourse on Metaphysics, 14, 15.
157. See also JAW, Relativity Notebook XVIII, p. 176.
158. Wheeler, "Superspace," 253.
159. JAW, Relativity Notebook XVII, pp. 5–6, quoting William Shakespeare's The Tempest, Act IV.

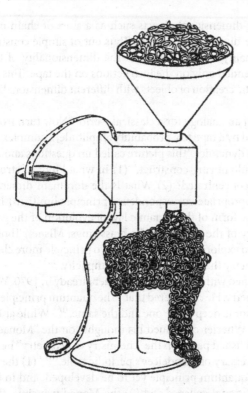

Figure 5.10. Misner, Thorne, and Wheeler's (1973, p. 1210) diagram of "Ten thousand rings." Republished with permission of Princeton University Press, from Charles W. Misner, Kip S. Thorne, and John Archibald Wheeler. *Gravitation*. Princeton, NJ: Princeton University Press, 2017; permission conveyed through Copyright Clearance Center, Inc.

In their mammoth, 1279-page, *Gravitation* (1973), Misner, Thorne, and Wheeler present "Wheeler's story about the vizier and what the vizier had to say about superspace [as] told at the May 18, 1970 Gwatt Seminar on the Bearings of Topology upon General Relativity," in Switzerland.[160] The story imagines a vizier commanding his subjects to assemble 10,000 brass rings and an apparatus "that will cut open a ring, loop it through another ring, and resolder the joint." This device, illustrated in figure 5.10, would take commands from a strip of computer instruction paper, like a teletype-reading Turing machine.[161] Depending on the instructions contained on the tape, the machine would produce different shapes of objects. With the simplest instructions a straight line of rings would be created (one-dimensional object). With a more complicated set of instructions a sheet may be made (two-dimensional object). Yet more complicated instructions might produce more

160. Misner, Thorne, and Wheeler, *Gravitation*, 1210; also JAW, B:W564, Relativity Notebook XVII, p. 200; cf. Kilmister, "King and the Vizier"; D. Kaiser, "Two Textbooks."
161. Compare: "Each automaton (that is, apparatus *cum* memory sequence)," DeWitt, "Everett-Wheeler Interpretation," 330.

complicated, three-dimensional objects such as a shirt of chain mail, as illustrated in the figure (three-dimensional object). Thus out of simple constituents, objects of varying dimensionality could be created. The dimensionality of the surface of the object created depended only on the instructions on the tape. This was a remarkable demonstration of the creation of objects with different dimensionalities from a simple base.

But this was still an analogy for a classical world. "Now turn from a structure deterministically fixed by a tape to a probability amplitude, a complex number, ψ(tape)" that might rule the dynamics. This picture called up questions about the shape of the majority of the resultant ring-construct. "(1) In what kinds of structures is the bulk of the probability concentrated? (2) What is the dominant dimensionality of these structures in an appropriate correspondence principle limit? (3) In this semiclassical limit, what is the form of the dynamic law of evolution of the geometry." Here, at the end of the story of the vizier and the brass rings, Misner, Thorne, and Wheeler abruptly declined to explore these questions: "No principle more clearly rules out this model for pregeometry than the principle of simplicity."[162]

Perhaps dissatisfied with the story of the vizier already in 1970, Wheeler continued to explore pregeometry. He considered that "The quantum principle and pregeometry [were] perhaps in some deep sense one and the same."[163] While at High Island at the end of September, Wheeler continued his thoughts on the "Monad principle" while working on the published paper on the "chemistry of geometry" he delivered in 1969 for the 100th anniversary of Mendeleev's periodic table.[164] "(1) the new enlargement of our view of the quantum principle yet to be developed, and to be forced on one, we hope, by gravitational collapse and (2) the Monad principle, that physics at the very small must somehow replicate the physics of the universe." Later that evening he questioned both what he was looking for and his own capacity to find it: "the problem of what, in one word? Collapse? Particles and geometry? Pregeometry? The final capture of the quantum? 'What distinguishes the bird that flies'? The essence of nature? If I fail through lack of wisdom or lack of sufficient animal energy or through being taken before my time, I want to leave the most helpful record possible for those who follow."[165]

In the early morning of April 8, 1971, in bed, while visiting the Battelle Center in Seattle, Wheeler sought to "Capture the heart of the puzzle and lay it out in a few perfect words." He reflected that he was called "the poet of physics" by his late friend Leonard Schiff. "In this spirit one searches. It is difficult for me to believe that geometry, 'simple' 3 dim[ensional] geom[etry], is the foundation of it all. Rather, it is only [...] one way towards the goal. Geometry, e[lectricity]& m[agnetism], and all the USPS must be built of something far simple [sic]." Here "USPS" was a permutation of another (short-lived) slogan, "persistent uniform submerged structure." Geometry itself had now moved definitely out of focus in the ontological picture. Here Wheeler

162. Misner, Thorne, and Wheeler, *Gravitation*, 1210.
163. JAW, Relativity Notebook XVII, p. 20.
164. See chapter 34 of the draft typescript copies of *Gravitation*, Folder: Undated drafts of chapters, Box 1, Misner Papers; Wheeler, "Mendeleev's Atom."
165. JAW, Relativity Notebook XVIII, pp. 18, 20.

briefly alighted upon the idea of basing his pregeometry on binary elements. He was guided by his principle of simplicity. "If going for simplicity go the whole way. Build out of simplest of all, the YES-NO element. Rich Werner proposed this to me when at Cincinnati conf[erence] ~ 2 y[ea]rs ago I put the issue to him, as I do to so many."[166] However, it may have been that this was too simple—a binary calculus is a very restrictive language with a limited possibility of expression. Rather, Wheeler was led back to philosophy—though rather than turning to early modern metaphysics, he turned to modern, formal logic.

At 8 a.m. EST, Saturday, April 10, 1971, Wheeler was in flight from Chicago to New Jersey. He recored the first mention of the proposal for pregeometry he would pursue for the next phase of his career.[167] Pregeometry would be the calculus of propositions. He began,

> The simplicity and beauty of the universe are surely dazzling. How get at this dazzling central machinery? More than anything the right philosophical principle should be the guide today, as equivalence principle was earlier. "Collapse" tells one, it can be believed, that (1) space can come down to Planck size (there is no such thing as spacetime or time[)] (3) [*sic*] even space geometry itself cannot be more than an approximate idea (4) coming out of gravitation and e[lectricity]& m[agnitism] and spin 1/2 *v*'s should be signal enough that one is building up in the right way from pregeometry, without at start trying to calculate particle properties (5) one might speak of a "logical" structure of the universe, in the sense that yes-no does it all. "Logic" means calculus of propositions, too.

Wheeler was searching for a philosophical principle, and he was guided by the physics of the collapsing star. The creation of a black hole tells one that physics must follow down to the tiniest dimensions. At these dimensions, fluctuations of geometry annihilate our concepts of space, time, and spacetime, such that even geometry must in some sense be a smooth approximation of a complex structure. If Wheeler was on the right track his central concerns since the 1950s—geometry, electromagnetism, and the non-classical two-valued neutrino—should emerge from pregeometry. Building from his earlier speculation that the answer may be found in binary statements, Wheeler realized that

> "logic" also included the richer field of propositions. Is there an isomorphism between pregeometry and the calculus of propositions? [...] I have often spoken of the idea of a floor covered with white paper and ruled off in 1-foot squares and someone down on his hands and knees writing in each square a "set of equations for the universe" and the question, what makes one set of equations put on wings are become airborne while all the others lie on the ground, dead. But the calculus of propositions is marvelous, it is not only pregeometry, it is also prenumber. If one is looking for something beautiful and uniquely right, what else could there be but this? Then

166. JAW, Relativity Notebook XVIII, p. 176–78. Cf. Wheeler, "Particles and Geometry."
167. In 1989, Wheeler would return to the idea of a binary calculus in his slogan "it from bit," Wheeler, "Information, Physics, Quantum."

even the quantum principle would be seen as a part of a larger whole. The worlds of mathematics and of physics would be one in a flash.[168]

Wheeler had seemingly worked his way back to a nearly century-old philosophers' dream of reducing mathematics to logic.[169]

But "How trace this idea out and access it?" Wheeler did not turn to the formalism of quantum geometrodynamics, as he had previously. He renounced that formalism as he was confronted by his, and his community's, inability to use it to make useful calculations. His next task was to learn more about logic. "Need to look at structure of the calculus of propositions. Is the word the justified in this sentence?" Was there a singular calculus of propositions? What about the mathematician's conceptual device, the Turing machine? "Turing machine does what? What would be meaning of undecidable proposition in its physical transcription?" Wheeler was sure whatever came out of his thinking would be abstract—but useful!—connections. "Relation between Calc[ulus] of Prop[ositions] and physics (pregeometry and geometry) might be considered to be as abstract, as dualistic, as relation between a function and its Fourier transform, or even more abstract. Notion of generator function come in." What was needed in the face of these abstractions? "an approximation principle, a correspondence principle, to serve as a guide." Already Wheeler was thinking of what such an approximation principle would approximate over: the size of propositions.

> Does one in calc. of prop. get to such large numbers (and whatever in that context is one anyway to make out of the phrase "large numbers") that any sort of smooth approximation works, analogous to JWKB or Fermi-Thomas or correspondence principle in general? —Summing up, the first question now is, is there any such thing as THE calculus of propositions? [...] Landing.[170]

As his flight ended, Wheeler sought analogies to approximation methods he knew from other areas of physics: the Jeffrys-Wentzel-Kramers-Brillouin approximation from quantum mechanics; or the Fermi-Thomas approximation (section 2.3.3); or Bohr's correspondence principle, cast in its limit-to-classical-physics guise. Even as he searched for a fundamental "magic structure," he was aiming for a theory that replicated the successful theories of his day: abstract generalizations that connected to experiment through approximations, idealizations, and perturbation techniques. Even at this abstract and conjectural level, perturbation and approximation structured Wheeler's research, as it had in scattering theory in the 1930s.

This chapter began with Wheeler's Relativity Notebook I. Wheeler's Relativity Notebook XVIII continued with his new interest in the calculus of propositions. Then the series ends, to be continued by a new series of "Pregeometry" notebooks, and while there is more of a story to tell, Wheeler eventually left the vacuum behind.[171]

168. JAW, Relativity Notebook XVIII, pp. 178–79; in electrical engineering, a circuit could be considered "isomorphic to a proposition," Misner, Thorne, and Wheeler, *Gravitation*, 1211.
169. Tennant, "Logicism and Neologicism."
170. JAW, Relativity Notebook XVIII, pp. 178–79.
171. Misner, Thorne, and Wheeler, *Gravitation*, §§44.5–6; Wheeler, "Relativity"; Wheeler, "Pregeometry."

Near the end of his last Relativity notebook, before his customary collection of clippings, he wrote an "Assessment of structure of physics": "If space not really a God given perfection, does space really have any status at all. Space only a reflection of something deeper."[172]

5.8 Conclusion

On April 1, 1974, some of Wheeler's former students—probably including Kip Thorne and Charles Misner—published a parody of Wheeler's thought: "Rasputin, Science, and the Transmogrification of Destiny" by John Archibald Wyler. The title was a play on some of his most recent writing and talks such as his work for the periodic table anniversary "Mendeléev and the chemistry of geometry."[173] This chapter has examined material from Wheeler private notebooks. This complements other work on Wheeler's extensive pedagogical impact.[174] However this parody demonstrates the degree to which Wheeler shared his notebook-thoughts with his students and colleagues. One must know whereof one speaks.

Wheeler's parodists imagined a fictional walk through the woods with the Russian mystic Rasputin. They lambasted Wheeler's propensity for expounding principles and paradoxical mottos. "What shall be our motto for Grigory Efimovich Rasputin? He sought to the end to see the fantastic wealth of scientifice [sic] facts all as transient consequences of one Universal Principle of Vacuity. How much more briefly can we state this theme than 'Not Much out of a Little.'" Matter falling into a "black whole," figure 5.11, caused "the present crisis that one can name in the history of the universe: the transmogrification of Destiny."[175] Into this "whole" fell the great names of philosophy and physics, as well as more pedestrian objects. It was characterized by Void, Cataclysm, and Annual Repetition, "thus transcending any particular conversation, including leprechaun and baron." (One of Wheeler's abiding concerns was that particles that fall into black holes may violate lepton and baryon conservation laws.) "Transmogrification" also parodied Wheeler's use of lists, and his reach toward Leibniz:

In the history of "the world" and its epilog of "the monad," six lessons stand out of relevance to the crisis of the transmogrification of destiny ("a lot of nothing"). First, not much emerges out of a little as the salamanders and grub-worms populate the littoral. Replace fins with legs. Gills without gills.

Last in this list of lessons, "Wyler" conjures Merlin.

172. 7:25 a.m., Tuesday, September 29, 1971. JAW, Relativity Notebook XVIII, p. 252.
173. See chapter 34 of the draft typescript copies of *Gravitation* found in loose pages at the Charles W. Misner Papers, Special Collections, University of Maryland Libraries. And ch. 19 of Rees, Ruffini, and Wheeler, *Black Holes*.
174. Christensen, "John Archibald Wheeler."
175. Wyler, "Rasputin," 175–79.

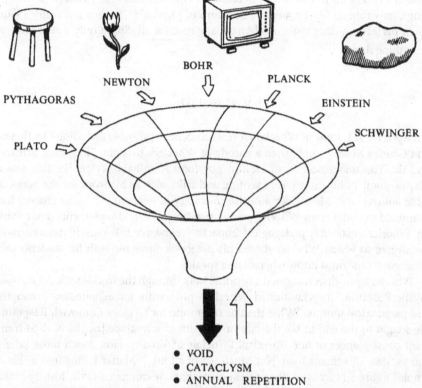

Figure 1 - A black whole reduces whatever matters in particular to a system of void, cataclysm, and annual repetition,
thus transcending any particular conversation, including leprechaun and baron.

Figure 5.11. Wyler's (1974, p. 176) black whole. Reprinted by permission from Springer Nature: John Archibald Wyler. "Rasputin, Science, and the Transmogrification of Destiny." *General Relativity and Gravitation* 5:175–82, copyright (1974).

Sixth, like Merlin the magician, protean prophet, the principle of transmogrification kept changing its shape as it was pursued. Examine the digression. Seek to build a catalog of improbable generalization.[176]

This chapter has tried to trace how the principle of transmogrification changed its shape as it was pursued—how Wheeler's abiding epistemological commitments to follow theories to their simplest foundations shaped his ontological picture. Wheeler was no Merlin, however. He was a philosopher-physicist.

176. Wyler, "Rasputin," 175–79.

6

Roger Penrose's Impossible Diagrams

6.1 Introduction

Chapter 2 discussed P. A. M. Dirac's picture of the vacuum: a sea of strange negative-energy electrons. It was a calm sea, with bubbles manifesting as positive particles. Dirac's pencil sketch of his vacuum (figure 2.10) resembled a cross section of a seascape, as if a waterproof camera were just coming to surface. John Archibald Wheeler (chapter 5) picked up Dirac's nautical analogy, but for Wheeler the vacuum was a heaving sea, with breaking waves, and quantum foam opening and closing connections between bubbles. This was in the context of General Relativity, but Wheeler's work was focused on the small scale of quantum phenomena. Following on Albert Einstein and Nathan Rosen's "throats," Wheeler's "wormholes" in spacetime were candidates for electron-positron pairs. For Wheeler, black holes were fascinating because they were relativistic objects at the atomic scale. In his description, a pilot flying over the vacuum sea swooped so low that a microscope could be used to examine foaming shapes. The pilot did not look up to the horizon.

The horizons of empty space are the subject of this chapter. They formed the global shape of the vacuum, the forms of universes on the page. These horizons were not restricted to the micro-world of quantum theory. Rather, they connected the fields of particle physics, general relativity, and cosmology. And in their history, the horizons of the vacuum connect the history of physics to the history of art and psychology.

These horizons were drawn and formalized by British mathematician and physicist Roger Penrose (b. 1931). During his graduate work at the University of Cambridge, he moved between the mathematics and physics communities (PhD Mathematics, 1957). He worked to apply modern mathematical methods, particularly geometry, to relativity. Penrose's diagrams—now eponymous—were quickly taken up by fellow "relativists" after their introduction in 1962. In the late 1960s Penrose collaborated with a young member of Dennis Sciama's group at Cambridge, Stephen Hawking (1942–2018). Together they applied this mathematics and diagrammatics to the singularities of gravitational collapse—black holes. In 1968 they submitted a draft of this work to the Gravity Research Foundation essay competition, which was founded by industrialist Roger Babson in Salem, Massachusetts. A diagram from this paper is reproduced in figure 6.1. This diagram is in Hawking's hand.[1] As Helene Mialet has shown, these diagrams dominated Hawking's later practice of physics;[2] this is an early and important example of his thinking with them. This work was published

---

1. Roger Penrose, personal communication, December 2015. Hawking's ALS symptoms had not yet claimed his ability to draw. From Hawking and Penrose (1968) "On Gravitational Collapse and Cosmology," AIP Gravity Research Foundation, Box 16, Folder "1968 Competition."
2. Mialet, *Hawking Incorporated*, ch. 3.

More than Nothing. Aaron Sidney Wright, Oxford University Press. © Oxford University Press 2024.
DOI: 10.1093/oso/9780190062804.003.0006

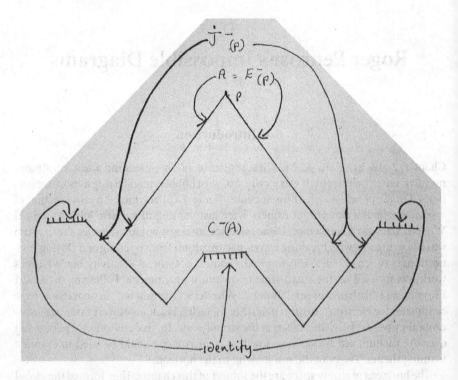

Figure 6.1. Penrose diagram in Stephen Hawking's hand. Courtesy of AIP. Gravity Research Foundation essays and abstracts, 1949–2019, Niels Bohr Library & Archives, American Institute of Physics, College Park, MD USA.

in 1970, and presents what are now called the Penrose-Hawking black hole theorems.[3] Figure 6.2 shows a related diagram to that in the earlier prize essay. If Wheeler's work (and his students') argued that black holes were at the center of physics, Hawking and Penrose's theorems established the *inevitability* of black hole formation.[4] They argued that under certain physical assumptions, a large collapsing star would *always* collapse to a singularity.[5] They provided the certainty many physicists needed to embrace Wheeler's advocacy for black hole physics.

This chapter's narrative stops just short of the black hole theorems. Rather, it focuses on the origins and early development of Penrose's geometry and diagrams of empty spacetime. It begins with the spacetime diagrams that Hermann Minkowski introduced to Special Relativity in 1907. And it explores an additional, unexpected, genealogy. In 1954, Penrose traveled from Cambridge to Amsterdam for the International Congress of Mathematicians, where he visited the neighboring M. C. Escher exhibition. There is a well-known connection between Penrose and Escher, which

3. Hawking and Penrose, "Singularities of Gravitational Collapse and Cosmology."
4. Penrose, "Gravitational Collapse and Space-time Singularities"; cf. Hawking and Ellis, *Large-scale Structure*.
5. Earman, "Penrose-Hawking Singularity Theorems."

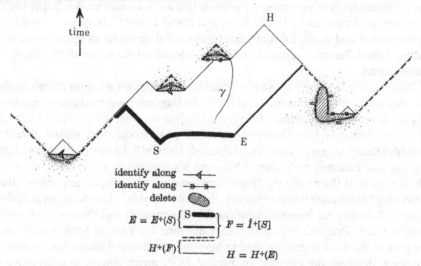

FIGURE 1. A future-trapped set S, together with the associated achronal sets $E = E^+(S)$, $F = \dot{I}^+[S]$, $H^+(F)$, $H = H^+(E)$. (For the proof of lemma (2.12).) The figure is drawn according to the conventions whereby null lines are inclined at 45°. The diagonally shaded portions are excluded from the space-time and some identifications are made. The symbol ∞ indicates regions 'at infinity' with respect to the metric. A future-inextendible timelike curve $\gamma \in D^+(E)$ is depicted, in agreement with the conclusion of lemma (2.12).

Figure 6.2. Penrose diagram showing a future-inextendible timelike curve. Republished with permission of The Royal Society (UK), from Stephen W. Hawking and Roger Penrose. "The Singularities of Gravitational Collapse and Cosmology." *Proceedings of the Royal Society of London A*, 314, no. 1519 (1970): 529–48; permission conveyed through Copyright Clearance Center, Inc.

Penrose himself has elaborated.[6] Escher's tiling birds and fish sparked Penrose's investigations in aperiodic tilings, now called Penrose tiles. He eventually patented the designs.[7] And Penrose's "impossible objects" were incorporated into Escher's work, such as "Waterfall" (1961). Despite their eventual use in architecture and crystallography, the tilings were not a central part of Penrose's scientific life. In contrast, I argue here that the play of perspectives in Penrose's impossible objects is the same sort of play—the same sort of thinking—that underlies his relativistic diagrams.[8] This biographical study shows the development of Penrose's visual skills and reasoning.[9] To establish the connection between visual reasoning and relativity, this chapter next turns to the first diagrams Penrose introduced in the early 1960s. The early history of these diagrams supports a central argument of this book: that physicists' epistemic tools and ontological commitments were intertwined. Rather than

6. Penrose, "Escher and the Visual Representation of Mathematical Ideas."
7. Penrose, Set of tiles for covering a surface.
8. See also Wright, "Origins of Penrose Diagrams."
9. As shown for chemists by Rocke, *Image and Reality*.

the epistemic tools of experimental science, this book focuses on the "paper tools" of theory, as developed by Ursula Klein and David Kaiser.[10] As Penrose honed his mathematical and graphical tools, the fundamental character of space, time, and matter shifted. Penrose's epistemic strategies determined the nature of his objects of investigation.

This is not an argument about physicists' explicit ontological commitments, in the way that some physicists asserted that Feynman diagrams supported a basic ontology of particles rather than fields.[11] Relativists like Penrose were not convinced enough of the correctness of their theory to make such an inference. They looked toward a quantum theory of gravity that would supersede General Relativity (GR). Rather, I am arguing that Penrose's tools shaped his—and his community's—*objects of theory*.[12] These objects of theory are the "paper tool" analogs of the "epistemic things" that Hans-Jörg Rheinberger found in microbiology laboratories.[13] They were not unreal— General Relativity has been well-confirmed by experiment—but they were not *stable* objects, either. Penrose's objects of theory did form the basis of further work, and are part of the models that are used to interpret gravitational wave observations.[14] However, detailing the eventual connection of Penrose's objects to observational evidence is beyond the scope of this chapter. Instead, it focuses on the early influences and development of Penrose diagrams, and how they reconfigured the objects of GR.

Penrose was not alone in his research. In addition to Hawking, the diagrams were quickly put to use and adapted by Rainer Sachs and Brandon Carter.[15] Penrose was engaging in an existing debate about how to understand radiation in General Relativity that had been reinvigorated by Hermann Bondi in 1957, and in collaboration with Felix Pirani.[16] The goal here is not to map out this community in detail.[17] Penrose's new techniques were immediately communicated at conferences, in publications, and at the Les Houches summer school for graduate students. This propagation of theoretical tools contributed to the "Renaissance" of General Relativity as a discipline.[18] David Kaiser has argued for the mutual embrace of diagrams in physics and the postwar boom in physics trainees.[19] Here the focus is less on the social function of diagrams, and more on the intellectual history of Penrose himself, the better to observe how the diverse contexts of mathematics and physics; science and art; and academia and family, came together in the crafting of a paradigmatic vision of the vacuum.

10. U. Klein *Experiments, Models, Paper Tools*; D. Kaiser, *Drawing Theories Apart*.

11. Schweber, "Feynman and Visualization"; D. Kaiser, "Stick-Figure Realism."

12. D. Kaiser, "Whose Mass Is It?"

13. Rheinberger, *History of Epistemic Things*.

14. Kennefick, "Star Crushing," esp. 14–17 on the Spatial Conformal Flatness assumption.

15. Sachs, "Gravitional Radiation"; Carter's work was so influential that the diagrams are sometimes called Carter-Penrose diagrams. Carter, "Complete Analytic Extension"; Carter, "Black Hole Equilibrium States."

16. Thorne, "Gravitational Radiation"; Kennefick, *Speed of Thought*.

17. Colleagues here included Rainer K. Sachs, Robert Geroch, Ezra Newman, Roy Kerr, Joshua Goldberg, and others. See Blum, Lalli, and Renn, "Reinvention of General Relativity."

18. Wright, "Bringing Infinity to a Finite Place."

19. D. Kaiser, *Drawing Theories Apart*, esp. ch. 3 and ch. 7.

6.2 Spacetime Diagrams

Physics has been partly constituted by drawn figures for as long as there has been inquiry that fits the label; and in England, back through nineteenth-century Mixed Mathematics; and seventeenth- and eighteenth-century Natural Philosophy. Isaac Newton (1642–1727) fashioned himself—and his geometric diagrams—as the heir to the mathematics of antiquity.[20] Andrew Warwick connected the origin of a problem solving style among students of Cambridge's mathematical tripos to (in part) the introduction of pens and paper around the turn of the nineteenth century. Aspiring wranglers would begin each examination problem by drawing a diagram.[21] In my own experience, this strategy is currently ubiquitous. Diagrams evolved alongside drawings of apparatuses or of nebulae; graphs of data; of spectra; and of chemicals.[22] While the diversity of these figures and their interactions resists easy generalization, some historians have argued that science includes multiple visual cultures.[23]

Relativistic spacetime diagrams have a history dating back to German mathematician Hermann Minkowski (1864–1909) in the years following Einstein's 1905 Special Relativity paper.[24] Peter Galison has detailed how Göttingen's communities of electrodynamical and mathematical researchers provided fertile ground for Minkowski's application of geometrical ideas to physics. Minkowski saw that Einstein's restructuring of simultaneity in terms of rod-and-clock reference frames was incomplete. More than just the single time dimension was relative; the spatial dimensions of length, width, and height were also. Together they formed a four-dimensional "world-picture." Galison argues that Minkowski's merger of three-dimensional space with time was connected to ontological commitments beyond the rejection of Euclid's and Kant's space and time. Minkowski related his world-picture to Hendrik Lorentz's electromagnetic worldview,[25] and a Leibnizian pre-established harmony between mathematics and the natural world.[26]

Penrose's diagrams were not the first spacetime diagrams to engender ontological shifts. Minkowski's 1908 lecture on *Space and Time* includes the diagrams in figure 6.3.[27] On the left, a *Weltlinie* (worldline) snakes upward, tracing the past and future positions of an observer at *P*, and intersecting with the dotted curvature hyperbola. On the right, two worldlines, l_1 and l, are drawn in rough parallel. A system of coordinates, x, y, t, for l allow the construction of different measures of distance between the two lines. Alongside the four-dimensional approaches of Henri Poincaré and Arnold Sommerfeld, Minkowski's four-dimensional world of words, symbols, and images propagated throughout modern physics.[28] Minkowski diagrams were a

20. Guicciardini, *Isaac Newton*.
21. Warwick, *Masters of Theory*, ch. 3.
22. Nasim, *Observing by Hand*; Franklin, *Shifting Standards*; Hentschel, *Mapping the Spectrum*; U. Klein *Experiments, Models, Paper Tools*.
23. Tucker, "Historian, Picture, Archive"; Hentschel, *Visual Cultures*, ch. 2.
24. Einstein, *Einstein's Miraculous Year*, 123–60.
25. Hirosige, "'Lorentz' Theory of Electrons"; Darrigol, "Electron Theories."
26. Galison, "Minkowski's Space-Time."
27. Minkowski, *Raum und Zeit*; Minkowski, "Time and Space."
28. Walter, "Breaking in the 4-vectors."

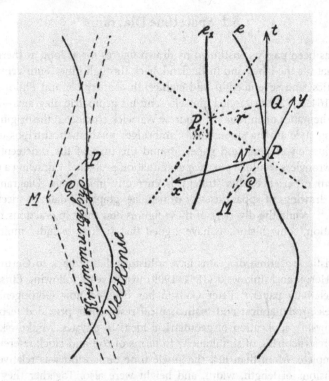

Figure 6.3. Hermann Minkowski, *Raum und Zeit* 1909, 12.

foundation of Richard Feynman's diagrammatic thinking, and how other physicists reacted to his work.[29] John Archibald Wheeler was known for his extensive used of diagrams in geometrodynamics (chapter 5).[30]

A more modern Minkowski diagram is shown in figure 6.4. In a modern diagram, time runs along the vertical axis and one spatial dimension runs along the horizontal axis. This leaves two spatial dimensions "suppressed." (Compare figure 6.3 right, in which Minkowski used perspective to show a *y*-axis pointing into the page. There, only the *z*-axis is suppressed.) The paths of "observers," O, through space-time—whether they be persons, or galaxies, or experimental apparatuses—trace out "world-lines" on the diagram. There is a powerful convention in spacetime diagrams that the time axis is understood to be multiplied by the speed of light, *c*, resulting in a diagram with units of spatial-extent on all axes. This means that Einstein's postulate that light travels at *c* in all frames of reference is rendered graphically by always drawing rays of light at 45° angles. Then the requirement that nothing travels faster than light becomes the rule that observers' world lines cannot have a slope of less than 1; that is, the possible paths of all observers are confined to their "light-cone" with borders at 45°.

29. Gross, "Pictures and Pedagogy"; D. Kaiser, *Drawing Theories Apart*, esp. chs. 5 and 6.
30. Galison, "Re-reading the Past"; Hentschel, *Visual Cultures*, §3.4.

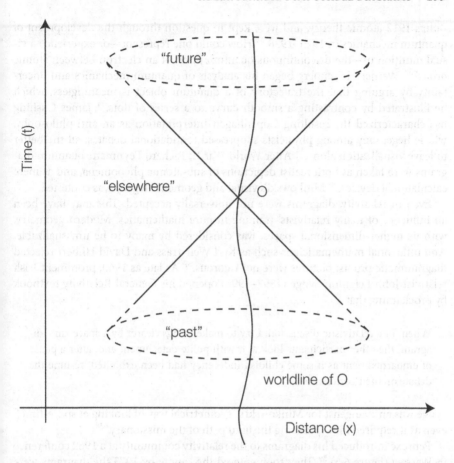

Figure 6.4. A standard Minkowski diagram with observer *O*, and Penrose's "three regions."

In the long history of philosophical debates over time—and in many contemporary discussions—the issue is framed in terms of the reality and structure of the past, present, and future. For St. Augustine, there was only the present; contemporary proponents of the "block universe" assert that all of the past, present, and future are real. Minkowski's work revealed that relativity invoked a fourth possibility: "The territory between the cones," or what Penrose called "elsewhere."[31] Elsewhere was a region of spacetime that an observer has never seen and can never reach. In addition to Minkowski's more elaborate philosophical connections, this sort of basic refiguring of the structure of space and time shows how new mathematical and graphical insights were tied to metaphysics.

However, even as Minkowski's four-dimensional framework was integrated into a wide swath of physics, the images were not always brought along. The propriety and possibility of visual representations of atomic phenomena were put in flux by Niels

31. Minkowski, "Raum und Zeit," xxvii; others called it simply the "present": Synge, *Relativity: The Special Theory*, §1.8, but "elsewhere" may not have been original to Penrose.

Bohr's 1912 atomic theory, and were kept in question through the development of quantum mechanics in 1925–1928.[32] How could one represent—or experience a visual intuition of—the discontinuous quantum jump of an electron between atomic orbits?[33] Werner Heisenberg began his analysis of quantum mechanics and uncertainty by arguing that the trajectory of a quantum object is meaningless, which he illustrated by contrasting a smooth curve to a series of dots.[34] James Cushing has characterized the resulting Copenhagen Interpretation as an anti-philosophy, whose hegemony among physicists suppressed foundational debates; all the better to leave visualization alone.[35] After World War II, Richard Feynman's quantum diagrams were taken as both realist depictions of sub-atomic phenomena, and as mere calculational devices.[36] Minkowski's visual and geometric legacy was contested.

Even in relativity, diagrams were not universally accepted. This may have been an influence of many relativists' training in pure mathematics. Modern geometry, with its higher-dimensional spaces, was considered by many to be unvisualizable. And influential mathematicians such as Karl Weirstrass and David Hilbert rejected diagrammatic proofs: pictures were not rigorous.[37] As late as 1960, prominent Irish relativist John Leighton Synge (1897–1995) opened his General Relativity textbook by proclaiming that

> When, in a relativistic discussion, I try to make things clearer by a space-time diagram, the other participants look at it with polite detachment and, after a pause of embarrassment as if some childish indecency had been exhibited, resume the debate on their own terms.

Synge was an evangelist for Minkowski's "geometrical way of looking at space-time," even as it required him "pursuing the hard path of the missionary."[38]

Penrose introduced his diagrams to the relativity community at a 1962 conference in Warsaw (figure 6.5).[39] They soon entered the canon: by 1973 the diagrams were included in influential monographs such as Hawking and George Ellis's *The Large-Scale Structure of Space-Time* and Charles Misner, Kip Thorne, and John Archibald Wheeler's *Gravitation*.[40] However, even though now standard, diagrammatics did not become universal in relativity. Steven Weinberg's 1972 textbook *Gravitation and Cosmology* contains almost no diagrams of any sort, and no Penrose diagrams. But Weinberg's choice to elide the diagrams and their accompanying mathematics circumscribed the topics he was able to cover (though he also attributed this to a

32. Beller, "Birth of Bohr's Complementarity."

33. For an overview, see Kragh, *Quantum Generations*, ch. 11.

34. Heisenberg, "Über den anschaulichen Inhalt," 173; Wheeler and Zurek loosely translate "*Bahnkurve*" as "worldline," which is usually given as *Weltlinie*. Heisenberg, "Physical Content."

35. Cushing, *Quantum Mechanics*.

36. D. Kaiser, "Stick-Figure Realism."

37. Galison, "Suppressed Drawing," 151–54, 161–62; Mancosu, "Visualization in Logic and Mathematics"; Barrow-Green and Gray, "Geometry at Cambridge."

38. Synge, *Relativity: The General Theory*, IX.

39. Robinson, Schild, and Schücking, eds, "Relativistic Theories of Gravitation," 20; Demianski, "Jablonna Conference."

40. Hawking and Ellis, *Large-scale Structure*; Misner, Thorne, and Wheeler, *Gravitation*.

J. A. Wheeler (Princeton) and R. Penrose (Austin)

Figure 6.5. Photograph of J. A. Wheeler and R. Penrose. Reproduced from Ivor Robinson, Alfred Schild, and Engelbert Schücking. "Relativistic Theories of Gravitation." *Physics Today*, 16, no. 8 (1963): 17–20, with the permission of the American Institute of Physics.

lack of space). This speaks to the importance of the diagrams, and their inseparability from their associated formalism and results. "The nongeometrical approach taken in [Weinberg's] book has, to some extent, affected the choice of the topics [...] covered. [...] [He] regret[s] the omission here of a detailed discussion of the beautiful theorems of Hawking and Penrose on gravitational collapse." In the body of Weinberg's text, he included citations to the major developments in black hole physics, even those by Penrose and by Wheeler's group that saw the role of geometry in relativity as more than an "analogy." His brief section on "Gravitational Collapse" gave details for only the simplest calculations of the duration of collapse, and the behavior of light escaping a star's surface.[41] Unlike for Synge, after Hawking and Penrose's work, those who eschewed geometry and diagrams could not simply resume debate on central issues in relativity on their own terms. Penrose's conformal techniques became a sort of obligatory passage point for black hole physics.[42]

In order to understand how this shift in General Relativity occurred, I now turn from spacetime diagrams to a second lineage of Penrose's diagrams, in art and the psychology of perception.

6.3 Impossible Objects

While Penrose was an undergraduate mathematics student at University College London in the early 1950s, he traveled to visit his older brother Oliver, who was a

41. Weinberg, *Gravitation and Cosmology*, "approach" on viii, "Analogy" on 147–48, references on 350–54, section 11.9, 342–49.
42. Cf. Latour, *Science in Action*, 245.

research student at Cambridge. Oliver brought Roger out to lunch at the Kingswood with a fellow research student, the cosmologist Dennis Sciama (1926–1999).[43] Roger had listened to the cosmologist Fred Hoyle's BBC Radio broadcasts, in which he presented his Steady State cosmological model of a universe with no beginning or end, and continuous creation of matter-out-of-nothing.[44] Hoyle (1915–2001) was a lecturer at St. John's College, Cambridge, and by the end of the decade would be appointed Plumian Professor of Astronomy and Experimental Philosophy. In his radio program, he addressed an old puzzle in cosmology called Olbers' paradox: if the universe is infinitely old, and has an infinite number of stars, why is the night sky mostly black? Shouldn't the sky be filled with the light from all the infinite stars? Hoyle's answer involved the expansion of the universe. Galaxies could recede from our own faster than the speed of light, and then disappear.[45] "Indeed if it [a galaxy] is far enough away the light never reaches us at all because its path stretches faster than the light can make progress."[46]

In a 1989 interview with Alan Lightman, Roger Penrose recalled that

Oliver told me that Dennis was a cosmologist. So I said to Dennis, "Well, you're a cosmologist aren't you. Can you explain this to me about galaxies disappearing off the edge? It doesn't make any sense." I drew a picture and said, "Surely, you can always see the galaxy. You draw back your light cones, and any galaxy crosses the cone." Apparently he hadn't thought about this before and was very impressed.

Oliver brought this to Hoyle, who agreed. In reflection, Roger continued with a characterization of attitudes toward drawing that is consistent with Synge's comments (above):

I don't think they (Hoyle and Sciama) were used to drawing pictures; whereas, that was the sort of thing I did naturally, being geometrically minded.[47]

Upon finishing his mathematics B.Sc., Penrose joined St. John's College (Hoyle's college), Cambridge, for a PhD in mathematics. He initially worked with the geometer W. V. D. Hodge. However, he maintained his interest in physical science, and took courses in General Relativity with Hoyle's Steady State compatriot Hermann Bondi, and in Quantum Theory with Dirac.[48] In this way he maintained connections with his brother's friends (including Felix Pirani) past Oliver's graduation in 1953.[49]

This confluence of family, drawing, and formalism repeated itself soon after. This section sets out Penrose's encounter with M. C. Escher, and his work with his father on mathematical puzzles and the psychology of impossible objects. There is

43. Lightman, "Interview of Roger Penrose."
44. Hoyle, *The Nature of the Universe*; Kragh, *Cosmology and Controversy*, chapter 4.
45. Similar ideas of objects dimming as they receded were common; see Milne, *Relativity*, Plate 1.
46. Hoyle, "Man's Place in the Expanding Universe," 420.
47. Lightman, "Interview of Roger Penrose."
48. Garcia-Prada, "Interview Penrose, Pt. 1," 18.
49. Oliver Penrose's career has been at Imperial College, London, and the Open University. Currently, he is Professor Emeritus of Mathematics at Heriot-Watt University, Edinburgh.

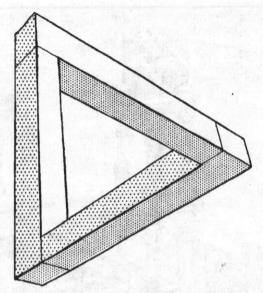

Fig. 1. Perspective drawing of impossible structure.

Figure 6.6. "Perspective drawing of impossible structure." from L. S. Penrose and R. Penrose, "Impossible Objects" 1958, 31. Copyright British Psychological Society.

a well-known connection between Penrose and Escher, in which, first, Escher's tiling patterns of birds and fish sparked Penrose to investigate the aperiodic tilings now called Penrose tiles; and, second, Penrose and his father's "impossible objects" (figure 6.6) were integrated into Escher's artwork, in prints such as figure 6.7, "Waterfall" (1961).[50] Penrose actually sent Escher a copy of his 1958 "impossible objects" paper.[51] In an interview, Penrose recalled that tiling was recreational: "Tiling problems have always been a doodling side interest of mine, just for fun; if I got bored with what I was doing I'd try and fit shapes together, for no particular scientific reason."[52] These shapes have found some application in science and mathematics (crystallography, for example) but they were not a central aspect of Penrose's scientific life. In this section, I will argue that some of the plays of perspective of looking at—and drawing—impossible objects are the same as looking at and drawing Penrose diagrams in General Relativity. As such, I claim that this art, psychology, and recreational mathematics sits squarely in the lineage of contemporary understandings of spacetime and Relativity.

In September 1954, as a research student at St. John's, Penrose attended the International Congress of Mathematicians meeting in Amsterdam. In association with the

50. Penrose, "Escher and the Visual Representation of Mathematical Ideas"; Hofstadter, *Gödel, Escher, Bach*, 13.
51. Penrose and Penrose, "Impossible Objects"; Garcia-Prada, "Interview with Penrose, Pt. 2," 12.
52. Garcia-Prada, "Interview with Penrose, Pt. 2," 12.

Figure 6.7. M. C. Escher's "Waterfall" © 2021 The M. C. Escher Company–The Netherlands. All rights reserved. www.mcescher.com

Congress, the Stedelijk Museum staged an exhibition of work by Dutch artist Maurits Cornelis Escher (1898–1972).[53] There he would have seen plane-filling lithographs of birds, fish, and reptiles that fueled his doodling into aperiodic tilings. He also

53. Escher, *[Tentoonstelling/ Exhibition]*.

would have seen prints that played with perspective, such as Escher's famous (1953) "Relativity." In this print, three different axes of gravitational attraction pull perpendicular to three staircases. What counts as "up" and "down" shifts dramatically as one's eye roams across the print. It is this play of perspective that I see manifest in Penrose's work in General Relativity. But rather than a direct reference from Escher's art to science, Penrose first developed these concepts in collaboration with his father, in different contexts.

In the December 25, 1958, issue of the weekly magazine *New Scientist*, Roger and his father Lionel S. Penrose (1898–1972) compiled a spread of "Puzzles for Christmas." Lionel was the Francis Galton professor at University College, London, and had a strong amateur interest in mathematics.[54] He is best known for establishing the correlation between parental age and Down syndrome.[55] "Puzzles" was a visual array spread across two pages that invited readers to trace paths through mazes, consider the geometry of billiard tables, and tile a plane with irregular shapes. It included a "Staircase for lazy people" and "Parking a car," which featured an "impossible" staircase and ramp (figure 6.8). "The first problem" on the staircase "is to go from A to B mounting the fewest number of steps. [...] The further problem is to find out why" it is possible to go one fewer step than seems necessary, "or, rather, why the staircase is impossible." The staircase is for lazy people because it appears that the same height could be climbed with one fewer step. A similar set of stairs is the main feature of Escher's "Ascending and Descending" (1960). For the car park, Penrose and Penrose note that "the principle of impossible objects can be extended to cover situations in which perspective produces peculiar effects because size relationships are rendered inconsistent." If one imagines the car driving around the ramp, it must shrink down with the roadway, and be able to fit out the small gate. According to the answer key published at the end of the issue, "Puzzles 2 and 3 only require the comment that they are constructed on the principle that, by itself, each part of the drawing is logical but that the parts are connected connected in such a way that the whole figure is inconsistent."[56]

Compare this to Escher's famous print, "Relativity." Obviously, staircases were in common. But the use of visual perspective was more sophisticated in Penrose and Penrose's puzzles, in which scale was at play. Escher's staircases were disjointed, whereas Penrose and Penrose's connected objects heightened the sense of illusion. "Relativity"'s interest did not arise from changing visual perspective; rather, it arose from the different physical orientations of the three stairs and walking figures. Both sets of images fell under Penrose and Penrose's "solution" to their puzzles: just like the impossible staircase and car park, Escher's stairs were logical when taken individually, but as a whole the three were inconsistent. On Penrose and Penrose's own understanding, their work functioned in the same way as Escher's "Relativity," as a conflict between parts and wholes.

Penrose and Penrose further developed their work and play with impossible objects, and the experience of looking at them and thinking with them. Also in 1958, they published an article entitled "Impossible Objects: A Special Type of Visual

54. Lightman, "Interview of Roger Penrose."
55. Kevles, *Eugenics*, ch. 10.
56. Penrose and Penrose, "Puzzles for Christmas."

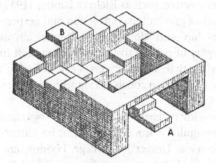

FIGURE 2.

2.—Staircase for lazy people

The first problem in Figure 2 is to go from *A* to *B*, mounting the fewest number of steps. If we turn to the right after four steps and then to the left we climb eleven steps before reaching the top. Similarly, by turning to the left after four steps and to the right after four more, eleven steps are climbed. But *B* can be reached by mounting only ten steps and also descending three on the way. This is done by turning to the right after four steps and then to the right again and then crossing the bridge. The further problem is to find out why this is possible or, rather, why the staircase is impossible.

3.—Parking a car

The principle of impossible objects can be extended to cover situations in which perspective produces peculiar effects because size relationships are rendered inconsistent. In Figure 3, the driver of a motor van wishes to exit through the archway at *A*. At present his van is too large to get through but, by circling the track, never going up hill, in an anticlockwise direction, he can reduce his size sufficiently to get out. The track might be used for making cars smaller so that a great number could be conveniently parked in a confined space. If the reader is not able to follow this argument he should not waste time considering it but should at once proceed to the next puzzle.

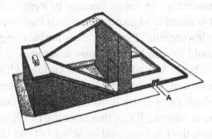

FIGURE 3.

Figure 6.8. "Staircase for lazy people" and "Parking a car," from Penrose and Penrose, "Puzzles for Christmas," 1580. (c) 1958 New Scientist Ltd. All rights reserved. Distributed by Tribune Content Agency.

Illusion," in the *British Journal of Psychology*.[57] It cited the 1954 catalog of Escher's exhibition at the Stedelijk Museum that R. Penrose had visited.[58] They first presented a perspective drawing (figure 6.6), each vertex of which could be accepted as representing a three-dimensional rectangular structure. (While looking at figure 6.6, cover any one corner and note the perceived direction of the bars.) The bars in the drawing were, however, connected in such a manner as to produce an impossibility. As the eye pursued the lines of the figure, sudden changes in the interpretation of distance of the object from the observer were necessary. Taken as a whole, the image was inconsistent.

Despite the presentation of these objects as "impossible," it is important to note that this did not mean they were not real. The complications of the figures "leads to the

57. Penrose and Penrose, "Impossible Objects."
58. Escher, *[Tentoonstelling/ Exhibition]*.

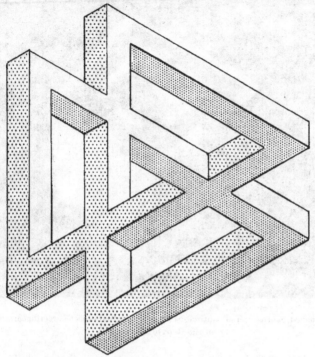

Fig. 2. Diagram of structure with multiple impossibilities.

Figure 6.9. "Diagram of structure with multiple impossibilities," from Penrose and Penrose, "Impossible Objects," 32. Copyright British Psychological Society

illusory *effect* of an impossible structure," but "[a]ctual objects suitably designed [...] can give exactly the same impressions as inconsistent drawings."[59] This point is reinforced by the final figure in the article, which presents an "[u]ntouched photograph" of a physically constructed "continuous staircase" (figure 6.10). It was a fantastic image, and Penrose and Penrose emphasized that it was "untouched": not a photo-collage or other deception. This caution may reflect the concern for scientific fraud that was raised by the recent unraveling of the Piltdown Man hoax in the *Times*.[60] The authors note that "[a]ctually the extreme right hand step was much nearer to the camera and on a much higher level than the step which appears to be just above it." Sketches and photographs for this paper (and into 1962) are included in L. Penrose's archive at University College London.[61] The photograph of the model appears to be taken at the Galton Laboratory.

59. Emphasis added, Penrose and Penrose, "Impossible Objects," 31, 32.
60. Museums Correspondent, "Piltdown Man Forgery."
61. UCL Special Collections, L. S. Penrose Papers, PENROSE/1/7/1, "Penrose Stairs." See also PENROSE/1/7/3, "Puzzles." Images available at https://wellcomecollection.org/works/quedyxap.

Fig. 4. Untouched photograph of continuous steps: the actual object is rectilinear and the surfaces of the steps are horizontal.

Figure 6.10. "Untouched photograph of continuous steps: the actual object is rectilinear and the surfaces of the steps are horizontal" from Penrose and Penrose, "Impossible Objects," 33. Copyright British Psychological Society

Penrose and Penrose cited only one source beyond the Escher catalog: Frederick P. Kilpatrick's introduction to his edited volume on perception experiments at Princeton University and the Navy's Institute for Associated Research in Hanover, New Hampshire.[62] During his PhD studies at Princeton, Kilpatrick used experimental devices invented by Adelbert Ames Jr in New Hampshire.[63] Ames was a founding member of the Dartmouth Eye Institute and developed his interest in visual perception and the arts with his sister, the artist and suffragist Blanche Ames Ames.[64] Ames's work was generally known in the psychology literature, but it is unclear how Penrose and Penrose found this somewhat obscure US Navy volume. Most likely, Lionel Penrose discovered this research program while he (and his family) lived in North America during World War II. L. Penrose was the director of psychiatric research at a hospital in London, Ontario.

Ames's apparatuses were aimed at understanding perception through illusion. The "Ames window" (figure 6.11) is well known, and the "Ames room" is used in contemporary filmmaking to give the impression of vastly different scales in the same image (the heights of wizards and hobbits, say). The point was to show how different

62. Kilpatrick, *Human Behavior.*
63. Behrens, "Adelbert Ames, Jr."
64. Behrens, "Ames and Ames."

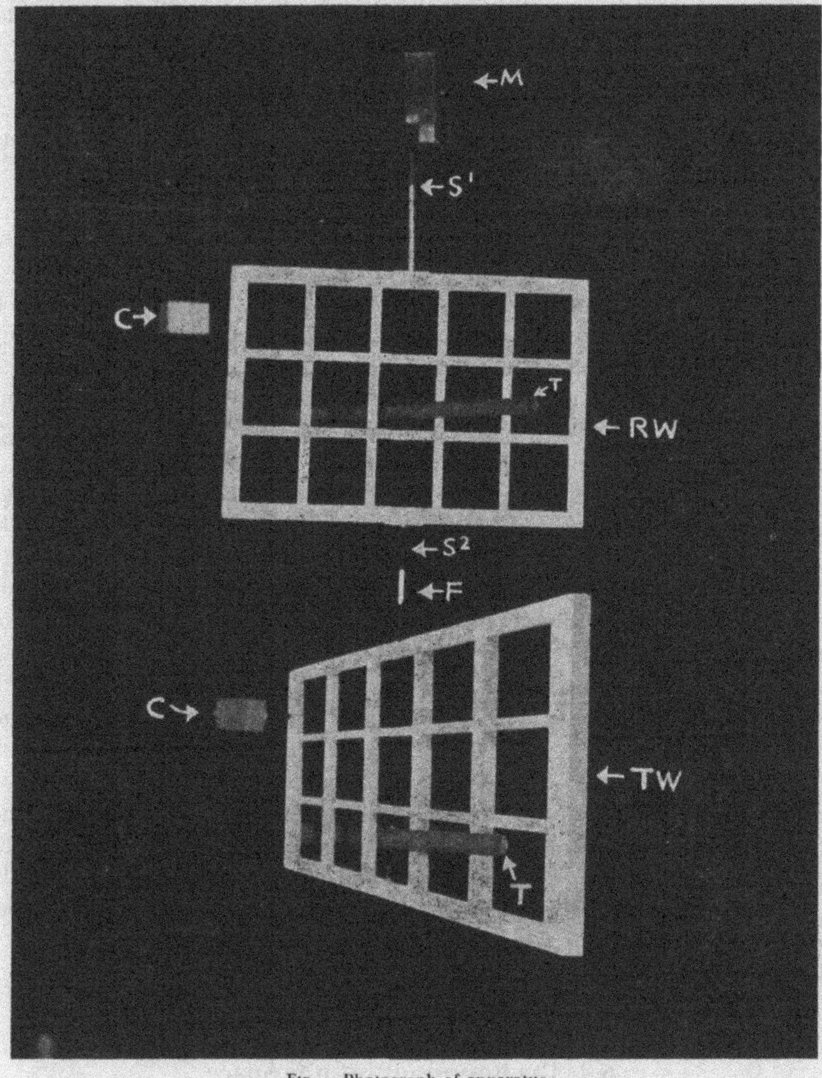

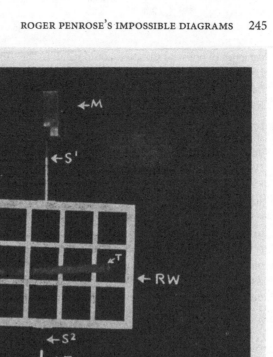

FIG. 1. Photograph of apparatus

The photograph was taken with the surfaces of the rectangular window and the trapezoidal window in the same plane, which was perpendicular to the optical axis of the camera. The dimensions of the rectangular window are: length, 23½"; height, 16½"; thickness, ¼"; width of outside frame, ½"; width of mullions, ¼". The outside dimensions of the trapezoidal window are: length, 19½"; height of long side, 23⅝"; height of short side, 12½"; thickness, ⅛".

Figure 6.11. A labeled photograph of Ames's window apparatus. "The photograph was taken with the surfaces of the rectangular window [RW, top] and the trapezoidal window [TW, bottom] in the same plane, which was perpendicular to the optical axis of the camera." From Ames (1951), 2.

varieties of physical situation could result in the same sense perception. For example, I find it quite difficult to accept that the trapezoidal window in figure 6.11, bottom, is flat in the same plane as the rectangular window and this sheet of paper (or screen). Rather, it looks like it is not a trapezoid, but a rectangular window held at an angle. My perception is in a sense underdetermined by multiple possible interpretations. More sophisticated illusions result from introducing motion into the scene. The two windows rotate, and an Ames room can make it seem as if water runs uphill (cf. Escher's "Waterfall").[65]

Kilpatrick, Ames, and co-workers such as Hadley Cantril called their work "transactional" psychology. Their larger analytic claim was that because perception was meaningfully underdetermined, the meaning or sense of our perceptions must be influenced by our pre-existing mindset, past experience, or unconscious assumptions. The intellectual lineage here did not lead back to the scientific British idealism of Arthur Eddington or Karl Pearson; rather, Ames drew on American pragmatism in the work of Percy Bridgman and John Dewey (for whom Ames brought demonstrations to New York City).[66]

These drawings and photograph, in addition to emphasizing the thought Penrose and Penrose devoted to perspective, reveal a second genealogy for R. Penrose's conformal diagrams. Not only did they draw upon physicists' tradition of drawing space-time diagrams, dating back to Minkowski and before, they also built upon a local, familial, tradition of rendering "impossible objects" in sophisticated ways. As Penrose and Penrose wrote of their "impossible objects": "As the eye pursues the lines of the figure, sudden changes in the interpretation of distance of the object from the observer are necessary." As I will now show, this is what happened when one looked at a Penrose diagram; the scale changed rapidly from some normal measure in the center of the diagram to the edges, where distances on the page were imagined to be infinite. Penrose diagrams were considered a tool to investigate "global" properties of spacetimes, and contradictions between pieces of Penrose diagrams were manifest. This part-whole contradiction also characterized the impossible triangles and staircases. This second genealogy places art and the psychology of perception squarely within the constituents of Penrose's new diagrams and their role in relativity's "renaissance."

6.4 The Bounds of Infinite Space

Roger Penrose submitted his first publication on relativity to the *Proceedings of the Cambridge Philosophical Society* in the months between his psychological and recreational articles with his father. "The Apparent Shape of a Relativistically Moving Sphere" gave three arguments for a seemingly paradoxical result of relativity combined with visual perspective.[67] According to the FitzGerald-Lorentz contraction

---

65. Ames's article is reprinted with modifications in Kilpatrick's volume. Ames, "Rotating Trapezoidal Window"; Kilpatrick, *Human Behavior*.
66. Ames, "Rotating Trapezoidal Window," 26; Behrens, "Adelbert Ames, Jr.," 278.
67. Penrose, "Apparent Shape."

formula—which Einstein integrated into special relativity—any object, such as a measuring rod, should appear shortened to a stationary observer if the object moved past the observer at a velocity close to the speed of light. To a driver stopped at a rail crossing, a train moving at relativistic velocity appears shortened. However, "[i]t turns out," Penrose argued, "that the appearance of a sphere, no matter how it is moving, is always such as to present a *circular* outline to any observer."[68] Physically, he argued this happens because light from different parts of the sphere reach an observer at different times, and the contracted sphere moves out of the way of light that would normally be blocked by itself.

Penrose's analysis depended on precisely specifying the different perspectives of physical observers and relativist/cosmologist analysts. He wrote that "an instantaneous photograph of a rapidly moving sphere has the same outline as that of a stationary sphere." He provided a note to "photograph," specifying that this was "not a 'snapshot' in the sense of Synge," rather he was "concerned, here, with a world-picture rather than a world-map" in the sense of Edward Arthur Milne (1896–1950).[69] This was a delicate comment, because it is evident that Synge had misunderstood Milne's analysis.[70] He also credited one line of argument to a suggestion of Synge himself. Milne's distinction between world-picture and world-map accounted for the finite velocity of light: we might mathematically construct a model of an expanding universe, and be able to survey it; but any actual image we have of distant stars captures light that traveled over time to us. So, by the time we see the light from a distant star, it has receded even farther from us. In Milne's terminology, our world-picture is always smaller than a world-map. Like an atlas of the globe, a world-map used mathematics to display regions that a single observer could not observe.

Milne wrote in a cosmological context in which the relevant motions were of stars increasing in radial distance from observatories on Earth. Receding stars may have their apparent diameter reduced, but would remain circular. But Penrose considered the more general case, including objects moving latitudinally and longitudinally across the sky. For Penrose, if a relativistically moving sphere were depicted on a world-map, it would appear compressed. However, as seen in a world-picture, as if an observer were taking a photograph, it would appear with a circular outline. Mathematically, this is because the transformation between stationary and FitzGerald-Lorentz contracted frames of reference for an observer is a stereographic projection, which distorts many shapes, but keeps circles as circles. It is a special case of conformal transformations (see below). From the beginning of Penrose's engagement with relativity, he employed careful attention to visual perspective.

6.4.1 Boundary Conditions across Borders

Penrose had received his PhD in mathematics in 1957, after which he began a peripatetic series of positions on both sides of the Atlantic. He was briefly an

---

68. Penrose, "Apparent Shape," 137.
69. Penrose, "Apparent Shape," 137.
70. Milne, *Relativity*, 120; Synge, *Relativity: The Special Theory*, 107.

Assistant Lecturer in Pure Mathematics at Bedford College, London, before returning to St. John's College, Cambridge, for a three-year term as a Research Fellow. However, before that term was up he won a NATO Research Fellowship which sent him to Princeton and then to Syracuse University (1959–1961). At Princeton he joined John Archibald Wheeler's group at the height of his research on geometro-dynamics and the possibility that matter is composed of curled-up empty space (see section 5.4). At Syracuse, he joined Peter Bergmann (Einstein's former assistant), whose group had a strong focus on gravitational radiation. There he met a future coauthor, Ezra Newman.[71] Returning to England, Penrose spent the following two years, 1961–1963, as a Research Associate at King's College, London, only to head back to the United States in 1963 for one year as a Visiting Associate Professor at the University of Texas at Austin. Relativist Alfred Schild (1921–1977) had recently moved from the UT Department of Mathematics to Physics, and helmed its new Center for Relativity. These names and places mapped a substantial fraction of the world's major centers for relativistic research.[72] This helps explain Penrose's influences, and also how his work became influential. Penrose had been enticed to Texas at a relativity conference the year before at another center of relativity research, Warsaw, at which he introduced his diagrams.[73]

The International Conference on Relativistic Theories of Gravitation was held in Warsaw and the nearby Jabłonna Palace of the Polish Academy of Science in July 1962. It was organized by Leopold Infeld (1898–1968), a Jewish Polish relativist who fled fascism and became Einstein's assistant at the Institute for Advanced Study, in Princeton. Their work on the problem of the motion of bodies in general relativity, with Banesh Hoffman, is regarded as a classic.[74] In 1939, Infeld took a position at the University of Toronto, but harassment by conservatives and anti-Semites drove him back to Poland in 1950 (well before Khrushchev's revelations of Stalin's crimes).[75]

The conference spoke to the ambitions of the international community of relativists. It was the third in a series of meetings that accompanied the post–World War II "renaissance" of general relativity.[76] Infeld's introductory remarks self-consciously announced that both senior and junior scholars were invited and given speaking time. And he spoke to scientific internationalism. Infeld welcomed the assembled scientists to a city destroyed by the Nazis—and rebuilt. Though power failed one day at Jabłonna, which disrupted the recording of sessions. In the year after the construction of the Berlin Wall, this conference had 114 participants, from 22 countries, on both sides of Churchill's iron curtain.[77] This conference demonstrates that scientific internationalism remained alive in the community of relativists, even if

---

71. E. Newman "Description of GR"; Kennefick, *Speed of Thought*, 209–10
72. See Blum, Lalli, and Renn, "Reinvention of General Relativity," Table 1, p. 614.
73. Schücking, "First Texas Symposium," 47.
74. Einstein, Infeld, and Hoffmann, "Gravitational Equations I."
75. Infeld, *Why I Left Canada*; Sketches of Infeld and his relationships with Synge, Schild, and others are in Kennefick, *Speed of Thought*, ch. 9.
76. The first was organized to mark the fiftieth anniversary of special relativity in 1955 at Berne, Switzerland. It also coincided with the year of Einstein's death. The second was at Royaumont, France, in 1959. Bergmann, "Fifty Years of Relativity"; Lichnerowicz and Tonnelat, *Théories relativistes de la gravitation*.
77. Infeld, *Conférence internationale*, VIII–X.

Eddington's 1919 eclipse expedition had failed to "heal the wounds of war."[78] The attendees span multiple chapters of this book: Dirac, Feynman, Wheeler, and Charles Misner.

At the conference, Penrose gave a short paper entitled "The Light Cone at Infinity." Its elements were characteristic of his developing relativistic research: it was geometrical, and incorporated his interest in spinors; it was diagrammatic; and it aimed toward a quantum theory of gravitation. This last element may be surprising, but it was far from unique. At the Warsaw conference, quantum gravity was much more a focus than gravitational experiments or astronomical observations. The prevalence of quantum gravity in the Renaissance of Relativity challenges interpretations in the historiography of physics that privilege new observations and experiments as the cause of the field's growth.[79] More broadly, this supports a central claim of this book, that Thomas Kuhn's picture of normal science does not hold up well against much of modern physics.[80] For Kuhn, science was mostly "puzzle solving" "normal science"; as unsolvable experimental "anomalies" accrued, the tinder for a revolution caught fire; soon a new paradigm was established, and normal science resumed. While most debate over Kuhn's work has been over "paradigm shifts" and "incommensurability" across revolutions, here I am challenging a narrow reading of Kuhn's concept of "normal science."[81] The first sentence of chapter IV of *The Structure of Scientific Revolutions*—"Normal Science as Puzzle-solving"—reads:

Perhaps the most striking feature of the normal research problems we have just encountered [in nineteenth-century mechanics] is how little they aim to produce major novelties, conceptual or phenomenal.[82]

In contrast, over decades, Penrose and his fellow relativists (not in the sense Kuhn worried about!) aimed to remake quantum theory, relativity, or both.

While he aimed at the major shift to a quantum gravity, Penrose's Warsaw paper was framed by "infinity." Why infinity? To understand the importance of infinity to physics, it is necessary to understand the role of "boundary conditions" applied to physicists' equations. When describing the electromagnetic field surrounding a charge, for example, one can calculate the magnitude of the field at any distance from the charge, from a centimeter to a kilometer and beyond. However, the laws of electrostatics state that the field falls off as the inverse of the square of the distance from the charge ($\propto 1/r^2$), so that as distance increases the strength of the field approaches zero. But if one wants to know the total energy of the field in a mathematically tractable form, one assumes that the field is *actually* zero "at infinity"; this even though "at infinity" makes little physical sense. Alongside political

78. Forman, "Scientific Internationalism and the Weimar Physicists"; Stanley, "Eddington as Quaker Adventurer."

79. As expressed in Will, "Renaissance of GR"; I develop this point further in Wright, "Bringing Infinity to a Finite Place"; see also Blum, Lalli, and Renn, "Reinvention of General Relativity."

80. See Warwick and Kaiser, "Kuhn, Foucault," 396–98; Seth's "problems-principles" axis is a thoughtful alternative: Seth, "Crisis and Construction," esp. 49–50.

81. For a broader interpretation of normal science, see Mody, "What Do Scientists and Engineers Do?"

82. Kuhn, *Structure*, 35.

boundaries, boundary conditions were an active object of discussion at Warsaw. For example, Dennis Sciama's contribution connected the boundary conditions of electromagnetic radiation to the boundary conditions of expanding universes.[83] This work in relativity was in the lineage of Julian Schwinger's work in Quantum Electrodynamics (chapter 4). Schwinger's formalism specified the boundary conditions for electromagnetic processes on slices of "flat" spacetime in special relativity.[84]

Penrose began his paper:

> Questions concerning radiation [...] involve statements about events in the "neighbourhood of infinity." It would appear [...] that some deeper understanding of the mathematical nature of this "infinity" might be of great conceptual value to physics.[85]

To this end, he introduced the idea of applying a conformal transformation to the spacetime. This was in Cambridge's tradition of projective geometry tracing back through Dirac, the reception of special relativity, and Maxwellian electrodynamics.[86] A conformal transformation is one that preserves the angles between objects but distorts their size; the Mercator projection of the globe of the Earth onto a flat page is a conformal transformation. For Penrose, "[t]he idea [was] that if space-time is considered from the point of view of its conformal structure only, points at infinity can be treated on the same basis as finite points."[87]

Penrose discussed some geometrical aspects of conformal transformations and a conformally-transformed Minkowski spacetime, $\mathscr{M}$: this manifold "contains ∞^4 points and ∞^5 null (straight) lines [...] ∞^9 *space-like circles* and ∞^9 *time-like circles.*" He then gave his audience instructions for picturing $\mathscr{M}$: "[i]magine the whole of two-dimensional Minkowski space-time to be mapped continuously onto the interior of a square," directing them to figure 6.12, left. The convention that light moves at $45°$ dictates that Penrose's square be on point. The hatching within the square indicates that as one moves out from the center of the square, the distance scale shrinks so that "infinity is represented by the sides of the square."[88] This pattern is also found in Escher's "Circle Limit" lithographs.[89] The instructions to "identify" different edges of the diagram invite the reader to mentally manipulate the figure and attach the edges such that the "resultant compact manifold is topologically a torus" (like a doughnut).[90] This active engagement of viewer and image—this mental animation—recalls the "transactional" view of perception Penrose and his father cited for "Impossible Objects." Here, the object of perception was not the diagram, or an image on the audiences' retinas. It was a mental creation that engaged with the empirical stimulus.

---

83. Sciama, "Retarded Potentials and the Expansion of the Universe."
84. Bryce DeWitt adapted this work for GR. See also, Schweber, *QED*, ch. 7.
85. Penrose, "Light Cone at Infinity," 369.
86. Warwick, *Masters of Theory*, 416–24.
87. Penrose, "Light Cone at Infinity," 369.
88. Penrose, "Light Cone at Infinity," 369–371.
89. Coxeter, "Non-Euclidean Symmetry"; Coxeter et al., *M. C. Escher*, Fig. 22, p. 389.
90. Penrose, "Light Cone at Infinity," 371.

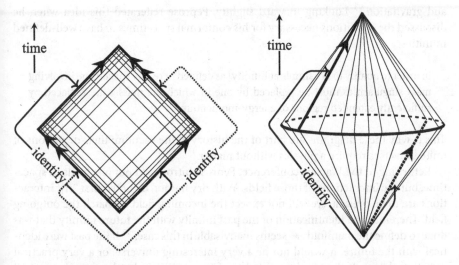

Figure 6.12. Illustrations of conformally transformed Minkowski manifolds, after Penrose's "The Light Cone at Infinity" (1964). Credit: author.

This mental manipulation had consequences: the conformal transformation allowed infinity to be placed on the page, but at the price of a change in topology from a "flat" Minkowski spacetime to a torus. This change in topology reconfigured the basic concepts of relativistic space. After the transformation "the three regions 'past,' 'future,' 'elsewhere' into which a [light-] cone divides normal Minkowski space [see figure 6.4] are *connected* to each other."[91] In the conformally transformed space, there was no certain distinction between past and future, or between regions an observer could travel to, and regions that can only be reached by traveling faster than light. "There [was], thus, no invariant distinction between a space-like or a time-like separation for two general points in" the conformally transformed manifold.[92] These were the type of major conceptual shifts Penrose's treatment engendered—infinity was made tractable and place-able at the cost of reconfiguring the basic concepts of relativistic space and time.

This mathematics could be connected to physics because *only* zero-rest-mass fields can be interpreted as being conformally invariant. Only these objects could be passed between Penrose's "physical" and "unphysical" metrics, so that conclusions drawn in the conformal space might apply to physical spacetime. But zero-rest-mass fields do not describe matter like electrons; rather, they describe fields like light and gravitation. These fields were the object of Weyl's and Einstein's quantum gravity investigations, and they were the "nothing" that John Wheeler attempted to make into "everything" (chapter 5). In his GR textbook, Weyl suggested that mass was just an accumulation of field energy of electromagnetism

91. Penrose, "Light Cone at Infinity," 371.
92. Penrose, "Light Cone at Infinity," 371.

and gravitation.[93] Looking forward slightly, Penrose reiterated this idea when he discussed the conditions necessary for his conformal spacetimes to have well-defined infinities:

> in order to qualify to be [simple at infinity] as defined here, any space-time involving mass singularities must be replaced by one in which the masses are replaced by sufficiently extended regions of energy-momentum.[94]

This is why these diagrams are part of the history of the vacuum; their depictions of finite universes require a physics without matter.[95]

Returning to the Warsaw conference, Penrose introduced not just fields in space-time, but the interactions of those fields. With dry humor, he noted that "[i]f interactions are to be present, we will not expect the incoming field to match the outgoing field. Therefore, the identification of the past infinity with the future infinity that was done to define the manifold $\mathscr{M}$ seems inadvisable in this case."[96] If the past were identical with the future, it would not be a very interesting universe, or a very practical radiation theory. If the topological twists of tori and Klein bottles from figure 6.12 are undone, Penrose could compare incoming and outgoing fields at past infinity and future infinity, as if they were normal points. In the final words of his presentation, he pointed his audience to the dominant tool of Quantum Field Theory predictions: "The possibility of an S-matrix theory for gravitation suggests itself."[97] Of course, Penrose was the one doing the suggesting.

6.4.2 An Atlas of Spacetime

Soon after the 1962 Warsaw conference, Penrose submitted a paper to *Physical Review Letters* (PRL) refining the conceptual shifts of his conformal techniques and their dia-grammatic accompaniments.[98] PRL was the flagship journal of the American Physical Society—as a "letters" journal, its brief reports were supposed to make the most important new research in physics accessible to the widest audience of English-reading physicists. In PRL, Penrose no longer insisted that his mathematical treatment demanded "identifying" the pasts and futures of conformally transformed Minkowski spacetime. Rather, he noted that one can simply "remove" (i.e., ignore) a light-cone of points in $\mathscr{M}$ that severs the earlier identifications. For the physicist-audience of PRL, there could be distinctions between "past," "future," and "elsewhere."

However, the concepts of past and future were not returned to their previous state. Rather, the concept of "infinitely far away" was broken up into five pieces, according to whether "far away" was in the future, past, or "elsewhere," or was accessible only to

93. Weyl, *Space-Time-Matter*, 200.
94. Penrose, "Zero Rest-Mass Fields Including Gravitation," 183.
95. Penrose's adaptation of Minkowski space also brings in the vacuum. Minkowski space is "flat," and according to GR, flat space is empty of particulate matter.
96. Penrose, "Light Cone at Infinity," 372.
97. Penrose, "Light Cone at Infinity," 373.
98. Penrose, "Asymptotic Properties of Fields and Space-times."

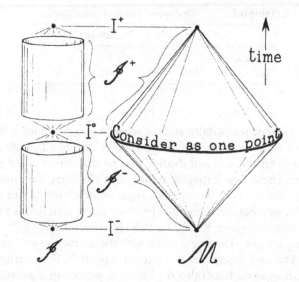

Figure 6.13. Penrose's "Conformal structure of infinity" for Minkowski spacetime. Reprinted figure with permission from Roger Penrose, "Asymptotic Properties of Fields and Space-Times." *Physical Review Letters* 10, no. 2: 66-68. https://doi.org/10.1103/PhysRevLett.10.66, copyright (1963) by the American Physical Society.

particles traveling at the speed of light. Penrose relayed the "basic idea" of his work as arising from the puzzle that (as above) he wanted to be able to deal with quantities at infinity even though "there is no such thing as a point <u>at</u> infinity [...].

> But if we think only in terms of <u>conformal</u> structure of space-time (only <u>ratios</u> of neighboring infinitesimal distances are to have significance), then infinity can be treated as though it were simply an ordinary three-dimensional boundary $\mathscr{I}$ [pronounced "scry"] to a finite four-dimensional conformal region.[99]

Penrose's $\mathscr{I}$ was the boundary of Minkowski space, $\mathscr{M}$, in words, symbols, and figures. Scry was a concatenation of *script I*. "$\mathscr{I}$ can be separated into five distinguishable disjoint parts" depicted in figure 6.13: three infinite distances for objects not traveling at the speed of light, future infinity I^+, the past infinity I^-, and the "spatial" infinity corresponding to the "elsewhere" direction in figure 6.4, I^0; and two infinite surfaces, $\mathscr{I}^+$ and $\mathscr{I}^-$, representing the future and past infinities for light, respectively (see table 6.1).

These figures were the material basis for the physical concepts that Penrose analyzed—they were material on the page, to be drawn as redrawn in notebooks and on blackboards.[100] And they used the conventions of realistic drawing to more strongly evoke the sense of engaging with a physical object. Figure 6.13 was drawn in

---

99. Penrose, "Asymptotic Properties of Fields and Space-times," 66.
100. For analysis of mathematicians' use of inscriptions, see Barany and MacKenzie, "Chalk."

Table 6.1. The five disjoint parts of infinity, $\mathscr{I}$.

	Past	Elsewhere	Future
Massive fields	Γ^-	I^0	I^+
Zero rest-mass fields	$\mathscr{I}^-$	—	$\mathscr{I}^+$

perspective and shaded realistically, as if the reader were confronted with a spinning top. Rather than Warsaw's mathematical "generators" (figure 6.12, right), in PRL the cones were drawn opaque and had shading that became denser at the edges, which placed an inferred light-source roughly at the reader's location. The same was true of the cylinders on the left side. These details are seemingly irrelevant to the scientific content of the image—but this irrelevance precisely marks their role as a sort of visual "reality effect," to borrow from Roland Barthes's literary theory.[101]

Text is closely integrated into figure 6.13, with the instruction to "Consider as one point" wrapped around the equator of the cones, labeled I^0. This instruction was not so bizarre as it may seem. Recall that the Mercator projection is a conformal transformation; we know that the south pole of the Earth is a point on the globe, but this point is stretched into a line stretching across the bottom of an atlas. I^0 was drawn on the left side of the figure as a point, as were I^+ and I^-. On the left then, the figure showed the pieces of $\mathscr{I}$, three points and two cylinders, and on the right the figure showed the "manifold" of Minkowski space, $\mathscr{M}$, with its boundaries connected together. This collection of pieces of spacetimes is another sense in which Penrose's diagrams were akin to atlases.[102] Together, and in parts, these figures gave a definite place—an objecthood, a materiality—to the mathematical entities that were the formal objects of GR.

I argue that creating this new objecthood and materiality of spacetimes required the same sort of visual techniques as those Penrose deployed in the contexts of recreational mathematics and psychology (above). And not only in Penrose's draftsmanship. "Parking a Car" was an image "in which perspective produces peculiar effects because size relationships are rendered inconsistent."[103] Inconsistent size relationships are the purpose of conformal transformations. Returning to the example of the Mercator projection, the relative sizes of Greenland and the African continent are dramatically misrepresented, but compass bearings from port to port are not. For Penrose the challenge was the distance between mathematical points: "there is no such thing as a point at infinity, since such a point would be an infinite distance from its neighbors."[104] In his diagrams, the distances between points changed across the figure. According to Penrose and Penrose, their "impossible" triangle (figure 6.6) "produce[s] an impossibility": "As the eye pursues the lines of the figure, sudden changes in the interpretation of distance of the object from the observer are

---

101. Barthes, "Reality Effect."
102. In contemporary GR, physicists speak of multiple "charts" that cover a manifold.
103. Penrose and Penrose, "Puzzles for Christmas," 1597.
104. Penrose, "Asymptotic Properties of Fields and Space-times," 66.

necessary."[105] Physicists looking at the pages of PRL were presented with figure 6.13, in which visual perspective shading gave an impression of an object at a fixed distance. But according to the mathematics, the physical spacetime underlying the diagram should extend infinitely back into the page. There were two competing perspectives at play. The first took the diagram to represent regular physical objects such as a spinning top. In this case, the edge of the top is a definite distance from the implied viewer. The second, less visual, perspective required physicists to understand that the edge of the top is at an infinite distance. In related work on gravitational collapse, Penrose showed how important it was for relativists to be able to shift their perspective as they considered different areas of a diagram. Figure 6.14 shows how an "outside observer"—drawn as an eye—would perceive a collapsing star: they "will always see matter [...], the collapse through $r = 2m$ to the singularity at $r = 0$ being invisible to him."[106] Only an observer destined to be destroyed by the singularity could see it.

The illusions of the "Staircase for Lazy People" (figure 6.8) and the model staircase (figure 6.10) were explained in terms of part-whole inconsistencies. "Each part of the structure is acceptable as representing a flight of steps but the connexions are such that the picture, as a whole, is inconsistent."[107] This is distinct from looking at traditional perspective drawings, in which a single perspective—painting as window or painting as lens of the eye—dominates the entire ground.[108] Penrose's relativistic diagrams shared the conflicting perspectives of his recreational drawings. This is why the viewers of the relativistic diagrams are instructed to "consider [a circle] as one point," and see a spinning-top-shape space time as composed of cylinders. This instruction told viewers to reconsider the spinning top as a combination of visually disjoint elements. For example, the upper half cone of the top represents the future. From the cone's shape, it looks like objects that are far apart in the present will become closer and closer together as time goes on, farther to the top of the cone. But this is not at all what Penrose meant. The cone must be considered jointly with the cylinder on its left, which represents the topology of the future. The cylinder does not converge to a point, and as time passes, objects distributed around its circumference maintain their distance. This demonstrates that competing perspectives and part-whole inconsistencies were required to interpret both Penrose's recreational drawings and relativistic diagrams. And this shows, I believe, the wholeness of Penrose as a thinker, and the importance of considering the intellectual context of physics within broad and flexible frames.

Within relativity, one frame for the history of Penrose's early work is the history of black holes and gravitational wave research, culminating in the LIGO experiment's search for gravitational radiation emanating from colliding black holes. Penrose's and similar efforts to establish conceptual foundations for General Relativity have indeed played this role. But Penrose's work, the Arnowitt-Deser-Misner formalism, and Bryce DeWitt's numerical relativity grew out of foundational work that was aimed

105. Penrose and Penrose, "Impossible Objects," 31.
106. Penrose, "Gravitational Collapse and Space-time Singularities."
107. Penrose and Penrose, "Impossible Objects," 31.
108. For Ptolemaic windows and Kepler's lenses, see Alpers, *Art of Describing*; see also Elkins, *Poetics of Perspective*.

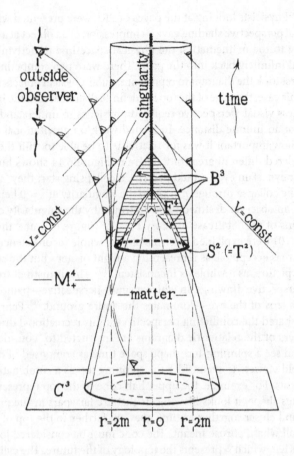

Figure 6.14. Penrose's 1965 diagram of gravitational collapse and singularity creation. Reprinted figure with permission from Roger Penrose, "Gravitational Collapse and Space-time Singularities." *Physical Review Letters* 14, no. 3: 57–59. https://doi.org/10.1103/PhysRevLett.14.57, copyright (1965) by the American Physical Society.

at quantum gravity.[109] Peter Bergmann expressed an attitude of the main thread of workers on quantum gravity: that quantized geometry led to quantized distance between points, which led to the possibility that spacetime points themselves would have to be abandoned. In order to quantize gravity, new basic objects of theory would need to be found.[110] This is how reworking foundations was connected to the sharpest edge of research. When Charles Misner presented the results of his doctoral work, he could give only a "broad outline [of] the principal features of a quantum theory of general relativity." He diagnosed the cause of this limitation: "If rigor could be supplied we would have a theory rather than an approach to one."[111]

109. Arnowitt, Deser, and Misner, "Dynamics of General Relativity"; D. Kaiser, "Cold War Curvature."
110. Bergmann, "Fifty Years of Relativity," 492.
111. Misner, "Feynman Quantization," 497.

Penrose's methods went some way to provide this rigor. At Warsaw, Misner reported on his work with Stanley Deser and Richard Arnowitt. Like Penrose, he too began with "the neighborhood of infinity" and ended with a "brief mention of quantum theory."[112] During the "renaissance of General Relativity," a large part of relativists' mainstream research was not rote puzzle solving; rather it aimed at fundamental change.[113]

But this work addressed long-standing problems: gravitational radiation, and electromagnetic radiation in curved space, had been debated since GR's interwar years. The efforts of Penrose and others to remake foundations spoke to both traditional debates in relativity, and to the aspirations of the community for a future solid area of research. Daniel Kennefick has charted the conflict over the reality of gravitational waves, including Einstein and Nathan Rosen's 1936 paper "Do Gravitational Waves Exist?"—answer: no—and Einstein's indignation at the editor of the *Physical Review*'s request for changes. But Hermann Bondi's later skepticism was closer to home for Penrose; he had taken Bondi's relativity course at Cambridge.[114] Looking beyond Penrose himself to the broader community of relativists, Kennefick argues that the power of Penrose's new tools (together with his work with Newman[115]) drew in both sides of philosophical conflicts; and that once critics and proponents were using the same tools, their disagreements dissipated.[116] This suggests that—beyond the context of individual results—Penrose's new paper tools shaped the ontology of space, time, and vacuum at the level of communal agreement.

6.4.3 Quantum/Relativistic/Cosmological/Astronomical Spacetimes

More than debates about gravitational waves, Bondi was engaged with cosmological debates over the shape, history, and evolution of the universe. Einstein preferred a cosmology with no beginning or end; the three dimensions of space were arranged like a filled-sphere with time proceeding as an unending line, together forming a four-dimensional cylinder. After Edwin Hubble's 1929 observation of the redshift of cosmological light sources, relativists such as Georges Lemaître, Alexander Friedmann, Howard Robertson, and Arthur Walker developed models of an expanding universe with a definite beginning. Rather than a cylinder, this model was a cone. After World War II, Bondi was part of a trio—with Fred Hoyle and Thomas Gold— who reinvigorated the possibility of a steady-state universe, without a cataclysmic

---

112. Misner, "Waves, Newtonian Fields," 189.
113. This is consistent with recent work in the history of GR. For example, Alex Blum's presentation to the 7th Annual Meeting of the European Society for the History of Science, "A farewell to speculation: How quantum gravity drove the renaissance of general relativity." Approximately one-third of the papers at the Warsaw conference worked along these lines. Even some research which did not share, say, Wheeler's speculative attitude directed its careful mathematical explication toward quantum gravity. See Ray Sachs's report at Warsaw.
114. Kennefick, *Speed of Thought*, ch. 3, ch. 10.
115. Their work on "asymptotically flat" spacetimes is another manifestation of the role of the vacuum in this work. Flat space was empty space. Newman and Penrose, "10 Exact Gravitationally-conserved Quantities"; Newman and Penrose, "New Conservation Laws."
116. Kennefick, *Speed of Thought*, 210.

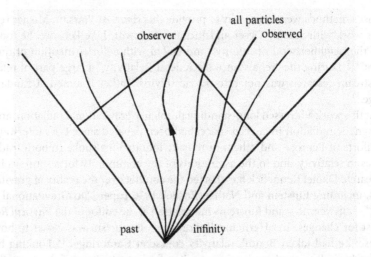

Figure 6.15. Penrose's diagram of a cosmology without a particle horizon, "past infinity is null." Reprinted from "Cosmological Boundary Conditions for Zero Rest-mass Fields," by Roger Penrose, from *The Nature of Time*, edited by Thomas Gold and Hermann Bondi. Copyright (c) 1967 by Cornell University. Used by permission of the publisher, Cornell University Press.

origin. They had met during the war under Hoyle's leadership in the Royal Admiralty Radar Group. Their proposals invoked the continuous creation of matter in an expanding universe. As part of this program, Gold and Bondi organized a conference at Cornell University, May 30–June 1, 1963, on "The Nature of Time."[117]

Penrose was invited to Cornell and began the work of extending his conformal methods to cosmological spacetimes. Until this point, he had analyzed the phenomena of gravitational waves, but only in the background of "flat" Minkowski spacetime. This was not a problematic restriction for most discussions of radiation: first because any Earth-bound experiment would not experience strong gravitational fields; and second, because classical relativity assumes that at any small-enough scale, even the strongest curvature becomes flat. Topographical maps of Ithaca, New York, need not factor in the curvature of the Earth. Turning to the shape of the universe, at Cornell he investigated the "Cosmological Boundary Conditions for Zero Rest-Mass Fields," and applied his techniques to contrast the Einstein–de Sitter cylinder cosmology with the steady-state adaptation of an expanding universe.[118] Again he focused on boundary conditions, and again he worked with a sort of empty space with only zero-rest-mass fields.

In his presentation, Penrose reached across disciplinary boundaries. In the language of quantum field theory, he spoke of "source free equations" and zero-rest-mass fields in spacetime according to their spin: "neutrino, spin 1/2 [...], photon, spin 1,"

---

117. Kragh, *Cosmology and Controversy*, Hubble, etc., chs. 1–2; steady-state trio, ch. 4; conference, p. 372.
118. Penrose, "Cosmological Boundary Conditions for Zero Rest-mass Fields."

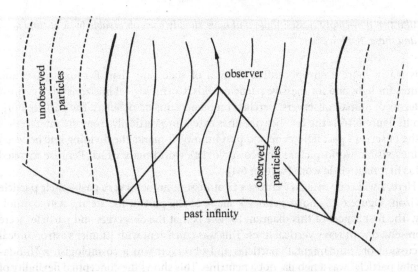

Figure 6.16. Penrose's diagram of a cosmological particle horizon, "past infinity is space-like." Reprinted from "Cosmological Boundary Conditions for Zero Rest-mass Fields," by Roger Penrose, from *The Nature of Time*, edited by Thomas Gold and Hermann Bondi. Copyright (c) 1967 by Cornell University. Used by permission of the publisher, Cornell University Press.

and "graviton, spin 2."[119] He applied his conformal methods to the massless objects of relativity, such as the Weyl curvature tensor.[120] And Penrose reached to the conceptual apparatus of astronomy by focusing on observable "horizons." For this, he drew on Wolfgang Rindler's 1956 paper in the *Monthly Notices of the Royal Astronomical Society*. Rindler defined "horizon" as "*a frontier between things observable and things unobservable. (The vague term things is here used deliberately.)*" He then divided horizons into two types:

> An event-horizon, for a given fundamental observer A, is a (hyper-) surface in space-time which divides all events into two non-empty classes: those that have been, are, or will be observable by A, and those that are forever outside A's possible powers of observation.

In contrast,

> [a] particle-horizon, for any given fundamental observer A and cosmic instant t_0 is a surface in the instantaneous 3-space $t = t_0$, which divides all fundamental particles

119. Penrose, "Cosmological Boundary Conditions for Zero Rest-mass Fields," 42.
120. Even if the steady-state theory in a sense rejected Einstein's field equations, they retained the geometry of Robertson-Walker line elements. See, e.g., Hoyle and Narlikar, "Mach's Principle"; Kragh, *Cosmology and Controversy*, 173–86.

260 MORE THAN NOTHING

into two non-empty classes: those that have already been observable by A at time t_0 and those that have not.[121]

This was a more anthropocentric division of spacetime than Penrose's pasts and futures for light and for massive particle-fields. Cosmological horizons were defined in terms of human observers, perched on planets, inside nebulae. The particle horizon in figure 6.16 is the two heavier lines extending vertically from the intersection of the observer's past light cone and past infinity. It marks the dividing line between Milne's visible world-picture and cosmologist's world-map, which Penrose invoked in his first relativistic work (section 6.4).

Here, I will only discuss Penrose's treatment of an observer's past—their particle horizon. Figure 6.15 shows Penrose's shape of the past in the steady-state cosmology. The first aspect of this diagram to note is that the observers and particles were represented with wavy vertical lines. This was consistent with Rindler's astronomical discussion of "fundamental" particles and observers—to a cosmologist, a "fundamental particle" was a nebula, not a neutrino. This shows the conceptual flexibility of Penrose's diagrammatic approach—when words and formulae spoke of massless neutrino fields, diagrams incorporated galaxies. However, since massless fields should travel only on 45° lines, these vertical lines were inconsistent with the diagrams' framing. Strictly speaking, in a conformal diagram the paths of entities through spacetime should never be vertical. There is a long-standing exception in physics for "test particles": an electromagnetic field could be mapped out with a small charged metal probe, with the agreement that the probe's own contribution to the field can be ignored. This came out in the questions after Penrose's talk.

> Dennis Sciama: The analysis applies only for a finite number of particles. Do you suppose this treatment could be extended to cover the case of an infinite number of particles?

> Penrose: I feel sure that this type of approach would be valuable, but the detailed physical questions soon begin to arise; these worry me. I am trying only to construct conditions which make mathematical sense to me. The situation might be changed if the universe is treated as full of particles; I do not know. My suggestions hold if there are just a few particles with a cosmological background.[122]

My point here is that in his drawings of "fundamental observer" galaxies, Penrose crossed the line between discountable-test-particles and particles that really affect the cosmological spacetime.[123] Here, Penrose made good use of Rindler's intentionally vague "things." Without engaging in a full social analysis, this interpretive flexibility, perched between scientific communities, suggests that Penrose diagrams served as

121. Rindler, "Visual Horizons in World Models," 663; later, Rindler and Penrose became collaborators: Penrose and Rindler, *Spinors and Space-time*.
122. Penrose, "Cosmological Boundary Conditions for Zero Rest-mass Fields," 54.
123. I presented this material and made this point in conversation with Penrose at the Max Planck Institute's Centenary of General Relativity conference in Berlin in December 2015.

"boundary objects."[124] The diagrams were a tool with which scientists from differing traditions could work together.

The second important aspect of this diagram is that it shows that all the particles in the universe are visible to an observer in the steady-state cosmology. This is in contrast to the Einstein–de Sitter cosmology (figure 6.16), in which an infinite expanse of particles is outside the reach of an observer. At Cornell in 1963—though he did not express it in these terms—Penrose had returned to his initial question about modern cosmology, the question he asked Dennis Sciama at lunch with his brother at the Kingswood restaurant. According to Fred Hoyle's BBC Radio broadcasts, shouldn't the night sky be as bright as the Sun? Neither the steady-state cosmology nor Einstein's static universe offered convincing answers to the brightness of the stars in one's particle horizon.[125]

The shape of empty space continued to hold Penrose's attention. He presented this material to advanced graduate students at the 1963 summer school at Les Houches, France, organized by Cécile and Bryce DeWitt.[126] The detailed formal presentation of his conformal methods was communicated to the *Proceedings of the Royal Society* by Hermann Bondi in June 1964 and was published in February 1965.[127] It included conformal techniques and diagrams, and it was aimed toward a "quantization programme," or S-Matrix theory of gravity.[128] In order to relate his work to discussions of asymptotically flat cosmological spacetime by Bondi and his students, Penrose integrated Einstein's cosmological constant into his analysis.[129] The cosmological constant was a contested object in relativity. Some, like Vladimir Fock, rejected it; while others, like Yakov Zel'dovich, interpreted it as a minimum energy for empty space, and made connections to particle physics (see chapter 8).

Bondi and the steady-state theory were not the only motivation for Penrose's work in cosmology. Schild had convinced a number of relativists to visit Texas at the Warsaw meeting in 1962. An occasion to bring them there was found in Maarten Schmidt's 1963 announcement that the 200-inch telescope at the Mount Palomar observatory showed only one visible object near the already-known strong radio source 3C 273. The redshift of the frequencies of this light made it the farthest known object from Earth, and meant that it was 100 times brighter than other X-ray sources.[130] This "quasi-stellar object" offered the energies at which relativity would become relevant to astronomy. Schild, and his junior colleagues Ivor Robinson and Engelbert Schücking, invited relativists, cosmologists, and astronomers to Dallas for a conference that December. (The conferences have continued since. They have done more than a little to define "relativistic astrophysics" as a field of study.[131]). Penrose attended the "First

124. Star, "Not a Boundary Object."
125. Modern explanations do not assume an infinitely old universe, and argue that the expansion of the universe causes starlight to be "redshifted" toward lower wavelengths, below the threshold of human sight. For a contemporary discussion, see Peebles, *Principles of Physical Cosmology*, esp. 124–31.
126. Penrose, "Conformal Treatment of Infinity."
127. Penrose, "Zero Rest-Mass Fields Including Gravitation."
128. Penrose, "Zero Rest-Mass Fields Including Gravitation," 161, 200.
129. Penrose, "Zero Rest-Mass Fields Including Gravitation," 185 et seq.
130. Schmidt, "3c 273: A Star-like Object with Large Red-shift."
131. Robinson, Schild, and Schücking, *Quasi-stellar Sources*; Schücking, "First Texas Symposium."

Texas Symposium on Relativistic Astrophysics" but did not speak.[132] However, exactly one year later—back in England after his Visiting Assistant Professorship with Schild at Austin—he applied his conformal methods to gravitational collapse and published the first expression of his theorems on the inevitability of singularities.[133]

6.4.4 Narratives of Creation and Collapse

Penrose integrated the physical possibility of gravitational collapse and "big bang" cosmology into his overall striving for the next theory, after general relativity. He expressed this in his 1965–1966 Adams Prize–winning paper "An Analysis of the Structure of Space-time." Cambridge University's Adams Prize in mathematics was among the most prestigious prizes at the university, and the award would be financially significant to a young researcher. Today the value is £14,000, divided with the winner's academic institution. The same year, a student of Sciama's at Cambridge—Stephen Hawking—was awarded an auxiliary prize, for an essay deeply influenced by Penrose's work.[134] Penrose's substantial paper—138 typescript pages—was circulated as a preprint by Wheeler's group at Princeton in 1967, and eventually much of it was included in the proceedings of a conference organized by Wheeler and DeWitt-Morette that aimed to connect physicists and mathematicians (section 5.6).[135]

The essay began with a quasi-historical analysis of theories of spacetime, from Aristotle to Einstein,[136] that then broadened to consider contemporary attempts to think quantum mechanics and spacetime together. There was a resonance here between Penrose's work and his rhetoric. At Warsaw, he had cut toroidal spacetimes to create the possibility of chronologies and the difference between pasts and futures. In this essay he refined this into a mathematical definition of "causal connection" that remains a central aspect of relativistic research.[137] Temporal orders were developed in both scientific content and in rhetoric.

Contemporary, Einsteinian, spacetime theories took nature's spacetime to be represented by a mathematical continuum of points, called a manifold. Even quantum theories that might do away with spacetime—such as Geoffrey Chew's S-Matrix program—relied on the continuum in some other mathematical space.[138] But Penrose was not content with the continuum's complicated hierarchy of infinities of points between zero and one. "It is a belief in the ultimate simplicity of nature, then, which, is to me an almost compelling reason for rejecting the differential manifold

132. Schücking, "First Texas Symposium," 50.
133. He cites both the Texas Symposium and the Les Houches lectures. Penrose, "Gravitational Collapse and Space-time Singularities."
134. Hawking, "Singularities and the Geometry of Spacetime."
135. Penrose, "Analysis of the Structure of Space-time"; Cécile DeWitt-Morette also signed her work Cécile DeWitt and Cécile Morette. Penrose, "Structure of Space-time."
136. A narrative not dissimilar to contemporary stories, e.g., Huggett, *Space*.
137. Kronheimer and Penrose, "Structure of Causal Spaces"; Hawking and Ellis, *Large-scale Structure*, ch. 6.
138. Cushing, *Theory Construction and Selection*; D. Kaiser, *Drawing Theories Apart*, ch. 9.

concept as an 'ultimate' view of space-time."[139] This aesthetics paralleled Dirac's faith in the power of mathematical beauty (chapter 2). But this was not a purely mathematical and aesthetic paper. Collapse and cosmology played a similar role for Penrose as it did for Wheeler: these were physical situations that made quantum gravity more than idle speculation.

> the quasi-topological arguments used in section VIII [Collapse and Cosmology] are largely directed at trying to ascertain whether effectively singular states exist in the universe, at which one might expect any breakdown in the manifold structure to be of real significance.

Another parallel between the two men: collapse and cosmology were avenues to new ontological pictures. For Wheeler, wormholes led to quantum foam. For Penrose, singularities in spacetime pointed to a place where a breakdown occurs in the standard mathematics of a manifold of spacetime points. Both aesthetic and more-physical motivations combined to send Penrose into investigations of "quantized directions"; of breaking field tensors into more-basic spinors; and of a "twistor calculus" that made light-cones more fundamental than spacetime points. He did not take the speculative leaps that characterized Wheeler's work, but no-less-radical ontological changes were the goal of Penrose's "reformulations and 'ways of looking at'" relativity.[140]

Here, I would like to focus on only one page of "An Analysis of the Structure of Space-time" (figure 6.17). This grid shows a sort of visual history of the Schwarzschild solution to Einstein's Equations that describe the spacetime surrounding a star.[141] These equations relate the curvature of spacetime to the distribution of mass-energy in the universe, and are expressed in the language of Riemannian geometry.[142] For a given distribution of matter, the goal was to find a geometry, in the form of a metric tensor or line element. Schwarzschild was deployed on Germany's Eastern Front in 1916 when he turned his astronomical training toward solving Einstein's Equations for a spherical distribution of matter, surrounded by vacuum—a universe with a lone star.[143] His work was cut short when he contracted an infection in the trenches and died later that year.

Schwarzschild's solution was expressed as a line element, ds^2, that determines how distances between objects in a spacetime are measured. (A line element is a curved-space generalization of the Pythagorean theorem for the length of the hypotenuse of a triangle, $c^2 = a^2 + b^2$.) In a spherical coordinate system, (t, r, θ, ϕ), the solution is:

$$ds^2 = \left(1 - \frac{2m}{r}\right) dt^2 - \frac{dr^2}{(1 - 2m/r)} - r^2(d\theta^2 + \sin^2\theta \, d\phi^2). \tag{6.1}$$

139. Penrose, "Analysis of the Structure of Space-time," 592.
140. Penrose, "Analysis of the Structure of Space-time," 598–99.
141. For this history pre–World War II see Eisenstaedt, "Histoire et singularités de la solution de Schwarzschild"; Eisenstaedt, "Trajectoires et impasses."
142. The Einstein Equations are $R_{\mu\nu} - \frac{1}{2}g_{\mu\nu}R = \kappa T_{\mu\nu}$. Penrose's zero rest mass fields were pieces of $T_{\mu\nu}$.
143. Schemmel, "Classical and Relativistic Cosmology."

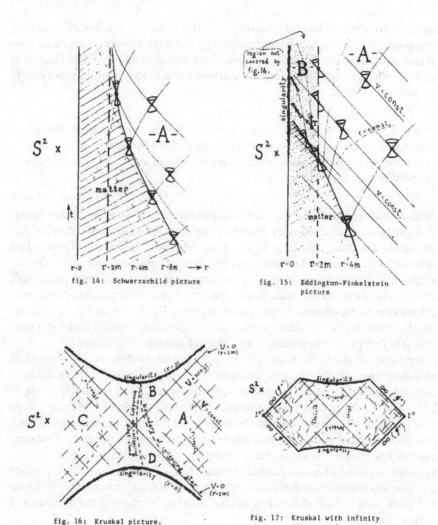

fig. 14: Schwarzschild picture

fig. 15: Eddington-Finkelstein picture

fig. 16: Kruskal picture.

fig. 17: Kruskal with infinity represented.

Figure 6.17. Four manifestations of the Schwarzschild solution. Roger Penrose, "An Analysis of the Structure of Space-time," Copyright Oxford University Press. Reproduced with permission of the Licensor through PLSclear.

Here, m is the mass of the star, r is the radial distance from its center, and θ and ϕ are longitudinal and latitudinal directions. There are two problematic terms in this line element, when a fraction can have a denominator of zero: at the center of the coordinate system, $r = 0$, and at the "Schwarzschild radius" when the radius equals twice the mass of the star, $r = 2m$. As the line element approaches these points, its value goes to infinity, and the curvature of spacetime is infinitely strong—they are singularities. In figure 6.17 top left, Penrose illustrated this picture. From left to right: $S^2 \times$

signals that the two radial coordinates are suppressed, and so each point in this diagram should be considered as a sphere; time is on the vertical axis and radial distance on the horizontal axis; the solid line at $r = 0$ is one singularity, and the shaded region represents the interior of a collapsing star; the dashed vertical line at $r = 2m$ is the second singularity at the Schwarzschild radius; a line marking the boundary of the star as seen from an outside observer is marked by light cones as they are distorted by the changing curvature; and the area –A– is the empty spacetime surrounding the star.

From afar, an observer would never see either singularity; a thinning sliver of star remains just outside $r = 2m$. One could, perhaps, ignore everything farther in toward $r = 0$. However, the mathematics indicated a problem: the "proper time" for an observer at the surface of the star—not one watching from far away—to reach the singularity is "<u>finite</u> [...t]hus, an observer who follows the star inwards must [...] either of necessity be destroyed [...] or else, find himself in a portion of the universe not covered by the coordinates of [equation 6.1]."[144] For a distant observer, the singularity was hidden. But new coordinates were needed to trace an adventurous male observer at the surface of the star.

Following the work of Arthur Eddington, David Finkelstein addressed this problem by introducing a new time parameter $v = t + r + 2m \log(r - 2m)$. This was a type of "analytic extension" that smoothed over the infinite spike of the singularity. The Schwarzschild metric became

$$ds^2 = \left(1 - \frac{2m}{r}\right) dv^2 - 2\, dv\, dr - r^2(d\theta^2 + \sin^2\theta d\phi^2). \tag{6.2}$$

This accorded to the second diagram (figure 6.17, top right), in Eddington-Finkelstein coordinates.[145] It was now possible to picture an observer on the surface of the star falling through the point at $r = 2m$ and reaching the singularity at the center of the coordinate system, $r = 0$. Area "–B–" was the "region not covered" by the Schwarzschild picture. Note that the light cones representing observers are elongated as in the first diagram but now gradually tilt into the singularity. The cones along the $r = 2m$ line with the rightmost future-directed (up) boundary on the line are at the point of no return from the collapsing star. Even light could not escape. Penrose recalls that as a student, he traveled from Cambridge to London to see Finkelstein give a talk. Finkelstein presented a diagram of spaces behind singularities, and this sparked Penrose's research on black holes.[146]

As with the $r = 2m$ case in Schwarzschild coordinates (equation 6.1), the proper time for an observer to reach the singularity at $r = 0$ is finite. Could another set of coordinates be found to carry an observer past this singularity, too? Could one travel to the left of the vertical line at $r = 0$? No, "our observer who successfully followed the star through r = 2m must now be torn to pieces by the infinite tidal forces at the <u>true</u> singularity at r = 0."[147] Because the singularity at the Schwarzschild radius could

---

144. Penrose, "Analysis of the Structure of Space-time," 686.
145. Finkelstein, "Past-future Asymmetry."
146. Lightman, "Interview of Roger Penrose."
147. Because "the curvature scalars constructed from the Weyl tensor [...] tend to infinity as r approaches zero." Penrose, "Analysis of the Structure of Space-time," 688.

266 MORE THAN NOTHING

be smoothed over with a different coordinate system, it was not a "true singularity." The puzzle of what features of a geometry were physically significant motivated a substantial portion of relativistic research.

Penrose then integrated this picture of collapsing stars with the expanding universe cosmology of the big bang. Connecting this diagram to Oppenheimer and Snyder's 1939 discussion of a collapsing dust cloud,[148] Penrose observed that the area inside the collapsing cloud—hashed and labeled "matter" in the top-right diagram—"turns out to be nothing other than (a portion of) the Friedmann universe," a cosmological model of a universe beginning from a singularity, expanding, and then contracting back to a singularity.[149] For Penrose this motivated "the essentially identical nature of the situation of a final singularity in gravitational collapse and of a final (or initial) singularity for a relativistic cosmology."[150] This may have been Wheeler's influence, as he had spoken about "Geometrodynamics and the Issue of the Final State" at Les Houches.[151]

In slightly anachronistic terminology, Penrose observed that a black hole was only a white hole in reverse, and a white hole—out of which matter can leave but never enter—was an interesting model of the big bang. Princeton mathematician Martin Kruskal had developed a time-symmetric picture of the Schwarzschild solution by changing both the time and space coordinates, U, V.[152] Then the Kruskal metric became

$$ds^2 = \left(\frac{32m^3}{r}\right) r^{-\frac{r}{2m}}\, dU\, dV - r^2(d\theta^2 + \sin^2\theta d\phi^2). \qquad (6.3)$$

Notice that there is only one place where a fraction could have a denominator of zero. This gave the "Kruskal picture" (figure 6.17, bottom left) that shows an empty, infinite, complement –C– to the region –A–. An analytic extension had conjured an entire new universe. The diagram is of the UV-plane, and Penrose has labeled some lines to indicate how these relate to the original Schwarzschild coordinates. The $U = 0$ and $V = 0$ diagonal lines are the "Schwarzschild radius" depicted as upright lines in the top diagrams. Neither U nor V can properly be called a time coordinate, but the diagram was (conventionally) oriented so that the future is up. The difference between a collapsing star and a collapsing cloud of dust was labeled with dotted lines. As in the Eddington-Finkelstein case, the star began in the distant past in region –A– and crossed the Schwarzschild radius, $U = 0$ or $r = 2m$, to enter region –B– and ended at the future-singularity at $r = 0$. The dust cloud, however, began as it ended—at a singularity—emerging from the past singularity at $r = 0$ into –D– passing briefly into –A– across $V = 0$ and then crossing $U = 0$ into –B– and ending at the future $r = 0$ singularity.

148. Oppenheimer and Snyder, "On Continued Gravitational Contraction."
149. Penrose, "Analysis of the Structure of Space-time," 689; Kragh, Cosmology and Controversy, 24–27.
150. Penrose, "Analysis of the Structure of Space-time," 689.
151. Wheeler, "Geometrodynamics and the Issue of the Final State."
152. Kruskal, "Maximal Extension of Schwarzschild Metric," $V/U = -e^{t/2m}$, $VU = e^{r/2m}(1 - r/2m)$.

ROGER PENROSE'S IMPOSSIBLE DIAGRAMS   267

At this stage in Penrose's paper, he had already introduced his conformal methods, and briefly remarked that it will be useful to have a version of the diagram with
infinity represented as finite lines for a future discussion. He referred to the last diagram (figure 6.17, lower right), with infinities bounded by bold lines labeled I^0, ∞,
and $\mathscr{I}$.[153]

Every other diagram in this series had open edges, their universes implicitly
spilling off the page in an infinite expanse. This is the visual correlate of the
mathematical ambiguity Penrose's conformal methods resolved. If a physicist were
asked to mark the extent of the gravitational field of the Eddington-Finkelstein
star in figure 6.17, top right, where would they put their pencil? In the original
Schwarzschild picture (figure 6.17, top left), where are the boundaries needed to place
an electric field's boundary conditions? Penrose gave a solution to these problems,
with his "fictional" conformal spacetime. He bounded infinity, and gave universes
something they had never had before: a place on the page. He gave these universes,
all vacuum save for singularities, a shape.

6.5 Conclusion

In chapter 4 I argued that Julian Schwinger's reformulation of Quantum Field
Theory put the vacuum at the heart of the practice of quantum calculations.
Each quantum process was understood as a transition from vacuum to vacuum.
Penrose's conformal techniques effected a similar state within General Relativity. He
participated in an effort to understand processes in GR as taking place within "asymp-
totically flat" spacetimes—universes with complex middles, but simple, flat, *empty*,
Minkowski-space beginnings and ends. (This effort is maintained in contemporary
GR observations, such as LIGO[154]) In this way, an S-Matrix theory of gravitation
could begin the same way Schwinger's QFT did: with fields specified on a surface in
Minkowski space. Schwinger, and other quantum theorists, had only to keep track
of the many possible interactions of fields in flat spacetime; relativists like Penrose
would have to understand these interactions, and also the complex interactions of
these fields with spacetime itself. As of yet, there is no successful quantum theory
of gravity. Relativists in the 1950s and 1960s were working just to reach the starting
point of Schwinger's calculations. Nevertheless, one important goal of this chapter
has been to explicate the development of diagrammatic, conceptual, and mathematical tools that were at the center of research in General Relativity in the twentieth
century—and that centered empty space within this research.

Among the small group of relativists who were remaking these foundations of their
field, Penrose enlisted the vacuum in another, deeper, way. His analysis of asymptotically flat spacetimes, cosmologies, and gravitational collapse took place within the
context of his conformal picture of spacetime. But the only objects in relativity that
are left unchanged by conformal transformations were insubstantial zero-rest-mass
fields. For Penrose, the stars and clouds of dust that collapse into black holes were

---

153. Using coordinates $p = \tan^{-1} \sinh^{-1} V$, $q = \tan^{-1} \sinh^{-1} U$. Penrose, "Analysis of the Structure of
Space-time," 690.
154. Kennefick, "Star Crushing," esp. 14–17.

not clouds of hydrogen gas undergoing fusion reactions—electrons and nuclear particles had rest-mass. In addition to his recreational mathematics, Penrose's pictures may have implicitly relied on the charisma of Wheeler and Misner's geometrodynamics, and their effort to describe regular matter as nothing more than curled up empty space (chapter 5). Penrose did spend time with them in Princeton during his NATO fellowship in 1959–1960. More than just establishing empty spacetime as the boundary for relativistic calculations, Penrose's conformal methods made the stuff of relativity massless. More so than Schwinger, Penrose's paper tools required physicists to work with empty space. These methods remain at the center of experimental and theoretical studies of black holes.[155] In this way, Penrose gave universes a shape that was defined by vacuum edges, but also defined his universes as empty.

155. Wald, *General Relativity*, esp. chs. 7–12.

7
Sidney Coleman's False Vacuum

7.1 Introduction: Story and Symmetry

Physicist Sidney Coleman (1937–2007) cared deeply about reading and writing stories. Here is a story he wrote in the spring of 1961, on a skipping typewriter borrowed from his roommate, during his doctoral studies at the California Institute of Technology. It was part of a letter to his mother, Sadie, and younger brother, Robert, at home where he grew up outside Chicago:

About a month ago Shelly Glashow and I exchanged a series of insults concerning each other's poor physical conditions that ended with a dare to jointly climb Mt. Wilson [...]. Now, this is a 22 mile round trip, and the greatest exercise I had had for a month preceding was peristaltic action, so perhaps this was not the wisest challenge for me to accept, but, once the deed was done, there was no room to retreat, especially since the plans for the venture became widely known in the Dept. and the subject of much comment of the "It'll never work!" variety, familiar to Galileo, the Wright brothers, and Allen Dulles.

Fred Zachariasen: "I hear Shelly is going to be your Sherpa guide."

Myself: "No, I believe in type casting. Shelly is going to be the abominable snowman."

Anyway we left early Sunday morning, with encouraging cries of "The exercise will be good for our health." "Excelsior" and "Why are we climbing Mt. Wilson? Because it's there!"

Well, as the Victorian novelists used to say, let us draw the curtain of oblivion over scenes too painful for delicate maternal eyes to see, and dissolve (mixing a metaphor in the process) to Monday morning and my bedroom, where we see, in my bed, a large angry red object, roughly eight feet in diameter. It is a mass of pain. I am dimly visible in its interior. [...] After a while I crawl out of bed, dress myself, and move, in tiny two-inch steps, towards the physics dept. On the way I meet my friend Paul Levine. He says, "My God, what's happened to you?" "Good health," I reply, "a bad case of good health."

But, like the Bourbons, I have forgotten nothing and learned nothing, for the next weekend, I try the same feat again, by a slightly different route, and find that it results in slightly less agony. Currently, I am well on my way to becoming a mountain goat, having climbed trails that go up, around, and across Mt. Wilson no less than five times. [...]

Of course, there is little change in my overt appearance. To the casual eye I present the usual configuration of sallow skin, baggy eyes, poor posture and pipestem arms. But from the waist down, what a transformation! I have the legs of an athlete: thighs of iron, calves of steel. (Feet of clay.)

More than Nothing. Aaron Sidney Wright, Oxford University Press. © Oxford University Press 2024.
DOI: 10.1093/oso/9780190062804.003.0007

The paper of which I sent you the preprint was accepted and appeared in Physical Review Letters, April 15.[1]

This story is revealing in both its form and content.

Coleman's story reveals several continuities in his life. He felt connected to his family (and his Jewish petit bourgeois upbringing). He traveled in elite scientific circles (if not from his undergraduate years at Illinois Institute of Technology in Chicago, certainly from his arrival at Caltech in Pasadena). Fred Zachariasen's cameo was the voice of a faculty member in theoretical particle physics. Sheldon (Shelly) Glashow was a postdoctoral fellow at Caltech and became a second mentor to Coleman after Coleman's thesis advisor, Murray Gell-Mann. The April 15 paper was a coauthored project of Coleman and Glashow that worked out implications and modifications of Gell-Mann's brand new "Eightfold Way" classification of elementary particles.[2] Coleman was in the right time and place to be at the leading edge of research: Gell-Mann's preprint is dated March 15, Coleman and Glashow's letter was received by *Physical Review Letters* (PRL) on March 20.[3] According to Glashow, he and Coleman became "traveling disciples" of Gell-Mann's symmetry scheme, lecturing across Europe and North America in the early 1960s at High Energy Physics meetings and at summer schools.[4] Glashow and Coleman remained friends for decades, including many other hiking expeditions in the United States and Europe.

Coleman could be very funny, both in one-liners and in fuller comic scenarios. He made classic literary references, such as to Victorian literature; the Bourbon restoration (in a line attributed to Talleyrand); and a dream of Nebuchadnezzar. He made effective use of contemporary references and juxtaposition. For example, "Galileo, the Wright brothers, and Allen Dulles," in which a pattern is established with modern geniuses who overcame their doubters only to crash into a modern fool who was given all-too-much trust.[5] Coleman tempered moments of self-assurance with self-deprecation. He did something healthy, but it almost killed him. He mastered the mountain, but remained unattractive. His thighs were stronger, calves were stronger, but—repeating the structure of his earlier juxtaposition—he had feet of clay. These elements came together in a *narrative*. More than a simple recounting of events, it had a recognizable beginning, middle, and end. Its plot was romantic. Our hero began with a natural challenge (Mt. Wilson) and social conflicts (Glashow, the physics department). They completed a climb, but was it only a Pyrrhic victory? Things got painful. But the end was happy: the mountain was truly mastered, and the social conflicts dissolved. In fact, there may have been two mountains in this story. The second was the research the two hikers published in PRL, the most prestigious physics journal.

1. Many of Coleman's letters have been published in Coleman, Kaiser, and Wright, *Theoretical Physics in Your Face!*, Coleman to home, undated [after 15.4.1961]. I have corrected minor spelling errors. SRC.
2. Coleman and Glashow, "Electrodynamic Properties of Baryons"; Gell-Mann, *Eightfold Way*; Pickering, *Constructing Quarks*, ch. 3.
3. Gell-Mann circulated a pre-preprint in January, Gell-Mann, *Eightfold Way*, 46n6.
4. Sheldon Glashow (private communication), 2014.
5. Dulles, director of the CIA, was a central figure in the disastrous US Bay of Pigs Invasion of Cuba, which unfolded from April 17 to 20, 1961.

This well-crafted story fit with Coleman's lifelong love of writing, in particular of science fiction (SF). As an undergraduate in Chicago he fell in with members of the University of Chicago's Science Fiction Club, including George Price and Earl Kemp. In 1955, Kemp proposed that Price, Coleman, and four others found Advent:Publishers, with the goal of publishing SF *criticism*, in particular Damon Knight's *In Search of Wonder: Essays on Modern Science Fiction* (1956).[6] Advent was a break-even operation.[7] Coleman attended SF conventions, and he contributed to amateur fanzines such as Terry Carr's *Lighthouse* and the private-circulation *Lilapa*. An initial—later abandoned—doctoral thesis project had strong resonance with SF: "acausality" in classical field theory. On April 17, 1960, he wrote home to his mother and brother:

> Thesis work (maybe) going along at a wonderful rate. I now believe that in certain
> cases, events may <u>precede</u> their causes. The times involved, however, are on the
> order of .00000000000000000000001 seconds. Not very useful for beating the horse
> races.[8]

In the 1970s, he wrote book reviews for his friend Avram Davidson, the editor of the *Magazine of Fantasy and Science Fiction*.[9] Coleman assigned one of these book reviews to students in Harvard's Program in Expository Writing in 1975.[10] SF was a strong presence throughout Coleman's career. He served as a bridge between elite circles of high-energy physicists and the US counterculture.[11] Through to the 1990s, his instructions for the staff at Harvard to forward his mail while visiting CERN, in Geneva, included forwarding "mimeographed or xeroxed amateur or semi-professional publications associated with science fiction."[12]

It is natural to ask about Coleman's view of the relation of science and SF. At least, it was a natural question for the editor of *Vertex* magazine. *Vertex* interviewed Coleman and his friend, fellow-physicist, and science fiction author Gregory Benford, in Laguna Beach, California, for their December 1973 issue. They were asked "How do you, as scientists, feel about science and SF?"

> COLEMAN: There are a lot of things to say about science and SF, but I think the
> important thing is that they are very very different sorts of things. [...] I think there
> is a terribly naive belief, a really awful error, that in some sense the science is in SF
> like a passenger is in a car—that just ain't so.

6. The founders were Robert Briney, Sidney Coleman, Earl Kemp, James O'Meara, George Price, Jon Stopa, and Ed Wood. Over the years, Price did the lion's share of the editorial work.
7. Coleman to Ted Cogswell, 4.1960, Excerpted by Earl Kemp, in *eI* 10, http://efanzines.com/EK/eI10/index.htm.
8. Coleman to Sadie, 17.4.1960. See also Coleman, "Acausality"; and Coleman, *Classical Electron*; years later, when philosopher Adolf Grünbaum asked for a reprint, Grunbaum to Coleman 20.5.1976, he referred him to the "Acausality" Erice lectures, Coleman to Grunbaum 25.5.1976 SRC; on re-expressions of classic results, see Galison, "Re-reading the Past."
9. Coleman, "Books (1970)"; Coleman, "Books (1973)"; Coleman, "Books (1974)."
10. Larry Weinstein to Sidney Coleman, 2.9.1975, Sidney Coleman to Edward Ferman, publisher of the *Magazine of Fantasy and Science Fiction*, 15.8.1975, SRC.
11. D. Kaiser, *Hippies Saved Physics*, ch. 8.
12. HUHEPL::COLEMAN to BARRETT (via bitnet), 21.6.1991 SRC.

He wanted nonscientists to understand that what they read in SF was not really science, even if, as Benford replied, science formed part of the "mechanism" of SF. There was little connection to scientific facts or theories.[13] But Coleman saw other strong connections between them. *Vertex* asked: "But isn't that one of the major functions of SF? To blow the mind, to really expand all the senses."

> COLEMAN: That's another correlation between science and SF. Some of the pleasures that you get out of science you also get out of SF. This is something probably very special to me and Greg and other people who are actually working scientists; but, I assure you, one of the reasons for doing science, especially the kind of science I do (elementary particle physics, high energy theory) is that it makes your head feel funny, Goddamned strange. That's also the feeling I get out of SF.

Interestingly, neither physicist was wedded to so-called hard science fiction; both rated Ursula Le Guin as the best contemporary SF author.[14] Coleman felt an *aesthetic* and *affective* similarity between his science and SF—both elementary particle physics and SF produced the same "makes your head feel funny" pleasure.

In this chapter, Coleman's unified emotional and aesthetic sensibility of SF and of science motivates an inquiry that is the inverse to the natural question of the role of science in science fiction stories. The chapter is about the role of stories in Coleman's science. To ask about stories in theoretical physics might seem out of place. It works against a widely accepted opposition between the change over time a narrative represents and the deeply held conviction that physical truths are timeless.[15] Coleman was not a mathematician, but he lived and breathed among the mathematical concepts, practices, and formalisms of theoretical physics. The facts of physics are not taken to be products of change over time in the way that, for example, the facts of zoology are taken to be the product of evolutionary processes. And yet there are (at least) four ways in which narratives enter into the calculations and results of theoretical physics: (1) A calculation or proof may have literary qualities, at the level of both sentences and document *structure*. Proofs may be more or less "vivid."[16] Peter Galison shows how John Archibald Wheeler (chapter 5) conceived of bits of formalism as machine components *in action*.[17] And mathematical works may be emplotted in literary form, as comedies or epics.[18] (2) Physicists often used *analogies* to motivate, guide, illustrate, or interpret their work.[19] (3) Stephan Hartmann argues that some stories physicists deploy are not optional supplements to formal reasoning and interpretation. He observed that in doctoral qualifying examinations, students are not considered to have demonstrated their understanding of a subject until they

13. On such boundary work, see Milburn, "Modifiable Futures," 562.

14. Turner, "Vertex Roundtable Interview." Coleman wrote angrily to Donald Pfeil, editor of *Vertex*, about this interview. He was upset about careless copy editing, and the magazine's description of Coleman. He did not object to the content of his replies. SC to Pfeil, 14.11.1973, reply 19.11.1973, SC to Pfeil 26.11.1973, SRC.

15. Traweek, *Beamtimes and Lifetimes*, epilogue.

16. Gowers, "Vividness in Mathematics and Narrative."

17. Galison, "Structure of Crystal."

18. Netz, "Aesthetics of Mathematics."

19. A classic text is Hesse, *Models and Analogies*.

produce a story to accompany their calculations.[20] (4) Finally, a story may be central to deriving a result. M. Norton Wise gives the example of the heat diffusion equation (see appendix B).[21] Coleman's physics embodied all four of these aspects of narrative. He deployed literary devices, used illustrative analogies, told explanatory stories, and wrapped stories into derivations. But Coleman was a connoisseur, not an enthusiast. He did not deploy narrative evenly across his entire body of work.

Within particle physics, Coleman was known for sophisticated deployments of the mathematics of symmetry and group theory. In the practice of theoretical physics, symmetry groups provided *structure*: they established boundaries between acceptable and unacceptable proposals and guided theorists in constructing theories and models. Some of this can be understood by analogy to architecture. An architect plans the footprint of a building on a spacious lot. Initially, any free sketch could become a model and building. But she has been told the building should be mirror-symmetrical along its longer axis. The range of acceptable sketches is much reduced; perhaps she concentrates on rectangles and ovals. If an additional symmetry requirement is imposed—mirror symmetry along long and short axes—rectangles and ovals will be restricted to squares and circles. As the amount of symmetry increases, the architect's range of choice is restricted. So it was for physicists. The structure provided by symmetry groups provided restrictions, but—as in art and architecture—the structure also supported creativity and discovery. Much of Coleman's early work was devoted to elaborating and understanding this structure.

Symmetry and group theory are, if anything, *anti*-narrative. The diffusion equation is part of thermodynamics and fluid mechanics. As a *dynamical* equation, it describes matter in motion. It is part of a broader class of *differential* equations that represent change, often change in time. As a representation of the change in time of physical bodies, the diffusion equation has a similar structure to a story or narrative.[22] Symmetry and group theory are different. They are more about *invariance* than variation and more about the structures of mathematics itself than about representing the natural world. Historians of mathematics recognize the invention of group theory in the early nineteenth century as an inflection point. It marks mathematicians' great turn away from representations of the natural world, inward to the invention of an autonomous "pure" mathematics.[23] In his 1962 dissertation on strong interaction symmetries, Coleman claimed: "Our conclusions, like all conclusions drawn from symmetry considerations alone, are independent of detailed dynamical assumptions."[24] The difference between mechanical analysis and symmetry groups is similar to a classic distinction between prose and pictures. In the eighteenth century, G. E. Lessing famously distinguished between the linear and sequential experience of reading a text, word by word, and the immediate, simultaneous, perception of

20. Hartmann, "Models and Stories."
21. Wise, "Science as (Historical) Narrative."
22. I use story and narrative interchangeably. Minimally, a story must contain a recounting of events and must give an interpretation or structure to the events (classically: beginning, middle, and end). See White, *Metahistory*, ch. 1.
23. Gray, *Plato's Ghost*; Alexander, *Duel at Dawn*; cf. Hacking, *Philosophy of Mathematics*, §5.A.
24. Coleman, "Structure of Strong Interaction Symmetries," 6; cf. "There is more in the world than algebra; for example, there is dynamics." Coleman, "All Possible Symmetries," 399.

paintings.[25] For Lessing, texts were extended in time while pictures were extended in space, and each had a distinct mode of perception. The analogy is: mechanical analysis is to prose as symmetry and groups are to pictures.

Coleman's use of narrative was at odds with his use of symmetry and group theory. They were, however, not at odds with the personality of a man who once puckishly described himself as "an esthete."[26] Each of these features of his physics was rich enough for aesthetic appreciation. This chapter will show how Coleman held narrative and symmetry in tension—and how the work that flowed from this tension shaped and reshaped the vacuum.

7.2 Partial Symmetries

Traditionally, "being symmetric" in quantum theory was not taken to admit of degrees; either a theory had a given symmetry or it did not. By the late 1950s, as the explosion of new particles continued, theorists pushed for more freedom, more room to speculate, in developing new theories.[27] (This was a less radical approach than John Wheeler's geometrodynamics, discussed in chapter 5.) This section lays out the "partial symmetry" scheme developed by Sheldon Glashow and Murray Gell-Mann that provided the background for Coleman's early work and for the development of spontaneous symmetry breaking.

In 1959, Glashow had defended his PhD at Harvard, under Julian Schwinger, and was a postdoctoral fellow at the Institute for Theoretical Physics at Copenhagen (later the Niels Bohr Institute). He proposed that Lagrangian QFTs could exhibit "partial" symmetry (also called "approximate" symmetry). This formal freedom allowed him to speculatively unite the electromagnetic and weak interactions in one partially symmetric model.[28] Similarly partial was Gell-Mann's (and independently, Yuval Ne'eman's) $SU(3)$-symmetry proposal that Glashow and Coleman evangelized in the early 1960s. It was not that $SU(3)$ was an *exact* symmetry in nature. Nor was it the case that the mathematical representations of the elementary particles were *exactly* $SU(3)$-symmetric.[29]

Formally, Gell-Mann represented approximate symmetries in the following manner. He wrote down a Lagrangian, L, which described the motion and interactions of the baryons in a simplified model (due to Sakata Shoichi). In the Sakata model, all particles that interacted through the strong nuclear force were composed of collections of protons, nutrons, Λ particles, and their anti-particles (table 7.1). He divided the Lagrangian into three parts:

$$\bar{L} + L' + L'', \tag{7.1}$$

25. In spacetime diagrams, see Wright, "Bringing Infinity to a Finite Place," 136.

26. Coleman to Cogswell, 4.1960, http://efanzines.com/EK/eI10/index.htm.

27. Gell-Mann and Rosenfeld, "Hyperons and Heavy Mesons," 408; Pickering, *Constructing Quarks*, ch. 3.

28. Glashow, "Renormalizability," §3; Glashow, "Partial-symmetries"; cf. Pickering, *Constructing Quarks*, 166–67; Mehra and Milton, *Climbing the Mountain*, 429–33.

29. Their Lagrangian changed when multiplied by the group generator.

Table 7.1. The elementary particles ca. 1960 (Barkas and Rosenfeld 1961, 5). Antiparticles are not listed. 1,000MeV is about the mass of a Hydrogen atom, or $\sim 1.8 \times 10^{-27}$kg.

	Particle	Spin	Mass (MeV)	Stable or Unstable
Photon	γ	1	0	S
Leptons	ν	1/2	0	S
	$e^{\mp}$	1/2	0.510976 ± 0.000007	S
	$\mu^{\mp}$	1/2	105.655 ± 0.010	U
Mesons	π^+	0	139.59 ± 0.05	U
	π^0	0	135.00 ± 0.05	U
	$K^{\pm}$	0	493.9 ± 0.2	U
	K^0, K^1, K^2	0	497.8 ± 0.6	U
Baryons	p	1/2	938.213 ± 0.01	S
	n	1/2	939.507 ± 0.01	U
	Λ	1/2	1115.36 ± 0.14	U
	Σ^+	1/2	1189.40 ± 0.20	U
	Σ^-	1/2	1195.96 ± 0.30	U
	Σ^0	1/2	1191.5 ± 0.5	U
	Ξ^-	?	1318.4 ± 1.2	U
	Ξ^0	?	1311 ± 8	U

where $\bar{L}$ included all the terms *except* the masses, L' was proportional to the masses of the basic particles,

$$L' \sim m_{(p,n)} + m_{(\Lambda)},$$

and L'' was proportional to the *difference* of the masses of the basic particles,

$$L'' \sim m_{(p,n)} - m_{(\Lambda)}.$$

This division of the Lagrangian allowed Gell-Mann (and others) to identify precisely what part of a model broke the symmetry. "If we now consider the Lagrangian with the mass-splitting term L'' omitted, we have a theory that is completely symmetrical in p, n, and Λ." If $SU(3)$ symmetry were exact, the masses of p, n, and Λ would be identical. Formally, the combination $\bar{L} + L'$ would remain unchanged (be invariant) when acted on by the generators of the symmetry group. Proceeding one step further, if L' were absent, $L = \bar{L}$ was massless. Just as for Roger Penrose (chapter 6), the model would be conformally invariant. Gell-Mann considered this "situation, still more remote from reality," and did not pursue its consequences.[30] Figure 7.1 is a schematic of

---

30. Gell-Mann, "Symmetries of Baryons and Mesons," 1074; the conformal invariance was pointed out to Gell-Mann by Walter Thirring, son of the relativist Hans Thirring, 1076 and n28.

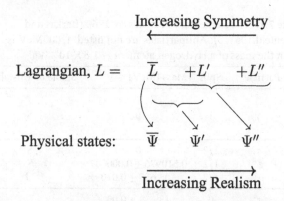

Figure 7.1. Explicit, or manifest, symmetry breaking by adding terms in the Lagrangian (or Hamiltonian).

this situation. States, $\overline{\Psi}$, built from only $\overline{L}$ are the most remote from reality; the most realistic states, Ψ'', are built from the entire Lagrangian, L. (Of course, constructing such states would only be a first step in comparing theory with experiment.) Gell-Mann did not ascribe a mechanism to the symmetry breaking—symmetry was meant to be distinct from dynamics.

For Coleman, approximate symmetries were connected to speculative worldmaking. At the 1967 "Ettore Majorana" Summer School in Erice, Italy, he replied to a question:

> What do I mean by a good approximate symmetry? To me, this means that the real world doesn't possess this symmetry but that there exists an imaginary world which does have this exact symmetry and the predictions in this imaginary world describe well a certain set of events in the real world.[31]

Coleman was by no means the only physicist to write of alternate worlds, though it is unlikely that the representativeness of his "imaginary world" picture can be assessed. Also at Erice, Sheldon Glashow gave "frankly speculative" lectures on symmetry breaking (see section 7.4). The discussion session suggests that he spoke of a "'good' world" contrasted with a "'perfect' world." But no traces of these worlds survive in the published lecture notes.[32] Perhaps the real world of physics publication was not good enough to accept these imaginings, but these imaginings were put into international circulation at summer schools like Erice.

7.3 Spontaneous Symmetry Breaking

The vacuum was put at the center of broken symmetries by the invention of "Spontaneous Symmetry Breaking" (SSB) in 1960 and its application in the "Higgs

---

31. Coleman, "All Possible Symmetries," 399–400.

32. E.g., "Glashow: In the 'good' world the Lagrangian is symmetrical, but the vacuum is not invariant." Glashow, "Breaking Chiral SU(3) ⊗ SU(3)," 83 and 133.

mechanism" in 1964.[33] These developments have been analyzed by historians and by some of the protagonists.[34] Nambu Yoichiro (1921–2015) brought SSB into particle physics in 1959. He framed his work as importing the techniques of solid-state physics into the relativistic realm, showing that "this interaction of the two branches of physics can be reciprocal [...]."[35] Chapter 2 showed that Paul Dirac was the first to make this maneuver, and suggests an explanation for how Nambu's advances were possible. In this section, I will lay out some of the most salient aspects of the story, especially the return of Douglas Hartree's many-electron approximation.

Nambu received his doctorate at the Imperial University in Tokyo in 1952 and worked in Japan until he became a fellow at Oppenheimer's Institute for Advanced Study in Princeton, NJ. In 1954 he joined the University of Chicago, where he spent the rest of his academic career. In 1956, John Schrieffer, a graduate student at the University of Illinois, Urbana, gave a seminar at Chicago on a new theory of superconductivity. Superconductivity occurs when material, such as liquid helium, is super-cooled close to absolute zero—the electrical resistance disappears, allowing electric currents to circulate perfectly without dissipating. John Bardeen, Leon Cooper, and Schrieffer (BCS) made a daring proposal to explain this: at low temperatures, the electrons formed "Cooper pairs," held together by phonon (vibration) interactions. Nambu was impressed, though he was concerned that the BCS proposal seemed to break electrodynamic gauge invariance. By the summer of 1959, Nambu had written a complex paper analyzing the BCS proposal and showing that in fact, gauge invariance (equivalently, charge conservation) was preserved. The dynamics of broken symmetries was not explicit in Bardeen, Cooper, and Schrieffer's paper; Nambu both defined the puzzle and pointed the way to a solution.[36] Also that summer, at the Ninth International Conference on High Energy Physics in Kiev, USSR, Nambu suggested that "there seems to be a possibility of learning something about the elementary particles by studying superconductivity."[37]

Nambu's proposal aimed at the heart of the problem of dozens of elementary particles. In the spirit of Dirac and Wheeler before him, in 1960 he declared: "A unified understanding of all the baryons and leptons will be of course our ultimate goal [...]." He aimed to achieve this by postulating the existence of an underlying massless fermion-field. In "a mathematical analogy" to the BCS superconductor theory, elementary fermions would pair into massive "quasi-particles," which would represent the baryons. Further "collective excitations" of these pairs would form a boson-field

33. A note on names. Uncharacteristically for a physicist, Peter Higgs downplays his priority claim to his eponymous mechanism: "perhaps it should be called 'the ABEGHHK'tH...mechanism' after all the people (Anderson, Brout, Englert, Guralnik, Hagen, Higgs, Kibble, 't Hooft) who have discovered or rediscovered it!" Here, I use the standard label. Higgs, "SBGT and All That," 159.

34. Pickering, *Constructing Quarks*, ch. 6; Nambu, "Gauge Principle"; Brown et al., "Panel Session"; Brown and Cao, "Spontaneous Breakdown of Symmetry"; Cao, *Conceptual Developments*, chs. 9–11; Stöltzner, "Higgs Models"; Schweber, "Shelter Island Conferences Revisited," 103–7.

35. Nambu, "Dynamical Theory Suggested by Superconductivity," 858.

36. Nambu, "Quasi-Particles and Gauge Invariance"; Hoddeson et al., "Collective Phenomena," 541–64.

37. Touschek, "Neutrino Gauge Group," 122.

which would represent the force-carrying mesons (see table 7.1).[38] But before such ambitious theories could be articulated, Nambu and a Chicago postdoctoral fellow, Giovanni Jona-Lasinio (b. 1932), built unrealistic models "to show that a new possibility exists for field theory to be richer and more complex than has been hitherto envisaged [...]."[39] In this latter goal at least, they succeeded.

Three decades after Dirac applied the Hartree approximation in his "hole theory," Nambu "recognize[d] [the BCS theory] as a generalized Hartree-Fock approximation."[40] (Vladimir Fock improved on Hartree's procedure by accounting for electrons' antisymmetric wavefunction in 1930.[41]) Hartree's approximation had several elements. In the Hamiltonian representing a complex atom, Hartree replaced *holistic* multi-electron expressions, $\psi(1, 2, 3 \ldots)$, with linear products of *individual* electron wavefunctions, $\psi(1)\psi(2)\psi(3) \ldots$. (Field theorists usually made the same approximation.) In an atom, each electron moved in the electric field of the central nucleus and the fields of every other electron. Hartree vastly simplified this picture by assuming that each electron interacted with the nucleus and an *average field* of the other electrons. The average field was defined by the following constraint: the average field for one electron should be the same as for any other, that is, the average field for $\psi(1)$ equaled the average field for $\psi(2)$: $(\psi(2), \psi(3), \ldots)_{AV} = (\psi(1), \psi(3), \ldots)_{AV}$. Because of this, Hartree's approximation was often called the "self-consistent" approximation that used "self-consistent" fields. In Nambu's field-theoretic notation, these averaged were written as expectation values, $\langle \psi(1)\psi(2) \rangle$.[42] The self-consistent fields were inserted into the Hamiltonian, replacing the complex electron-to-electron interactions. With the new self-consistent Hamiltonian, standard tools of calculation were applied, such as perturbation analysis. The act of inserting an expectation value into the Hamiltonian (or Lagrangian) was to become the central act of SSB. In 1929, Dirac's application of many-electron methods to relativistic quantum mechanics radically transformed his picture of the vacuum. Nambu followed the same path.

In 1961, Nambu and Jona-Lasinio published their "Dynamical Model of Elementary Particles Based on an Analogy with Superconductivity, part I." They applied Nambu's Hartree-Fock analysis of the BCS model to a model Lagrangian, and found the following consequences. Their model included *two* vacua, one with massive particles, $\Omega^{(m)}$, the other massless $\Omega^{(0)}$. But the two vacua were mathematically orthogonal—they had no overlap. "Thus the two 'worlds' based on $\Omega^{(0)}$ and $\Omega^{(m)}$ are physically distinct and outside of each other."[43] Nambu and Jona-Lasinio's "world" language followed Werner Heisenberg. Because the vacuum state "form[ed] the

38. Nambu, "Dynamical Theory Suggested by Superconductivity," 859, 858; this ontological vision was shared by J. D. Bjorken (b. 1934). Bjorken extended Nambu's methods to electromagnetism. His first sentence referred readers to "earlier work" on "manufacturing the electromagnetic field from other fields" in E. T. Whittaker's *A History of the Theories of Aether and Electricity*! Bjorken, "Dynamical Origin for the Electromagnetic Field," 174n1.

39. Nambu and Jona-Lasinio, "Dynamical Model. I," 358.

40. Nambu also used the recent work of Soviet mathematician Nikolai Bogoliubov on superconductivity; Nambu, "Quasi-Particles and Gauge Invariance," 648; cf. Nambu and Jona-Lasinio, "Dynamical Model. I," 345; Hoddeson et al., "Collective Phenomena," 560–61.

41. Fock, "Approximate Method."

42. Nambu, "Quasi-Particles and Gauge Invariance," 649.

43. Nambu and Jona-Lasinio, "Dynamical Model. I," 349.

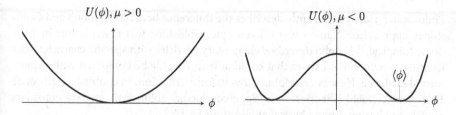

$U(\phi), \mu > 0$ $U(\phi), \mu < 0$

$\langle\phi\rangle$

ϕ ϕ

Figure 7.2. "Wine-bottle" potential for illustrating Spontaneous Symmetry Breaking. In a normal theory, at left, a ball rolling down the curve will stop at a single lowest-point. In SSB, at right, a ball released at the center of the potential can roll to multiple lowest-points, such as $\langle\phi\rangle$. In particle physics language, the ground state of the ball is infinitely degenerate.

substrate for the existence of elementary particles," Heisenberg preferred to call it a "world state."[44] Nambu and Jona-Lasinio argued that $\Omega^{(m)}$ was the "'true' ground state."[45] $\Omega^{(m)}$ had a second consequence for the vacuum. The initial massless Lagrangian L was invariant under chiral symmetry transformations;[46] but neither $\Omega^{(m)}$ nor its states, $\Psi^{(m)}$, were invariant. They showed that their theory implied $\Omega^{(m)}$ was really an infinite collection of vacua, one for each chirality, α, which in turn meant infinite worlds. "Thus there is an infinity of equivalent worlds described by $\Omega_\alpha^{(m)}, 0 \leq \alpha < 2\pi.$"[47] The conflict between an invariant initial Lagrangian, L, and a non-invariant vacuum $\Omega_\alpha^{(m)}$ led, dramatically, to the existence of *new* zero-mass particles in the theory. Within Nambu and Jona-Lasinio's ambitious scheme for explaining the elementary particles, they identified this zero-mass particle as a new bound state of their fundamental field.[48] (These are now called Nambu-Goldstone bosons, or in the period, Goldstone bosons, "zerons," and other names.) However, in order to make their discussion more realistic—no such particles existed—they introduced an additional conventional symmetry breaking term in their Lagrangian. Rather than "broken," they described the overall situation in which the vacuum is degenerate with respect to a symmetry of the Lagrangian as a "masked" symmetry.[49] This complicated and ambitious paper aimed to point the way to a unified picture of the elementary particles, in the course of which it discovered new depths of vacuum structure.

This structure was clarified and developed by British physicist Jeffrey Goldstone (b. 1933). He was well placed for this work. His 1958 doctoral dissertation at Cambridge University was on the application of field theory techniques—including Feynman diagrams—for solid-state physics, including the Hartree-Fock approximation.[50]

44. Quoted in Brown and Cao, "Spontaneous Breakdown of Symmetry," 222.
45. Nambu and Jona-Lasinio, "Dynamical Model. I," 349.
46. Chiral symmetries involve the 4×4 matrix γ^5. Recall that Dirac's electron theory introduced six 4×4 matrices, section 2.2.3. Physicists later found that the Dirac equation could be written with only four 4×4 matrices labeled γ^0 to γ^3. By definition, $\gamma^5 = i\gamma^0\gamma^1\gamma^2\gamma^3$.
47. Nambu and Jona-Lasinio, "Dynamical Model. I," 350–51.
48. Nambu and Jona-Lasinio, "Dynamical Model. I," 350–51.
49. Nambu and Jona-Lasinio, "Dynamical Model. II," 253.
50. Goldstone, "Brueckner Many-Body Theory"; D. Kaiser, *Drawing Theories Apart*, 237–42; Schweber, "Hacking the Quantum Revolution," §3.2; cf. Nambu, "Quasi-Particles and Gauge Invariance," 649n5

Philosopher Tian Yu Cao aptly describes the difference between Nambu's and Goldstone's approaches: Nambu was microscopic, Goldstone was macroscopic or phenomenological.[51] Nambu described elementary particles with specific characteristics forming specific bound states that could at least roughly be compared with experimental evidence. He was describing a *dynamical mechanism*. In contrast, Goldstone's 1961 paper, "Field Theories with 'Superconductor' Solutions," "merely considers models and has no direct physical applications [...]."[52]

The SSB procedure is shown schematically in figure 7.3. The key action for most practitioners was replacing an expression for a quantum field with a *c*-number, classical, expression of its vacuum expectation value, $\phi \rightarrow \langle \phi \rangle = \lambda$. Physicists acknowledged the connection to Hartree's earlier work by name (by Nambu and Jona-Lasinio) and by terminology such as "self-consistent" (Goldstone and many others).[53] Compared to Glashow's "partial" symmetries depicted in figure 7.1, Nambu, Jona-Lasinio, and Goldstone's technique *rotated* the symmetry-realism axis. This was a *more abstract* analysis of broken symmetry, in the sense that there was an additional layer of mathematical structure between the initial Lagrangian and the physical states. This may be why Goldstone warned his readers: "In all these theories the relation between the boson field and the actual particles is more indirect than in the usual perturbation type solutions of field theory."[54] Goldstone's indirectness was in fact part of a trend in particle physics, which also included "Euclidean" field theory (section 7.5.3). The trend was toward more "remote" connections between the fundamental and the phenomenological.[55] SSB was Janus-faced. From one direction, SSB allowed physicists to construct "broken" theories with a closer relationship to experimental measurements; simultaneously, from the other direction, they allowed physicists to build "symmetric" theories that were ever more remote from the world of experiments.

Compared to Nambu and Jona-Lasinio, Goldstone was more explicit about how the vacuum was *used* in the practice of symmetry breaking. His analysis of these solutions was "based on the idea that in the vacuum state the expectation value of the boson field is not zero." This was a radical departure from standard practice in particle physics.[56] (Though it was more acceptable in Goldstone's many-particle

---

51. Cao, *Conceptual Developments*, 316n22.

52. Goldstone, "Field Theories," 154.

53. In addition to work already mentioned, see "self-consistent vacuum," Englert and Brout, "Broken Symmetry," 322; "Hartree conditions," Kibble, "Symmetry Breaking," 1559; another particle physics application of the Hartree approximation was Martin and Schwinger, "Many-Particle Systems." N.B. this was distinct from the S-matrix "self-consistent bootstrap" theory and distinct from efforts to make "self-consistent," that is, non-divergent, renormalization schemes.

54. Goldstone, "Field Theories," 155.

55. Consider Schwinger, in 1959: "the observed physical world is the outcome of the dynamical play among underlying primary fields, and the relationship between these fundamental fields and the phenomenological particles can be comparatively remote, in contrast to the immediate correlation that is commonly assumed." Schwinger, "Euclidean Quantum Electrodynamics," 398.

56. This was ruled out in QED by Wendell Furry in 1937 (section 3.3). Goldstone invoked reflection symmetry, though the vanishing was more commonly attributed to charge conservation. Rohrlich, *Applied Quantum Electrodynamics*, 60; Jauch and Rohrlich, *Theory of Photons and Electrons*, §8-4. See also Schweber's and Nambu's assessments in Brown and Cao, "Spontaneous Breakdown of Symmetry," §5.

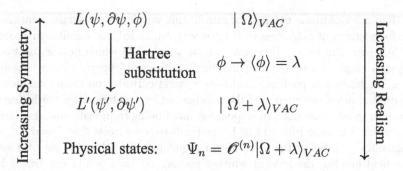

Figure 7.3. Schematic of Spontaneous Symmetry Breaking. For Nambu and Goldstone, the physical states Ψ include a massless boson. Coleman's "hidden symmetry" view considered only the different symmetries of L and Ψ.

theory, in which the vacuum was replaced by the ground state of a nucleus.[57]) Goldstone illustrated this vacuum contribution with a tadpole-shaped diagram. It represented a particle interacting with a vacuum fluctuation and scattering away.[58] The diagram represented energy coming from nothing (or dissipating into nothing). Goldstone implied that the vacuum expectation value of the current was not zero, $\langle 0|j_\mu|0\rangle_0 \neq 0$. Chapter 4 showed how central the *vanishing* of this current was to Julian Schwinger's renormalized QED. However, in 1957 Schwinger himself explored the consequences of a scalar field with non-vanishing vacuum expectation value, $\langle \phi_\mu \rangle_0 \neq 0$. Such a "mass generating" term would make the field, ϕ_μ, massive and violate the symmetries of the strong interactions.[59] In the same period, symmetry-breaking, mass-generating, tadpoles entered particle physics, though in a less-radical way.[60] Goldstone's "superconductor" theory made a small but radical step toward attributing positive physical qualities to the vacuum.

57. However, there were doubts even in the many-particle context. Bethe commented on Goldstone's vacuum matrix element: "This obviously would have no physical meaning, but would serve merely to give a definition of the self-consistent potential." The matrix element was $(n_i|V|n_i)$ for n_i unoccupied, Bethe, "Nuclear Many-Body Problem," 1360; compare to Goldstone, "Brueckner Many-Body Theory," fig. 5(c); Goldstone did in fact use the vacuum contribution, Δ, to establish self-consistent conditions. Goldstone, "Field Theories," 158; cf. Goldstone, Salam, and Weinberg, "Broken Symmetries," 966.

58. Goldstone, "Field Theories," 157, 159; Goldstone, "Brueckner Many-Body Theory," 271, figs. 3, 5, and 13; D. Kaiser, *Drawing Theories Apart*, 239–43; compare Bjorken's "Hartree bubbles," Bjorken, "Dynamical Origin for the Electromagnetic Field," 175–76.

59. Schwinger, "Theory of the Fundamental Interactions," 416–17, 423; Mehra and Milton, *Climbing the Mountain*, 418–28.

60. This work was independent of SSB, but was later integrated with it. Summarizing a 1960 conference, Abdus Salam remarked: "tadpole graphs are perfectly respectable graphs." At least for the particle "which carries the same quantum numbers as the vacuum," Salam, "Some Speculations on the New Resonances," 428; for a photograph of Paul Matthews drawing tadpole diagrams, see Brown, "Tyger Hunting in the Everglades." This was a development of Gregor Wentzel's aptly named "spurion" proposal, in which "spurious particles" carry quantum numbers into the vacuum, Brown and Cao, "Spontaneous Breakdown of Symmetry," 221. Tadpoles became associated with Coleman and Glashow through an influential paper that proposed that tadpole interactions broke strong interaction symmetries: Coleman and Glashow, "Departures from the Eightfold Way." In 1964, Peter Freund and Nambu named vacuum particles "zerons," Freund and Nambu, "Mass and Coupling Constant Formulas"; in 1966, Peter Higgs synthesized SSB and tadpoles. He described the work of Schwinger, Salam (with John Clive Ward), Freund, Nambu, Coleman, and Glashow as about symmetry-breaking "vacuons," Higgs, "SSB without Massless Bosons," 1156.

How did Goldstone use the vacuum in this way if his procedure contradicted the foundations of field theory? If rigor was demanded, the calculation probably did *not* pass. But formal rigor was not the standard to which he—or most other theorists—aspired. Goldstone argued that his abnormal "superconductor" solutions were "established as plausibly as the more usual perturbation theory solutions."[61] Theoretical physicists often used the standard set by paradigmatic perturbation calculations to relativize rigor. In response to questioning from mathematician Arthur Wightman, Coleman insisted that his methods were no worse than "good old wave mechanics [...]. So, at the level, not of a quantum field theorist, but of a quantum field butcher, the level at which I operate, so far, I would say, things look good."[62] Goldstone and Coleman appealed to what Thomas Kuhn called paradigmatic problems, or exemplars: established puzzles and methods of solution which were learned by all members of the community of expert practitioners. In twentieth-century physics, one of the important *uses* for paradigms was to calibrate communal epistemic standards—such as what mathematical operations were rigorous *enough*.

Perhaps it did not matter whether Goldstone's methods were rigorous or whether they described physical processes. He concluded his paper: "A method of losing symmetry is of course highly desirable in elementary particle theory [...]." He certainly provided a *method*, and in clearer terms than Nambu and Jona-Lasinio. Goldstone continued: "[...] but these theories will not do this without introducing non-existent massless bosons [...]." This became known as the Goldstone Theorem.[63] But there was a problem. As Goldstone, Abdus Salam, and Steven Weinberg wrote in 1962, SSB predicted "a spinless particle of mass zero, which almost certainly does not exist."[64]

In 1962, Marshall Baker and Sheldon Glashow christened Nambu and Goldstone's methods "Spontaneous Symmetry Breaking" because solutions that break a symmetry were produced without "without the introduction into the Lagrangian of any symmetry-breaking terms," as in section 7.2.[65] The final development to sketch is the results of efforts to evade Goldstone's theorem. That is, to have a theory *with* SSB and *without* massless, spinless, bosons in the "physical" states. This was accomplished in the mid-1960s by the work of several physicists, most prominently: Philip Anderson, Robert Brout, François Englert, Gerald Guralnik, Richard Hagen, Peter Higgs, and Tom Kibble. They managed to build models with SSB and without massless bosons by putting SSB in a new formal context: gauge theory.[66] The simplest gauge theory was

---

61. Goldstone, "Field Theories," 161. If one started from the position Nambu and Goldstone tried to establish—that superconductor solutions existed—then it was possible to construct more rigorous proofs of Goldstone's theorem: Jona-Lasinio, "Relativistic Field Theories." Also, Higgs's discussion of these proofs valued *use-value* over rigor. Sophisticated axiomatic proofs "do not seem relevant to the possible usefulness of such theories in generating zeron-free models of spontaneous symmetry breakdown," Higgs, "SSB without Massless Bosons," 1157.

62. Specifically, the WKB approximation. Coleman, "Uses of Instantons," 926; cf. Coleman and Weinberg, "Radiative Corrections," 1904.

63. Goldstone, "Field Theories," 163–64.

64. Goldstone, Salam, and Weinberg, "Broken Symmetries," 965.

65. Note that this definition of "spontaneous" is purely formal. Baker and Glashow, "Spontaneous Breakdown," 2463.

66. How? "The Goldstone theorem is avoided [in the "Goldstone-Higgs phenomenon"] because it is impossible to quantize the gauge fields in a gauge in which all of the [standard field theory] axioms are obeyed simultaneously." Coleman, "No Goldstone Bosons," 260.

QED, and section 4.3 discussed how Julian Schwinger tamed its complexities in 1951. Following Nambu and Goldstone's work, the vacuum played a central role in symmetry breaking, including Higgs et al.'s gauge-theory calculations. For Higgs, "symmetry is broken by the nonvanishing of the vacuum expectation value"; Englert and Brout used them synonymously, referring to "degenerate vacuum (broken symmetry)"; Guralnik, Hagen, and Kibble worked in "the usual broken-symmetry theories in which a degenerate vacuum is the sole symmetry-breaking agent [...]."[67] But viewing the vacuum as the *agent* of symmetry breaking was not uncontested.

7.4 Secret Symmetry

Coleman objected to the view that the vacuum *acted* to create SSB and massless bosons. He saw Goldstone's theorem as—like other group-theoretic results—independent of a dynamical mechanism. In the same 1967 "frankly speculative" Erice lectures discussed above, Glashow regularly referred to the vacuum as an agent—for example, he described the "vacuum breaking of SU(3)" symmetry. During the discussion session, Coleman objected:

> Coleman: I have a comment on the Goldstone theorem. I want to stress that it is not necessarily connected with a symmetry breakdown. [... he gave an example.] Thus, although the Goldstone theorem applies to spontaneous symmetry breakdown, it also applies to situations where the spontaneous symmetry breakdown never enters the picture.[68]

In 1967, Coleman argued *against* narrative and mechanism.[69] He used counterexamples to try to convince his peers that Goldstone's theorem was true—as a statement about symmetries—but that the vacuum as an agent was not needed to break the symmetry. In fact, he would argue that the symmetries were not really "broken."

In Coleman's 1971 Erice lectures on "Dilatations" (scale invariance), he distinguished between "manifest" symmetry and "hidden" symmetry. (Recall Nambu and Jona-Lasinio's "masked" symmetry.[70]) For manifest symmetry, "The ground state of the theory (the vacuum) is invariant under the symmetry group." Hidden symmetry "is usually called 'spontaneously broken symmetry', but this terminology is deceptive; the symmetry is not really broken [...] it is just that the consequences of the symmetry are less obvious than in the other case."[71] These lectures are also a fine example of how Coleman gave his technical writing narrative form. The lecture headings were

67. Higgs, "Broken Symmetries, Massless Particles," 133; Englert and Brout, "Broken Symmetry," 321; Guralnik, Hagen, and Kibble, "Global Conservation Laws," 586.

68. In work with Steven Weinberg, Charles Sommerfield, and Howard Schnitzer, Glashow, "Breaking Chiral SU(3) $\otimes$ SU(3)," 101, 133–34; see also Coleman's comments to Nicola Cabibbo, "Spontaneous Breakdown," 391.

69. Rather than "symmetry breaking" from one Lagrangian to another, he considered single theories with, as it were, misaligned Lagrangians, currents, and states. Still, the vacuum was at center stage. Coleman, "Invariance of the Vacuum"; see Aitchison, *Gauge Field Theories*, §6.1.

70. Nambu and Jona-Lasinio, "Dynamical Model. II," 253.

71. Coleman, "Dilatations," 369–70.

as follows: §1. "Introduction." §2. "The formal theory of broken scale invariance." §3. "The death of scale invariance," in which the students learned of "A disaster in the deep Euclidean region," and that *the entire theoretical structure of Sect. 2 is a lie!*" In the final subsection of §3. Coleman proclaimed "We are lost in a dark wood."[72] But hope was on the horizon, with §4. "The resurrection of scale invariance." §5. "Conclusion and questions" showed the ironic detachment Coleman maintained, even while telling a romantic tale. "The formal theory of scale invariance is a pack of lies, hopelessly afflicted with anomalies." Only one set of "true equations" had been found, and even these were unsatisfying.[73] Coleman narrated many of his papers and lectures using classic narrative form. But he was a man of his time. Even in a didactic context, he deployed irony to maintain a critical remove.

Coleman returned to Erice in 1973 and delivered an influential series of lectures on "Secret Symmetry." As Sociologist Andrew Pickering has explained, SSB and the Higgs mechanism really entered the mainstream of theoretical physics only in 1972. In 1971–1972 the Dutch physicist Martinus Veltman and his graduate student Gerard 't Hooft demonstrated that gauge theories of the weak and strong interactions were renormalizable. From Veltman's demonstration for *massless* theories, 't Hooft used the mass-generating Higgs mechanism to demonstrate the renormalizability of *massive* (and more realistic) theories.[74]

Connecting SSB to renormalizability changed the landscape. It provided another reason to build theories with SSB. Renormalizability was a delicate thing to specify. The basic idea was that it should be possible to apply (a version of) the algorithm for removing infinite singularities from QED, which was developed by Freeman Dyson.[75] A contemporary definition for renormalizability was: "the requirement that it should be possible to eliminate all infinities in a quantum field theory by a redefinition of a small number of physical parameters."[76] *Non*renormalizable theories would require an infinite number of redefinitions, that is, their predictions would never be finite. Even with finite predictions, a theory with an infinite number of free parameters would not be predictive. Following this logic, in the 1950s, Hans Bethe elevated renormalizability into a guiding principle for relativistic researchers.[77] Through the 1970s, S. Weinberg used "renormalizable" synonymously with a "satisfactory" theory, though he admitted that many physicists might disagree with him.[78] Coleman agreed with this prescription.

In fact, Coleman saw achieving renormalizability as the *purpose* of SSB. He did work with non-renormalizable "phenomenological" Lagrangians—not to get closer

72. Coleman, "Dilatations," 375–76, 379.

73. The true equations were the Callan-Symanzik equations. Coleman, "Dilatations," 392; compare: "We are in trouble. We should be, for we have been foolish." Callan and Coleman, "False Vacuum. II," 1764.

74. 't Hooft, "Renormalizable Lagrangians," §3; Pickering, *Constructing Quarks*, §6.3; for a contemporary recognition of the role of asymmetric vacua, see "it was shown by 't Hooft that the previous mechanism [i.e., "the presence of non-symmetric vacuum"] allows to construct renormalizable theories of massive vector particles." Fubini, *Present Trends*, 8.

75. Schweber, *QED*, §§9.11, 9.14.

76. Weinberg, "Search for Unity," 33.

77. Cao and Schweber, "Conceptual Foundations," 43; cf. D. Kaiser, *Drawing Theories Apart*, 203–4; Schweber, "Hacking the Quantum Revolution," §4.

78. Weinberg, "Search for Unity," 33; cf. Galison, *Image and Logic*, 643–44.

to experiment, but because they might "offer clues to general theorems" or "offer counter-examples to wrong arguments [...]." But he found them "hard to take seriously as physical theories [...]."[79] He expressed stronger views in "Secret Symmetry": "For nonrenormalizable theories, in the common sense (i.e. those with an infinite number of counterterms), there are an infinite number of free parameters, which is why these theories are commonly (and properly, I think) considered disgusting." For Coleman, renormalizability was akin to comprehensibility. He suggested that renormalizable theories "may exhaust the set of theories that make sense."[80] In the ensuing discussion, Frank Wilczek (who met Coleman when the latter spent the spring 1973 semester at Princeton) pressed on this issue. "Does this suggest to you that renormalizability is essentially an artifact of perturbation theory, or do you think it has intrinsic meaning?"

> Coleman: I think renormalizability has a profound meaning outside of perturbation theory. My guess is that there is no way to make sense out of a non-renormalizable theory. [He discussed studies of divergences.] Therefore, I think that non-renormalizable theories are absolute nonsense, not merely perturbative nonsense. This statement of course is sheer chutzpah.[81]

't Hooft used SSB to open the door to renormalizable theories of the weak and strong interactions. For Coleman, this opened up new landscapes of theories that could be taken physically seriously. In his "Secret Symmetry" lectures, he asked:

> All of this is very pretty, but what does it buy us? What is the practical use of the idea of spontaneous symmetry breakdown, even by the generous standards of practicality current among high-energy theoreticians?

In his next lecture, he showed that SSB would not affect the renormalization procedure. He then answered his question: "I asked, 'What does it buy us?' We now have the wonderful answer: secret symmetry buys us secret renormalizability."[82] Physicists presented a series of purposes for SSB, all of which had ontological, formal, and aesthetic aspects. Nambu made the strongest ontological proposal. The purpose of SSB was to achieve a unified picture of elementary particles as bound states of a single primitive field. Goldstone and Coleman were more formal and aesthetic. Goldstone abjured physical consequences and explored a method for breaking symmetries. Coleman presented the purpose of SSB as exquisitely formal—to apply an algorithm, to achieve comprehensibility.

The usefulness of SSB was reflected in the vivid descriptions contained in "Secret Symmetry." From Coleman's introduction: "The standard example [of SSB] is the Heisenberg ferromagnet," a model of magnetic material. "Even though the Hamiltonian is rotationally invariant, the ground state is not;" as a cool (i.e., ground-state)

79. Coleman, "Dilatations," 370.
80. Coleman, "Secret Symmetry," 157.
81. Coleman, "Secret Symmetry," 221; attitudes changed so radically by the 1990s that Coleman had difficulty recalling his earlier views; Deser et al., "Panel Discussion," 369.
82. Coleman, "Secret Symmetry," 154, 166; Smeenk, "False Vacuum," 235.

bar of metal, the microscopic magnetic dipoles are all aligned to produce an overall magnetization in a particular north-south direction (i.e. not the same in all directions). This example was suggested by Nambu and Heisenberg in 1959.[83] Coleman made it vivid. "A little man living inside such a ferromagnet would have a hard time detecting the rotational invariance of the laws of nature; all his experiments would be corrupted by the background magnetic field."[84] Coleman's parable of the little man inside a magnet became a paradigmatic description of SSB.[85] The remoteness Goldstone saw between symmetric theories and physical states manifested for Coleman as the impossibility of measuring the hidden symmetry.[86]

This tension between vivid description and tenuous connection to physical phenomena also held in Coleman's discussion of the Higgs mechanism. Coleman memorably described Higgs et al.'s manipulation of gauge parameters to create massive particles as a carnivorous process. "The vector meson has eaten the Goldstone boson and grown heavy."[87] This was not the first description of particles eating one another, but Coleman strikingly developed the image.[88] In the following discussion session, he was asked about models of composite Higgs bosons.[89] The carnivorous became cannibalistic:

> What we imagine is that the Yang-Mills fields create a force which binds two fermions into a composite Goldstone boson, which the Yang-Mills then eat, like ancient gods devouring their children.[90]

The students were unlikely to forget this description, and were thankful, perhaps, that there was no way to observe such devouring.

Eating is a paradigmatic physical and embodied process. Yet Coleman also described the "Higgs phenomenon" as a ghostly, formal process, a "magic trick." The gauge fields used by Higgs et al. were partly *excess* mathematical structure, which was eliminated in the end of all calculations by "fixing" the gauge. Coleman pointed out that one way to fix a gauge is to cancel out one of the gauge fields.

> The reason the Goldstone boson disappears in the gauge-invariant theory is that it was never there in the first place; the degree of freedom that would be associated with the Goldstone boson is a mere gauge phantom, an object that can be gauged away.[91]

83. Nambu, "Dynamical Theory Suggested by Superconductivity," 866.
84. Coleman, "Secret Symmetry," 141; the little man flipping magnets was developed into a general model of quantum measurement. See Bell, "Wave Packet Reduction."
85. Coleman, "1979 Nobel Prize"; Pickering, *Constructing Quarks*, 169.
86. Cf. "we can not do experiments that directly see the hidden symmetry," Coleman, "Secret Symmetry," 216–17.
87. Coleman, "Secret Symmetry," 149.
88. For eating and tadpoles, see Gell-Mann and Lévy, "Axial Vector Current," 719; see also Llewellyn Smith, *Renormalizable Theories of Weak Interactions*, 19.
89. See the lovely paper, Borrelli, "Composite Higgs," though I suspect that the dominance of non-rigorous "hybrid theoretical constructs" she ascribed to physics only since the late 1970s can be traced back to, say, the Maxwellian era of Oliver Heaviside.
90. Coleman, "Secret Symmetry," 222.
91. Coleman, "Secret Symmetry," 150.

To Coleman, "the Higgs phenomenon is an explanation, not a phenomenon. [...] Unlike the Goldstone phenomenon, the Higgs phenomenon leaves no footprints."[92] In my view, Coleman's evocative language was not a description of a *physical* phenomena among elementary particles. Much like other twentieth-century American references to "ancient gods," this was not an exercise in realism. Coleman described a *theoretical* phenomenon—by which I do *not* mean a postulated experimental phenomenon. He described a phenomenon theorists enacted on paper and in lecture, at the blackboard.[93]

7.5 Vacuum Stories

Sidney Coleman's research on the "false vacuum" brought together several threads from earlier chapters of this book. As for Paul Dirac's "hole" theory (chapter 2) and Roger Penrose's conformal relativity (chapter 6), Coleman's "false vacuum" was shaped by his paper tools, the techniques, and methods he applied in his work. He began his 1977 Erice lectures (which included the false vacuum):

> In the last two years there have been astonishing developments in quantum field theory. We have obtained control over problems previously believed to be of insuperable difficulty and we have obtained deep and surprising (at least to me) insights into the structure of the leading candidate for the field theory of the strong interactions, quantum chromodynamics. These goodies have come from a family of computational methods that are the subject of these lectures.[94]

These were "semi-classical" methods, which built directly on Julian Schwinger's use of classical "external" fields (section 4.3.1). These tools produced results that drove particle physicists to reconsider the classical limits of their quantum models, as had John Wheeler beginning in the 1950s (section 5.4). Coleman appealed to Weisskopf's inherent vacuum fluctuations when he applied his semi-classical tools in the cosmological arena. Progress in theoretical physics came when physicists applied new tools to the vacuum and found surprising, even astonishing, results.

---

92. Coleman, "Uses of Instantons," 928–29; ghosts and phantoms were again prominent because of Veltman and 't Hooft's renormalization demonstration. The the scalar fields ϕ and the vacuum-shifted gauge triplet $A_a(x)$ "must all be considered as ghosts." 't Hooft, "Renormalizable Lagrangians," eq. 3.5, 174; Richard Feynman's quantum gravity (section 4.4) was another source of ghosts, Cao, *Conceptual Developments*, 303–4, 317n40; cf. Cao and Schweber, "Conceptual Foundations."

93. Further examples of "theoretical phenomenon" are: the phenomenological "hard to take seriously as real physical theories" Lagrangians that "embody the Higgs phenomenon," and "fields acquir[ing] anomalous dimensions," Coleman, "Dilatations," 370, 378; "Accidental symmetry and related phenomena" occurred when renormalization constraints changed a model's symmetry group. Compare the "asymptotic phenomenon" of Bjorken scaling, which was about mathematical functions—but was open to experimental test. Coleman, "Secret Symmetry," §§3.8, 6.1, 198; cf. Cao, *Conceptual Developments*, §10.5.I.

94. Coleman, "Uses of Instantons," 805.

7.5.1 Classical Technique

Coleman was a master of technique. Demand for his unpublished Erice lectures was high and Coleman was anxious for their publication in the (usually late) edited volumes. He repeatedly wrote to his friend Antonino Zichichi. Zichichi (b. 1929) was a prominent experimentalist at Fermilab, CERN, and other laboratories, who championed the establishment of an international center for physics in his native Sicily. Coleman wrote to Zichichi, the Erice proceedings' editor, on November 15, 1972: "Books were set in print faster in the days of hand-set wooden type." On April 18, 1975: "Where is the 1973 book? Once it is an accident; twice is a policy." On April 4, 1977: "Whatever happened to the '75 book? Our [National Science Foundation] contract is bleeding to death paying for preprint requests: it costs us at least two dollars to print and mail each copy of 'Classical Lumps.'" Italian theorist Sergio Fubini wryly acknowledged Coleman's influence in his summary remarks at the 1974 International Conference on High Energy Physics. He listed four "requirements for an acceptable theoretical approach" to new "supersymmetric" theories, which concluded: "4) The whole scheme should not make Sidney Coleman unhappy."[95] In his early career, he deployed a physicists' understanding of group theory to high energy theory. By the 1970s, his toolbox enlarged to include "semi-classical" methods.

Coleman used "semi-classical" in three related senses. First, semi-classical described the use of a "generating functional" and "effective potential" formalism. This formalism made extensive use of classical "external" fields. Second, it described the use of the classical analogue of a quantum system as a basic description, to which quantum corrections are added. And third, it described a perturbation expansion in which the energy-scale (action) of the phenomena was large compared to Planck's constant $\hbar$. (For classical phenomena, $\hbar$ was irrelevant, effectively zero; for quantum systems, the action was about the same size as $\hbar$.) These took their cue from Julian Schwinger's action principle, which began from the calculation of the expectation value of the vacuum state, in the infinite future $\langle 0^+|$ and infinite past $|0^-\rangle$, perturbed by a classical external source, $J(x)$ (section 4.3.2),

$$e^{iW(J)} = \langle 0^+|0^-\rangle_J, \tag{7.2}$$

where in this formalism the generating functional $W(J)$ was used to calculate the dynamics of a theory.[96] Why functional methods? The Lagrangian function was already considered to be a "functional," meaning a function that takes as arguments other

95. Fubini, *Present Trends*, 11; Further: Gian-Carlo Wick wrote, asking for a preprint: "Dear Sidney, According to [Kurt] Symanzik, you 'explained brilliantly' something in Heidelberg," Wick to Coleman, undated [1970] and 8.1.1973. Later, Wick, Nina Byers and others sought out Coleman to explain topology, Wick to Coleman, undated [10.1975?] and Byers to Coleman 22.5.1976. In 1977, Antonino Zichichi awarded him the Ettore Majorana prize for the best lecturer in the history of the Erice school, Zichichi to Coleman 9.3.1976. SRC. For decades, Coleman's unpublished lecture notes informed QFT pedagogy; they were recently published as a textbook: Coleman, *Lectures on QFT*.

96. The effective potential Γ was defined in terms of a classical field, $\phi_c(x) = \delta W(J)/\delta J(x)$, and a Legendre transform, $\Gamma(\phi_c) = W(J) - \int dx_\mu J(x)\phi_c(x)$. Coleman and Weinberg, "Radiative Corrections," 1890; Schwinger, "Gauge Invariance and Mass II," 2425; for the debt to Schwinger, see Coleman and Jackiw, "Dilatation Generators," 587n18.

functions. But when physicists considered only one Lagrangian at a time, the general characteristics of functionals were less relevant. With SSB, *multiple* Lagrangians were at play in a single "theory" or "model," and physicists like Giovanni Jona-Lasinio argued that functional techniques were the appropriate methods for analyzing this situation.[97] In the 1960s, Jona-Lasinio and others developed this technique around the generating functional and the "effective potential," which was the collection of all terms in a Lagrangian that did not contain derivatives.[98] (If a simple Lagrangian was composed of kinetic energy terms and potential energy terms, $L = T - V$, the effective potential was V.)

In 1973, Coleman applied these methods in his most influential paper, "Radiative Corrections as the Origin of Spontaneous Symmetry Breaking," written with his graduate student, Erick Weinberg. They proposed that "the driving mechanism of the instability [of the vacuum] is not a negative mass term in the Lagrangian," as in Nambu's and Goldstone's approach, "but certain effects of higher-order processes involving virtual photons." This effort proposed a dynamical mechanism for SSB.[99] In another author's words: "The radiative corrections de-stabilize the vacuum and trigger a Higgs mechanism."[100]

Coleman and E. Weinberg developed their methods pedagogically, adapting paradigmatic cases to new circumstances.[101] They explained, "the virtue of this method is that it enables us to compute higher order corrections while still retaining a great advantage of the semiclassical approximation: the ability to survey all possible vacua simultaneously."[102] The functional, semi-classical, approach allowed the authors to simultaneously survey *both* the symmetric, $\langle \phi \rangle = 0$, theory *and* the asymmetric, $\langle \phi \rangle \neq 0$, theory. In Nambu and Jona-Lasinio's terms (section 7.3), they could survey all the "worlds" conjured up by the multiple vacua. These methods were immediately applied in Nobel Prize–winning work by Coleman's student, H. David Politzer (see below), and in a flurry of work on "dynamical" symmetry breaking.[103]

Semiclassical methods led Coleman to consider vacuum decay in 1973, three years before he undertook his well-known false vacuum research. At Erice in 1973, immediately after declaring that "secret symmetry buys us secret renormalizability," Coleman developed an analogy to classical physics to establish "the physical meaning of the effective potential," namely, that it was an energy density. Through his classical analogy,

97. Jona-Lasinio, "Relativistic Field Theories"; similarly, in this period "renormalization group" methods viewed QFTs in a functional space, see Cao, *Conceptual Developments*, §8.8.

98. Jona-Lasinio, "Relativistic Field Theories"; others included: Goldstone, Salam, and Weinberg, "Broken Symmetries"; Coleman and Weinberg, "Radiative Corrections," §II.

99. Coleman and Weinberg, "Radiative Corrections," 1889; according to particle physicists' citation tracking system, inSPIRE, this paper has been cited more than 4,000 times, at time of writing, and 836 times within ten years of publication, https://inspirehep.net/literature/81406. By comparison, Goldstone's "Superconductor Solutions" has 1,773 citations at time of writing, https://inspirehep.net/literature/12289.

100. Iliopoulos, "Progress in Gauge Theories," 105.

101. It is remembered as "enormously influential as a handbook," in which "Coleman pulled together all the most useful techniques," Georgi, "Sidney Coleman, 1937–2007," 8.

102. Coleman and Weinberg, "Radiative Corrections," 1889; much of the work in this paper was done by the authors' insistence that a renormalization mass counterterm be added to the Lagrangian, despite the lack of a normal mass term, see 1892; cf. Cao, *Current Algebra*, 163–65.

103. Borrelli, "Composite Higgs."

Coleman showed that his interpretation of the effective potential produced an understanding of the physical meaning when the effective potential had an imaginary part: it was the decay probability of the *ground state* of the system. By analogy, in field theory, it was the decay probability for the *vacuum* of the system. In a footnote, Coleman remarked:

> This phenomenon is well-known; it occurs in the classic calculation of the effective action for electrons coupled to a constant external electromagnetic field. (W. Heisenberg and H. Euler, ["Consequences of Dirac Theory of the Positron"]); J. Schwinger, ["On Gauge Invariance and Vacuum Polarization"]).) This is real for a magnetic field, but imaginary for an electric field; the reason is that, in the presence of a constant electric field, the vacuum is unstable and decays into electron-positron pairs.

Physicists returned, again and again, to paradigmatic vacuum calculations, and reimagined them with new methods. Like Rudolf Peierls and Paul Dirac (chapter 2), Coleman imagined the vacuum as a dielectric material. When a dielectric insulator breaks down, it allows current to propagate, reducing the field strength. In an exchange in the discussion session, Hagen Kleinert asked about the consequences of vacuum decay: "Where does the extra energy go when the false vacuum collapses?"

> Coleman: In the case of the atom, a transition from the 2s "false ground state" to the 1s "true ground state" is accompanied by the production of photons. In our case, the transition is accompanied by particle production.[104]

The analogy to a well-understood system provided a reply: when an atom decays from a higher-energy state to a lower-energy state, it emits photons (i.e., light). In the context of strong nuclear forces, this would become the emission of other particles. This discussion raises two historical questions. Why, if Coleman discussed vacuum decay as early as 1973, did he not publish on the "false vacuum" until 1977? And at that later date, why did he come to different answers?

Coleman's different answers in 1973 and 1977 are best explained by the different methods he applied, and by his comments on related work by other physicists. I will return to these questions in section 7.5.3. Here, I will outline the development of semi-classical methods in QFT, leading up to the false vacuum. One driving force was the discovery of "asymptotic freedom" in 1973 by two groups influenced by Coleman. At Harvard, Coleman's student Politzer developed Coleman and E. Weinberg's methods for the case of "non-Abelian" gauge theories; Coleman spent the 1973 spring term visiting Princeton, and was an interlocutor for David Gross and his student Frank Wilczek as they came to similar results. Asymptotic freedom was the phenomenon in which, in the limit of very high energy ("asymptotic"), a theory of interacting particles behaved as if there were no interactions ("freedom"). This was important because in 1969 theorist J. D. Bjorken made a phenomenological proposal to explain and predict

104. Coleman, "Secret Symmetry," 213–14, 218; cf. Coleman, Jackiw, and Politzer, "SSB in the $O(N)$ Model," 2498; Coleman, "Massive Schwinger Model," 242; Coleman, "Uses of Instantons," 922–23.

Figure 7.4. Sidney Coleman (left) speaking at Erice, July 1975. Seated are (l-r) Antonino Zichichi, Isidor I. Rabi, and unknown (Paul Frampton?). The blackboard reads "$\lim_{t \to \infty}$(Quality of life) = Science," the subject of Rabi's remarks. By permission of Diana Coleman.

results of electron-proton scattering experiments called "scaling." Richard Feynman showed that scaling implied that when a high-energy electron smashed into a proton, it interacted with the proton's constituents as if they were free, non-interacting, particles. In 1973, non-Abelian gauge theories became highly regarded: 't Hooft and Veltman showed they were renormalizable, asymptotic freedom connected them to phenomenology and thus, at least tentatively, to experiment. The leading candidate non-Abelian gauge theory for the strong interactions was Quantum Chromodynamics (QCD), an adaptation of the quark model with the addition of a new fundamental property, "color," and new strong force-carrying particles "gluons."[105] While some physicists worked to solidify the connection between QCD and high-energy scattering, attention rapidly turned to an unfamiliar subject: low-energy phenomena. Could quarks and gluons explain the behavior of protons and neutrons in atoms, at rest? One challenge bridged the energy scales: if quarks existed, why were isolated quarks never observed? Theorists wanted an explanation of how quarks could be "confined."[106]

Perhaps classical *particle* limits would be fruitful. Starting from bosonic fields, in the 1970s, attention turned to "soliton" solutions to classical field equations. Solitons—coherent, non-dissipative waves—had been studied since the nineteenth

105. 't Hooft reported the same result at a 1972 conference. See Pickering, *Constructing Quarks*, ch. 7; Cao, *Conceptual Developments*, §10.2.
106. Pickering, *Constructing Quarks*, 224–26, 310; Cao, *Current Algebra*, §8.3.

century, prompted by waves that maintained their shape over time as they propagated down canals. They suggested a bridge between continuous field theories and discrete particle theories. Solitons formed the bulk of Coleman's lectures at the 1975 Erice Summer School. Figure 7.4 shows Coleman addressing the audience.[107] Coleman recognized that particle physicists were not respecting the precise mathematical definition of "soliton"; he preferred to call these particle models "classical lumps," the title of his lectures.[108] Perhaps stable, *classical* particles—like the molecules and atoms of classical statistical mechanics—were composed of "lumps" of relativistic fields. Following the logic of semi-classical approximations, quantum phenomena would be added as corrections.[109] Coleman found a promising result—albeit in an unrealistic, two-dimensional, "sine-Gordon" model—that bosonic solitons could behave like fermions.[110] In their search for correct *quantum* theories of matter, Coleman and many of his fellow particle physicists turned to re-evaluations of *classical* particles.

Soliton studies echoed John Archibald Wheeler's inquiry into *gravitational* particle models in the 1950s, which were discussed in chapter 5. A central pillar of the new classical-particle explorations was the mathematics of topology. This provided a direct link between Wheeler's geometrodynamics and the mainstream of particle physics. In 1959, David Finkelstein and Charles Misner showed how the topology of a field theory—its being a flat sheet, or a sphere, or a torus, etc.—was associated with conservation laws and conserved quantities. They related their topological objects to Wheeler's space theory of matter, "already unified" field theory, and geons. Finkelstein and Misner's conserved quantities came in integer units, like electric charge, which provided particle physicists like Coleman with a possible basis for "classical lumps."[111] (Finkelstein's topological research was also an important influence on Roger Penrose's diagrams, chapter 6, which built on the "Finkelstein-Eddington" coordinates for black holes.) In a somewhat pessimistic 1977 paper, Coleman argued that "There Are No Classical Glueballs"; that is, in a classical version of QCD, gluons could not form an all-gluon particle. He developed this result with a postdoctoral fellow at Harvard, Larry Smarr. Smarr had studied black hole collisions under Bryce DeWitt at the largest US center for relativistic research, the University of Texas at Austin. If there were no classical "glueball" particles, could there be in classical QCD *impermanent*-but-long-lasting energy concentrations? Coleman and Smarr looked

107. For help with photo identification, I thank Greg Good, Paul Guinnessy, Samantha Holland, and Charles Day. For an account of the meeting, see L. Fermi, "'Ettore Majorana' Centre."

108. Coleman, "Classical Lumps," 299.

109. This calculation was a non-trivial task. At Erice in 1975, Michael Peskin doubted that soliton solutions would survive renormalization. Coleman also remarked: "It would be very nice if someone could compute the first order correction for 't Hooft's monopole, to see that nothing goes wrong. Everybody agrees that this should be done by someone, but everybody also agrees that someone should be somebody else!" Coleman, "Classical Lumps," 414–15.

110. This paper established a "duality, or nuclear democracy in the sense of Chew," between two theories in which the composite particle of one theory was the composite particle of the other. Coleman, "Quantum sine-Gordon Equation," 2096; on nuclear democracy, see D. Kaiser, *Drawing Theories Apart*, 306–17.

111. Finkelstein and Misner, "New Conservation Laws," 241–42; later, Finkelstein developed non-wormhole objects called "kinks," Finkelstein, "Kinks"; cited in Coleman, "Classical Lumps," 403n8 and 404–5n20; on topology and particle physics more generally, see Schweber, "Shelter Island Conferences Revisited," 118–21.

to geometrodynamics to ask "Are There Geon Analogues in Sourceless Gauge-field Theories?" Their answer was negative, but they gained some understanding of the *differences* between nuclear (Yang-Mills) theories and General Relativity.[112] Recall from section 4.4 that the *similarity* of Yang-Mills theories and GR motivated Richard Feynman's attempt at quantum gravity.[113] Coleman was moving into a middle ground. Nevertheless, he was not trying to quantize gravity, and he did not identify as a quantum gravity researcher.[114] However, in July 1979, he attended a symposium at the University of Durham on "Geometry and Physics."

The Durham conference marks two important developments in Coleman's personal life. He traveled to England with Diana Teschmacher, his future wife. And he wrote to the symposium organizer about his diabetes.

> I am an insulin-dependent diabetic and thus have to keep down my consumption of sugars and starches [...]. Fats are no problem, so I am usually able to work my way through any meal with minor adjustments (bread and butter, hold the bread), but if you are planning a pasta and trifle feast, give me advance warning.[115]

At Durham, Coleman spoke on renormalization and SSB; Penrose spoke on spacetime structure and twistor theory; Stephen Hawking spoke on spacetime foam.[116] Particle physics and general relativity came closer together in the mid-1970s: intellectually, in concepts like geons; practically, with semi-classical methods; and institutionally, in conferences such as "Geometry and Physics."

Physicists, sociologists, and historians usually attribute the origin of particle cosmology to the advent of "Grand Unified Theories." GUTs predicted particles so heavy that no accelerator could observe them. Studying GUTs required the heat of big bang cosmology. Historian David Kaiser effectively challenges this view and relocates the explanation to structural changes in physics as a discipline, particularly changes to pedagogy and advanced training. In the early 1970s, underemployed postdocs poured out of particle physics into neighboring subfields, including cosmology.[117] Kaiser's macro-scale arguments from government funding and citation statistics are convincing for US physicists. But philosophy also played a role. Both Kaiser and Andrew Pickering acknowledge that Soviet physicists did not have the same sharp disciplinary boundaries, and that prominent particle cosmology work was done in the 1960s and

---

112. Namely: gravitation always attracts, so Wheeler and Misner's geons are possible; however, like Yang-Mills charges repel, making gauge geons impossible: Coleman, "There Are No Classical Glueballs"; Coleman and Smarr, "Are There Geon Analogues?," they thanked Charles Misner. On UT Austin, see D. Kaiser, "Whose Mass Is It?," 549

113. Coleman may have attended Feynman's GR lectures at Caltech. At Harvard, he taught a Feynmanesque GR course in the late 1960s and early 1970s. D. Kaiser, "A ψ is just a ψ," 331–33.

114. Coleman to Christopher Isham, 12.7.1978, SRC.

115. Diana Coleman believes that Sidney's diabetes was a very important part of his professional and personal life that should be generally known (personal communication). Coleman to R. S. Ward, 1.6.1979. SRC.

116. Hawking attempted to solve the central mathematical difficulty of Wheeler's geometrodynamics, which motivated the flight to superspace: how to describe functions on changing topologies. Hawking, "Space-Time Foam."

117. D. Kaiser, "Whose Mass Is It?," 539-50, for previous accounts, see 556n13; cf. Pickering, *Constructing Quarks*, §10.2; Smeenk, "False Vacuum."

1970 by Lev Landau, Andrei Sakharov, Yakov Zel'dovich, and others.[118] This unified approach was exemplified in the 1930s by Vladimir Fock, who had cosmological concerns for Dirac's "hole" theory (section 2.3). The official philosophy of science of the USSR, dialectical materialism, held that each science was a unified whole.[119] Early US advocates for combining QFT and GR, such as Wheeler and S. Weinberg, had strong commitments to *unity*. This suggests that philosophical commitments, or Gerald Holton's unity *thema*, were conditions of possibility for rapid growth of particle cosmology in the 1970s.[120]

My discussion of semi-classical methods adds a finer-grained understanding of the dynamics of disciplinary re-formation. Particle cosmology was shaped by *practices*. Here, I have discussed the practice of making semi-classical approximations. This is a good example of theorists' use of "paper tools," in just the way Ursula Klein suggested was analogous to laboratory tools.[121] Recall Coleman's influential "sine-Gordon" paper on bosons and fermions. When he taught this material at Erice, in 1975, he presented more than his boson-fermion finding. He researched and taught this material because "the sine-Gordon equation is a marvelous theoretical laboratory for checking various approximation schemes."[122] Semi-classical methods bridge the two explanations of the rise of particle cosmology. They produced key intellectual achievements, such as asymptotic freedom, which are emphasized in standard accounts. They also address a failing of the high-energy GUTs explanation, namely, the central place of low-energy, semi-classical, theorizing. On the other hand, practices like semi-classical approximations were part of the pedagogical infrastructures that Kaiser elucidates. Practices answer a question here, too: if particle theorists were so hyper-specialized, how were they *able* to transition to neighboring subfields? Because their training included trans-disciplinary tools.

In the mid-1970s, these tools radically reshaped the QCD vacuum. Semi-classical methods introduced the possibility that transitions were possible between the "worlds" Nambu and Jona-Lasinio had previously held separate (section 7.3). Solitons opened up new vacuum physics.[123] At Erice in 1977, Coleman's "goodies [which] have come from a family of computational methods" included his false vacuum research.[124] There, semi-classical methods bridged gauge theory and cosmology.

7.5.2 True Fathers of the False Vacuum

Since Nambu's 1960 "superconductor" field theory, particle physicists found richer and richer structure in the vacuum. The achievements of Goldstone, Higgs et al., and

118. D. Kaiser, "Whose Mass Is It?," 540; Pickering, *Constructing Quarks*, 397–98n12–13; motivations inlcuded eliminating Big Bang singularities, see Smeenk, "False Vacuum."
119. Graham, *Science, Philosophy, and Human Behavior*, ch. 2, on Fock esp. 367–78, on unification: 378–79.
120. Holton, "Role of Themata."
121. U. Klein, *Experiments, Models, Paper Tools*.
122. Whence followed a table with the results of five approximation schemes. Coleman, "Classical Lumps," 380.
123. Callan, Dashen, and Gross, "Gauge Theory Vacuum," 334; Cao, *Current Algebra*, 194; for the related "quark-gluon plasma," see Rugh and Zinkernagel, "Quantum Vacuum," §3.3.
124. Coleman, "Uses of Instantons," 805.

't Hooft and Veltman made this vacuum structure central to achieving major theoretical desiderata: theories that were symmetric, unified, and renormalizable. It was common to ascribe agency to the vacuum, to describe it as changing, and to distinguish a "true" vacuum from other possibilities. As discussed in section 7.5.1, at Erice in 1973 Coleman was prompted to consider the possibility of vacuum decay from a higher-energy "false" vacuum to a lower-energy "true" vacuum. But at the time he did not pursue this as a research question. Others, however, did publish on the false vacuum. Their work, and Coleman's reactions to it, shows how vacuum theories interacted with so-called epistemic virtues, the norms and values that animate knowledge production.[125]

The most important early work on the false vacuum came from the Soviet Union.[126] In 1972, David A. Kirzhnits and his graduate student Andrei D. Linde extended the analogy between solid-state and relativistic physics. They asked whether superconductor theories—like Steven Weinberg's model of leptons—had a Curie Point. A magnetized bar of iron will demagnetize when heated to its Curie point; the preferred direction marked by its magnetic field disappears and the electromagnetic field becomes more symmetrical. At a high enough temperature, would SSB models like S. Weinberg's have their symmetries restored? The required temperatures were only available in the immediate aftermath of the big bang. In January 1974, in another magnetic analogy, Yakov Zel'dovich, Igor Kobzarev, and Lev Okun argued that a cooling universe would, like a magnet, divide into "domains" with different symmetry properties. They "consider[ed] the 'collapse of a bubble'" of symmetric matter surrounded by asymmetric matter.[127] Six months later, Kobzarev, Okun, and Mikhail Voloshin submitted their paper on "Bubbles in Metastable Vacuum." They analyzed a skewed version of Goldstone's double-well potential, figure 7.5, following a paper by Lee Tsung-Dao (T. D.) and Gian-Carlo Wick. Recall Coleman's allegory of the "little man" inside an infinite magnet. He argued that the "hidden" symmetries of nature would never be observable to the little man because it was impossible for a finite being to change the magnetization of an infinite number of elementary magnets. Lee and Wick argued that the domain structure of magnets should guide particle theorists: was it "experimentally possible in a limited domain in space to 'excite' (flip) the ordinary vacuum to an abnormal one"? In heavy-nucleus collisions, and contrary to Coleman, they answered "yes." For Lee and Wick, vacuum excitations might be new models for particles with confined quarks. They offered a *more dynamical* "physical picture that the vacuum more resembles a medium whose properties can be *changed* [emphasis added]."[128] Kobzarev, Okun, and Voloshin extended this work to cosmology, including the classical mechanics of bubble growth and collapse. They found that the "upper vacuum," ϕ_+, was "metastable," meaning approximately stable, or temporarily stable until a decay and dissolution. (In fact, earlier in 1961 Nambu and Jona-Lasinio considered that a higher-energy vacuum would "correspond to a

125. See Daston and Gailson, *Objectivity*, esp. 39–41.
126. For an excellent discussion, which however omits Coleman's false vacuum, see Smeenk, "False Vacuum"; cf. Kragh and Overduin, *Weight of the Vacuum*.
127. Kirzhnits and Linde, "Macroscopic Consequences"; Zel'dovich, Kobzarev, and Okun, "Cosmological Consequences," bubble: 2, skewing: 3.
128. Lee and Wick, "Vacuum Stability," figs. 1a and 7, 2292, 2310.

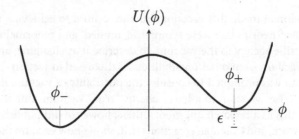

Figure 7.5. The effective potential, $U(\phi)$, for a theory of a field, ϕ, with a false vacuum. The true vacuum, ϕ_-, is the absolute minimum of $U(\phi)$; the false vacuum, ϕ_+, is a local minimum of $U(\phi)$ and has ϵ higher energy. (Credit: author)

metastable world."[129]) Kobzarev, Okun, and Voloshin considered the probability of bubble decay and formation from fluctuations and from particle collisions, and the effects of bubble walls colliding with particles. They suggested that we live in the "lower vacuum," ϕ_- (figure 7.5), for two reasons: because of the current small value of the cosmological constant, and because in a big bang cosmology, the initial temperature would have destroyed the "upper vacuum," ϕ_+.[130] This was a small but creative vacuum literature that reimagined both nuclear matter and cosmological models.

Much later, in 1983 *Discover* magazine profiled Coleman's false vacuum work. Coleman wrote to the editors to correct some oversights. They neglected his coauthor Curtis Callan,

> and, well before us, the subject was founded in a brilliantly original paper by M. B. Voloshin, I. Yu. Kobzarev, and L. B. Okun, working in the Soviet Union. These men are the true fathers of the false vacuum, and should receive due credit.[131]

In the 1970s, Coleman's reaction to two further developments them reveals some of the epistemic virtues he upheld. In 1976, Michael Stone, a research student at Cambridge, analyzed Lee and Wick's vacuum excitations using Coleman's functional techniques, focusing on the two-dimensional "sine-Gordon" model.[132] In summer 1977, Stone wrote asking Coleman "to quench [his] incipient paranoia." Earlier, Stone had sent Coleman a copy of his paper, but Coleman had not cited it in "The Fate of the False Vacuum." Coleman's reply highlights how he evaluated the physics literature and underscores the importance of priority among physicists.[133] Why no citation? "The reason is simple: I'm an idiot.

> To give more details, I had read and admired your first paper when it came out as a preprint, but had missed (or forgotten) its last two paragraphs, which clearly

129. Nambu and Jona-Lasinio, "Dynamical Model. II," 249.
130. Voloshin, Kobzarev, and Okun, "Bubbles in Metastable Vacuum," fig. 3; cf. Coleman, "False Vacuum. I," fig. 1, skewing was created by adding a linear or cubic term to the Lagrangian.
131. Coleman to the Editors, *Discover* magazine, 11.5.1983. SRC.
132. Stone, "Lifetime and Decay of 'Excited Vacuum' States."
133. See Baldwin, "'Keeping in the Race.'"

indicate that you knew the essential outlines of the semi-classical computation in 3+1 dimensions. At any rate, I remembered your paper as one that studied vacuum decay in a special model by methods that had no obvious generalization to more general theories, and I thus didn't bother to cite it.[134]

Simply treating the same subject was not enough to result in a citation. Coleman looked for models that were not narrowly crafted to exhibit particular features, but that spoke to models and QFTs more generally.

Coleman saw similar limitations, not due to misreading, in the work of his peer, Paul Frampton. Frampton, based at UCLA, attended the Aspen Center for Physics with Coleman, Curtis Callan, Gerard 't Hooft, and others in the summer of 1976. He analyzed a false vacuum in the Weinberg-Salam electroweak model, and alerted the world to Coleman and Callan's forthcoming publications.[135] At the new year 1977, Soviet physicist Andrei Linde (b. 1948) wrote to Coleman from Moscow. He read Frampton's article, and requested a preprint of Coleman's forthcoming publications. Coleman replied with a preprint on January 12, 1977. Teaching freshman physics prevented Coleman from writing up last summer's manuscript until that week. Linde made his own analysis of the Weinberg-Salam scenario. He sent a preprint to Coleman on September 9 with results "that differ strongly from those obtained by Frampton." Coleman replied: "I'm not surprised that you find Frampton wrong—his paper is a third rate job."[136] From Coleman's (and Linde's) perspective, Frampton's papers had three failings. First, they were an analysis of a special model, which did not emphasize connections to general field theories. Second, Linde argued that Frampton failed to consider important aspects of the physical model, such as the energy of bubble walls. Third, Linde argued that Frampton failed to keep his conclusions within the bounds of his approximations.[137] Working with the proper bounds of approximations was a central feature of Coleman's research (for which the sine-Gordon equation was a marvelous theoretical laboratory). These virtues he looked for in the work of others were also be the virtues he strove to achieve in his own work: generality, insight, and control over approximations.[138]

134. Stone to Coleman, undated; Coleman to Stone 26.9.1977. SRC.

135. Frampton, "Vacuum Instability"; Frampton, "Consequences of Vacuum Instability"; cf. Coleman, "False Vacuum. I," 2935n2.

136. Linde, "Vacuum Instability"; Linde to Coleman, 31.12.1976; Coleman also asked "a favor: please tell me if there are any Soviet papers on this subject (other than the work of Kobsarev et al. and references cited within) that I should cite. My ignorance of Russian keeps me at least a year behind on the Soviet literature." Coleman to Linde 12.1.1977. Linde to Coleman 9.9.1977, Coleman to Linde 29.9.1977. SRC; cf. Linde, "Vacuum Instability," 120; he also found Lee and Wick "not quite reliable," Krive and Linde, "Vacuum Stability in the σ Model," 268; Coleman affirmed Linde's critiques in Coleman, "Uses of Instantons," 915n41.

137. Fairness may dictate a fuller discussion of Frampton's work, but space restrictions preclude this.

138. E.g., "It is instructive to compute a slightly more general object than is needed for our immediate purposes." Coleman, "Quantum sine-Gordon Equation," 2092; for insights see §V; for "mistrust" of results "outside the domain of validity of the approximation," see Coleman and Weinberg, "Radiative Corrections," 1898.

7.5.3 Fate of the False Vacuum

From 1977 to 1980, Coleman published five times on the false vacuum. In 1977, "Fate of the False Vacuum: Semiclassical Theory" (with an erratum) and a sequel with Princeton physicist Curtis Callan, "Fate II. First Quantum Corrections." His 1977 Erice lectures "The Uses of Instantons" incorporated material from these two papers. With Andre Martin and Vladimir Glaser in 1978, he strengthened the mathematical basis for these calculations. Finally, in 1980, with his doctoral student Frank De Luccia, he considered the effects of gravitation on vacuum decay.[139] This body of work aimed to exemplify the epistemic values of generality, insight, and control over approximations. It was driven by Coleman's ongoing development of semi-classical approximation methods. It was also characterized by narrative, on the four levels this chapter has introduced: in literary qualities, in analogies, as supplemental stories, and as central to deriving physical results. This section shows how instantons provided a new mechanics for vacuum tunneling, how it was related to quantum fluctuations. Finally, Coleman's encounter with cosmology brought vivid eschatological narratives into his physics. Unlike cosmologists' historical narratives, Coleman told stories about the future.

The false vacuum was one of the "uses of instantons." It was a phenomenon that was apt for exploration and analysis by Coleman's semi-classical approximation methods. Recall that solitons were particle-like solutions to classical field equations with a finite action (section 7.5.1). In 1974, Alexander Polyakov and, separately, Gerard 't Hooft generalized these objects to *four*-dimensional solutions, making them finite in space and in time; Polyakov called them "pseudoparticles," 't Hooft referenced the temporal snapshot with "instanton."[140] Explaining how instantons had "finite action" will exhibit some of Coleman's storytelling and demonstrate how much his false vacuum inquiry was driven by technique. Technically, Polyakov and 't Hooft worked in "Euclidean" spacetime, not Minkowski spacetime. (It was usually, in principle, possible to transform mathematical results from one spacetime to the other.) In Euclidean spacetime, the effective potential in figure 7.5 was flipped, as shown in figure 7.6.[141] Coleman imagined that this was the potential, in classical mechanics, for a particle climbing the hills of figure 7.6. If a particle began to the left of the peak at ϕ_-, and was released it would continue gaining energy without limit as it fell. If its energy were integrated over an infinite amount of time, it would have infinite action. Similarly, if it began slightly to the right of ϕ_-, it would cross the valley, pass over ϕ_+, and fall infinitely downward. However, Polyakov and 't Hooft pointed out that there were other mathematically possible solutions. As Coleman described it, if the particle was

---

139. Coleman, "False Vacuum. I"; according to particle physicists' citation tracking system, inSPIRE, "Fate" has been cited more than 1,900 times, at time of writing, and 295 times within ten years of publication, https://inspirehep.net/literature/118681; Coleman, "Erratum: False Vacuum I"; Callan and Coleman, "False Vacuum. II"; Coleman, "Uses of Instantons"; Coleman, Glaser, and Martin, "Action Minima"; Coleman and De Luccia, "Gravitational Effects."

140. Cao, *Current Algebra*, §8.5.

141. After Coleman, "False Vacuum. I," fig. 3; Hermann Minkowski himself introduced the conversion between the metric of special relativity, $-dt^2 + d\bar{x}^2$, to an all-positive metric, like standard Euclidean geometry, $dT^2 + d\bar{x}^2$, in which t is replaced by $it = T$. This was elaborated in QFT by Schwinger in the later 1950s, M. Miller, "Euclidean Green's Functions."

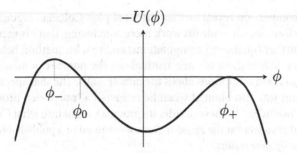

Figure 7.6. *Minus* the effective potential, $U(\phi)$, depicted in figure 7.5. Instantons that start at ϕ_0, roll to the right, and "bounce" at ϕ_+. (Credit: author)

released at ϕ_0, "the particle moves quickly through the valley in [figure 7.6], and slowly comes to rest at ϕ_+ at time infinity." He called this the "bounce."[142] In this case, even after integrating over all times, the action was finite. Why include instantons in gauge theories? In 1975, Polyakov and others hoped that they would produce long-range forces and lead to quark confinement, without adding new parameters to a theory.[143] While instanton confinement worked in some unrealistic models, by 1977 Coleman rejected the possibility: "Whatever confines quarks, it's not instantons." Instead he put forward a formal justification. "The only reason we are interested in configurations of finite action is that we are interested in doing semiclassical approximations [...]."[144] Rather than valuable objects of study in themselves, instantons were a (toy?) system for the application and development of paper tools.[145]

Coleman's efforts to achieve his epistemic virtues can be illustrated briefly. His false vacuum work aimed at generality by including pedagogical discussions of methods, and by crafting results (with Callan) that "applie[d] to a general renormalizable field theory."[146] The success of this strategy is indicated by the spread of these methods in both QCD and cosmology.[147] Coleman repeatedly kept the bounds of his approximations in view of his readers.[148] Coleman also distinguished his initial paper from the existing work of Kobzarev, Okun, and Voloshin (with considerable deference).

142. Coleman, "False Vacuum. I," 2933.
143. Polyakov, "Compact Gauge Fields."
144. Coleman, "Uses of Instantons," 858, 829; these methods were put to good use resolving a theoretical puzzle about $U(1)$ symmetry. In a paper written while visiting Harvard and which thanks Coleman, 't Hooft interpreted instantons as creating "a certain tunneling effect from one vacuum to a gauge-rotated vacuum." 't Hooft, "Four-dimensional Pseudoparticle," 3432. Coleman and 't Hooft exchanged five letters in early 1975 arranging the visit. SRC.
145. In his 1975 lectures on solitons, he asked: "Is there any chance that the lump will be more than a theoretical toy in our field?" He was pessimistic about non-toy possibilities in both QCD "bags" and magnetic monopoles. Coleman, "Classical Lumps," 391; contrary to my assessment, Schweber suggests that Coleman aimed at understanding the Higgs mechanism in GUTs. Schweber, "Shelter Island Conferences Revisited," 109. That may have been how Coleman's work was used by others.
146. Coleman, "False Vacuum. I," §§II–III; Callan and Coleman, "False Vacuum. II," 1762.
147. See Witten, "Instability of the Kaluza-Klein Vacuum," and "The Coleman-Callan method," applied in Guth, "Inflationary Universe," 351.
148. E.g., on the justifiability of thin-wall approximation, Coleman and De Luccia, "Gravitational Effects," 3310.

The terms of comparison reveal further virtues at play. Coleman argued that he used fewer assumptions, which made his work more convincing; that his method was better suited to further (quantum) computations; and that his method better accounted for symmetries. While there is some truth about the number of assumptions, both papers made ad hoc assumptions about the bubble walls. The sharpness of Coleman's symmetry claim was also blunted when he retracted a "purported proof" about the symmetry of "bounces."[149] These desiderata provide insight into what Coleman—and the editors and reviewers at the *Physical Review*—valued in a publishable paper, *apart from substantially new results.*

One way narrative entered this work was through analogies to classical (non-relativistic, non-quantum) physics. The most important such analogy came from the thermodynamics of superheated liquids.[150] When liquids such as water were heated very carefully, their temperature must rise above their boing point in order to boil (i.e., to change phase from liquid into vapor). Near the boiling point, bubbles of vapor could appear and then collapse back to liquid. This happened because of the surface tension around the bubble wall. When—because of thermodynamic fluctuations—a bubble was created with enough energy to convert liquid into vapor *and* to overcome the surface tension, it rapidly grew and converted the entire container of liquid to vapor. According to Coleman:

> An identical picture describes the decay of the false vacuum, with quantum fluc-tuations replacing thermodynamic ones. Once in a while, a bubble of true vacuum will form large enough so that it is classically energetically favorable for the bubble to grow. Once this happens, the bubble spreads throughout the universe converting false vacuum to true.[151]

As for Victor Weisskopf in the 1930s (section 4.2), Coleman's picture assumed that quantum fluctuations simply permeated all space, without underlying cause. This was a disanalogy with the thermodynamic case, in which fluctuations indicated the existence of an underlying molecular reality (see appendix B).

Weisskopf simply assigned fluctuations to a value in his "hole" theory formalism. Julian Schwinger and Richard Feynman took steps to eliminate divergent quan-tities and disconnected bubble graphs from their formalism. Coleman's case was somewhat different. He did not consider pure vacuum-to-vacuum transitions—at least, transitions that started and ended in the *same* vacuum. *Despite* calculating only vacua, he was considering transitions between *different* states. For this pur-pose, Coleman adapted a standard piece of semi-classical quantum technique, the 1926 Wentzel-Kramers-Brillouin approximation. The WKB approximation allowed

149. As shown by his graduate student, Ian Affleck, Coleman, "Erratum: False Vacuum I."
150. Bubble chambers used superheated hydrogen. Physicist Gary Feinberg wrote to Coleman with false vacuum questions, including: "Is our universe a giant bubble chamber, created by transcendent beings to detect the presence of intelligent life, through its production of vacuum bubbles by high energy physics experiments?" Coleman replied: "Dear Gary: Good questions. Do you have answers also?" Feinberg to Coleman, 23.5.1977; Coleman to Feinberg 13.6.1977. SRC. On Feinberg, see D. Kaiser, *Hippies Saved Physics*, ch. 8; Milburn, "Ahead of Time."
151. Coleman, "False Vacuum. I," 2929.

quantitative evaluation of a classical particle "tunneling" through a potential barrier that would be impossible in classical mechanics. In the late 1920s, George Gamow made this a key element of his model of radioactive—that is, metastable—nuclei.[152] In a pedagogical section of "Fate of the False Vacuum," titled "Barrier Penetration in Field Theory," Coleman adapted physicists' paradigmatic treatment of metastable entities into the new instanton formalism. This section deployed Coleman's particle climbing a hill narrative, figure 7.6, for describing instanton solutions. After making another key assumption—that the walls of the bubble were thin—Coleman was able to describe bubbles of false vacuum. He began with the particle description of balls rolling down hills. A particle released at ϕ_0 in figure 7.6 stayed near ϕ_- until a very large time R, and then moved rapidly across the valley, coming to rest at ϕ_+ (the "bounce").

Translating from mechanical language back into field theoretic language, the bounce looks like a large four-dimensional spherical bubble of radius R, with a thin wall separating the false vacuum without from the true vacuum within.[153]

The fate of the false vacuum was a narrative told with vocabularies from particle and field descriptions, and from quantum and classical mechanics.

Coleman's semi-classical narrative marked a shift from the rejection of spacetime descriptions of quantum jumps by Niels Bohr and Werner Heisenberg in the early twentieth century.[154] In Coleman's analysis, a classical field sits at the false vacuum, ϕ_+ in figure 7.6. In the WKB picture, a classical entity simply jumps onto the other side of the potential, near ϕ_-, after which it moves classically again. Coleman used instantons—less particles than events—to describe the quantum jump. Not only did instantons describe the previously indescribable, they did so with only *classical* motion (but, motion in Euclidean spacetime). Tunneling through the peak in figure 7.5 became traversing the valley in the upside-down figure 7.6.[155]

The classical growth of vacuum bubbles provided the profit of Coleman's semi-classical computations. "[T]he thing to compute is the probability of decay of the false vacuum per unit time per unit volume, Γ/V. Of course, such a computation would be bootless were it not for cosmology. [After the big bang,] As the universe expanded and cooled down, it might well have settled into a false vacuum instead of a true one." Cosmology provided conditions for the relevance of vacuum bubbles. But Coleman was not focused on recreating the past history of the cosmos, as were cosmologists and increasing numbers of particle physicists, such as Steven Weinberg. Coleman was interested in predicting the future. He quantified this in terms of Γ/V. "The relevant

152. Jammer, *Conceptual Development*, 277–80; Stuewer, "Gamow's Theory of Alpha-decay," this was also tied to Coleman's interpretation of the imaginary part of an effective potential, section 7.5.1.
153. Coleman, "False Vacuum. I," 2933.
154. See Beller, *Quantum Dialogue*, ch. 5.
155. While visiting Harvard, 't Hooft used the classical solutions and tunneling interpretation in a landmark instanton paper, citing, among other sources, private communication from Coleman. 't Hooft, "Four-dimensional Pseudoparticle," 3432, 3450n3; Linde attributed this to Coleman; Linde, "Vacuum Instability," 306; for vacuum tunneling "events," see Callan, Dashen, and Gross, "Gauge Theory Vacuum," 334, WKB on 335; cf. Stone, "Semiclassical Methods for Unstable States"; Polyakov attributed "tunneling" to Vladimir Gribov, Polyakov, "Quark Confinement," 457.

parameter *for describing future events* is that cosmic time for which the product of Γ/V and the four-volume of the past light cone becomes of order unity" [emphasis added]. When would enough time have passed since the big bang so that vacuum decay was almost certain? "If this time is on the order of milliseconds, the universe is still hot when the false vacuum decays, even on the scale of high-energy physics, and the zero-temperature computation of Γ/V is inapplicable." Unlike his predecessors (section 7.5.2), Coleman did not predict that vacuum bubbles would form just after the big bang—when the universe was hardly a vacuum. "If this time is on the order of years, the decay of the false vacuum will lead to a sort of secondary big bang with interesting cosmological consequences." This was a recognizably cosmological observation, and where most of his peers concentrated their attention. However, Coleman continued: "If this time is on the order of 10^9 yr [i.e., the present], we have occasion for anxiety."[156] It would be the end of the world.

Coleman was less interested in the cosmological case, but even there he told an eschatological narrative. In the discussion after his "Instantons" lecture at Erice in 1977, Robert Budney asked: "Could it be that the big bang was a bubble?"

COLEMAN: Not as I've described it. [...] The idea is—and this is not serious physics—that you have some universe, maybe with some matter in it, maybe not, and it bubbles. [...] Eventually the universe consists of the remnants of the clashing bubble walls. The walls just don't go through each other, that much I know. So this is the matter we see. We live in the debris of the catastrophe.[157]

The clashing bubble walls created particles—not the tunneling or expansion of the bubbles. "This refutes the naive expectation that the decay of the false vacuum would leave behind it a roiling sea of mesons. In fact, the expansion of the bubble leaves behind only the true vacuum."[158] This naive expectation was *Coleman's* original expectation in 1973 (section 7.5.1); it was dashed by Voloshin, Kobzarev, and Okun.[159] Matter must be generated another way. Adding to the pessimistic tone of his cosmogenesis, Coleman told Budney that he did not know how to calculate the details.

On the other hand, Coleman told a personalized and darkly comic narrative about the *future* possibility of vacuum decay. While he labeled his cosmogenesis "not serious physics," predicting the future was closer to the standard activity of predicting the outcome of new experiments.[160] This case was when the time for likely vacuum decay was "on the order of 10^9 yr"—the current age of the universe. "Fate of the False Vacuum"'s introduction advertised this as cause for anxiety. But bubbles expanded at nearly the speed of light. "As a consequence of this rapid expansion, if a bubble were expanding toward us at this moment, we would have essentially no warning of its approach until its arrival." This had apocalyptic consequences. Once an observer

156. Coleman, "False Vacuum. I," 2929–30.
157. Coleman, "Uses of Instantons," 937.
158. Coleman, "False Vacuum. I," 2935.
159. They argued that all the transition energy went to accelerating the bubble wall. Voloshin, Kobzarev, and Okun, "Bubbles in Metastable Vacuum," 644.
160. On speculation and cosmology, see Kragh, "Contemporary History."

sees an expanding bubble, within a time "on the order of $10^{-10} - 10^{-30}$ sec. later, he is inside the bubble and dead." At Erice, he elaborated

(In the true vacuum, the constants of nature, the masses and couplings of the elementary particles, are all different from what they were in the false vacuum, and thus the observer is no longer capable of functioning biologically, or even chemically.)

Vacuum decay offered a mechanism for catastrophic changes in the basic constants of nature.[161] But even after his warnings, Coleman darkly suggested we should not worry after all. "Since even 10^{-10} sec. is considerably less than the response time of a single neuron, there is literally nothing to worry about; if a bubble is coming toward us, we'll never know what hit us."[162]

Physicists both accepted and challenged this story. At Erice, Nicola Cabibbo asked, "Assuming we are living in a false vacuum, are there some things we should not do?" But Coleman thought there was no risk from inducing vacuum decay from, for example, particle accelerators.[163] Coleman's pessimism was challenged by Michael Peskin, a postdoc at Harvard's Society of Fellows. In the postscript to a twelve-manuscript-page-long letter applying the—unfamiliar to Coleman—statistical mechanics of droplets to the false vacuum, Peskin "offer[ed] one unrelated complaint about [Coleman's] lectures on the demise of the false vacuum." Considerations from standard statistical mechanics, "leads to the conclusion that, if the vacuum decay has not already occured [sic], we are, in all probability, safe for another few billion years! We have enough to worry about already, Sidney, for you to encourage us also to worry about this unlikely circumstance." Coleman replied: "Dear Dr. Peskin: I love you."[164]

At Erice in 1977, Coleman gave preliminary results of calculations with his student Frank De Luccia on vacuum decay, now considering gravity. Their paper was submitted for publication in early 1980. He echoed Feynman's dismissal of the relevance of quantum gravity (section 4.4): "for any conceivable application, these [gravitational] effects are too small to worry about." But, technique drove the work: "Nevertheless, when we discovered it was within our power to compute them, we were unable to resist the temptation to do so."[165] If anything, these computations intensified Coleman's pessimism. In a marked departure from Feynman's rejection of dynamic geometry, Coleman and De Luccia found that "the nonzero energy density inside the bubble

---

161. Some historians see this as marking a fundamental shift away from the conviction that the laws of nature are timeless, and toward the idea that the laws of nature change over time. Schweber, "Historical Perspective," 670; Schweber, "Shelter Island Conferences Revisited," 107–13; for a different view, see Wise, "Science as (Historical) Narrative," 361n12; Dirac was a predecessor of Coleman, see Kragh, "Cosmo-Physics."

162. Coleman, "Uses of Instantons," 937–38; Coleman, "False Vacuum. I," 2934.

163. Coleman, "Uses of Instantons," 935.

164. Peskin to Coleman, 20.3.1977, Coleman to Peskin 23.3.1977, SRC. See Callan and Coleman, "False Vacuum. II," 1768n2.

165. Coleman, "Uses of Instantons," 938; Coleman and De Luccia, "Gravitational Effects," 3306; Dan Freedman wrote to Coleman from SUNY Stony Brook that the "Gravitational Effects" preprint "was immediately stolen from our preprint shelves. Please send me a copy." Freedman to Coleman, 14.4.1980. SRC.

distorts its geometry."[166] Unfortunately, the spacetime inside an expanding bubble of true vacuum was *unstable* and "inevitably suffers gravitational collapse," like the inverse of a big bang.

> This is disheartening. The possibility that we are living in a false vacuum has never been a cheering one to contemplate. Vacuum decay is the ultimate ecological catastrophe; in a new vacuum there are new constants of nature; after vacuum decay, not only is life as we know it impossible, so is chemistry as we know it. However, one could always draw stoic comfort from the possibility that perhaps in the course of time the new vacuum would sustain, if not life as we know it, at least some structures capable of knowing joy. This possibility has now been eliminated.[167]

This was a true "physical eschatology"—a scientific vision of the end times that allowed for the possibility, at least, of "structures" experiencing something of the human spirit, even after the apocalypse.[168]

Coleman's ecological vision resonated with the linked-developments of political events and literary trends. Throughout the 1970s, US physicists experienced major funding cuts and underemployment of PhDs; particle physicists experienced the sharpest decline.[169] Facing shortfalls in their NSF grants, Coleman and other particle theorists at Harvard reduced their salaries and sought summer support abroad.[170] The 1979 oil crisis and partial meltdown of the Three Mile Island nuclear plant darkened attitudes to the Earth's ability to sustain modern civilization, and humanity's technological capacity to adapt. In science fiction, the 1970s saw dozens of works of ecological catastrophe. This included Coleman's friend and fellow interviewee (section 7.1) Gregory Benford's apocalyptic novel *Timescape* (1980).[171]

The "structures capable of knowing joy" may have referred to philosophical theologies that connected the capacity to attain happiness with being human (as in Augustine's *Summa theologiae* or Moses Mendelssohn's *Jerusalem*). But perhaps the source was from the author Coleman and Benford agreed, in 1973, was head-and-shoulders above the field: Ursula Le Guin. In a 1975 address to a science fiction convention, widely available by 1979, Le Guin established her purpose: "We're here to enjoy ourselves, which means we are practicing the most essentially human of all undertakings, the search for joy."[172] In her novel *The Dispossessed*, the anarchist philosopher Laia Asieo Odo connects the capacity for joy to joyful work:

> A child free from the guilt of ownership and the burden of economic competition will grow up with the will to do what needs doing and the capacity for joy in doing it.

166. Coleman and De Luccia, "Gravitational Effects," 3310.

167. They acknowledged that this was a "rhetorical excess," Coleman and De Luccia, "Gravitational Effects," 3314.

168. For later physical eschatology, see Kragh, "Physics and Cosmology," 915.

169. Kevles, *Physicists*, XXV; D. Kaiser, "Whose Mass Is It?"

170. Among many letters: Coleman to Harvard Dean Henry Rosovsky, 6.5.1976; Coleman to William Wallenmeyer (AEC), 17.6.1974; Coleman to Daniel Amati (CERN) 18.4.1974.

171. Benford, *Timescape*; Stableford, *Science Fact and Science Fiction*, s.v. "ecocatastrophe."

172. Le Guin, "Stone Ax," 226.

Figure 7.7. Sidney and Diana Coleman with Stephen Hawking in Cambridge, UK, ca. 1980s. By permission of Diana Coleman.

This was a utopic vision of personal development, free from economic distortions, in which the capacity for joy was produced by one's upbringing, and which was activated by meaningful work.

> It is useless work that darkens the heart. The delight of the nursing mother, of the scholar, of the successful hunter, of the good cook, of the skillful maker, of anyone doing needed work and doing it well—this durable joy is perhaps the deepest source of human affection, and of sociality as a whole.[173]

Le Guin wrote of the joy of craftwork, of method freely applied. It was a joy I believe drove Coleman's physics.

7.6 Conclusion

From 1960 to 1980, particle physicists defined and redefined the vacuum, giving it new roles in both formal manipulation and physical interpretation. These

173. Le Guin, *Disposessed*, 247.

transformations were driven by physicists' practices, including the practice of storytelling. These methods made distinctive use of classical objects. "Superconductor" solutions were found by means of Douglas Hartree's approximation procedure in which a quantum variable was replaced by its classical average. Earlier, Paul Dirac, Julian Schwinger, and Richard Feynman had insisted that the vacuum be characterized by zero—by the unobservability of the electron sea, by eliminating vacuum currents, and by discarding "bubble" diagrams. By giving the vacuum a non-zero value, Spontaneous Symmetry Breaking gave the vacuum a new physicality.[174] By way of concluding this chapter, I will discuss two critical aspects of this history: physicists' use of classical entities (such as fields, solutions, numbers) and Sidney Coleman's storytelling.

Coleman embraced SSB as a topic for advancing field theory. However, he critiqued the standard discussion of SSB and the "Goldstone phenomena" as a *diachronic* story with a standard beginning-middle-end structure. Initially the vacuum was ordinary, $\langle \phi \rangle = 0$; then a non-zero value for $\langle \phi \rangle = \lambda$ was chosen; and then the theorist would say that $\langle \phi \rangle$ broke a symmetry, and that massless particles appeared. Coleman's doctoral work with group theory taught him that *solid* mathematical results about symmetries should be "independent of detailed dynamical assumptions."[175] At Erice summer schools, he denied that his colleagues symmetry-breaking stories were necessarily connected to Goldstone's result. Instead he advocated a *synchronic* mismatch of Lagrangians and physical states; not a brea*king* process, but a hidden state of affairs. Despite his critiques of, and alternatives to, the standard SSB methods, Coleman provided them with exceptionally vivid descriptions. His critical distance from the "Higgs phenomenon" did not prevent him from deploying carnivorous and even cannibalistic metaphors. What student could forget the mechanism of dynamical symmetry breaking, cast in the role of Cronos devouring his children? Coleman and others' "semi-classical" methods reshaped gauge theory, and opened new depths of complexity in vacuum physics. Classical solutions to field equations provided particle models, and became tunneling events between the infinite vacua of Quantum Chromodynamics. In 1978, Gerard 't Hooft gave a seminar at Harvard. He wrote to Coleman: "I could speak about: 'Confinement and topology of gauge fields.' Privately, we could babble about: 'The vacuum is more complicated than we think.'"[176] These tools became perhaps the most important means for connecting quantum theory, gravitation, and cosmology. When Coleman applied these methods to the "false vacuum" he discovered new apocalyptic visions and, at the same time, provided tools for a new generation of particle cosmologists to remake our understanding of the universe.

What should be said about the connections between Coleman's physics and science fiction? I will first sketch a parallel between this case and Roger Penrose's visual thinking. Then I will consider connections between Coleman and recent discussions of physics, SF, and affect.

174. Brown and Cao, "Spontaneous Breakdown of Symmetry"; Brown et al., "Panel Session"; Cao, "Englert-Brout-Higgs Mechanism." The emphasis on the vacuum was somewhat retrospective.
175. Coleman, "Structure of Strong Interaction Symmetries," 6.
176. 't Hooft to Coleman, 21.3.1978.

In chapter 6, I argued that the visual thinking Roger Penrose applied in his recreational mathematics and psychology was the same visual thinking he applied in his spacetime diagrams. A similar case can be made for Coleman. For example, the activity of writing SF book reviews was quite similar to the activity of preparing his Erice review lectures—including selecting texts, apt retelling, comparing, juxtaposing, judging logic and aesthetics (in both genres!).[177] An example closer to the content of Coleman's physics comes from time-travel narratives (and reaches just beyond the timeframe of this study). Recall Coleman's 1973 book review in the *Magazine of Fantasy & Science Fiction* in which he affirmed that "making your head feel funny is one of the main functions of science fiction." The review critiqued a time-travel novel, *The Man Who Folded Himself*, by David Gerrold, in which the protagonist can jump forward and backward in time and make changes to past and future. Coleman identified an inconsistency in the logic of the story: Gerrold insisted that the protagonist have a unique past, like a trunk on a branching tree of possibilities; Coleman argued that the unique past was inconsistent with the jumping.[178] Important here are the jumping forward and backward, branching, and multiple pasts.

Now jump ahead to 1988. The scene is similar to that which surrounded John Archibald Wheeler in the 1950s (chapter 5): wormholes, black holes, cosmology, and the foundations of quantum mechanics are actively debated (and are seen as more central to particle physics). Stephen Hawking proposed that wormholes resolved a puzzle in the foundation of quantum theory. Coleman contested this interpretation, depicting "Black Holes as Red Herrings." But he put his new wormhole techniques to astonishing use the same year, in "Why There Is Nothing Rather than Something: A Theory of the Cosmological Constant."[179] The cosmological constant term, Λ, in the Einstein Field Equations in General Relativity was known to be zero by cosmological observation.[180] ("It can be thought of as the energy density of the ground state of quantum field theory, empty space."[181])

The central difference between Hawking's and Coleman's approaches, I suggest, was their view of looking back through wormholes, or the past. In the 1960s, Hawking and Penrose had perfected the classical theory of black holes as *one-way* objects. Stuff that went in did not come out. In the 1970s, Hawking proposed that semiclassically, black holes in fact radiated, which would cause them to shrink and evaporate. This left a question: when a black hole evaporated, what happened to the stuff? In the 1980s, Hawking answered: picture a black hole as an entrance to a wormhole, a tube,

177. See Benford, "Sid the Critic, Sid the Scientist."
178. Coleman, "Books (1973)," 26–27.
179. While Wheeler's wormholes were three-dimensional, Coleman et al.'s were four-dimensional. The foundational puzzle was decoherence. For an accessible overview and references, see Schwarzschild, "Cosmological Constant"; Coleman, "Black Holes as Red Herrings"; Coleman, "Why There Is Nothing." These papers were discussed inside and outside Coleman's circle. See Coleman to Hawking 5.2.1988; 't Hooft to Coleman, 20.4.1988, and reply, 27.4.1988; Linde to Coleman, 26.5.1988. In 1988, Coleman gave seminars on "Red Herrings" at Yale, Boston University, Princeton, Rockefeller, and Pennsylvania, Coleman, "Information for the 1987–88 Annual Report." "Why There Is Nothing" became a Science Citation Index "Hot Paper," Angela Martello to Coleman, 8.1.1990; all SRC.
180. In 1996, Coleman's wormhole work was described in the New York *Times* as "ironic science" with untestable conclusions. He replied that his theory of a zero cosmological constant was very much in danger of being refuted. Coleman, "Not 'Ironic Science.'"
181. Coleman to Angela Martello, 15.1.1990.

in spacetime; at evaporation, the contents of the wormhole are left in the tube, which becomes a distinct "baby universe." To Coleman, however, wormholes were much more like tunnels (or instantons). Tunnels were characteristically *two-way* objects. Just as in the SF book review, to Coleman two-way time travel meant not only multiple futures, but also multiple pasts. If Hawking was right that wormholes made the future a mess of wormholes, Coleman insisted there should also be multiple wormhole pasts.[182] Further, there was always the possibility of making contact with a "separate" universe by tunneling. (Technically, Coleman's other universes were colder but not in the "past" of our universe, because they were a different spacetime. He thought this time-free thinking should not be so strange, for "as Augustine observed, the past is present memory."[183]) Wormholes gave Coleman the opportunity to explain "Why There Is Nothing Rather then Something," that is, why the cosmological constant, Λ, was zero. He considered existing explanations to rely on "prearrangement," or fine-tuning: the initial conditions of the cosmological equations had to be arranged with spectacular precision to produce predictions aligned with the observational fact $\Lambda = 0$. Coleman personified the constants of nature and described them, just after the big bang, as *looking* through wormholes to "see the cosmological constant. Prearrangement is replaced by precognition."[184] In this sense, Coleman applied the same logic to his analysis of time travel stories as to the semiclassical worlds of quantum field theory in curved spacetime.

Time travel was also a convergence of science and SF for one of the things Coleman treasured most: an affect, a feeling. David Kaiser has shown how, in the 1960s and 1970s, a group of "hippies" made physics "fun" by exploring foundational philosophical questions and connections between physics, literature, religion, and more.[185] Literature scholar Colin Milburn has analyzed the ways in which scientists have adapted aspects of science fiction narratives into scientific, laboratory, practices: as blueprints, as supplements to possible applications, and as speculations of future developments.[186] None of these seems to characterize Coleman's science. Instead, I suggest that Coleman thought and felt the same way about (some of) his science and his science fiction. Milburn recently drew attention, beyond the laboratory uses of SF, to physicists who saw physics and SF as aligned. As Murray Gell-Mann famously commented in his "quark" paper: "It is fun to speculate [...]."[187] In this respect, my analysis of Coleman agrees with Milburn. In his 1973 interview with *Vertex* magazine, Coleman insisted: "I assure you, one of the reasons for doing science, especially the kind

---

182. In "Red Herrings," Coleman insisted on an initial state that included wormholes, which Hawking rejected. Hawking, "Wormholes in Spacetime," 908; Hawking may have been influenced by Fred Hoyle and Jayant Narlikar's bubble universes or Linde's chaotic inflation. See Kragh, "Contemporary History," 536.

183. Coleman, "Why There Is Nothing," 652; similarly for "future," Coleman to Arvind Borde, 5.12.1988, SRC.

184. Coleman, "Why There Is Nothing," 646. Though all sides agreed the others' calculations were technically correct, Coleman was largely regarded as victorious in this dispute. A possible explanation as to why is that insisting on time-symmetry was consistent with the practice of particle physics and scattering theory. Hawking, trained as a relativist, was arguing against the grain.

185. D. Kaiser, *Hippies Saved Physics*.

186. Milburn, "Modifiable Futures."

187. Milburn, "Ahead of Time," 222–23, cf. fn150.

of science I do [...] is that it makes your head feel funny, Goddamned strange."[188] He described the effect of good SF in the same terms.[189] In his Erice lectures, Coleman's students learned that highly valued results were "remarkable theoretical surprises."[190] He characterized a favorite paper as "surprising," "striking," and containing "unexpected results."[191] Coleman shows us what it *felt like* to be a physicist, and a motivation for doing the kind of physics he did.

This was a very different motivation from producing new industrial or military technology. It was closer to an affirmation of science as the free exercise of curiosity, science for its own sake. Humanists and social scientists repeatedly noted the dissonance between the image of the pure scientist and the actual practice of Cold War scientists, who often preached of lofty ideals while also designing weapons or providing military strategic analysis. Nevertheless, Jessica Wang recently argued, we should be careful not to hastily reject an ideal of disinterested inquiry. In contemporary debates over, for example, cuts to public funding for science without immediate commercial or military application, the ideal of pure science plays an important role.[192]

In this spirit, it is worth noting that striving to feel "goddamned strange" was unlike both military-industrial pursuits and the drive for the pure intellectual mastery of nature. In the 1950s, just before Coleman arrived, the Harvard physics department quashed philosophical, foundational questions. The attitude was "shut up and calculate."[193] This ethos aligned with the standard framing of science—and physics— as essentially masculine.[194] As a doctoral student in theoretical physics at Harvard, Evelynn Fox Keller, felt herself an "anomaly" as a woman.[195] Coleman's disarmingly simple and subjective desire did not respect the stereotype of impersonal science and the division of feeling and reason.[196] "Feeling funny" is not a heroic posture, and Coleman did not rush to name and claim (priority or particles).[197] He was not, in the period under study here, particularly interested in the foundational questions that Keller observed were particularly silenced by physicists' ideology.[198] But Coleman pursued problems in which solutions were found by both calculation *and* flash of insight, something beyond the daily grind.[199] I would like to suggest that seeking a strange feeling was a motivation different from the traditional characterizations of efficacy and aesthetics. It was fun, but institutionally and intellectually at the center

188. Turner, "Vertex Roundtable Interview."

189. Coleman, "Books (1973)," 26–27.

190. Coleman, "Dilatations," 359; cf. "as surprising as it is beautiful," Coleman, "Renormalization and Symmetry," 616; "deep and surprising (at least to me) insights," Coleman, "Uses of Instantons," 805.

191. Coleman, AIP Interview; Coleman, "Quantum sine-Gordon Equation," 2088; Coleman, "Classical Lumps," §5.1.

192. Wang, "'Broken Symmetry,'" 42–44.

193. As remembered by Mermin, "Could Feynman Have Said This?," cf. D. Kaiser, *Hippies Saved Physics*, ch. 1.

194. See Keller, *Gender and Science*; Traweek, *Beamtimes and Lifetimes*; Milam and Nye, "Scientific Masculinities."

195. Keller, "Anomaly of a Woman."

196. On such oppositions, see Keller, *Gender and Science*, 7–12.

197. See Oreskes, "Objectivity or Heroism?," for SF being used for "territorial" naming and priority, see Milburn, "Ahead of Time," 217.

198. Keller, "Cognitive Repression in Contemporary Physics."

199. Unlike the working problems of S. Weinberg's and Glashow's students. Coleman, AIP Interview.

of mainstream physics. Let me return to my reading of Coleman's "false vacuum" through the work of Ursula Le Guin. Coleman found joy in the freely directed labor of theoretical physics. A great diversity of "structures" seem to be capable of knowing such joy.[200]

The vacuum was a fruitful object for engendering funny feelings. As for Heisenberg, Nambu, and many others, for Coleman the properties of the vacuum were the properties of the world.[201] Defining new vacua was akin to world-making. Vacuum decay was a quantum leap from one world to another. The vacuum also threw up paradoxes and Coleman embraced the pleasant feeling of paradox. His 1980 "false vacuum" paper, with Frank De Luccia, concluded with comments on gravitation and the zero value of the cosmological constant. This was critically important in the context of a false vacuum, because "even the slightest negativity would be enough to initiate catastrophic gravitational collapse." Coleman worried that there was no *reason* for the cosmological constant to vanish. "Normally, when something is strictly zero, there is a reason for it."[202] This was a strange inversion of the classic rationalist maxim, the principle of sufficient reason. As Baruch Spinoza wrote in 1663, "Nothing exists of which it cannot be asked, what is the cause (or reason), why it exists." For Coleman, *nothing* needed a reason. It's enough to make your head feel funny.

200. My reading of science through SF is indebted to Haraway, *Primate Visions*, ch. 16; further institutional and personal context will be found in Coleman, Kaiser, and Wright, *Theoretical Physics in Your Face!*
201. E.g., Coleman, "Invariance of the Vacuum."
202. Coleman and De Luccia, "Gravitational Effects," 3314–15; for Guth it was a "deep mystery," Guth, "Inflationary Universe," 355n11.

8
Epilogue

Since 1980, the vacuum has retained its central place in theoretical physics. Some developments have led to new observational and experimental discoveries. Others have challenged the epistemological foundations of physical science.

Relativists maintain their focus on vacuum solutions to Einstein's gravitational field equations (though few follow John Wheeler's program of reducing matter to empty space).[1] Roger Penrose's conformal diagrams are standard tools of teaching and research.[2] Penrose's (and Ted Newman's) techniques became an "indispensable" part of relativists' toolbox, including in work on gravitational waves.[3] In 2017, the Nobel Prize in physics was awarded to three members of the LIGO collaboration—Rainer Weiss, Barry Barish, and Kip S. Thorne (Wheeler's student)—for observing gravitational waves, which are often described as ripples in the fabric of spacetime.

Quantum theorists manipulate vacuum states, vacuum expectation values, and creation and annihilation operators in much the same way as in the "hole" theory and QED of the 1930s and 1940s. Victor Weisskopf's concept of primitive vacuum fluctuations has been widely assumed to be correct. His "hole"-theoretic picture of vacuum polarization was eclipsed by other field-theoretic pictures. In 1954, Weisskopf's most famous student, Murray Gell-Mann, and Francis Low published a landmark paper describing vacuum polarization and the shape of the "cloud of electrons and positrons" that surrounded a "bare" electric charge in the vacuum.[4] Gell-Mann and Low abjured both "fluctuation" and "virtual" labels for their vacuum pairs. But describing vacuum polarization in terms of a cloud of virtual particles became standard.[5] This is a strong continuity with the earlier pictures developed by Paul Dirac, Rudolf Peierls, Weisskopf, and others. The vacuum was a dielectric medium which screened charges, reducing their apparent strength (renormalizing them). A version of Vladimir Fock's objection to "hole" theory applies ninety years later: what about the mass of all these polarized particles? Perhaps the answer lies in an as-yet-unknown theory of quantum gravity. In particle physics, the Standard Model is based on a collection of gauge fields that undergo Spontaneous Symmetry Breaking.[6] The 2012 discovery of the (post-SSB, massive) Higgs boson at CERN was heralded as proving "the power of empty space."[7]

---

1. Stachel, "Rise and Fall of Geometrodynamics"; Kiefer, "Quantum Geometrodynamics."
2. Carroll, *Spacetime and Geometry*, ch. 6.
3. Kennefick, *Speed of Thought*, 210; for a sociological view, see Collins, *Gravity's Shadow*.
4. The description was quite neutral, the cloud was not described as arising from fluctuations or as virtual. Gell-Mann and Low, "Quantum Electrodynamics," 1301.
5. Peskin and Schroeder, *Introduction to QFT*, fig. 7.8, 255; Itzykson and Zuber, *QFT*, 326–28; for a fluctuation-focused discussion, see Zee, *QFT in a Nutshell*, §III.7.
6. Schweber, "Quantum Field Theory."
7. Randall, *Higgs Discovery*.

More than Nothing. Aaron Sidney Wright, Oxford University Press. © Oxford University Press 2024.
DOI: 10.1093/oso/9780190062804.003.0008

Two technological innovations spurred new vacuum theorizing, beginning in the 1980s and 1990s: the development of lasers and new techniques of micromanufacturing. Chapters 3 and 4 discussed Albert Einstein's theory of "spontaneous emission" of electromagnetic radiation by atoms, beginning in 1916. Since the development of lasers, the rate of spontaneous emission of atoms in a cavity can be manipulated by "squeezed vacuum" techniques.[8] Optically manipulating the quantum vacuum entered non-relativistic theory in the 1981 and, by 1997, experiments. The squeezed vacuum is produced as a matter of course in efforts to improve LIGO instrumentation and measure EPR violations.[9] Micromanufacturing gave the Casimir Effect new life. In 1948, Dutch physicists Hendrik Casimir and Dirk Polder developed a new theory of the Van Der Waals forces between electrically neutral molecules. Casimir recalls that discussions with Niels Bohr prompted him to rethink the nature of vacuum fluctuations between neutral objects, which he idealized as tiny cavities or metal plates.[10] During Robert Oppenheimer's address at the 1948 Solvay Congress (discussed in chapter 4), Casimir insisted that vacuum fluctuations were real and measurable—and contradicted Julian Schwinger's insistence that the vacuum current be zero. However, Casimir's comment produced no recorded reaction.[11] The Casimir effect attracted scattered attention for the next forty years.[12] Experimentalists made several attempts at measuring Casimir attraction or repulsion, but precise results arrived only in the late 1990s.[13] The existence of the Casimir effect is often cited as independent evidence for the reality of quantum vacuum fluctuations, though both physicists and philosophers have challenged that conclusion.[14]

The central narrative of *More than Nothing* ends in 1980, just at the cusp of two developments that reshaped theoretical physics and its epistemological foundations: inflationary cosmology and string theory. I will discuss them in turn. On January 31, 1980, Sidney Coleman was on sabbatical at the Stanford Linear Accelerator (SLAC) in California. He called in a memo to his colleagues in the theory group at Harvard, recommending that Alan Guth (b. 1947), a research associate at SLAC, be invited to Harvard to give a seminar on his "extremely interesting work."[15] Guth integrated Coleman's false vacuum work (among other sources) with relativistic cosmology to solve a conflict between astronomical observations and particle physicists' Grand Unified Theories. He proposed that after the big bang, the universe was in a false vacuum that expanded incredibly rapidly before tunneling into our current, true,

8. Bromberg, *Laser in America*.

9. Roughly speaking, "squeezing" the vacuum means manipulating the spectrum of vacuum fluctuations so that some modes have lower fluctuations than a standard vacuum. Walls and Milburn, *Quantum Optics*, esp. §§3.7.2, 5.1.2, 8.1.6, 8.3.1, 13.

10. Kragh and Overduin, *Weight of the Vacuum*, ch. 8.

11. Institut Solvay, "Electron Theory," 285–86.

12. Exceptions included Schwinger's source theory analysis, topological studies in GR, and the MIT "bag" model of hadrons. See Mehra and Milton, *Climbing the Mountain*, ch. 15; Birrell and Davies, *Quantum Fields*, ch. 4; Milonni, *Quantum Vacuum*, ch. 10.

13. Klimchitskaya, Mohideen, and Mostepanenko, "Casimir Force."

14. Rugh and Zinkernagel, "Quantum Vacuum"; Saunders, "Zero-Point Energy"; an attractive moderate position is that "the Casimir effect gives no more (or less) support for the reality of the vacuum energy of fluctuating quantum fields than any other one-loop effect in quantum electrodynamics," Jaffe, "Casimir Effect," 1.

15. Coleman to Howard Georgi, Sheldon Glashow, and Steven Weinberg, 31.1.1980, SRC.

vacuum. Guth catalyzed a huge literature, including many variants. Perhaps the most influential was "slow roll" inflation, which shows in its name the continuing influence of Coleman's "bounce" techniques (section 7.5.3).[16] Inflation has become a standard part of modern cosmology.[17] It also fomented challenges to traditional understandings of physical science in the form of anthropic reasoning and the multiverse. For example, in 1982 Andre Linde proposed a "chaotic" variant of inflation in which, after the initial expansion is complete, the universe settled into an infinite collection of distinct bubble-universes. As historian Helge Kragh has elaborated, this poses a problem for those who think that physical science necessarily restricts itself to empirically verifiable (or falsifiable) statements. Distinct universes are distinct from ours, and there is no possibility to learn about the others. Eventually, inflation was connected to anthropic reasoning. An, as it were, spiritual forefather of anthropic reasoning was John Wheeler, in his puzzling about the role of the observer in quantum theory and cosmology, in the years just after the events of chapter 5. Australian relativist Brandon Carter formalized an anthropic argument in 1974: among an infinite collection of (possible) worlds, our universe is singled out by its having the conditions necessary for the existence of human life. This was also a challenge to a traditional view of physical science as producing facts and theories that are independent of humans. Both the multiverse and anthropic reasoning began as marginal, idiosyncratic, views and became more influential in the twenty-first century.[18]

As Kragh remarks, acceptance of the multiverse was encouraged by the rise of string theory. String theory grew out of S-matrix nuclear physics in the 1960s and might have become detached from the central role of the vacuum in field theory. However, as string theory blossomed into a leading area of theory in the middle 1980s, vacuum configurations of strings and curled-up extra dimensions became central to attempts to connect string theory to empirical reality.[19] Even these attempts, however, were qualitatively different from the standard approach of theoretical physicists. In the 1970s, Coleman was aware that his work was distant from experimental results. At Erice in 1975, his remarks reflected an institutional deference to experiment: "This has been a long series of physics lectures with no reference whatsoever to experiment. This is embarassing [sic]."[20] Things were worse for string theorists. Many believed that even if string theory was correct, it would never be open to test by particle physics experiments. String theorists did not accept embarrassment. As historian Peter Galison has shown, in the 1980s, string theorists attempted to change the meaning of physical science—to shift the epistemic values of their community—away from experiment and toward mathematical and conceptual desiderata.[21] String theory and the multiverse meet in the "landscape" of string theory vacua. The landscape is an enormous

16. The conflict was: GUTs predicted the observation of magnetic monopoles, which were not observed. Guth, "Inflationary Universe"; in 1979, Alexei Starobinsky made a similar cosmological proposal. Smeenk, "False Vacuum"; Schweber, "Shelter Island Conferences Revisited," 110–13.

17. For general and critical reviews, see Guth and Kaiser, "Inflationary Cosmology"; Earman and Mosterin, "Inflationary Cosmology."

18. Kragh, "Contemporary History."

19. Rickles, String Theory, 170n4.

20. Of course, he gave the lectures regardless. Coleman, "Classical Lumps," 390; cf. Schweber, "Empiricist Temper."

21. Galison, "Mirror Symmetry"; Rickles's "phase 3," Rickles, String Theory, 8 and n15.

314   MORE THAN NOTHING

number (up to 10^{500}) of possible ways for string theory's extra dimensions to be curled up ("compactified"). Much in the way that a vacuum state defined a quantum field theory, each compactification defined a possible string theory. At the beginning of the twenty-first century, Leonard Susskind popularized the landscape as a challenge for string theorists. If there are 10^{500} sets of string theory predictions, all equally valid, some critics argue that string theory predicts everything and therefore explains nothing.[22] The development of the multiverse and string landscape are rooted in the story of theoretical physics told in this book. However, as they engage in novel forms of reasoning and epistemic virtues, they also mark a stopping place for the central narrative. Nevertheless, in the twenty-first century, the false vacuum propels inflation and string theorists survey a landscape of vacua. In its explanations of the smallest and the largest scales, theoretical physics has kept the vacuum at center stage.

22. Kragh, "Contemporary History"; each compactification supported a matter-free solution to the Einstein Field Equations, which was taken to be the low-energy limit for the string theory. Rickles, *String Theory*, 191–94, §10.3.

9

The Void

9.1 Overview

In the introduction to this book I asked a question: "How did the vacuum become a central object for theoretical physics?" My overarching answer is that this happened because of theorists' practices. We might think of theorists' practices as a dynamical system, consisting of multiple interacting elements. Here, I provide an overview of how the interacting elements of these practices shaped the vacuum, and how the vacuum shaped them in turn. A partial list of these elements include: individual creativity; reaction to experiment; idealization; formalization; and the application of paper tools.

Creativity. There is always a difficult-to-analyze element of singular personality in the history of science. This element is included in the meaning of "scientific practice," when it is contrasted to ahistorical, formal representations of science and scientific theory. The history of the vacuum was driven by physicists who applied the full capacity of their creativity and personality to their work. Physics has been driven by beauty, diagrams, and aesthetics; has been propelled by philosophical, political, and religious commitments; and has been narrated like a good science fiction story. Even the most mathematical, formal, and rigorous scientific work is essentially human. This is a basic finding by historians of science, whatever era, geography, or area of natural inquiry they study. I emphasize it here, in particular, for readers of this book who are students of science. But there is another sense of "scientific practice" that draws on associations with repetitive behavior and routines, such as bicycling to and from an office each workday. These practices, like commuting, are often communal and social. Arguably, scientific practice should be conceived of as a "creative practice," like in art. In scientific practice, repetitive procedures—like pipetting biological material or performing relativistic calculations—produce creative advances.[1] John Archibald Wheeler's contributions to General Relativity show the importance of both individual personality and shared practices (chapter 5). His research was characterized by recursively returning to a set of philosophical questions and goals, including the drive to identify a single universal substance underlying creation. This was on the one hand a product of his individual personality, and resonated with his Unitarian Christianity. On the other hand, this was a product of his note-taking practices, which created the physical record of his research to which he returned. His notebooks began in the context of military research and collaboration with experimenters (with their rigid note-making

---

1. Knorr Cetina, "Objectual Practice."

More than Nothing. Aaron Sidney Wright, Oxford University Press. © Oxford University Press 2024.
DOI: 10.1093/oso/9780190062804.003.0009

discipline), and he instructed his students on keeping their own series of notebooks. Similarly, Roger Penrose's diagrams (chapter 6) were a product of lifelong practices of drafting and recreational mathematics. These practices also exhibit deep connections between theoretical physics and other aspects of culture and society, including the fine arts, the military, and religion. These practices were a learned and a shared context for creating and communicating scientific novelty. Indeed, fortune favors prepared minds.

Experiment. Paul Dirac transformed the vacuum in the context of justifying his marvelous relativistic equation for electrons (chapter 2). Ivar Waller had thrust Dirac onto the horns of a dilemma: accept strange "negative-energy" states of motion as states real electrons could occupy, or have his equation cut off from experimental phenomena. Dirac saved the phenomena: to keep its contact with experiment, he modified his theory, and transformed the vacuum. Through the 1930s, experimental evidence from scattering experiments and cloud chamber photographs convinced many physicists, if not of the reality of Dirac's "holes," at least of the necessity of a new vacuum electrodynamics. In California, J. Robert Oppenheimer's group was in close contact with experimentalists, rapidly modeling new experimental phenomena. However, in the chaotic 1930s, theorists also selectively downplayed the importance of inconvenient experiments (chapter 3). Immediately after World War II, increasingly precise atomic measurements contributed to the new "renormalized" Quantum Electrodynamics, with its vacuum loops and fluctuations (chapter 4). In General Relativity, Wheeler was initially closely engaged with experiment and observation. He synthesized cosmological theory and brand-new neutrino physics in early geometrodynamics. But, by the 1960s, experiment and observation—including of new quasi-stellar objects—served for relativists as inspiration more than anything else (chapters 5 and 6). By the 1970s, Sidney Coleman's research was self-consciously removed from experimental phenomenon (chapter 7). The "measurement" Coleman's false vacuum envisaged—like relativists' narratives about falling into black holes—was a catastrophe. Theorists' interactions with experiment ranged widely and resist easy generalization. They proposed new measurements and phenomena flippantly (such as Dirac's anti-electron) and also precisely (such as Edwin Uehling's polarization effect). Theorists felt tightly constrained by experimental work but at other times dismissed inconvenient experiments. Among the theorists studied here, one pattern that emerges is that engagement with experiment lessened over time.

Idealizations. Dirac engaged self-consciously in theory building, as indicated by the titles of his scientific publications. He made suppositions and conjectures and reasoned toward their consequences. His most daring conjecture was that the vacuum was filled with a sea of electrons in negative-energy states, and from this he reasoned toward anti-electrons and modifications to Compton scattering. If this story is begun with Dirac's conjecture, it fits a very traditional understanding of scientific theorizing: the hypothetical-deductive method, in which a "higher," general, and abstract theory is specialized to specific physical circumstances via idealized models that are compared to experiment. But the story began earlier. I argued in chapter 2 that the form of Dirac's general theory was in fact determined by idealizations and approximations: Dirac's non-relativistic many-electron formalism and Douglas Hartree's atomic model.[2] A fundamental theory about the elementary constituents of nature was directed by idealizations that—despite their beauty—were aimed at overcoming contingent, human, difficulties. (Difficulties such as the inability to solve classes of mathematical equations.) Similarly, Roger Penrose's *rigorous* solution to the radiation problem in GR was premised on the idealization that the universe contained only zero-rest-mass fields. He did not seem to be motivated by philosophical or religious commitment to simplicity and unification, as was Wheeler. Like Dirac, Penrose used his idealization because no one knew how to solve the equations for general or less-idealized models of phenomena. Idealization was combined with mathematical approximation in the paper tool of perturbation analysis. Perturbations structured QED from 1927 to 1980 and beyond (chapters 2, 3, 4, and 7). Making idealizations is a practice of theory, one which is both routine and generative of scientific novelty. These idealizations are essentially pragmatic, in the sense that they are responses to specific, contingent, human limitations.[3]

Formalism. In many ways, theorists engaged with mathematical formalisms, the collections of mathematical symbols and rules for their manipulation that were a necessary condition for calculations to proceed. There were continuities among the different vacuum formalisms. They should be able to represent a state of zero particles and zero energy. This common core was articulated differently according to the context. For example, Dirac's "sea" was a state of zero *observable* particles.

Theorists espoused varying degrees of commitment to the accuracy of their formalisms. Were the electrons in Dirac's sea merely mathematical fictions or were they real electrons? Dirac argued that they were real. But others espoused a range of subtle views. Wolfgang Pauli was willing to use Dirac's formalism without endorsing its correctness. Igor Tamm accepted that Dirac's sea was the best picture to accept around 1930, but expected a better picture to emerge later on. If different formalisms described different pictures of the vacuum, why did theorists agree that they were discussing the same object? Why was it accepted, for example, that Pauli and Weisskopf's wave-based calculations were relevant to Dirac and Rudolf Peierls's particle-based calculations? According to a traditional view of scientific objectivity, this intersubjective agreement—this invariance between different perspectives—constituted the objectivity of the vacuum.[4] In part, this was due to theorists' communal agreement on core vacuum concepts and the qualities of their formalisms. (These included the capacity to represent a zero-particle state, e.g.) This book shows that the conditions of possibility for much of this agreement was common *practices*, including the use of paper tools. One such practice was theorists' recalculation of published results, as performed by Furry (chapter 3) and Weisskopf (chapter 4). By working through one another's formalism, theorists were able to see their object of study as if through

2. A body of philosophical work—produced by Mary Morgan, Margaret Morrison, Nancy Cartwright, and others—argues that in an important sense lower-level models of phenomena are *independent* from higher-level, general, theory. They argue that the traditional priority accorded to theory over models be erased. Morgan and Morrison, *Models as Mediators*; Suárez and Cartwright, "Theories"; I take my analysis to demonstrate that, in some important cases, the traditional priority was in fact reversed.

3. Following Rudolf Carnap's discussion of pragmatics, semantics, and syntax. Quoted in Winther, "Structure of Scientific Theories," §1.

4. Daston and Gailson, *Objectivity*, ch. 5.

new lenses.[5] In chapter 3 and appendix A, I argue that theorists' agreement about the new vacuum electrodynamics was established by physicists' common use of a paper tool, perturbation analysis. The argument is that perturbation analysis applied to any of the quantum vacuum formalisms in the literature produced non-empty vacua. The use of a common paper tool produced a common result that formed the basis for a common understanding of the vacuum.

Julian Schwinger, Freeman Dyson, and others developed a successful formalism for QED built on vacuum states (chapter 4). Schwinger's vacuum expectation values became part of the standard theorists' toolkit. In analogy to experimental research, once their properties were well understood, objects of theory became part of the infrastructure that supported studies of new epistemic objects. Vacuum expectation values became analogs to the chemical reagents or model organisms that laboratory managers order from a catalog. But the progression from a proposed formal system to wide acceptance was not simple. Establishing the validity of a formalism was a difficult task, judging from some prominent failures. Wendell Furry and J. Robert Oppenheimer's "pair theory" formalism—which represented electron-positron pairs by a single wavefunction—failed to gain acceptance (chapter 3). Wheeler struggled for decades to establish a formalism for quantum geometrodynamics, without success (chapter 5). Even within a mathematical world on paper, theorists were not completely free to invent.

At important junctures, physicists *chose* to center their work on the vacuum. In the 1940s, Wheeler had failed to revive action-at-a-distance electrodynamics and overthrow field theory. After surveying the landscape, he threw himself "whole hog" into zero-rest-mass fields.[6] Coleman rather freely embraced the pleasure of vacuum paradoxes and narratives. In different ways, both Wheeler and Coleman were attracted to the vacuum's *fecundity*, its capacity to generate scientific novelty and surprise. As emphasized in chapter 3, the vacuum became a resource from which physicists could develop solutions to theoretical problems. Vacuum expectation values and scattering between vacuum states were paper tools too powerful to ignore. Penrose's paper tools drew relativists into agreement.[7] In large part, the vacuum became central to theory because it was so *useful* to theorists. More than this, however, some theorists felt forced to accept the active vacuum into their physics. (Wheeler, having made his choice, was very attuned to this feeling. He worked hard to convince his peers that geometrodynamics was inevitable.) In the 1930s, Walter Heitler was less enthused than Weisskopf about the Dirac sea. But Heitler thought there *must* nevertheless be a new vacuum electrodynamics (chapter 3). Richard Feynman disliked the loop diagrams he needed to compute vacuum polarization. He wanted to cut them out of his QED.

5. I thank David Kaiser for the suggestion that theorists' congruences here may be a parallel to Allan Franklin's discussion of experimentalists' "calibration." Franklin, "Avoid the Experimenters' Regress."

6. In 1961, Feynman recognized the connection between Wheeler's radical hypotheses and the generation of research questions. "I now realize that there is much to be said for considering theoretically the possibility that Q.E.D. is correct, although incomplete. This assumption may be wrong, but it is precise and definite, and suggests many things to study theoretically, while the other negative assumption, (that it fails somehow) is not enough to suggest definite theoretical research. This is Wheeler's principle of 'radical conservatism.'" FP Box 17, Folder 2, Draft report to 1961 Solvay conference, p. 34.

7. Kennefick, *Speed of Thought*, 210.

But he felt *compelled* to keep the loops by tradition and by conservation laws (chapter 4). Experimenters may experience the world *pushing back* at them through their instruments when they are unable to make some laboratory phenomena disappear; in turn, theorists feel constrained by their theories.[8] Analyzing the vacuum produced surprising results such as modifications to Compton scattering or "mass without mass." It was well enough understood for two physicists to argue about the vacuum's properties, and to convince physicists to accept bizarre new conclusions. It was mysterious enough to remain an object of research.

Theorists' practices installed the vacuum at the center of two pillars of theoretical physics, Quantum Field Theory and General Relativity. These practices were not ephemeral. Long after their first application, theorists' practices continued to shape the vacuum, its ontology, and its place in physics. In turn, paper tools based on vacuum physics continued to shape theorists' study of elementary particles and gravitation. The vacuum made meaning, and was made meaningful, in theorists' mathematical form of life.

9.2 The Momentary Vacuum

As an object of theory, the vacuum can be analyzed according to multiple temporalities: synchronously, a snapshot of a single moment; and in two orientations diachronically, through time considering the past and looking toward the future. Considering the vacuum in these three aspects provides a historically informed perspective on questions about the reality the vacuum, and of other objects of theory. How is it that some theorists experience a strong feeling of the reality of their research objects?

Synchronously, we approach the vacuum at a particular year, month, and day. Perhaps on this day John Wheeler is in dialogue with his notebook, groping toward a new understanding of the vacuum. Here, at the leading edge of research, the vacuum is the most personal and subjective. The boundary between idea and object is indistinct. Much of this book's narrative has followed this avant-garde, in moment-by-moment reconstructions of the elements of theorists' practices that were sketched in section 9.1. In these moments, it is hardest to say that an object of research is an object at all.

And yet, many scientists express realist attitudes toward these ill-defined objects. Realist commitment is a paradigmatic feature of scientific investigation. But is realism not, in the moment, premature? It comes before the subsequent scientific communication, replication, verification, or any of the myriad scientific processes that are extended socially and in time. What is the origin of this realistic commitment? In experimental sciences, Hans-Jörg Rheinberger points to the unusual temporality of scientific research, in which a (diachronic) consideration of the future is pulled into a scientists' (synchronic) present. Drawing on the work of Michael Polanyi and

8. Compare the discussions of "resistance" the world offers, in opposition to "plasticity": Pickering, *Mangle of Practice*; Galison, *Image and Logic*, §9.5; on Karen Barad's reading, theorizing is a material-discursive practice, Barad, *Meeting the Universe Halfway*, esp. ch. 3.

Gaston Bachelard, Rheinberger connects the sense of reality of a hazy research object to scientists' expectations of the object's future performance as a generator of novelty, of an endless potential for investigation. Polanyi wrote: "To trust that a thing we know is real is to feel that it has the independence and power for manifesting itself in yet unthought of ways in the future."[9]

Reading with Polanyi in mind, John Wheeler's notebooks reveal the convergence in theoretical research of surprise, a presumed future of investigation, and realist commitment. In 1956, Wheeler outlined a scientific question: "How come mass of vacuum vanishes so exactly." Many physicists took for granted that the vacuum would be massless. But Wheeler recast this quotidian fact as surprising: "Seems fantastic this should come out!" He considered two possible directions for investigation: "Is it defined as the answer? Or calculable [from quantum geometrodynamics]?" His note then flowed to a projection of future successes. "Would be an enormous triumph to get a simple proof. THE VACUUM HAS NO MASS!"[10] The confidence expressed by his capital-letter slogan was, perhaps, a product of his pulling a subjunctive "triumph" backward into his present. Wheeler's notebooks are filled with juxtapositions of the subjunctive and indicative moods, of imagined futures and confident declarations of fact. Wheeler's object of theory felt real in the concatenated moment of the contact of pen and notebook with an imagined future.

There is another synchronous aspect of objects of theory to consider: the existence, at a certain moment, of conflicting representations of an object. Given two or more conflicting representations of an object of theory, there are a range of possible responses. One might hold that one or the other representation is (more) correct, or one might hold that given the conflict, neither representation should be trusted. One might dissolve the conflict by disarticulating the representations by, for example, limiting the domain of application of each representation until they do not overlap.[11] Or one might adopt pluralism about the object. I will return to some of these questions below. For the moment, I want to stay with Polanyi's and Rheinberger's *positive* analysis of surprise and unknowability. This is a radical departure from the Cartesian heritage in philosophy that values clear and distinct ideas and logical precision. Rheinberger locates the existence of epistemic objects at the concatenation of multiple, overlapping, and perhaps inconsistent representations.[12] In my reading, this is a straightforward generalization from laboratory practices, in which genes or electrons are known and manipulated by multiple instruments that produce different kinds of traces for analysis. In Rheinberger's study of the history of protein synthesis, these representations include inscriptions, physical marks made in the laboratory. We might say that scientists' feeling toward their research objects of an unthought future potential is *produced* by their epistemic situation. That is to say, laboratory scientists produce multiple, perhaps inconsistent, representations of their object of research as a matter of course. (Consider DNA as a stick-and-ball wooden model,

9. Quoted in Rheinberger, "Cytoplasmic Particles," 272, cf. 273; Rheinberger, *History of Epistemic Things*, 23.

10. JAW, Relativity Notebook IV, p. 10.

11. Consider how relativistic and non-relativistic representations of electrons are distinguished according to their kinetic energy, so that each representation correctly applies in its distinct domain.

12. Rheinberger, *History of Epistemic Things*, 225–26; Rheinberger, "Cytoplasmic Particles," 274.

as a crystallographer's X-ray, as a chain of CGAT letters, or as a sediment in a test tube.) At least for the laboratory sciences Rheinberger analyzes, this situation is the rule, not the exception. If a laboratory scientist has a realist commitment to their object of study, that commitment will be a product of a research environment with multiple incompatible representations.[13]

Further developing the analogy between experimental and theoretical research: objects of theory are also posed between conflicting representations. This happened even within individual lines of research or publications. Consider first Furry and Oppenheimer's "pair" theory (section 3.3). On the one hand, they carefully arranged their formalism so that it was capable of representing an empty state with no light quanta, no electrons, and no "positives" (i.e., positrons). On the other hand, by way of "conceptual experiments," they showed that the vacuum could never be measurably empty. Each representation of the vacuum, in their view, followed from rigorous scientific reasoning. Each representation, simultaneously, was scientifically useful in their research. Using these representations, and referring to experimental evidence of pair creation, Furry and Oppenheimer made cautious claims about what was real. In Relativity, there is a well-known impossibility of visually representing an infinite four-dimensional spacetime on a finite sheet of paper. Roger Penrose's conformal diagrams were one method to overcome the physical and conceptual difficulties of representing spacetime. Multiple diagrams could be drawn for a single spacetime. But more than this, alternative forms of visualization of a spacetime were often deployed. These representations could be based on mutually inconsistent assumptions (such as making, or not making, a conformal transformation). Examples include conformal diagrams, Penrose's depiction of gravitational collapse (figure 6.14), and the "embedding" diagrams John Wheeler taught to generations of students.[14] The spacetime under study was located by the concatenation of its representations, the diagrams as well as formal representations in terms of tensors, or spinors. These were not contradictions to be reconciled. They were acknowledged to be partial representations of something, an object of theory, which eluded simple and clear expression. With the character of this epistemic situation in view, it is easier to see how vagueness and surprise could be built in to the development of research objects.

Let me summarize. I have considered two synchronous aspects of objects of theory, aspects that come into view at singular moments or snapshots. The first aspect is the future-directedness of research objects; the way in which a scientist may feel the reality of their object of study in its future potential. The second aspect is the location of research objects at the intersection of possibly incompatible representations. As an object of theory, the vacuum was represented and manipulated differently, even within a single body of research. It could not simply be identified with a single representation. This may be an origin of the ineffable nature of research objects Rheinberger identifies, and the capacity to surprise and generate novelty.

13. For Rheinberger, the feeling of the reality of objects *of* research is the feeling of their reality *as* objects of research, what Rheinberger sometimes refers to in inverted commas as the "scientific real." Rheinberger, *History of Epistemic Things*, 23. If these objects were not surprising, or vague, they would no longer be the *targets* of research. Clearly understood, unsurprising scientific objects may become technical objects, part of researchers' toolkit, and still real.

14. Misner, Thorne, and Wheeler, *Gravitation*, 613.

Theorists often address themselves to an entity that is not captured in any single representation. Electrons are posed between representations (relativistic and non-relativistic; before and after renormalization; before and after spontaneous symmetry breaking).[15] Theorists like Wheeler have felt and urged realist commitment to these entities. In the nineteenth century, James Clerk Maxwell was skeptical about his and other competing *models* of the luminiferous aether. Despite this, he had "no doubt" a material aether filled the spaces between the stars.[16] These synchronic considerations help us understand the process of research, how objects of theory are created, and how they change. Entities that are only partially understandable are entities that are capable of eliciting surprise. Rheinberger's analysis of novelty and surprise is in fact aligned with a traditional realist desire: to identify scientific entities that are independent of human consciousness. When we feel surprise, it is usually created by something independent of ourselves (say, a teacher's pop quiz). In this way, the synchronic, in-the-moment realist of (some) laboratory scientists and (some) theorists can be understood in parallel. However, as Maxwell's example illustrates, synchronic considerations may not be the right guide to which objects of inquiry will be reliable over time. Even research in so-called mature science, like physics, may be *im*mature in its moments of creation.[17] This calls for diachronic considerations.

9.3 The Real Vacuum

So far, I have explored an analogy between experimental systems and theoretical research, in order to shed light on the coming into being of new scientific objects and their change over time. I now want to consider the continuities objects of theory have over time. And, what should we, with a surveyable history of an object of theory, say about the reality of these objects. Is it possible to acknowledge something as deeply historical but that transcends its history? Some readers may be surprised to find here a discussion of what we should think of the vacuum, in addition to a discussion of what we should think of what *scientists* thought of the vacuum. It may be possible to write an historical ontology and remain philosophically neutral on the question of reality of the ontology under study.[18] But it would be naive to proceed as if the writing of history did not have implications for the coherence, centrality, or importance of that history's subject, whether a nation, or a social class, or a scientific discipline.[19] In my view, the humanistic study of the sciences has important continuities with the sciences themselves in seeking to understand the world and our place in it.[20] I am a realist, which means that I hold that scientific entities like electrons and spacetime exist.[21]

15. For earlier instances, see Arabatzis, *Representing Electrons*.

16. Maxwell, "Ether," 775; Wright, "Fresnel's Laws," 44.

17. For "immaturity," see Hacking, "Michel Foucault."

18. See Arabatzis, "Hidden Entities."

19. See Iggers, *Historiography in the Twentieth Century*; Graham, "Epilogue," esp. 293–95.

20. Examples of metaphysically engaged historical studies include, for anti-realism, Cushing, *Theory Construction and Selection*; for pluralism, Chang, *Is Water H$_2$O?*; and for realism, Graham, *Lysenko's Ghost*.

21. I take this to be the core realist commitment. Realists often hold several related positions. See Chakravartty, "Scientific Realism."

And I believe that realism matters outside narrow philosophical debates.[22] But I will not try to convince non-realists here. Rather, I address myself to others with realist leanings. The history of theory suggests severe limits to realist commitments. But it also offers new arguments for the existence of objects of theory. I will proceed by analogy to an analysis of experimental practices, Ian Hacking's "entity realism." Entity realism focuses on reliable experimental practices. This connects to my framing discussion of synchronic and diachronic temporalities. Reliability is an essentially *diachronic* characteristic that names a quality that exists over an extended period of time.

Hacking looks for scientific entities in the laboratory, far from theorists' notebooks. Within the laboratory, he does not focus on the objects of experimenters' research. Rather, Hacking is a realist about the entities experimenters *use* to do their research on other objects or phenomena. One of his examples is the PEGGY II particle physics experiment, which investigated symmetry breaking in the Weak interaction. PEGGY II included a kind of "electron gun" that created pulses of linearly polarized electrons that were directed via magnets into other parts of the experiment. Hacking argues that we hold electrons to be real when we manipulate them reliably enough to produce new knowledge about something else (in PEGGY II, the Weak interaction). Some amount of theory is involved in our understanding and manipulation of entities in tools and instruments, such as the mass, charge, and spin of electrons. Hacking calls these "home truths" and denies that characterizing them requires commitment to comprehensive theories. Experimenters use multiple, mutually inconsistent, theoretical models to understand electrons, such as classical models of electrons and quantum models of electrons. For Hacking, that is so much the worse for the theories. One need not have an exact characterization of an entity to hold that it is real.[23]

Compared to a traditional account of scientific discovery, Hacking's entity realism is a diachronically extended, retrospective account.[24] Several decades elapsed between the era in which the existence of electrons was actively debated by the scientific community (the 1890s to 1920s, say) and the era in which control of electrons[25] became a commonplace in scientific instruments such as electron microscopes

22. For example, Karen Barad's "agential realism" connects quantum mechanics to feminist social philosophy. Barad, *Meeting the Universe Halfway*. "Agential realism" does not, however, address historical challenges to realist commitments; in a different direction, Kim Fortun argues that robust realist commitment is needed to counter the creation of ignorance around toxic industrial sites: Fortun, "From Latour to Late Industrialism."

23. Hacking, "Experimentation and Scientific Realism"; Hacking, *Representing and Intervening*, ch. 16; Rheinberger appeals to Hacking for the idea that what is "real" is located at the intersection of multiple inconsistent representations. Rheinberger, *History of Epistemic Things*, 113; Rheinberger, "Cytoplasmic Particles," 274 fn8. However, Rheinberger picks up the deflationary aspect of Hacking's discussion of "real," rather than the entity realist aspect.

24. Michael Friedman's unification account is another diachronic argument for scientific realism, in which the passage of time has a "central importance." Friedman, *Foundations of Space-Time Theories*, 247.

25. According to recent science, all regular objects are partly composed of electrons. To prevent "manipulates electrons" from being trivially the case in all experiments, some specification must be made to "manipulates electrons, qua electron," that is, manipulate in a manner that involves the relevant causal properties and home truths of electrons.

(1950s onward).[26] Instrument makers necessarily consider their materials over time.[27] To establish reliability, a temporally extended series of events is required. For scientific entities to be real, they must have a history.

Hacking's entity realism is skeptical about realism about theory, in particular about theories that (aspire to) provide a complete and fundamental representation of a significant part of the world. Maxwell's aetherial electromagnetism is an example of such a theory. However, Hacking presents entity realism as the *best* argument for realism about scientific entities, not the only possible argument. Here, I sketch a complementary position to entity realism, based in the practice of theory and directed toward objects of theory.

In parallel to entity realism's focus on reliable experimental manipulation, we may ask: what objects of theory have been reliably used to produce new scientific knowledge? Attending to these objects will produce a patchwork of bits of theory, not cohesive theoretical world views like the electromagnetic theory of matter.[28] This patchwork will not consist of the most fundamental objects that are so often the primary focus of theoretical research. Wheeler's magic fundamental building material—superspace—drove the quantum geometrodynamics research program for several years; but it failed to become a reliable tool for creating lasting new knowledge. Better to attend to the uses of Schwinger's vacuum expectation values, with their placid, still, and empty vacuum states, $|0\rangle$. Whatever the passing objects of a field theorist's research—exotic quarks or quantum gravity—these reliable vacuum states continue to be enlisted to produce new results, and new knowledge.[29] Here is the pattern: examine the practices of theorists over time; identify the reliable practices for producing new knowledge; take the objects of theory that are used in these practices as candidates for realist belief.

This procedure might be seen as a practice-based adaptation of an influential proposal for scientific realism defended by Philip Kitcher. Kitcher considers a canonical argument against scientific realism from the history of electromagnetism and optics: Augustin Fresnel's 1818 theory of polarization, reflection, and refraction. Fresnel's theory was a marvelous success, and "Fresnel's equations" remain central to optical science and pedagogy. The difficulty for realists today is that Fresnel strongly believed that light was the vibrations of a molecular luminiferous aether. This example seems to block the realist inference from the *success* of a part of science to the *existence* of the entities invoked by that science. As it happens, despite Fresnel's genuine conviction in the reality of the aether, he did not actually *use* the aether in producing

26. See Rasmussen and Chalmers, "Role of Theory."
27. Hacking does in fact also frame entity realism in similar terms to the future-oriented present I discussed above. However, the imagined future is not one of unending novelty, but one of winning grant funding to build reliable new devices. E.g., Hacking, "Experimentation and Scientific Realism," 77.
28. In philosophical terms, this is called a "selective" strategy: Chakravartty, "Scientific Realism," §2.3.
29. I thank Gordon McOuat for suggestive comments on connections to the philosophy of biology. The vacuum states might be an example of "generative entrenchment." Soler, "Solidity of Scientific Achievements," 3; as happens to experimental epistemic objects, this placid vacuum-tool reemerged as an object of research. The new vacuum of spontaneous symmetry breaking can be interpreted as new research on this vacuum state, chapter 7. In GR, the Unruh effect is sometimes taken as a critique of the basic vacuum (or a critique of the concept "particle"). See Birrell and Davies, *Quantum Fields*, §§3.3 and 8.3.

his lasting results.[30] Kitcher defends realism on the basis that the lasting results were produced by the "working posits" of Fresnel's theory, rather than "idle posits" like the aether.[31] The challenge for Kitcher is to produce a criterion for distinguishing the "working" from the "idle" that is objective or principled enough to withstand charges of arbitrariness.[32] My suggestion here is to look, diachronically, for those objects of theory which have been reliably *used* to produce lasting new knowledge. Recall that an object of theory can just as well be a phenomenon as an object; for Fresnel's optics we might identify the phenomenon of light behaving with a specific form of wavelike motion.

The reliability of theoretical practices can be characterized at multiple levels. Within a particular synchronic moment, an object of theory might be reliably used across a range of applications. For example, Fresnel's characterization of wavelike optical phenomenon was applied to multiple contexts of measurement. Broadening the temporal horizon, Fresnel's object of theory can be characterized as reliable outside its original context, among other aether-theories, and eventually, around the turn of the twentieth century, to non-aetherial electrodynamics. Recall that Hacking advocates belief in the reality of experimental entities, even if those entities have multiple inconsistent material and theoretical representations. If theoretical representations disagree, so much the worse for theory. Nancy Cartwright pushes realists to commit to more than entities; she advocates realism about the causes that appear in the unique causal stories scientists use to explain phenomena. If these stories have conflicting representations in higher-level theory, so much the worse for higher-level theory.[33] Hacking and Cartwright focus their attention on the areas in which inconsistent theories nevertheless agree.[34] Considered diachronically, some objects of theory—such as Fresnel's wavelike behavior of light—occupy a similar position to Cartwright's causal stories.[35] They are theoretical but not fundamental. Fresnel's objects of theory reliably produced new knowledge in several versions of aetherial electrodynamics. This situation should increase our confidence in these objects of theory, even if we reject the conflicting, putatively more fundamental versions of the aether.

30. Here I follow Jed Buchwald: "although Fresnel undoubtedly did think from the very beginning of his research that the ether, which carries optical waves, consists of particles that interact with material molecules, and although that image was immensely important to him in suggesting the possibility and plausibility of a wave optics, he did not use it to produce the underlying principles and practices of his novel physics for light." Buchwald, "Ether Spawned the Microworld," 216.

31. Kitcher, "Real Realism."

32. Chakravartty, "Scientific Realism," §2.3.

33. Cartwright, *Dappled World*.

34. Part of the underlying reasons for this, and perhaps of the justifications for this, can be found in the "causal theory of reference." The causal theory of reference, advanced independently by Saul Kripke and Hilary Putnam, is a framework for understanding how the reference of linguistic terms, like "electron," can be stable even when the conceptual scheme for understanding the meaning of the term changes. See Hacking, *Representing and Intervening*, ch. 6. Putnam now takes varied attitudes toward representational conflicts. On the one hand, he accepts the reality of both "scientific image" and "manifest image" of reality. He takes a different position where there are technically equivalent but disagreeing scientific representations of a system. For example, where there are equivalent wave and particle representations of a quantum mechanical system Putnam now advocates realism about the *system*, despite the differing representations. Putnam, "Quantum Mechanics to Ethics."

35. Reciprocally, causal stories may count as objects of theory.

Why should the agreement of multiple *conflicting* theoretical structures increase our confidence in their areas of agreement? This follows from the epistemic value of *robustness*: that we should be more confident when we have diverse evidence to justify our beliefs.[36] The robustness of Fresnel's objects of theory was established when they were used to produce knowledge across later aether theories, and up to today.[37]

In the history of theoretical practices, some robust objects of theory have taken on a special additional significance: they become standards by which future theories are judged. Two such objects were discussed in chapter 3. First, Wendell Furry brought an abrupt end to some of his radiation theory calculations when he discovered that his result disagreed with the Klein-Nishina formula for X-ray scattering. The Klein-Nishina formula was not something with which he could disagree. Before knowing the precise nature of his error, Furry knew the error was his. The Klein-Nishina formula produces a graph that agrees with experimental results, and so its robustness may not be surprising. The second example is more abstract: Einstein's A and B radiation constants, which represented the effects of theoretical radiation processes (or causal stories). These basic processes—stimulated absorption, stimulated emission, and spontaneous emission—were interpreted and reinterpreted several times. After their introduction in 1916, new foundations were moved beneath them, and new consequences drawn from them. These activities included Paul Dirac's derivations in 1926 and 1927, and Victor Weisskopf's 1935 claim that "[s]pontaneous emission is thus an emission of a light quantum which is forced by the zero-point oscillations of empty space."[38] Reproducing Einstein's radiation processes became a standard to which new theories were held. The ability to increase understanding of these processes or put them to new uses was highly valued. (See sections 3.2 and 4.2.1.) Peter Galison has noted how the life of arguably obsolete physics has been extended by physicists through their selective "re-reading" of old results in new language. His example is Maxwell's equations as considered by John Wheeler, Charles Misner, and Kip Thorne. Galison suggests that the activity of re-reading has generated productive new research, from special relativity to modern gauge theory.[39] These special objects of theory may be strong examples of successful "working posits" that can establish a practice-based argument for (limited) scientific progress.[40]

36. Against the prevailing probabilistic model of robustness, an explanation-based model of robustness may be appropriate here. See Schupbach, "Robustness Analysis"; for an influential critique of probabilistic models of scientific theories, see Magnus and Callender, "Realist Ennui"; for entity realists, multiple means of intervening in biological specimens increases our confidence in microscopy: Hacking, *Representing and Intervening*, ch. 11.

37. This practice- and object-based picture may be usefully compared to Michela Massimi's rather historicist "perspectival realism." There, the success of theoretical propositions is determined by *multiple* "scientific perspectives" that change over time. Massimi, "Three Tales."

38. Weisskopf, "Über die Polarisation der Elektronen bei der Streuung an Kristallen," 365.

39. Galison, "Re-reading the Past," 48–49; for the retrospective reimagining of mass-generating fields, see D. Kaiser, "Whose Mass Is It?"

40. This would be consistent with the philosophy of "effective field theory," in which mid-level objects of theory, such as Maxwell's equations, are viewed as stable over time and through changes in higher-level (and higher-energy) theories. If physicists invent a Theory of Everything, whether it is a string theory or something else, effective field theorists are confident that Maxwell's equations will remain a useful part of the low-energy part of the theory. See Cao and Schweber, "Conceptual Foundations"; Schweber, "Hacking the Quantum Revolution."

The wave picture of light and Einstein's radiation processes are objects of theory that have close relationships with experimental practice. But as I discussed in chapter 1, experimental access to the vacuum is more difficult. One expression of this difficulty was in terms of the logic of measurement. The physicist Percy Bridgman asked: "how shall we establish that a purportedly empty space is really empty without going there with an instrument to prove it, and when we have introduced the instrument the space is no longer empty." If it was impossible to ground the concept "empty space" in a measuring procedure, Bridgman argued from the 1920s to the 1950s, it had no physical meaning at all.[41] The physicists discussed in this book rejected this analysis. Through their practices, they made the vacuum meaningful. One keen analyst in the 1980s rejected the demand that the physical vacuum should have to accord with our taken-for-granted assumptions about emptiness or nothingness. Yakov Zel'dovich argued that nature simply prevented us from using an intuitive concept of truly empty space:

> The conception of the vacuum cannot be simplified by a mere redefinition of words. One cannot say that the vacuum, i.e., empty space, is devoid of all properties "by definition" and that all the complexities are connected with something that should not be called "vacuum."
>
> We must define vacuum as space without any particles. Such a definition coincides with the condition of a minimum of the energy density in the given volume of space. [...] But the actual properties of the "minimal" state which is called the "vacuum" are dictated by the laws of physics, and we cannot insist that the minimum be zero or that the simplest possible situation be as simple as we wish.[42]

Zel'dovich thought it was reasonable to define the vacuum as a region without *particles*, but he insisted that this particle-free space was not, therefore, empty of energy. Physical concepts can only be as simple as the laws of nature allow, and for Zel'dovich, these laws ruled out truly empty space.

Zel'dovich gave a very general description of the vacuum, as the "minimal"-but-not-zero energy state. In this quotation, he used the language of particle physics, but he was also a respected relativist. His description of the vacuum holds in General Relativity, as well. At this level of generality, synchronic and diachronic continuities in physicists' vacua become salient across time periods and subfields.[43]

41. Bridgman, "Physical Concepts," 269; Bridgman, *Logic of Modern Physics*, 150–66; Jammer, *Concepts of Space*, 189–90.

42. Zel'dovich, "Vacuum Theory," 217; Zel'dovich shaped Hawking's picture of the vacuum during the latter's visit to the USSR in 1973, Hawking, *Brief History of Time*, 110–11.

43. Geoffrey Chew's S-Matrix theory abjured fields in favor of particle "trajectories"—but it retained a central role for the vacuum. In 1961 he wrote: "We have been emboldened to conjecture that a Regge pole for *the quantum numbers of the vacuum* is responsible" for some scattering behavior. He called it the "Pomeranchuk trajectory" or, as a vacuum particle the "Pomeron," after Soviet physicist Isaak Pomeranchuk. Chew, "S-matrix Theory," 398, emphasis added; and Eden et al., *The Analytic S-Matrix*, 127. The Pomeron remained as a difficult aspect of S-matrix theory through its decline in the late 1960s and into the 1990s. Cushing, *Theory Construction and Selection*, §§6.5, 6.8. As Kaiser indicates, the Pomeron-exchange technique far outlives its original context. D. Kaiser, *Drawing Theories Apart*, 385. For recent developments of this vacuum-object, see Donnachie et al., *Pomeron Physics and QCD*. It seems that it was easier to give up fields than to give up ascribing properties to empty space, and treating the vacuum on a par—not with fields, but— with particle trajectories.

In 2008, Yoichiro Nambu was awarded the Nobel Prize in Physics for his work on the theoretical mechanism of Spontaneous Symmetry Breaking, which was based on his solid-state representation of the vacuum. In his Nobel lecture, Nambu described the vacuum's degrees of freedom (loosely, a degree of freedom is a way of moving or acting): "the 'vacuum' is not void, but has many intrinsic degrees of freedom."[44] At this level of generality, the vacua described by quantum theory and by relativity agree.[45] Taking this agreement seriously would mean taking the separate views' disagreement's seriously. If the bubbling quantum fluctuations and the rippling relativistic fabric are meant to describe the same entity, then we should have confidence in neither picture. But perhaps there is a reality sitting at the intersection of these two incompatible representations. Perhaps we cannot say what the vacuum ultimately *is*, but we can say that the vacuum is more than nothing.

44. Nambu, "Nobel Lecture," 1015–16.
45. For example, Cao frames Einstein's view of the metric: "the metric field should also be allowed at [Einstein's] hands to cast off its chains under Mach, and step onto the stage of physics as a participant in its own right, with dynamic degrees of freedom of its own." Cao, *Conceptual Developments*, 98.

Appendix A
Perturbations

Perturbation theory is a strategy of idealization and approximation. It attempts to find close-enough models of physical phenomena for which close-enough solutions to mathematical equations could be constructed. Scientists might prefer exact solutions to realistic representations, but exact solutions to the equations of mathematical physics are rather rare. The classic example is from astronomy. The "two-body problem" is the problem of describing the motion of two bodies that experience mutual gravitational attraction: for any arbitrary values of the initial positions and velocities of the bodies, describe their subsequent motions. This problem can be solved. The "three-body problem" asks the same question for three bodies, such as a star and two planets. The mathematician and historian Edmund Taylor Whittaker found more than 800 attempts at the problem between 1750 to 1916. None succeeded. Mathematicians began questioning even the *possibility* of finding a solution to the three-body problem.[1] From the middle of the eighteenth century, perturbation theory set aside the question of an exact solution to the three-body problem—and more-complicated puzzles—and aimed at approximations that could be used by astronomers to calculate the predicted paths of planets, moons, comets, etc.[2] The basic idea was to begin with a known solution to a simple problem, and assume that the difference between the simple situation and the complicated reality had only a small effect on the resulting dynamics. For example, consider the three-body problem with the Sun, Earth, and Moon. A perturbative approach would first solve the two-body problem, Sun+Earth, and then add the effect of the Moon as a perturbation. At the turn of the twentieth century, the most influential practitioner of perturbation methods was the 3rd Baron Rayleigh, John William Strutt (1842–1919). His two-volume monograph *The Theory of Sound* was something of a pinnacle of Victorian mixed mathematics applied to vibrations and oscillations. For example, he treated the vibrations of an elastic string under the perturbing influence of a small bead fixed at one point. Rayleigh was a central figure for the Göttingen mathematician Richard Courant (1888–1972), and *The Theory of Sound* had a large impact on Courant's 1924 *Methoden der mathematischen Physik*.[3] For example, Courant updated Rayleigh's bead-on-a-string analysis.[4]

---

1. Whittaker, *Analytical Dynamics*, 339; Morris Klein, *Mathematical Thought*, ch. 21.
2. Wilson, "Perturbations and Solar Tables Pt. I"; Wilson, "Perturbations and Solar Tables Pt. II"; Garber, *Language of Physics*, 202–6, 221–22.
3. Schaffer, "Rayleigh"; Warwick, *Masters of Theory*, 276–77, 323–34; Strutt (Baron Rayleigh), *Theory of Sound*, vol. 1, 115–18; David Hilbert was a ceremonial second author. Reid, *Courant*, 70–71, 98.
4. Courant and Hilbert, *Methods of Mathematical Physics*, vol. 1, 348–50.

More than Nothing. Aaron Sidney Wright, Oxford University Press. © Oxford University Press 2024.
DOI: 10.1093/oso/9780190062804.003.000A

Perturbation methods were a part of quantum theory from its origins.[5] In wave mechanics, Erwin Schrödinger acknowledged Rayleigh explicitly.[6] In 1927, Schrödinger's assistant Erwin Fues wrote a *Handbuch der Physik* article on perturbation theory. He framed it in terms of "the transposition of astronomical methods to atomic physics," which was motivated by the planetary analogy of the atom.[7] Max Born knew Courant's work through his time in Göttingen, and he integrated perturbation theory into the formulation of quantum dispersion theory in 1925.[8] Born's mathematical approach to quantum mechanics was built in collaboration with the American mathematician Norbert Weiner.[9] In turn, Weiner had turned to Rayleigh during his reconstruction of harmonic analysis in the 1920s.[10] A paradigmatic example was the ubiquitous "Born approximation": an idealization of perturbing potential energies on atomic systems. Then a series approximation was applied. It was taken up in Hans Bethe's account of collision problems and *Bremsstrahlung* (section 3.5).[11] In QED, Dirac's perturbative approach was developed by Wolfgang Pauli, Werner Heisenberg, and Enrico Fermi, in particular.[12] More than a tool, perturbation methods structured the field. Walter Heitler organized his influential monograph *The Quantum Theory of Radiation* around the orders of perturbation theory.[13] Perturbation theory was the premier method for connecting quantum theory to experiment.[14]

A.1 A Definition

Perturbation theory (PT) is an algorithm, or procedure. To understand it, I will establish some terminology, following philosopher Margaret Morrison's discussion of modeling.[15] The application of PT to some physical phenomena begins with an *idealization*: a simplification of the physical circumstance. Once the physical phenomena was idealized, the second stage of PT was *abstraction*: the idealized system was represented mathematically. Ideally, the abstract equations representing this idealized system had exact solutions. Finally, some small *de*-idealization was introduced—the perturbation—that would create a small change in the solutions, and that could represent a less idealized, more realistic, physical system.

5. Darrigol, *c-Numbers to q-Numbers*, 128–32.

6. Schrödinger, "Quantisation as a Problem of Proper Values (part III)," 62–101; Mehra and Rechenberg, *HDQT*, vol. 5 §IV.5; cf. Kemble, "The General Principles of Quantum Mechanics. Part I," §4; Weyl, *Theory of Groups and Quantum Mechanics*, §II.9; Mott and Massey, *The Theory of Atomic Collisions*, ch. XIV; Jammer, *Conceptual Development*, 264–66; Pauli misattributed it to Dirac, Pauli, *General Principles of Quantum Mechanics*, 83; Pais makes the same error, Pais, *Inward Bound*, 329, 337. Dirac's 1927 papers on radiation theory did not claim novel results in perturbation theory, though he was characteristically slight in his citations.

7. Fues, *Calculation of Perturbation*, 4.

8. Hendry, "Bohr-Kramers-Slater," 206–9.

9. Born and Wiener, "A New Formulation of the Laws of Quantization of Periodic and Aperiodic Phenomena."

10. Masani, *Norbert Wiener, 1894–1964*, §9D.

11. Schweber, *Nuclear Forces*, ch. 4.

12. Darrigol, "Quantized Matter Waves," 226, 243–47; Schweber, "Enrico Fermi."

13. Heitler, *Quantum Theory of Radiation* (1936), 98 fn.

14. For modern references, see Goldstein, Poole, and Safko, *Classical Mechanics*, 12; Peskin and Schroeder, *Introduction to QFT*, §4.1; but see D. Kaiser, *Drawing Theories Apart*, ch. 8.

15. M. Morrison, *Reconstructing Reality*, 20–21.

Here are Dirac's introductory comments from the chapter on perturbation theory in *The Principles of Quantum Mechanics*:

Most quantum problems, however, cannot be solved exactly with the present re-sources of mathematics, as they lead to equations whose solutions cannot be expressed in finite terms with the help of the ordinary functions of analysis. For such problems one must use a perturbation method. This consists in splitting up the Hamiltonian into two parts, one of which must be simple and the other small. The first part may then be considered as the Hamiltonian of a simplified or unper-turbed system, which can be dealt with exactly, and the addition of the second will then require small corrections, of the nature of a perturbation, in the solution for the unperturbed system. If this second part contains a small numerical factor ϵ, we can obtain the solution of our equations for the perturbed system in the form of a power series in ϵ, which, provided it converges, will give the answer to our problem with any desired accuracy.[16]

The basic idea of a series expansion can be seen in Maclaurin's series of the exponential function, e, from elementary calculus:

$$e^x = \frac{x^0}{0!} + \frac{x^1}{1!} + \frac{x^2}{2!} + \frac{x^3}{3!} + \frac{x^4}{4!} + \cdots \qquad (A.1)$$

The exponential function was a ubiquitous basis for the central expressions in quantum theory: wavefunctions, operators, potentials, etc.[17]

Schematically, a series in the Hamiltonian was written:

$$H = \epsilon^0 H^0 + \epsilon^1 H^1 + \epsilon^2 H^2 + \epsilon^3 H^3 + \epsilon^4 H^4 + \cdots. \qquad (A.2)$$

Perturbations are required to not result in too drastic changes in the overall dynamics. Higher-order terms, such as $\epsilon^4 H^4$, have the potential to violate this requirement, and invalidate the analysis. In the case of QED, this worry is alleviated by the smallness of the electric charge or equivalently, the fine structure constant, α. A fourth-order term $\alpha^4 H^4$ gives a contribution:

$$\alpha^4 H^4 \sim \left(\frac{1}{137}\right)^4 H^4 = \left(\frac{1}{352275361}\right) H^4. \qquad (A.3)$$

H^4 might represent complex phenomena, such as Hans Euler and Werner Heisenberg's scattering of light by light (section 3.4). Because of the smallness of (their equivalent of) α^4, their calculations did not contradict their knowledge of the nor-mal, lower-order, phenomena of light scattering and dispersion. If the perturbation parameter is not small, or is overwhelmed by terms within H^4, the series was divergent and, arguably, meaningless.[18]

16. Dirac, *Principles* (1930), 167.
17. E.g., Heitler, *Quantum Theory of Radiation* (1936), ch. III§9, eq. 31.
18. Schweber, *QED*, §9.17; D. Kaiser, *Drawing Theories Apart*, 195–97, 228–29.

This definition of perturbation theory is quite capacious. It *excludes* idealizations that lead to exact solutions (such as the two-body problem).[19] It also excludes purely mathematical uses of approximations. It includes physicists' varied uses of perturbation methods and their evaluations of what counted as a "perturbative" calculations. This means that I count as perturbation theories some work which has been described as "non-perturbative" (discussed further in chapter 7). For example, Tomonaga, Schwinger, Feynman, and Dyson's initial versions of renormalized QED relied on series expansions in terms of the interaction between electrons and electromagnetism, $j_\mu(x)A^\mu(x)$, in which $A^\mu(x)$ was a classical perturbation.[20] The series was the object of "renormalization" procedure *and* of calculations of observable quantities, such as the Lamb shift. Beginning about 1950, Schwinger reformulated the "Theory of Quantized Fields" such that it did not rely on the interaction term in this way; it is often referred to as "nonperturbative."[21] When I describe Schwinger's later work as perturbative, I do not mean to question the integrity of this work, vis-à-vis the earlier approaches. However, Schwinger's TQF *was* perturbative in my sense. It relied heavily on the idealization of separate dynamical systems. And in fact Schwinger calculated many results "by means of a perturbation calculation."[22] A typical abstract from 1954 began: "The Dirac field, as perturbed by a time-dependent external electromagnetic field that reduces to zero on the boundary surfaces, is the object of discussion."[23] This paper was later referred to as "non-perturbative." My overall point is that when I describe a given calculation as "perturbative," I refer to my capacious definition. I do not deny the grounds on which the calculation may have been labeled differently. But I believe my usage is well-grounded in the historical record and physicists' practices.

A.2 An Example

Here is an example of a perturbation calculation, adapted from J. J. Sakurai's *Modern Quantum Mechanics*.[24] The physical situation is a vial of hot hydrogen atoms in an electric field, of the kind experimentalists used to investigate the splitting of spectral lines.[25] The idealization is to consider the "unperturbed" atom to be a single electron interacting with a fixed Coulomb potential (instead of a dynamic nucleus). The unperturbed—or zeroth-order—Hamiltonian function is

$$H^{(0)} = \frac{p^2}{2m} - \frac{e^2}{r}, \tag{A.4}$$

where p, m, and e are the momentum, mass, and charge of the electron, and r is the distance between the electron and the center of the potential. The external electric

19. Dirac, *Principles* (1930), chs. VII and VIII.
20. Schweber, *QED*, eqs. 9.9.5 and 9.9.10.
21. Mehra and Milton, *Climbing the Mountain*, ch. 9; Schweber, "Schwinger's Green's Functions."
22. Schwinger, "Gauge Invariance and Vacuum Polarization," 664.
23. Schwinger, "TQF5," 615.
24. Sakurai, *Modern Quantum Mechanics*, ch. 5.
25. Kragh, *Quantum Generations*, ch. 11.

field is the perturbation. Its interaction with the atom is represented by the interaction part of the Hamiltonian, $H^{(I)} = \lambda V$, where λ is a small constant and V is the electric potential. The total Hamiltonian is:

$$H = H^{(0)} + H^{(I)} = \frac{p^2}{2m} - \frac{e^2}{r} + \lambda V. \tag{A.5}$$

The dynamics of the unperturbed system are given by the Schrödinger equation for a given atomic state, Ψ_n,

$$-i\hbar\frac{\partial}{\partial t}\Psi^{(0)} = H^{(0)}\Psi_n^{(0)} = E_n^{(0)}\Psi_n^{(0)}, \tag{A.6}$$

which is an "eigenvalue" problem; here the eigenvalue, $E_n^{(0)}$, is the energy of the nth atomic state in the unperturbed system. The bounds of applicability of the perturbation analysis can be specified in terms of the unperturbed levels, say levels 1 and 2:

$$\lambda V_{12} \ll |E_2^{(0)} - E_1^{(0)}|, \tag{A.7}$$

which means that the magnitude of the perturbation should be much less than the difference of the two energies. If λV were large, the potential might eject the electron from the atom, drastically changing the physical situation, and invalidating the perturbation analysis.

The goal of a perturbation analysis is now to solve the eigenvalue problem for the Hamiltonian with perturbation and the more realistic state, $H\Psi^{(\lambda)}$ or

$$(H^{(0)} + \lambda V)\Psi_n^{(\lambda)} = E_n^{(\lambda)}\Psi_n^{(\lambda)}. \tag{A.8}$$

The quantities of interest are the eigenfunctions or wavefunctions, $\Psi_n^{(\lambda)}$, and the difference between the original and perturbed eigenvalues, $\Delta_n = E_n^{(\lambda)} - E_n^{(0)}$. ($\Delta_n$ is the analogue to the Lamb shift, chapter 4.) Plugging in Δ_n and rearranging terms gives

$$(E_n^{(0)} - H^{(0)})\Psi_n^{(\lambda)} = (\lambda V - \Delta_n)\Psi_n^{(\lambda)}. \tag{A.9}$$

This equation can be solved for Ψ_n by inverting the operator on the left-hand side and using the superposition principle to add an unperturbed solution:[26]

$$\Psi_n^{(\lambda)} = \Psi_n^{(0)} + \frac{1}{E_n^{(0)} - H^{(0)}}(\lambda V - \Delta_n)\Psi_n^{(\lambda)}. \tag{A.10}$$

The divergent integrals in Quantum Electrodynamics are due to integrating expressions such as equation (A.10) over all values of E_n, resulting in dividing by zero. To find an expression for Δ_n, note that from quantum theory,

$$\Psi_n^{(0)*}(\lambda V - \Delta_n)\Psi_n^{(\lambda)} = 0,$$

---

26. I suppress some subtleties. See Sakurai, *Modern Quantum Mechanics*, eqs. 5.1.21–35; cf. Schrödinger, *Collected Papers of Wave Mechanics*, 68, eq. 14.

and so

$$\Delta_n = \lambda \Psi_n^{(0)*} V \Psi_n^{(\lambda)}. \tag{A.11}$$

Equations (A.10) and (A.11) are mathematically exact. This discussion is approximate in the sense that the physical picture underlying it is an idealization, but no quantitative approximation procedures have been applied. This is the next step.

To find values for $\Psi_n^{(\lambda)}$ and Δ_n in equations (A.10) and (A.11), these functions are expanded in a series (see eq. (A.1)):

$$\Psi_n^{(\lambda)} = \lambda^0 \Psi_n^{(0)} + \lambda^1 \Psi_n^{(1)} + \lambda^2 \Psi_n^{(2)} + \cdots, \tag{A.12}$$

$$\Delta_n = \lambda^0 0 + \lambda^1 \Delta_n^{(1)} + \lambda^2 \Delta_n^{(2)} + \cdots, \tag{A.13}$$

where the exponent in parentheses indicates the first-, second-, etc., order correction. Plugging equation (A.13) into (A.11) gives:

$$\begin{aligned} \mathcal{O}(\lambda^1): \quad & \Delta_n^{(1)} = \Psi_n^{(0)*} V \Psi_n^{(0)} \\ \mathcal{O}(\lambda^2): \quad & \Delta_n^{(2)} = \Psi_n^{(0)*} V \Psi_n^{(1)} \\ \vdots \quad & \vdots \\ \mathcal{O}(\lambda^N): \quad & \Delta_n^{(N)} = \Psi_n^{(0)*} V \Psi_n^{(N-1)} \\ \vdots \quad & \vdots \end{aligned} \tag{A.14}$$

To find the states, $\Psi_n^{(\lambda)}$, substitute equation (A.12) into (A.10). The first order term is

$$\mathcal{O}(\lambda^1): \quad \Psi_n^{(1)} = \frac{1}{E_n^{(0)} - H^{(0)}} V \Psi_n^{(0)}. \tag{A.15}$$

This expression for $\Psi_n^{(1)}$ in terms of the known $\Psi_n^{(0)}$ is analogous to Victor Weisskopf's expression for his first-order wavefunction c_1 in terms of the vacuum state, c_0 (figure 3.5). It can be used to find $\Delta_n^{(2)}$ from equation (A.14),

$$\Delta_n^{(2)} = \Psi_n^{(0)*} V \frac{1}{E_n^{(0)} - H^{(0)}} V \Psi_n^{(0)}. \tag{A.16}$$

Continuing the pattern allows the next order of $\Psi_n^{(\lambda)}$ to be expressed in term of the unperturbed state as well:

$$\begin{aligned} \mathcal{O}(\lambda^2): \Psi_n^{(2)} = & \frac{1}{E_n^{(0)} - H^{(0)}} V \frac{1}{E_n^{(0)} - H^{(0)}} V \Psi_n^{(0)} \\ & - \frac{1}{E_n^{(0)} - H^{(0)}} \Psi_n^{(0)*} V \Psi_n^{(0)} \frac{1}{E_n^{(0)} - H^{(0)}} V \Psi_n^{(0)}. \end{aligned} \tag{A.17}$$

This procedure can be iterated as necessary—working back and forth between the expansions for $\Psi_n^{(\lambda)}$ and $\Delta_n^{(\lambda)}$—to find expressions for the perturbed system *in terms of the unperturbed system*, $\Psi_n^{(0)}$. Equations that express the complicated, perturbed, system in terms of the simple, known, unperturbed system, such as (A.17), are the desired result of perturbation theory.[27]

One further example is instructive: a physical system with only two possible states, or energy levels, subject to the same electric field perturbation as above. The expansion for the perturbed states, equation (A.12), can be written,[28]

$$
\Psi_n^{(\lambda)} = \lambda^0 \Psi_n^{(0)}
$$
$$
+ \lambda^1 \sum_{k \neq n} \Psi_k^{(0)} \frac{V_{kn}}{E_n^{(0)} - E_k^{(0)}}
$$
$$
+ \lambda^2 \left(\sum_{k \neq n} \sum_{l \neq n} \Psi_k^{(0)} \frac{V_{kl} V_{ln}}{(E_n^{(0)} - E_k^{(0)})(E_n^{(0)} - E_l^{(0)})} - \sum_{k-n} \Psi_k^{(0)} \frac{V_{nn} V_{kn}}{(E_n^{(0)} - E_k^{(0)})^2} \right)
$$
$$
+ \cdots.
$$

(A.18)

This expression has two interesting features. First, why specify in the summations that, for example, $k \neq n$ in the second line of equation (A.18)? If $k = n$, then

$$
\frac{1}{E_n^{(0)} - E_k^{(0)}} = \frac{1}{E_n^{(0)} - E_n^{(0)}} = \frac{1}{0},
$$

which is infinity (or, nonsense).[29] Removing the $k = n$ terms is analogous to removing the self-energy of the electron in QED. The second interesting feature of equation (A.18) is that the perturbation causes a *mixture* of the contribution of different energy levels. Say the state of interest is the first excited state, such as the $2S^2$ state of hydrogen in the Lamb shift. The expression for $\Psi_2^{(\lambda)}$ will involve the energy levels of the unperturbed state, E_2^0, but will also involve the other level, E_1^0. This sort of mixing was one reason why Heisenberg's perturbation analysis of "hole" theory resulted in new physical situations, such as the appearance of a light quantum and an electron-positron pair (section 3.4).

27. To give a sense of the correspondence between the simple example here and physicists' actual calculations, compare Hans Bethe's notes at the 1947 Shelter Island conference, under the heading "Weisskopf," which are available from the Cornell University Archives, https://ecommons.cornell.edu/handle/1813/53358, accessed September 20, 2020. In contrast to the current example of expansion of the energy level Δ and wavefunction Ψ, Weisskopf expanded in the Hamiltonian, H, and wavefunction, ψ. See also section 4.3.

28. Sakurai, *Modern Quantum Mechanics*, eq. 5.1.44.

29. See Dirac, "On the Annihilation of Electrons and Protons," eq. 20, 368.

Appendix B
Fluctuations

B.1 Mean-squared Fluctuations

In chapters 3 and 4, I described physicists Victor Weisskopf and Julian Schwinger as repeatedly *referring* to fluctuations, but not *calculating* fluctuations. This appendix frames fluctuation calculations as a theoretical tool; discusses Einstein's paradigmatic fluctuation analysis; and then discusses the example of vacuum fluctuations in QED.

The calculation of statistical fluctuations originated in the nineteenth century, in the theory of errors of observation. In reference to physical systems, early fluctuation calculations—for example, by J. W. Gibbs—focused on the fact that they were unimportant or imperceptible.[1] This orientation continued into the twentieth century. For example, Ralph Fowler expressed the usefulness of fluctuation calculations in his monograph, *Statistical Mechanics* (1929). After outlining the foundations of this field, and their logical and conceptual limits, Fowler remarked:

> One further step can be made, which, while it does not fill the logical gaps indicated above, yet goes a long way towards giving us confidence in our conclusions [...].
> This step is the calculation of *fluctuations*.

The center of Fowler's worries was that *mathematical averages* in the formalism of statistical mechanics were only *assumed* to correspond to the *actual average* states of physical systems. Fluctuation analysis showed that deviations from the average were "insignificant" for systems with large numbers of molecules. "This very greatly consolidates the whole theory."[2] Fowler applied fluctuation calculations to a *theoretical* puzzle, the lack of logical and physical sense in a theory. Like Gibbs, he assumed that fluctuations were imperceptible to human senses. Fluctuations were important because of their unimportance.

Fluctuations were given a more *positive* role in theoretical physics by Albert Einstein, most strikingly in his 1905 theory of Brownian motion. He argued that the random motion of motes of dust was not caused by collisions with individual molecules of air; rather millions of collisions from all directions occurred within the time of an observation. On average, the collisions canceled one another out. Einstein inverted Gibbs's attitude. To Einstein, observed Brownian motions were actually fluctuations from this average. To calculate the velocity fluctuations of a dust mote, or other Brownian particle, Einstein considered the average, $\langle \cdot \rangle$, of the velocity, v, imparted by a large number of collisions, N: $\langle v \rangle = (v_1 + v_2 + \cdots v_N)/N$. The amount an individual measurement, v, differed from the average was $\langle v \rangle - v$, which was

---

1. Brush, "Development of Kinetic Theory"; Porter, "Statistics and Physical Theories."
2. Fowler, *Statistical Mechanics*, 10, cf. 497.

More than Nothing. Aaron Sidney Wright, Oxford University Press. © Oxford University Press 2024.
DOI: 10.1093/oso/9780190062804.003.000B

written Δv. If the coordinate system was set so that the average was zero, these differences could be negative. To avoid that inconvenience, physicists worked with the differences squared, $(\Delta v)^2$. Finally, because physicists were not interested in the amount of difference of individual measurements, but rather in the regular behavior of the system over many measurements, they calculated the average of the squared differences: $\langle (\Delta v)^2 \rangle$.[3] (The square root of $\langle (\Delta v)^2 \rangle$ is often called the standard deviation.) This was the mean-square deviation referred to by Niels Bohr and Leon Rosenfeld (section 4.2.2), among others. By a theorem from the theory of probability, if the values of a general quantity $\mathcal{O}$ were distributed in an appropriate way, the following relation held:[4]

$$\langle (\Delta \mathcal{O})^2 \rangle = \langle \mathcal{O}^2 \rangle - \langle \mathcal{O} \rangle^2 . \tag{B.1}$$

Meaning, the fluctuations were equal to the average of the squared values of $\mathcal{O}$ minus the average values of $\mathcal{O}$ squared. This analysis was applied to a wide variety of physical systems, from atomic to galactic, and for a wide range of quantities.[5]

In 1909, Einstein applied his Brownian fluctuation formalism to Max Planck's formula for the "spectral intensity" of blackbody radiation (i.e., the brightness of each color in the spectrum of the radiated light). Einstein considered a small box with volume V, within a blackbody cavity. He calculated the fluctuations of energy, E, within a narrow band of frequencies, dv, in the box with the form:

$$\langle (\Delta E)^2 \rangle = hvE + \frac{c^3}{8\pi V v^2 dv} E^2 . \tag{B.2}$$

Einstein famously interpreted the two terms in this equation as arising from the dual particle and wave nature of radiation, respectively. Einstein's result was re-derived by Pascual Jordan in 1925, and set the stage for the incorporation of fluctuations into quantum mechanics.[6]

Unlike fluctuations of Brownian particles, fluctuations of blackbody radiation were not measured by experiment.[7] Experimentally, other errors were always far too prominent to allow radiation fluctuations to be measured.[8] In 1904, Einstein argued that radiation fluctuations were observable *in principle*, and proposed an outlandish situation: a blackbody cavity with dimensions of the maximum wavelength, λ_{max}, of the radiation itself.[9] At 50 Kelvin /–223 Celsius, λ_{max} is 58μm (half the width of a human hair). As late as 1914, the experimental situation was that Planck's equation worked better than its rivals, but was not itself confirmed.[10] I do not know of any

3. Martin Klein, "Fluctuations"; Kuhn, *Black-Body Theory*, 187; Renn, "Einstein's Invention of Brownian Motion"; Uffink, "Insuperable Difficulties."
4. I.e., for particularly well-behaved probability distributions such as the Gaussian distribution.
5. Yeang, "Two Mathematical Approaches."
6. Pais, *Subtle Is the Lord*, 404; Staley, *Einstein's Generation*, 389–95.
7. Kojevnikov, "Einstein's Fluctuation Formula," 189.
8. Kangro, *Planck's Radiation Law*, esp. ch. 8; Nye, *Before Big Science*, 116.
9. Kuhn, *Black-Body Theory*, 179; Uffink, "Insuperable Difficulties," 54; in addition to its outlandishness, the argument was also incorrect, Pais, *Subtle Is the Lord*, 67–8.
10. Coblentz, "Black Body," 4; Kuhn, *Black-Body Theory*, 187–96; cf. editorial essay, Einstein, *CPAE 2*, §V.

direct experimental test of Einstein's formula for fluctuations in a blackbody cavity.[11] This was likely due to the infinitesimal magnitude of the fluctuations for realistic cavities, $\langle (\Delta E)^2 \rangle \sim 10^{-19}$. A measurement of these fluctuations would require a precision 10^{17} times better than our metrological standard for blackbody radiation measurements.[12]

According to Martin Klein, "Once Einstein had seen the power of the fluctuation theory [as applied to Brownian motion], he made it one of his principal instruments for probing into the unknown." In radiation theory, fluctuations provided Einstein with "devices for thinking" with which he could "explore the nature of radiation."[13] Both mean-square fluctuations and perturbation theory (appendix A) were theoretical technologies or paper tools. Unlike perturbation theory, fluctuations were not an engine for connecting quantum theories to experiments. Mean-square fluctuations, applied to radiation, were a theoretical tool *for theory's sake*.

B.2 Relativistic Fluctuations

In chapter 4, I noted that in their landmark papers on the self-energy of the electron and vacuum polarization, Victor Weisskopf and Julian Schwinger did not apply a fluctuation analysis to the quantities they discussed. What I meant was that, in their published papers and in the archival records that I consulted, they did not apply a mean-squared fluctuation calculation, eq. (B.1). (The mean-squared calculation defined "fluctuation" for most physicists.[14]) Were Schwinger to have made the calculation of fluctuations of the current, j_μ, he would have formed an expression such as:

$$\langle (\Delta j_\mu)^2 \rangle = \langle j_\mu^2 \rangle - \langle j_\mu \rangle^2 . \tag{B.3}$$

The second term is zero by equation (4.6). Recall that j_μ was composed of pairs of field operators, ψ. Then j_μ^2 would be composed of quadratic terms, $\psi\psi\psi\psi$. There were several ways of parsing this expression. Schwinger actually did calculate one such term because it appeared in the calculation of the self-energy of the electron.[15] In general, this was a divergent function of the momentum of the particles. But fluctuation

11. In the 1950s, deviations from expectations in photon statistics reshaped our understanding of light: Silva and Freire, "Concept of the Photon"; Bromberg, "Explaining the Laser's Light."

12. Consider a gold blackbody cavity. The temperature is 1337.334 K, and wavelength of light 632.9nm or frequency 473.7×10^{12}Hz (a reddish orange). According to Planck's formula, the spectral radiance, E, is 0.0649×10^{-12}Watts per element of volume per Hz. Applying formula (B.2) gives $\langle (\Delta E)^2 \rangle \sim 4 \times 10^{-32}$. The ratio of fluctuations to the radiance is approximately 6×10^{-19}. The gold blackbody facilitates a comparison with blackbody measurements by the US National Institutes of Standards and Technology. Until recently, these measurements defined the SI system of temperature. The uncertainty of NIST's measurement of the spectral radiance of 632.9nm light was 9×10^{-2}%. Mielenz, Saunders, and Shumaker, "Freezing Temperature of Gold," table 7.

13. Martin Klein, "Fluctuations," 52–3; cf. "Jordan submitted his new formalism to the test of fluctuations," Darrigol, "Quantized Matter Waves," 231.

14. See, Fowler, *Statistical Mechanics*, ch. XX; Chandrasekhar, "Stochastic Problems in Physics and Astronomy"; Yeang, "Two Mathematical Approaches."

15. Schwinger, "QED2," 671–73.

problems were posed in terms of narrow bands of frequencies, and so it would have been possible to find an expression for these fluctuations—as long as the frequency band did not include the singular point. That singular point was zero frequency—the vacuum. In a follow-up publication, Schwinger calculated a similar four-field function. Perhaps characteristically, he argued that it was "strictly divergent," but that its value could be "unambiguously" found to be zero.[16] Simply following the path of previous QED fluctuation calculations would lead to zero fluctuations or infinite nonsense.[17] In fact, there was no accepted framework for doing *relativistic* statistical mechanics (quantum or not) until the 1960s.[18]

However, several attempts at calculations of relativistic quantum fluctuation phenomena were made by others. Before World War I, Einstein had calculated both fluctuations in the displacement of a Brownian particle, and also the statistical character of diffusion phenomena, such as dye spreading in water. This work was developed by Paul Langevin, Adriaan Fokker, Max Planck, and others.[19] In the early 1930s, Erwin Schrödinger took up the challenges facing the integration of relativity and quantum theory—in particular, studying the effects of the energy-time uncertainty relation on Lorentz transformations. As part of this underappreciated work, Schrödinger noted the formal identity of Fokker's diffusion equation and his eponymous equation. Perhaps the formalism of random motions could inform a relativistic account of uncertainty? It was not to be. Schrödinger's stumbling block was that in quantum theory the wave field, ψ, was a complex quantity. It could not be represented as an observable field strength.[20] Hendrik Kramers made fluctuation calculations in his relativistic precursor to renormalized QED, though he did not offer a strong physical interpretation.[21] From the middle 1950s, there was a surge of interest in the application of the tools of renormalized QED to the non-relativistic physics of solid matter (and which reciprocally influenced the Standard Model of particle physics). Both Schwinger and Feynman contributed to this work (with their students), but neither drew from it ontological consequences for relativistic QED.[22]

The strongest effort to connect fluctuations and relativistic QED was made by Theodore Welton during his work with Weisskopf at MIT. In 1948, he published "Some Observable Effects of the Quantum-Mechanical Fluctuations of the Electromagnetic Field." Like Weisskopf, Welton postulated the existence of electromagnetic fluctuations in the vacuum (but not charge fluctuations). Unlike Weisskopf, Welton calculated the effects of these fluctuations on the position, q, of the electron with a non-relativistic mean-squared formula, $\langle (\Delta q)^2 \rangle$. As in Weisskopf's 1939 "hole" theory analysis, Welton's electromagnetic fluctuations spread the density of the charge of the electron over a finite region (figure 4.3). With this analysis, Welton reproduced

16. Schwinger, "QED3," 815.
17. See Bjorken and Drell, *Relativistic Quantum Fields*, 35–37, 93.
18. Haar, "Foundations of Statistical Mechanics," §D; Hakim, "Relativistic Statistical Mechanics. I."
19. Maiocchi, "Brownian Motion"; Yeang, "Tubes, Randomness, and Brownian Motions"; Yeang, "Two Mathematical Approaches."
20. Schrödinger, "Théorie relativiste de l'électron," 296–306; Kragh, "Unifying Quanta and Relativity?," cf. Bitbol, *Schrödinger's Philosophy*, ch. 3.
21. Institut Solvay, "Correspondence Principle," 253; Dresden, *H. A. Kramers*, 353–54, cf. 375.
22. Martin and Schwinger, "Many-Particle Systems"; Schwinger, "Brownian"; Feynman and Vernon Jr., "Theory of a General Quantum System"; Mehra and Milton, *Climbing the Mountain*, 332–34.

Hans Bethe's non-relativistic calculation of the Lamb shift.[23] Weisskopf presented Welton's calculation in his influential 1949 review article.[24] And Willis Lamb presented Welton's "qualitative explanation" for the energy shift as a prelude to discussing Bethe's calculation and the relativistic results that followed.[25] In his influential lectures at Cornell, Freeman Dyson accepted Welton's fluctuations as a description of the physical origin of the Lamb shift.[26] In this way, a non-relativistic calculation became an accepted physical interpretation of an (otherwise) relativistic QED.

23. Welton, "Quantum-mechanical Fluctuations," §§ II and III; Schweber, *QED*, 336; cf. an idiosyncratic attempt to apply Nyquist's circuit noise: Callen and Welton, "Irreversibility and Generalized Noise."
24. Weisskopf, "Theory of the Electron," 313.
25. Lamb, "Anomalous Fine Structure," §4.4.
26. Dyson, *Advanced Quantum Mechanics*, II–80; D. Kaiser, *Drawing Theories Apart*, 188–95.

Bibliography

Aaserud, Finn. *Redirecting Science: Niels Bohr, Philanthropy, and the Rise of Nuclear Physics*. Cambridge: Cambridge University Press, 1990.

Achieser, Aleksander, and Isaak Pomeranchuk. "Kohärente Streuung von γ-strahlen an Kernen." *Physikalische Zeitschrift der Sowietunion* 11, no. 5 (1937): 478–497.

Aitchison, Ian J. R. *An Informal Introduction to Gauge Field Theories*. Cambridge: Cambridge University Press, 1982.

Alexander, Amir. *Duel at Dawn: Heroes, Martyrs, and the Rise of Modern Mathematics*. Cambridge, MA: Harvard University Press, 2010.

Alpers, Svetlana. *The Art of Describing: Dutch Art in the Seventeenth Century*. Chicago: University of Chicago Press, 1983.

Ames, Jr., Adelbert. "Visual Perception and the Rotating Trapezoidal Window." *Psychological Monographs: General and Applied* 65, no. 7 (1951): i–32.

Anderson, Carl D. "The Apparent Existence of Easily Deflectable Positives." *Science* 76, no. 1967 (1932): 238–239.

Arabatzis, Theodore. "Hidden Entities and Experimental Practice: Renewing the Dialogue between History and Philosophy of Science." In *Integrating History and Philosophy of Science: Problems and Prospects*, edited by Seymour Mauskopf and Tad Schmaltz, 125–139. Dordrecht: Springer Netherlands, 2012.

Arabatzis, Theodore. *Representing Electrons: A Biographical Approach to Theoretical Entities*. Chicago: University of Chicago Press, 2006.

Arnowitt, Richard, Stanley Deser, and Charles W. Misner. "Dynamical Structure and Definition of Energy in General Relativity." *PR* 116, no. 5 (1959): 1322–1330.

Arnowitt, Richard, Stanley Deser, and Charles W. Misner. "The Dynamics of General Relativity." In *Gravitation: An Introduction to Current Research*, edited by Louis Witten, 227–264. 1962. Reprint, New York: Wiley, 2004.

Ash, Mitchell G., and Alfons Söllner. *Forced Migration and Scientific Change: Emigré German-speaking Scientists and Scholars after 1933*. Washington: German Historical Institute, 1996.

Bacciagaluppi, Guido, Elise Crull, and Owen J. E. Maroney. "Jordan's Derivation of Blackbody Fluctuations." *SHPMP* 60 (2017): 23–34.

Bacciagaluppi, Guido, and Antony Valentini. *Quantum Theory at the Crossroads: Reconsidering the 1927 Solvay Conference*. Cambridge: Cambridge University Press, 2009.

Badino, Massimiliano, and Jaume Navarro, eds. *Research and Pedagogy: A History of Quantum Physics through Its Textbooks*. Berlin: Edition Open Access, Max Planck Research Library, 2013.

Baker, Marshall, and Sheldon L. Glashow. "Spontaneous Breakdown of Elementary Particle Symmetries." *PR* 128, no. 5 (1962): 2462–2471.

Baldwin, Melinda. "'Keeping in the Race': Physics, Publication Speed and National Publishing Strategies in *Nature*, 1895–1939." *BJHS* 47, no. 2 (2014): 257–279.

Barad, Karen. *Meeting the Universe Halfway: Quantum Physics and the Entanglement of Matter and Meaning*. Durham, NC: Duke University Press, 2007.

Barany, Michael J., and Donald MacKenzie. "Chalk: Materials and Concepts in Mathematics Research." In *Representation in Scientific Practice Revisited*, edited by Catelijne Coopmans, Janet Vertesi, Michael Lynch, and Steve Woolgar, 107–129. Cambridge, MA: MIT Press, 2014.

Barbour, Julian B. "Einstein and Mach's Principle." In Janssen, Norton, Renn, Sauer, and Stachel, *Genesis of General Relativity*, 3:569–604.

The Barchas Collection at Stanford University: A Catalogue of the Samuel I. and Cecile M. Barchas Collection in the History of Science and Ideas. Stanford, CA: Stanford University Libraries, 1999.

Barkas, Walter H., and Arthur Hinton Rosenfeld. *Data for Elementary-Particle Physics*. Technical report UCRL-8030 Rev. Berkeley: University of California Radiation Laboratory, 1961.

Barrow-Green, June, and Jeremy Gray. "Geometry at Cambridge, 1863–1940." *Historia Mathematica* 33, no. 3 (2006): 315–356.

Barthes, Roland. "The Reality Effect." In *The Rustle of Language*, translated by Richard Howard, 141–148. 1967. University of California Press, 1986.

Beck, Guido. "Hat das negative Energiespektrum einen Einfluß auf Kernphänomene?" *ZP* 83, no. 7 (1933): 498–511.

Bederson, Benjamin, and H. Henry Stroke. "History of the New York University Physics Department." *Physics in Perspective* 13 (2011): 260–328.

Behrens, Roy R. "The Artistic and Scientific Collaboration of Blanche Ames Ames and Adelbert Ames II." *Leonardo* 31, no. 1 (1998): 47–54.

Behrens, Roy R. "The Life and Unusual Ideas of Adelbert Ames, Jr." *Leonardo* 20, no. 3 (1987): 273–279.

Bell, John S. "On Wave Packet Reduction in the Coleman-Hepp Model." In *Speakable and Unspeakable in Quantum Mechanics: Collected Papers on Quantum Philosophy*, 45–51. 1975. Cambridge: Cambridge University Press, 1987.

Beller, Mara. "'Against the Stream'—Schrödinger's Interpretation of Quantum Mechanics." *SHPMP* 28, no. 3 (1997): 421–432.

Beller, Mara. *Quantum Dialogue: The Making of a Revolution*. Chicago: University of Chicago Press, 1999.

Beller, Mara. "The Birth of Bohr's Complementarity: The Context and the Dialogues." *Studies in History and Philosophy of Science Part A* 23, no. 1 (1992): 147–180.

Benczer-Koller, Noémie. "Chien-Shiung Wu." In Byers and Williams, *Out of the Shadows*, 272–281.

Benford, Gregory. "Sid the Critic, Sid the Scientist." *eI* 7, no. 1 (2008). http://efanzines.com/EK/eI36/.

Benford, Gregory. *Timescape*. New York: Simon & Schuster, 1980.

Bergmann, Peter G. "Fifty Years of Relativity." *Science* 123, no. 3195 (1956): 486–494.

Bergmann, Peter G. "Summary of the Chapel Hill Conference." *RMP* 29, no. 3 (1957): 352–354.

Bergmann, Peter G. "The Sandwich Conjecture." In Carmeli, Fickler, and Witten, *Relativity*, 43–53.

Bethe, Hans A. "Nuclear Many-Body Problem." *PR* 103, no. 5 (1956): 1353–1390.

Bethe, Hans A. "On the Annihilation Radiation of Positrons." *PRSLA* 150, no. 869 (1935): 129–141.

Bethe, Hans A. "The Electromagnetic Shift of Energy Levels." *PR* 72, no. 4 (1947): 339–341.

Bethe, Hans A., and Walter Heitler. "On the Stopping of Fast Particles and on the Creation of Positive Electrons." *PRSLA* 146, no. 856 (1934): 83–112.

Bethe, Hans A., and J. Robert Oppenheimer. "Reaction of Radiation on Electron Scattering and Heitler's Theory of Radiation Damping." *PR* 70, nos. 7–8 (1946): 451–458.

Bethe, Hans A., and Rudolf Peierls. *The Bethe–Peierls Correspondence.* Edited by Sabine Lee. Singapore: World Scientific, 2008.

Bethe, Hans A., and Edwin E. Salpeter. *Quantum Mechanics of One- and Two-electron Atoms.* New York: Academic Press, 1957.

Bhabha, Homi J. "The Creation of Electron Pairs by Fast Charged Particles." *PRSLA* 152, no. 877 (1935): 559–586.

Biagioli, Mario. *Galileo, Courtier: The Practice of Science in the Culture of Absolutism.* Chicago: University of Chicago Press, 1994.

Biagioli, Mario. *Galileo's Instruments of Credit: Telescopes, Images, Secrecy.* Chicago: Chicago University Press, 2006.

Birrell, N. D., and Paul C. W. Davies. *Quantum Fields in Curved Space.* Cambridge: Cambridge University Press, 1984.

Bitbol, Michel. *Schrödinger's Philosophy of Quantum Mechanics.* Dordrecht: Kluwer Academic, 1996.

Bjorken, J. D. "A Dynamical Origin for the Electromagnetic Field." *Annals of Physics* 24 (1963): 174–187.

Bjorken, J. D., and Sidney Drell. *Relativistic Quantum Fields.* New York: McGraw-Hill, 1965.

Blackett, Patrick M. S., and G. P. S. Occhialini. "Some Photographs of the Tracks of Penetrating Radiation." *PRSLA* 139, no. 839 (1933): 699–726.

Bloch, Felix. "Zur Bremsung rasch bewegter Teilchen beim durchgang durch Materie." *AP* 408, no. 3 (1933): 285–320.

Blum, Alexander, and Thiago Hartz. "The 1957 Quantum Gravity Meeting in Copenhagen: An Analysis of Bryce S. Dewitt's Report." *European Physical Journal H* 42 (2017): 107–157.

Blum, Alexander, and Christian Joas. "From Dressed Electrons to Quasiparticles: The Emergence of Emergent Entities in Quantum Field Theory." *SHPMP* 53 (2016): 1–8.

Blum, Alexander, Roberto Lalli, and Jürgen Renn. "The Reinvention of General Relativity: A Historiographical Framework for Assessing One Hundred Years of Curved Space-time." *Isis* 106, no. 3 (2015): 598–620.

Bohr, Niels. "Field and Charge Measurements in Quantum Theory." In *Foundations of Quantum Physics II (1933–1958).* Vol. 7 of *Niels Bohr Collected Works,* edited by Jørgen Kalckar, 195–209. 1937. Elsevier, 1996.

Bohr, Niels. "On the Application of the Quantum Theory to Atomic Structure: Part I: The Fundamental Postulates." In *The Correspondence Principle (1918-1923)*. Vol. 3 of *Niels Bohr Collected Works*, edited by Leon Rosenfeld and J. Rud Nielsen, translated by Leon Francis Curtiss, 456–500. 1923. Amsterdam: Elsevier, 1976.

Bohr, Niels. "Problems of Elementary-particle Physics." In *Foundations of Quantum Physics II (1933-1958)*. Vol. 7 of *Niels Bohr Collected Works*, edited by Jørgen Kalckar, 217–222. 1947. Elsevier, 1996.

Bohr, Niels. "The Quantum Postulate and the Recent Development of Atomic Theory." *Nature* 121 supp. (1928): 580–590.

Bohr, Niels, Hendrik A. Kramers, and John C. Slater. "The Quantum Theory of Radiation." In Waerden, *Sources of Quantum Mechanics*, 159–176.

Bohr, Niels, and Léon Rosenfeld. "On the Question of the Measurability of Electromagnetic Field Quantities." In *Foundations of Quantum Physics II (1933-1958)*. Vol. 7 of *Niels Bohr Collected Works*, edited by Jørgen Kalckar, 55–121. 1933. Elsevier, 1996.

Bokulich, Alisa. "Open or Closed? Dirac, Heisenberg, and the Relation between Classical and Quantum Mechanics." *SHPMP* 35, no. 3 (2004): 377–396.

Bonolis, Luisa. "Walther Bothe and Bruno Rossi: The Birth and Development of Coincidence Methods in Cosmic-ray Physics." *American Journal of Physics* 79, no. 11 (2011): 1133–1150.

Born, Max, Werner Heisenberg, and Pascual Jordan. "On Quantum Mechanics. II." In Waerden, *Sources of Quantum Mechanics*, 321–386.

Born, Max, and Norbert Wiener. "A New Formulation of the Laws of Quantization of Periodic and Aperiodic Phenomena." *Journal of Mathematics and Physics* 5, nos. 1–4 (1926): 84–98.

Borrelli, Arianna. "The Case of the Composite Higgs: The Model as a 'Rosetta Stone' in Contemporary High-Energy Physics." *SHPMP* 43, no. 3 (2012): 195–214.

Borrelli, Arianna. "The Emergence of Selection Rules and Their Encounter with Group Theory, 1913–1927." *SHPMP* 40, no. 4 (2009): 327–337.

Breit, Gregory. "An Interpretation of Dirac's Theory of the Electron." *PNAS* 14, no. 7 (1928): 553–559.

Breit, Gregory. "The Effect of Retardation on the Interaction of Two Electrons." *PR* 34, no. 4 (1929): 553–573.

Breit, Gregory, and John Archibald Wheeler. "Collision of Two Light Quanta." *PR* 46, no. 12 (1934): 1087–1091.

Bridgman, Percy W. *Dimensional Analysis*. New Haven, CT: Yale University Press, 1922.

Bridgman, Percy W. *The Logic of Modern Physics*. New York: Macmillan, 1927.

Bridgman, Percy W. "The Nature of Some of Our Physical Concepts. I." *BJPS* 1, no. 4 (1951): 257–272.

Brill, Dieter R., and Alexander Blum. "The Revival of General Relativity at Princeton: Daring Conservatism." *EPJ Web of Conferences* 168 (2018): 01013. doi:10.1051/epjconf/201816801013.

Bromberg, Joan Lisa. "Explaining the Laser's Light: Classical versus Quantum Electrodynamics in the 1960s." *AHES* 70, no. 3 (2016): 243–266.

Bromberg, Joan Lisa. *Fusion: Science, Politics, and the Invention of a New Energy Source*. Cambridge, MA: MIT Press, 1982.

Bromberg, Joan Lisa. "The Concept of Particle Creation before and after Quantum Mechanics." *HSPS* 7 (1976): 161–191.

Bromberg, Joan Lisa. "The Impact of the Neutron: Bohr and Heisenberg." *HSPS* 3 (1971): 307–341.

Bromberg, Joan Lisa. *The Laser in America, 1950–1970.* Cambridge, MA: MIT Press, 1991.

Brown, Laurie M., ed. *Renormalization: From Lorentz to Landau (and Beyond).* New York: Springer, 1993.

Brown, Laurie M. "The Compton Effect as One Path to QED." *SHPMP* 33, no. 2 (2002): 211–249.

Brown, Laurie M. "Tyger Hunting in the Everglades." *PT* 17, no. 4 (1964): 36.

Brown, Laurie M., Robert Brout, Tian Yu Cao, Peter Higgs, and Yoichiro Nambu. "Panel Session: Spontaneous Breaking of Symmetry." In Hoddeson, Brown, Riordan, and Dresden, *Rise of the Standard Model*, 478–522.

Brown, Laurie M., and Tian Yu Cao. "Spontaneous Breakdown of Symmetry: Its Rediscovery and Integration into Quantum Field Theory." *HSPBS* 21, no. 2 (1991): 211–235.

Brown, Laurie M., Max Dresden, and Lillian Hoddeson, eds. *Pions to Quarks: Particle physics in the 1950s.* Cambridge: Cambridge University Press, 1989.

Brown, Laurie M., Max Dresden, and Lillian Hoddeson. "Pions to Quarks: Particle Physics in the 1950s." In Brown, Dresden, and Hoddeson, *Pions to Quarks*, 3–39.

Brown, Laurie M., and Lillian Hoddeson. "The Birth of Elementary Particle Physics: 1930–1950." In *The Birth of Particle Physics*, edited by Laurie M. Brown and Lillian Hoddeson, 3–36. Cambridge: Cambridge University Press, 1983.

Brown, Laurie M., and Donald F. Moyer. "Lady or Tiger?—the Meitner-Hupfeld Effect and Heisenberg's Neutron Theory." *American Journal of Physics* 52, no. 2 (1984): 130–136.

Brown, Laurie M., and Helmut Rechenberg. "Paul Dirac and Werner Heisenberg—a Partnership in Science." In *Reminiscences about a Great Physicist: Paul Adrien Maurice Dirac*, edited by Behram N. Kursunoglu and Eugene Paul Wigner, 117–162. Cambridge: Cambridge University Press, 1987.

Brown, Laurie M., Michael Riordan, Max Dresden, and Lillian Hoddeson. "The Rise of the Standard Model: 1964–1979." In Hoddeson, Brown, Riordan, and Dresden, *Rise of the Standard Model*, 3–35.

Brush, Stephen G. "The Development of the Kinetic Theory of Gases." *AHES* 12, no. 1 (1974): 1–88.

Buchwald, Jed Z. "How the Ether Spawned the Microworld." In Daston, *Biographies of Scientific Objects*, 203–225.

Buchwald, Jed Z., ed. *Scientific Practice: Theories and Stories of Doing Physics.* Chicago: University of Chicago Press, 1995.

Buchwald, Jed Z. *The Creation of Scientific Effects: Heinrich Hertz and Electric Waves.* Chicago: University of Chicago Press, 1994.

Buchwald, Jed Z. "Why Hertz Was Right about Cathode Rays." In Buchwald, *Scientific Practice*, 151–169.

Buchwald, Jed Z., and Robert Fox, eds. *The Oxford Handbook of the History of Physics.* Oxford: Oxford University Press, 2013.

Buchwald, Jed Z., and Sungook Hong. "Physics: Its Methods, Practitioners, Boundaries." In *From Natural Philosophy to the Sciences: Writing the History of Nineteenth-Century Science*, edited by David Cahan, 163–195. Chicago: University of Chicago Press, 2003.

Buchwald, Jed Z., and Andrew Warwick, eds. *Histories of the Electron: The Birth of Microphysics*. Cambridge, MA: MIT Press, 2001.

Byers, Nina, and Gary Williams, eds. *Out of the Shadows: Contributions of Twentieth-century Women to Physics*. Cambridge: Cambridge University Press, 2006.

Cabibbo, Nicola. "Spontaneous Breakdown and the Weak Interaction Angle." In Zichichi, *Hadrons and Their Interactions*, 381–392.

Callan, Curtis G., and Sidney Coleman. "Fate of the False Vacuum. II. First Quantum Corrections." *PRD* 16, no. 6 (1977): 1762–1768.

Callan, Curtis G., Roger Dashen, and David J. Gross. "The Structure of the Gauge Theory Vacuum." *PL* B 63, no. 3 (1976): 334–340.

Callen, Herbert B., and Theodore A. Welton. "Irreversibility and Generalized Noise." *PR* 83, no. 1 (1951): 34–40.

Camilleri, Kristian. "Constructing the Myth of the Copenhagen Interpretation." *Perspectives on Science* 17, no. 1 (2009): 26–57.

Camilleri, Kristian. *Heisenberg and the Interpretation of Quantum Mechanics: The Physicist as Philosopher*. Cambridge: Cambridge University Press, 2009.

Canales, Jimena. *The Physicist and the Philosopher: Einstein, Bergson, and the Debate That Changed Our Understanding of Time*. Princeton, NJ: Princeton University Press, 2015.

Cantor, G. N., and M. J. S. Hodge, eds. *Conceptions of Ether: Studies in the History of Ether Theories, 1740–1900*. Cambridge: Cambridge University Press, 1981.

Cao, Tian Yu. *Conceptual Developments of 20th Century Field Theories*. Cambridge: Cambridge University Press, 1997.

Cao, Tian Yu. *From Current Algebra to Quantum Chromodynamics: A Case for Structural Realism*. Cambridge: Cambridge University Press, 2010.

Cao, Tian Yu. "The Englert-Brout-Higgs Mechanism—An Unfinished Project." *International Journal of Modern Physics A* 31, no. 35 (2016): 1630061.

Cao, Tian Yu, and S. S. Schweber. "The Conceptual Foundations and the Philosophical Aspects of Renormalization Theory." *Synthese* 97, no. 1 (1993): 33–108.

Carazza, Bruno, and Helge Kragh. "Heisenberg's Lattice World: The 1930 Theory Sketch." *American Journal of Physics* 63 (1995): 595–605.

Carmeli, Moshe, Stuart I. Fickler, and Louis Witten. "Preface." In Carmeli, Fickler, and Witten, *Relativity*, v–vi.

Carmeli, Moshe, Stuart I. Fickler, and Louis Witten, eds. *Relativity: Proceedings of the Relativity Conference in the Midwest, Held at Cincinnati, Ohio, June 2–6, 1969*. New York: Plenum, 1970.

Carroll, Sean M. *Spacetime and Geometry: An Introduction to General Relativity*. New York: Addison-Wesley, 2003.

Carson, Cathryn. *Heisenberg in the Atomic Age: Science and the Public Sphere*. Washington, DC: German Historical Institute, 2010.

Carson, Cathryn. "The Peculiar Notion of Exchange Forces—I: Origins in Quantum Mechanics, 1926–1928." *SHPMP* 27, no. 1 (1996): 23–45.

Carson, Cathryn. "The Peculiar Notion of Exchange Forces—II: From Nuclear Forces to QED, 1929–1950." *SHPMP* 27, no. 2 (1996): 99–131.

Carson, Cathryn, A. B. Kozhevnikov, and Helmuth Trischler, eds. *Weimar Culture and Quantum Mechanics: Selected Papers by Paul Forman and Contemporary Perspectives on the Forman Thesis*. London: Imperial College Press, 2011.

Carter, Brandon. "Black Hole Equilibrium States." Orig. pub. 1973, *GRG* 41, no. 12 (2009): 2873–2938.

Carter, Brandon. "Complete Analytic Extension of the Symmetry Axis of Kerr's Solution of Einstein's Equations." *PR* 141, no. 4 (1966): 1242–1247.

Cartwright, Nancy. *The Dappled World*. Cambridge: Cambridge University Press, 1999.

Cassidy, David C. "Cosmic Ray Showers, High Energy Physics, and Quantum Field Theories: Programmatic Interactions in the 1930s." *HSNS* 12, no. 1 (1981): 1–39.

Cassidy, David C. "Heisenberg's First Core Model of the Atom: The Formation of a Professional Style." *HSPS* 10 (1979): 187–224.

Cassidy, David C. *Uncertainty: The Life and Science of Werner Heisenberg*. New York: W. H. Freeman and Company, 1992.

Chakravartty, Anjan. "Scientific Realism." In *The Stanford Encyclopedia of Philosophy*, Spring 2014 ed., edited by Edward N. Zalta. 2014. http://plato.stanford.edu/archives/spr2014/entries/scientific-realism/.

Chandrasekhar, Subrahmanyan. "Stochastic Problems in Physics and Astronomy." *RMP* 15, no. 1 (1943): 1–89.

Chang, Hasok. *Is Water H_2O?: Evidence, Realism and Pluralism*. Dordrecht: Springer, 2012.

Chemla, Karine, and Evelyn Fox Keller, eds. *Cultures without Culturalism: The Making of Scientific Knowledge*. Durham, NC: Duke University Press, 2017.

Chevalley, Catherine. "Niels Bohr's Words and the Atlantis of Kantianism." In *Niels Bohr and Contemporary Philosophy*, edited by Jan Faye and Henry Folse, 33–55. Dordrecht: Kluwer, 1994.

Chevalley, Catherine. "Physics as an Art: The German Tradition and the Symbolic Turn in Philosophy, History of Art and Natural Science in the 1920s." In *The Elusive Synthesis: Aesthetics and Science*, edited by Alfred I. Tauber, 227–249. Dordrecht: Kluwer Academic, 1997.

Chew, Geoffrey F. "S-matrix Theory of Strong Interactions without Elementary Particles." *RMP* 34, no. 3 (1962): 394–401.

Christensen, Terry. "John Archibald Wheeler: A Study of Mentoring in Modern Physics." PhD diss., Oregon State University, 2009.

Christiansen, Jens Anton, and Hendrik Antoon Kramers. "Über die Geschwindigkeit chemischer Reaktionen." *Zeitschrift für physikalische Chemie* 104 (1923): 451–469.

Close, Frank E. *The Void*. Oxford: Oxford University Press, 2007.

Coblentz, William Weber. "Constants of Spectral Radiation of a Uniformly Heated Inclosure, or So-called Black Body." *Bulletin of the Bureau of Standards (United States)* 10, no. 1 (1914): 1–80.

Coleman, Diana, David Kaiser, and Aaron Sidney Wright, eds. *Theoretical Physics in Your Face! Selected Correspondence of Sidney Coleman.* Singapore: World Scientific, 2023.

Coleman, Sidney. "Acausality." In *Subnuclear Phenomena*, edited by Antonio Zichichi, vol. A, 282–325. New York: Academic Press, 1970.

Coleman, Sidney. "All Possible Symmetries of the S Matrix." In Zichichi, *Hadrons and Their Interactions*, 393–402.

Coleman, Sidney. "Black Holes as Red Herrings: Topological Fluctuations and the Loss of Quantum Coherence." *NP* B307, no. 4 (1988): 867–882.

Coleman, Sidney. "Books," 1970. Accessed 2019. http://efanzines.com/EK/eI36/.

Coleman, Sidney. "Books." *Magazine of Fantasy & Science Fiction* 45, no. 5 (1973): 22–29.

Coleman, Sidney. "Books." *Magazine of Fantasy & Science Fiction* 47, no. 2 (1974): 51–58.

Coleman, Sidney. *Classical Electron Theory from a Modern Standpoint.* Technical report RM-2820-PR. Santa Monica, CA: US Air Force Project RAND, 1961.

Coleman, Sidney. "Classical Lumps and Their Quantum Descendants." In *New Phenomena in Subnuclear Physics*, edited by Antonio Zichichi, 297–422. New York: Academic Press, 1977.

Coleman, Sidney. "Dilatations." In Zichichi, *Properties of the Fundamental Interactions*, 359–402.

Coleman, Sidney. "Erratum: Fate of the False Vacuum: Semiclassical Theory." *PRD* 16, no. 4 (1977): 1248.

Coleman, Sidney. "Fate of the False Vacuum: Semiclassical Theory." *PRD* 15, no. 10 (1977): 2929–2936.

Coleman, Sidney. *Lectures of Sidney Coleman on Quantum Field Theory.* Edited by Bryan Gin-ge Chen, David Derbes, David Griffiths, Brian Hill, Richard Sohn, and Yuan-Sen Ting. Singapore: World Scientific, 2018.

Coleman, Sidney. "More about the Massive Schwinger Model." *Annals of Physics* 101 (1976): 239–267.

Coleman, Sidney. "Not 'Ironic Science.'" *New York Times*, July 23, 1996, A18.

Coleman, Sidney. "Quantum sine-Gordon Equation as the Massive Thirring Model." *PRD* 11, no. 8 (1975): 2088–2097.

Coleman, Sidney. "Renormalization and Symmetry: A Review for Non-specialists." In Zichichi, *Properties of the Fundamental Interactions*, 605–624.

Coleman, Sidney. "Secret Symmetry: An Introduction to Spontaneous Symmetry Breakdown and Gauge Fields." In *Laws of Hadronic Matter*, edited by Antonino Zichichi, 139–224. New York: Academic Press, 1975.

Coleman, Sidney. "The 1979 Nobel Prize in Physics." *Science* 206, no. 4424 (1979): 1290–1292.

Coleman, Sidney. "The Invariance of the Vacuum Is the Invariance of the World." *JMP* 7, no. 5 (1966): 787–787.

Coleman, Sidney. "The Structure of Strong Interaction Symmetries." PhD diss., California Institute of Technology, 1962.

Coleman, Sidney. "The Uses of Instantons." In *The Whys of Subnuclear Physics*, edited by Antonino Zichichi, 805–941. New York: Plenum Press, 1979.

Coleman, Sidney. "There Are No Classical Glueballs." *CMP* 55, no. 2 (1977): 113–116.

Coleman, Sidney. "There Are No Goldstone Bosons in Two Dimensions." *CMP* 31, no. 4 (1973): 259–264.

Coleman, Sidney. "Why There Is Nothing Rather than Something: A Theory of the Cosmological Constant." *NP* B310 (1988): 643–668.

Coleman, Sidney, and Frank De Luccia. "Gravitational Effects on and of Vacuum Decay." *PRD* 21, no. 12 (1980): 3305–3315.

Coleman, Sidney, Vladimir Glaser, and Andre Martin. "Action Minima among Solutions to a Class of Euclidean Scalar Field Equations." *CMP* 58, no. 2 (1978): 211–221.

Coleman, Sidney, and Sheldon L. Glashow. "Departures from the Eightfold Way: Theory of Strong Interaction Symmetry Breakdown." *PR* 134, no. 3B (1964): 671–681.

Coleman, Sidney, and Sheldon L. Glashow. "Electrodynamic Properties of Baryons in the Unitary Symmetry Scheme." *PRL* 6, no. 8 (1961): 423–425.

Coleman, Sidney, and Roman Jackiw. "Why Dilatation Generators Do Not Generate Dilatations." *Annals of Physics* 67 (1971): 552–598.

Coleman, Sidney, Roman Jackiw, and H. David Politzer. "Spontaneous Symmetry Breaking in the $O(N)$ Model for Large N." *PRD* 10, no. 8 (1974): 2491–2499.

Coleman, Sidney, and Larry Smarr. "Are There Geon Analogues in Sourceless Gauge-field Theories?" *CMP* 56 (1977): 1–9.

Coleman, Sidney, and Erick Weinberg. "Radiative Corrections as the Origin of Spontaneous Symmetry Breaking." *PRD* 7, no. 6 (1973): 1888–1910.

Collins, Harry. *Gravity's Shadow: The Search for Gravitational Waves*. Chicago: University of Chicago Press, 2004.

Collins, Harry. "Lead into Gold: The Science of Finding Nothing." *Studies in History and Philosophy of Science Part A* 34, no. 4 (2003): 661–691.

Compton, Arthur H. "A Quantum Theory of the Scattering of X-rays by Light Elements." *PR* 21, no. 5 (1923): 483–502.

Compton, Arthur H. "Absorption Measurements of the Change of Wave-length Accompanying the Scattering of X-rays." *Philosophical Magazine* 46, no. 275 (1923): 897–911.

Compton, Arthur H. "Secondary Radiations Produced by X-Rays: and Some of Their Applications to Physical Problems." *Bulletin of the National Research Council* 4, no. 20 (1922).

Compton, Arthur H. *X-rays and Electrons: An Outline of Recent X-ray Theory*. New York: D. Van Nostrand, 1926.

Courant, Richard, and David Hilbert. *Methods of Mathematical Physics*. Translated by Peter Ceike, Ernest Courant, and Anneli Lax. Vol. 1. 1937. New York: Interscience Publishers, 1953.

Coxeter, H. S. M. "The Non-Euclidean Symmetry of Escher's Picture 'Circle Limit III'." *Leonardo* 12, no. 1 (1979): 19–25.

Coxeter, H. S. M., M. Emmer, Roger Penrose, and M. L. Teuber, eds. *M. C. Escher, Art and Science*. Amsterdam: North-Holland, 1986.

Crull, Elise, and Guido Bacciagaluppi, eds. *Grete Hermann—Between Physics and Philosophy*. Dordecht: Springer, 2016.

Crystal, Lisa. "Quantum Times: Physics, Philosophy, and Time in the Postwar United States." PhD diss., Harvard University, 2013.

Cunningham, Ebenezer. *Relativity and the Electron Theory*. London: Longmans, Green, & Co., 1915.

Cunningham, Ebenezer. *Relativity, The Electron Theory, and Gravitation*. 2nd ed. London: Longmans, Green, 1921.

Cunningham, Ebenezer. *The Principle of Relativity*. Cambridge: Cambridge University Press, 1914.

Cushing, James T. *Quantum Mechanics: Historical Contingency and the Copenhagen Hegemony*. Chicago: University of Chicago Press, 1994.

Cushing, James T. *Theory Construction and Selection in Modern Physics: The S matrix*. Cambridge: Cambridge University Press, 1990.

Dancoff, Sidney M. "On Radiative Corrections for Electron Scattering." *PR* 55, no. 10 (1939): 959–963.

Darrigol, Olivier. "Cohérence et complétude de la mécanique quantique: l'exemple de « Bohr-Rosenfeld »." *Revue d'histoire des sciences* 44, no. 2 (1991): 137–179.

Darrigol, Olivier. *Electrodynamics from Ampère to Einstein*. Oxford: Oxford University Press, 2000.

Darrigol, Olivier. *From c-Numbers to q-Numbers: The Classical Analogy in the History of Quantum Theory*. Berkeley: University of California Press, 1992.

Darrigol, Olivier. "Statistics and Combinatorics in Early Quantum Theory, II: Early Symptoma of Indistinguishability and Holism." *HSPBS* 21, no. 2 (1991): 237–298.

Darrigol, Olivier. "The Electron Theories of Larmor and Lorentz: A Comparative Study." *HSPBS* 24, no. 2 (1994): 265–336.

Darrigol, Olivier. "The Origin of Quantized Matter Waves." *HSPBS* 16, no. 2 (1986): 197–253.

Darrigol, Olivier. "Yukawa et Tomonaga: L' essor de la physique théorique Japonaise." *La Recherche* 23, no. 239 (1992): 42–50.

Darwin, Charles G. "Free Motion in the Wave Mechanics." *PRSLA* 117, no. 776 (1927): 258–293.

Darwin, Charles G. "The Dynamical Motions of Charged Particles." *PM* 39, no. 233 (1920): 537–551.

Darwin, Charles G. "The Electron as a Vector Wave." *Nature* 119, no. 2990 (1927): 282–285.

Darwin, Charles G. "The Wave Equations of the Electron." *PRSLA* 118, no. 780 (1928): 654–680.

Daston, Lorraine, ed. *Biographies of Scientific Objects*. Chicago: University of Chicago Press, 2000.

Daston, Lorraine, and Peter Gailson. *Objectivity*. New York: Zone Books, 2007.

Davies, P. C. W. *Superforce*. New York: Simon & Schuster, 1984.

Demianski, Marek. "The Jablonna Conference on Gravitation: A Continuing Source of Inspiration." *GRG* 46, no. 6 (2014): 1–10.

Deser, Stanley, Sidney Coleman, Sheldon L. Glashow, David Gross, Steven Weinberg, and Arthur Wightman. "Panel Discussion." In *Conceptual Foundations of Quantum Field Theory*, edited by Tian Yu Cao, 368–386. Cambridge: Cambridge University Press, 1999.

Deutsch, Martin. "Evidence for the Formation of Positronium in Gases." *Physical Review* 82, no. 3 (1951): 455–456.

DeWitt, Bryce S. "Dynamical Theory in Curved Spaces. I. A Review of the Classical and Quantum Action Principles." *RMP* 29, no. 3 (1957): 377–397.

DeWitt, Bryce S. "On the Application of Quantum Perturbation Theory to Gravitational Interactions." University of Texas, Austin, 1950. Accessed 2018. https://repositories.lib.utexas.edu/handle/2152/9620.

DeWitt, Bryce S. "Quantum Theory of Gravity. I. the Canonical Theory." *PR* 160, no. 5 (1967): 1113–1148.

DeWitt, Bryce S. "The Everett-Wheeler Interpretation of Quantum Mechanics." In DeWitt-Morette and Wheeler, *Battelle Rencontres*, 318–332.

DeWitt, Bryce S., and Cécile Morette DeWitt. "The Quantum Theory of Interacting Gravitational and Spinor Fields." *PR* 87, no. 1 (1952): 116–122.

DeWitt, Cécile Morette, and Bryce S. DeWitt, eds. *Relativity, Groups, and Topology: Lectures Delivered at Les Houches during the 1963 Session of the Summer School of Theoretical Physics.* New York: Gordon and Breach Science Publishers, 1964.

DeWitt-Morette, Cécile, ed. *Conference on The Role of Gravitation in Physics, Chapel Hill, North Carolina, January 18–23, 1957.* WADC technical report 57-216. Wright-Patterson Air Force Base, Ohio, Wright Air Development Center, Air Research / Development Command, US Air Force, 1957.

DeWitt-Morette, Cécile, and John Archibald Wheeler. "Battelle and Babel." In DeWitt-Morette and Wheeler, *Battelle Rencontres*, ix–xii.

DeWitt-Morette, Cécile, and John Archibald Wheeler, eds. *Battelle Rencontres: Lectures in Mathematics and Physics.* New York: W. A. Benjamin, 1968.

Dirac, P. A. M. "A New Notation for Quantum Mechanics." *MPCPS* 35, no. 3 (1939): 416–418.

Dirac, P. A. M. "A Theory of Electrons and Protons." *PRSLA* 126, no. 801 (1930): 360–365.

Dirac, P. A. M. "Developments in Quantum Electrodynamics." *Communications of the Dublin Institute for Advanced Studies. Series A* 3 (1946): 3–33.

Dirac, P. A. M. "Discussion of the Infinite Distribution of Electrons in the Theory of the Positron." *MPCPS* 30, no. 2 (1934): 150–163.

Dirac, P. A. M. "Note on Exchange Phenomena in the Thomas Atom." *MPCPS* 26, no. 3 (1930): 376–385.

Dirac, P. A. M. "On the Annihilation of Electrons and Protons." *MPCPS* 26, no. 3 (1930): 361–375.

Dirac, P. A. M. "On the Theory of Quantum Mechanics." *PRSLA* 112, no. 762 (1926): 661–677.

Dirac, P. A. M. "Quantised Singularities in the Electromagnetic Field." *PRSLA* 133, no. 821 (1931): 60–72.

Dirac, P. A. M. "Quantum Electrodynamics." Orig. Pub. 1943. *Communications of the Dublin Institute for Advanced Studies. Series A* 1 (1960): 1–36.

Dirac, P. A. M. "Quantum Mechanics and a Preliminary Investigation of the Hydrogen Atom." *PRSLA* 110, no. 755 (1926): 561–579.

Dirac, P. A. M. "Quantum Mechanics of Many-electron Systems." *PRSLA* 123, no. 792 (1929): 714–733.

Dirac, P. A. M. "Quelques problèmes de mécanique quantique." *Annales de l'Institut Henri Poincaré* 1, no. 4 (1930): 357–400.

Dirac, P. A. M. "Quelques problèmes de mécanique quantique." In *Collected Works*, 519–608. Originally published as "Quelques problèmes de mécanique quantique," *Annales de l'Institut Henri Poincaré* 1, no. 4 (1930): 357–400.

Dirac, P. A. M. "Recollections of an Exciting Era." In *History of Twentieth Century Physics: Proceedings of the International School of Physics "Enrico Fermi," Varenna 1972*, edited by Charles Weiner, 109–146. New York: Academic Press, 1977.

Dirac, P. A. M. "Relativistic Quantum Mechanics." *PRSLA* 136, no. 829 (1932): 453–464.

Dirac, P. A. M. "Relativity Quantum Mechanics with an Application to Compton Scattering." *PRSLA* 111, no. 758 (1926): 405–423.

Dirac, P. A. M. "Teoriya pozitrona." In *Collected Works*, 719–759.

Dirac, P. A. M. "The Basis of Statistical Quantum Mechanics." *MPCPS* 25, no. 1 (1929): 62–66.

Dirac, P. A. M. *The Collected Works of P. A. M. Dirac, 1924–1948*. Edited by R. H. Dalitz. Cambridge: Cambridge University Press, 1995.

Dirac, P. A. M. "The Compton Effect in Wave Mechanics." *MPCPS* 23, no. 5 (1927): 500–507.

Dirac, P. A. M. "The Conditions for Statistical Equilibrium between Atoms, Electrons and Radiation." *PRSLA* 106, no. 739 (1924): 581–596.

Dirac, P. A. M. "The Development of Quantum Mechanics." *Contributi del Centro Linceo Interdisciplinare di scienze mathematiche e loro applicazioni* 4 (1974): 1–11.

Dirac, P. A. M. "The Difficulties in Quantum Electrodynamics." In *Conference on Fundamental Particles and Low Temperatures*, vol. Vol. 1, 10–14.

Dirac, P. A. M. "The Effect of Compton Scattering by Free Electrons in a Stellar Atmosphere." *Monthly Notices of the Royal Astronomical Society* 85 (1925): 825–832.

Dirac, P. A. M. "The Fundamental Equations of Quantum Mechanics." *PRSLA* 109, no. 752 (1925): 642–653.

Dirac, P. A. M. "The Physical Interpretation of the Quantum Dynamics." *PRSLA* 113, no. 765 (1927): 621–641.

Dirac, P. A. M. *The Principles of Quantum Mechanics*. Oxford: Clarendon Press, 1930.

Dirac, P. A. M. *The Principles of Quantum Mechanics*. 4th ed. rev. Oxford: Oxford University Press, 1967.

Dirac, P. A. M. "The Proton." *Nature* 126, no. 3181 (1930): 605–606.

Dirac, P. A. M. "The Quantum Theory of Dispersion." *PRSLA* 114, no. 769 (1927): 710–728.

Dirac, P. A. M. "The Quantum Theory of the Electron." *PRSLA* 117, no. 778 (1928): 610–624.

Dirac, P. A. M. "The Quantum Theory of the Electron. Part II." *PRSLA* 118, no. 779 (1928): 351–361.

Dirac, P. A. M. "The Quantum Theory of the Emission and Absorption of Radiation." *PRSLA* 114, no. 767 (1927): 243–265.

Dirac, P. A. M. "Théorie du positron." In Institut International de Physique Solvay, *Structure et propriétés des noyaux atomiques*, 203–212.

Dirac, P. A. M. "Über die Quantenmechanik der Stoßvorgänge." *ZP* 44, no. 8 (1927): 585–595.

Dirac, P. A. M. "Über die Quantentheorie des Elektrons." In *Collected Works*, 352–357.

Dirac, P. A. M. "Zur Quantentheorie des Elektrons." In *Collected Works*, 360–379.

Dirac, P. A. M., and J. W. Harding. "Photo-electric Absorption in Hydrogen-like Atoms." *MPCPS* 28, no. 2 (1932): 209–218.

DiSalle, Robert. *Understanding Space-Time: The Philosophical Development of Physics from Newton to Einstein*. Cambridge: Cambridge University Press, 2006.

Dobkowski, Michael N. *The Tarnished Dream: The Basis of American Anti-Semitism*. Westport, CT: Greenwood Press, 1979.

Dongen, Jeroen van. *Einstein's Unification*. Cambridge: Cambridge University Press, 2002.

Donnachie, Sandy, Gunter Dosch, Peter Landshoff, and Otto Nachtmann. *Pomeron Physics and QCD*. Cambridge: Cambridge University Press, 2002.

Doxiadis, Apostolos, and Barry Mazur, eds. *Circles Disturbed: The Interplay of Mathematics and Narrative*. Princeton, NJ: Princeton University Press, 2012.

Dresden, Max. *H. A. Kramers between Tradition and Revolution*. New York: Springer, 1987.

Duncan, Anthony, and Michel Janssen. "From Canonical Transformations to Transformation Theory, 1926–1927: The Road to Jordan's Neue Begründung." *SHPMP* 40, no. 4 (2009): 352–362.

Duncan, Anthony, and Michel Janssen. "On the Verge of Umdeutung in Minnesota: Van Vleck and the Correspondence Principle. Part One." *AHES* 61, no. 6 (2007): 553–624.

Duncan, Anthony, and Michel Janssen. "Pascual Jordan's Resolution of the Conundrum of the Wave-particle Duality of Light." *SHPMP* 39, no. 3 (2008): 634–666.

Dyson, Freeman J. *Advanced Quantum Mechanics*. 2nd ed. Edited by Michael J. Moravcsik. 1951. 1957.

Dyson, Freeman J. "The Radiation Theories of Tomonaga, Schwinger, and Feynman." *PR* 75, no. 3 (1949): 486–502.

Dyson, Freeman J. "The S Matrix in Quantum Electrodynamics." *PR* 75, no. 11 (1949): 1736–1755.

Earman, John. "The Penrose-Hawking Singularity Theorems: History and Implications." In *The Expanding Worlds of General Relativity*, edited by Hubert Goenner, Jürgen Renn, J. Ritter, and Tilman Sauer, 7:235–267. Boston: Birkauser, 1999.

Earman, John. *World Enough and Space-Time: Absolute versus Relational Theories of Space and Time*. Cambridge, MA: MIT Press, 1989.

Earman, John, and Jesus Mosterin. "A Critical Look at Inflationary Cosmology." *Philosophy of Science* 66, no. 1 (1999): 1–49.

Eckert, Michael. *Arnold Sommerfeld: Science, Life and Turbulent Times, 1868–1951*. New York: Springer, 2013.

Eddington, Arthur Stanley. *Space, Time and Gravitation: An Outline of the General Relativity Theory*. Cambridge: Cambridge University Press, 1920.

Eddington, Arthur Stanley. "The Charge of an Electron." *PRSLA* 122, no. 789 (1929): 358–369.

Eddington, Arthur Stanley. *The Mathematical Theory of Relativity*. Cambridge: Cambridge University Press, 1923.

Eden, Richard John, Peter V. Landshoff, David I. Olive, and John Charlton Polkinghorne. *The Analytic S-Matrix*. Cambridge: Cambridge University Press, 1966.

Eden, Richard John, and John Polkinghorne. "Dirac in Cambridge." In *Aspects of Quantum Theory*, edited by Abdus Salam and E. P. Wigner, 1–6. Cambridge: Cambridge University Press, 1972.

Einstein, Albert. *Einstein's Miraculous Year: Five Papers That Changed the Face of Physics*. Edited by John Stachel. Princeton, NJ: Princeton University Press, 1998.

Einstein, Albert. "Ether and the Theory of Relativity." In *The Collected Papers of Albert Einstein*. Vol. 7, *The Berlin Years: Writings, 1918–1921*, edited by Michael Janssen, Robert Schulmann, József Illy, Christoph Lehner, and Diana Kormos Buchwald, 305–323. 1920. Princeton, NJ: Princeton University Press, 2002.

Einstein, Albert. "On the Electrodynamics of Moving Bodies." In *CPAE 2*, 275–310.

Einstein, Albert. "On the Method of Theoretical Physics." *Philosophy of Science* 1, no. 2 (1934): 163–169.

Einstein, Albert. "On the Quantum Theory of Radiation." In *CPAE 6*, 381–397.

Einstein, Albert. *Relativity: The Special and the General Theory*. 100th anniversary ed. Translated by Robert W. Lawson. Princeton, NJ: Princeton University Press, 2015.

Einstein, Albert. *The Berlin Years: Correspondence, January 1919–April 1920*. Vol. 9 of *The Collected Papers of Albert Einstein*, edited by Diana Kormos Buchwald, Robert Schulmann, József Illy, Daniel J. Kennefick, and Tilman Sauer. Princeton, NJ: Princeton University Press, 2004.

Einstein, Albert. *The Berlin Years: Writings, 1914–1917*. Vol. 6 of *The Collected Papers of Albert Einstein*, edited by A. J. Kox, Martin J. Klein, and Robert Schulmann. Princeton, NJ: Princeton University Press, 1996.

Einstein, Albert. "The Principle of Relativity and Its Consequences in Modern Physics." In *The Collected Papers of Albert Einstein*. Vol. 3, *The Swiss Years: Writings, 1909–1911*, edited by Martin J. Klein, A. J. Kox, Jürgen Renn, and Robert Schulman, translated by Edouard Guillaume, 130–176. Doc. 2. 1910. Princeton, NJ: Princeton University Press, 1994.

Einstein, Albert. *The Swiss Years: Writings, 1900–1909*. Vol. 2 of *The Collected Papers of Albert Einstein*, edited by John Stachel, David C. Cassidy, Jürgen Renn, and Robert Schulmann. Princeton, NJ: Princeton University Press, 1989.

Einstein, Albert, and Leopold Infeld. *The Evolution of Physics: The Growth of Ideas from the Early Concepts to Relativity and Quanta*. Cambridge: Cambridge University Press, 1947.

Einstein, Albert, Leopold Infeld, and Banesh Hoffmann. "The Gravitational Equations and the Problem of Motion." *The Annals of Mathematics*, Second Series, 39, no. 1 (1938): 65–100.

Einstein, Albert, and Nathan Rosen. "The Particle Problem in the General Theory of Relativity." *PR* 48, no. 1 (1935): 73–77.

Eisenstaedt, Jean. "Histoire et singularités de la solution de Schwarzschild (1915–1923)." *AHES* 27 (1982): 157–198.

Eisenstaedt, Jean. "Trajectoires et impasses de la solution de Schwarzschild." *AHES* 37, no. 4 (1987): 275–357.

Elkins, James. *The Poetics of Perspective*. Ithaca, NY: Cornell University Press, 1994.

Ellis, John, Nick E. Mavromatos, and Tevong You. "Light-by-light Scattering Constraint on Born-Infeld Theory." *PRL* 118, no. 26 (2017): 261802-6.

Englert, Francois, and Robert Brout. "Broken Symmetry and the Mass of Gauge Vector Mesons." *PRL* 13, no. 9 (1964): 321–323.

Enz, Charles P. "Heisenberg's Applications of Quantum Mechanics (1926–33) or the Settling of the New Land." *Helvetica Physica Acta* 56, no. 5 (1983): 993–1001.

Enz, Charles P. *No Time to Be Brief: A Scientific Biography of Wolfgang Pauli*. Oxford: Oxford University Press, 2002.

Escher, M. C. *[Tentoonstelling/ Exhibition]*. Cat. No. 118. Amsterdam: Stedelijk Museum, 1954.

Euler, Hans. "Über die Streuung von Licht an Licht nach der Diracschen Theorie." *AP* 418, no. 5 (1936): 398–448.

Euler, Hans, and Bernhard Kockel. "Über die Streuung von Licht an Licht nach der Diracschen Theorie." *Naturwissenschaften* 23, no. 15 (1935): 246–247.

Everett, Hugh. "'Relative State' Formulation of Quantum Mechanics." *RMP* 29, no. 3 (1957): 454–462.

Farmelo, Graham. *The Strangest Man: The Hidden Life of Paul Dirac, Mystic of the Atom*. New York: Basic Books, 2009.

Fermi, Enrico. "Fundamental Particles." In *Collected Papers*, 825–828.

Fermi, Enrico. "La théorie du rayonnement." *Annales de l'Institut Henri Poincaré* 1, no. 1 (1930): 52–74.

Fermi, Enrico. *Note e memorie (Collected Papers)*. Edited by Edoardo Amaldi, Herbert L. Anderson, Enrico Persico, Emilio Segrè, and Albert Wattenberg. 2 vols. Rome: Accademia nazionale dei Lincei, 1962–1965.

Fermi, Enrico. "Quantum Theory of Radiation." *RMP* 4, no. 1 (1932): 87–132.

Fermi, Enrico. "Über die Anwendung der statistischen Methode auf die Probleme des Atombaues." In *Collected Papers*, 1:291–304.

Fermi, Laura. "A Visit to the 'Ettore Majorana' Centre." *Physics Today* 29, no. 8 (1976): 9–13.

Feynman, Richard P. "Critical Comments." In DeWitt-Morette, *Role of Gravitation in Physics*, 149–53.

Feynman, Richard P. *Feynman Lectures on Gravitation*. Edited by Brian Hatfield. Reading, MA: Westview Press, 1995.

Feynman, Richard P. *Feynman's Thesis: A New Approach to Quantum Theory*. Edited by Laurie M. Brown. Hackensack, NJ: World Scientific, 2005.

Feynman, Richard P. "Quantum Theory of Gravitation." *Acta Physica Polonica* XXIV, no. 6 (12) (1963): 697–722.

Feynman, Richard P. "Relativistic Cut-off for Quantum Electrodynamics." *PR* 74, no. 10 (1948): 1430–1438.

Feynman, Richard P. "Space-time Approach to Non-relativistic Quantum Mechanics." *RMP* 20, no. 2 (1948): 367–387.

Feynman, Richard P. "Space-time Approach to Quantum Electrodynamics." *PR* 76, no. 6 (1949): 769–789.

Feynman, Richard P. "The Present Status of Quantum Electrodynamics." In *La théorie quantique des champs: douzième Conseil de physique, tenu à l'Université libre de Bruxelles du 9 au 14 octobre 1961*, 61–91. New York: International Publishers, 1962.

Feynman, Richard P., and Frank L. Vernon Jr. "The Theory of a General Quantum System Interacting with a Linear Dissipative System." In *Selected Papers of Richard Feynman, with Commentary*, edited by Laurie M. Brown, 257–312. 1963. River Edge, NJ: World Scientific, 2000.

Finkelstein, David. "Kinks." *JMP* 7, no. 7 (1966): 1218–1225.

Finkelstein, David. "Past-future Asymmetry of the Gravitational Field of a Point Particle." *PR* 110, no. 4 (1958): 965–967.

Finkelstein, David, and Charles W. Misner. "Some New Conservation Laws." *Annals of Physics* 6 (1959): 230–243.

Fock, Vladimir A. "An Approximate Method for Solving the Quantum Many-body Problem." In *Selected Works*, 137–163.

Fock, Vladimir A. "Geometrisierung der Diracschen Theorie des Elektrons." *ZP* 57, nos. 3–4 (1929): 261–277.

Fock, Vladimir A. "On Quantum Electrodynamics." In *Selected Works*, 331–368.

Fock, Vladimir A. "On Quantum Exchange Energy." In *Selected Works*, 263–278.

Fock, Vladimir A. "On the Notion of Velocity in the Dirac Theory of the Electron." In *Selected Works*, 95–108.

Fock, Vladimir A. *V. A. Fock—Selected Works: Quantum Mechanics and Quantum Field Theory*. Edited by L. D. Faddeev, L. A. Khalfin, and I. V. Komarov. Boca Raton, FL: Chapman & Hall/CRC, 2004.

Foote, Paul D., and F. L. Mohler. *Origin of Spectra*. New York: Chemical Catalog Company, 1922.

Ford, Kenneth W. *Building the H Bomb: A Personal History*. Singapore: World Scientific, 2015.

Forman, Paul. "Behind Quantum Electronics: National Security as Basis for Physical Research in the United States, 1940–1960." *HSPBS* 18, no. 1 (1987): 149–229.

Forman, Paul. "Scientific Internationalism and the Weimar Physicists: The Ideology and Its Manipulation in Germany after World War I." *Isis* 64, no. 2 (1973): 151–180.

Forman, Paul. "Weimar Culture, Causality, and Quantum Theory, 1918–1927: Adaptation by German Physicists and Mathematicians to a Hostile Intellectual Environment." In Carson, Kozhevnikov, and Trischler, *Weimar Culture and Quantum Mechanics*, 87–201.

Forman, Paul, John L. Heilbron, and Spencer Weart. "Physics circa 1900: Personnel, Funding, and Productivity of the Academic Establishments." *HSPS* 5 (1975): 1–185.

Fortun, Kim. "From Latour to Late Industrialism." *HAU: Journal of Ethnographic Theory* 4, no. 1 (2014): 309–329.

Fowler, Ralph H. "Statistical Equilibrium with Special Reference to the Mechanism of Ionization by Electronic Impacts." *PM*, 6th ser., 47, no. 278 (1924): 257–277.

Fowler, Ralph H. *Statistical Mechanics: The Theory of the Properties of Matter in Equilibrium.* Cambridge: Cambridge University Press, 1929.

Fowler, Ralph H., and E. A. Milne. "A Note on the Principle of Detailed Balancing." *PNAS* 11, no. 7 (1925): 400–402.

Fowler, Ralph H., and E. A. Milne. "The Intensities of Absorption Lines in Stellar Spectra, and the Temperatures and Pressures in the Reversing Layers of Stars." *Monthly Notices of the Royal Astronomical Society* 83, no. 7 (1923): 403–424.

Fowler, W. A., and Charles C. Lauritsen. "Pair Emission from Fluorine Bombarded with Protons." *PR* 56, no. 8 (1939): 840–841.

Frampton, Paul. "Consequences of Vacuum Instability in Quantum Field Theory." *PRD* 15, no. 10 (1977): 2922–2928.

Frampton, Paul. "Vacuum Instability and Higgs Scalar Mass." *PRL* 37, no. 21 (1976): 1378–1380.

Franklin, Allan. "How to Avoid the Experimenters' Regress." *Studies in History and Philosophy of Science Part A* 25, no. 3 (1994): 463–491.

Franklin, Allan. *Shifting Standards: Experiments in Particle Physics in the Twentieth Century.* Pittsburgh: University of Pittsburgh Press, 2013.

Freire, Olival Jr. *The Quantum Dissidents: Rebuilding the Foundations of Quantum Mechanics (1950–1990).* Heidelberg: Springer, 2015.

French, J. Bruce, and Victor Weisskopf. "The Electromagnetic Shift of Energy Levels." *PR* 75, no. 8 (1949): 1240–1248.

Freund, Peter G. O., and Yoichiro Nambu. "Mass and Coupling Constant Formulas in Broken Symmetry Schemes." *PRL* 13, no. 6 (1964): 221–225.

Friedman, Michael. *Foundations of Space-Time Theories.* Princeton, NJ: Princeton University Press, 1983.

Fubini, Sergio. *Present Trends in Particle Physics.* Technical report TH.1904. Geneva: CERN, 1974.

Fues, Erwin. *Calculation of Perturbation.* Technical report NASA-CR-110103. 1927. Greenbelt, MD: NASA, 1970.

Furry, Wendell H. "A Symmetry Theorem in the Positron Theory." *PR* 51, no. 2 (1937): 125–129.

Furry, Wendell H., and John Franklin Carlson. "On the Production of Positive Electrons by Electrons." *PR* 44, no. 3 (1933): 237–238.

Furry, Wendell H., and J. Robert Oppenheimer. "On the Limitations of the Theory of the Positron." *PR* 45, no. 12 (1934): 903–904.

Furry, Wendell H., and J. Robert Oppenheimer. "On the Theory of the Electron and Positive." *PR* 45, no. 4 (1934): 245–262.

Furry, Wendell H., and J. Robert Oppenheimer. "On the Theory of the Electron and Positive [II]." *PR* 45, no. 5 (1934): 343–344.

Fürth, Reinhold. "Über Strahlungsschwankungen nach der Lichtquantenstatistik." *ZP* 50, no. 5 (1928): 310–318.

Galison, Peter. "Feynman's War: Modelling Weapons, Modelling Nature." *SHPMP* 29, no. 3 (1998): 391–434.

Galison, Peter. *How Experiments End*. Chicago: University of Chicago Press, 1987.

Galison, Peter. *Image and Logic: A Material Culture of Microphysics*. Chicago: University of Chicago Press, 1997.

Galison, Peter. "Minkowski's Space-Time: From Visual Thinking to the Absolute World." *HSPS* 10 (1979): 85–121.

Galison, Peter. "Mirror Symmetry: Persons, Values, and Objects." In *Growing Explanations: Historical Perspectives on Recent Science*, edited by M. Norton Wise, 23–63. Durham, NC: Duke University Press, 2004.

Galison, Peter. "Re-Reading the Past from the End of Physics: Maxwell's Equations in Retrospect." In *Functions and Uses of Disciplinary Histories*, edited by Loren Graham, Wolf Lepenies, and Peter Weingart, 35–51. Dordecht: Springer, 1983.

Galison, Peter. "Structure of Crystal, Bucket of Dust." In Doxiadis and Mazur, *Circles Disturbed*, 52–78.

Galison, Peter. "The Discovery of the Muon and the Failed Revolution against Quantum Electrodynamics." *Centaurus* 50, nos. 1–2 (2008): 105–159.

Galison, Peter. "The Suppressed Drawing: Paul Dirac's Hidden Geometry." *Representations*, no. 72 (2000): 145–166.

Galison, Peter. "Theoretical Predispositions in Experimental Physics: Einstein and the Gyromagnetic Experiments, 1915–1925." *HSPS* 12, no. 2 (1982): 285–323.

Galison, Peter. "Theory Bound and Unbound: Superstrings and Experiments." In *Laws of Nature: Essays on the Philosophic, Scientific, and Historical Dimensions*, edited by F. Weinart, 369–408. Berlin: Walter de Gruyter, 1995.

Galison, Peter, and Barton Bernstein. "In Any Light: Scientists and the Decision to Build the Superbomb, 1952–1954." *HSPBS* 19, no. 2 (1989): 267–347.

Galison, Peter, and Bruce William Hevly, eds. *Big Science: The Growth of Large-scale Research*. Stanford, CA: Stanford University Press, 1992.

Galison, Peter, and Andrew Warwick. "Introduction: Cultures of Theory." *SHPMP* 29, no. 3 (1998): 287–94.

Gamow, George. "The Blegdamsvej Faust." In *Thirty Years That Shook Physics: The Story of Quantum Theory*, translated by Barbara Gamow, 174–218. 1932. Reprint, New York: Dover, 1966.

Garber, Elizabeth. *The Language of Physics: The Calculus and the Development of Theoretical Physics in Europe, 1750–1914*. New York: Springer, 1999.

Garcia-Prada, Oscar. "Interview with Sir Roger Penrose, Part 1." *European Mathematical Society Newsletter* 39 (2000): 17–21.

Garcia-Prada, Oscar. "Interview with Sir Roger Penrose, Part 2." *European Mathematical Society Newsletter* 38 (2001): 12–17.

Gaunt, John Arthur. "A Theory of Hartree's Atomic Fields." *MPCPS* 24, no. 2 (1928): 328–342.

Gaunt, John Arthur. "The Relativistic Theory of an Atom with Many Electrons." *PRSLA* 124, no. 793 (1929): 163–176.

Gavroglu, Kostas, and Ana Simões. *Neither Physics nor Chemistry: A History of Quantum Chemistry*. Cambridge, MA: MIT Press, 2012.

Gavroglu, Kostas, and Ana Simões. "Preparing the Ground for Quantum Chemistry in Great Britain: The Work of the Physicist R. H. Fowler and the Chemist N. V. Sidgwick." *BJHS* 35, no. 125 (2002): 187–212.

Gell-Mann, Murray. "Symmetries of Baryons and Mesons." *PR* 125, no. 3 (1962): 1067–1084.

Gell-Mann, Murray. *The Eightfold Way: A Theory of Strong Interaction Symmetry*. Technical report CTSL-20. Pasadena: Caltech Synchrotron Lab, 1961.

Gell-Mann, Murray, and Maurice Lévy. "The Axial Vector Current in Beta Decay." *INC* 16, no. 4 (1960): 705–726.

Gell-Mann, Murray, and Francis E. Low. "Quantum Electrodynamics at Small Distances." *PR* 95, no. 5 (1954): 1300–1312.

Gell-Mann, Murray, and Arthur H. Rosenfeld. "Hyperons and Heavy Mesons (Systematics and Decay)." *Annual Review of Nuclear Science* 7, no. 1 (1957): 407–478.

Georgi, Howard. "Sidney Coleman, 1937–2007." In *Biographical Memoirs*, 1–20. Washington, DC: National Academy of Sciences, 2011.

Geroch, Robert P. "Topology in General Relativity." *JMP* 8, no. 4 (1967): 782–786.

Gibbs, J. Willard. *Elementary Principles in Statistical Mechanics*. New York: Charles Scribner's Sons, 1902.

Gingras, Yves. "The Collective Construction of Scientific Memory: The Einstein-Poincaré Connection and Its Discontents, 1905–2005." *History of Science* 46, no. 1 (2008): 75–114.

Gingras, Yves. "What Did Mathematics Do to Physics?" *History of Science* 39 (2001): 383–416.

Glashow, Sheldon L. "Breaking Chiral SU(3) $\otimes$ SU(3)." In Zichichi, *Hadrons and Their Interactions*, 83–140.

Glashow, Sheldon L. "Partial-symmetries of Weak Interactions." *NP* 22, no. 4 (1961): 579–588.

Glashow, Sheldon L. "The Renormalizability of Vector Meson Interactions." *NP* 10 (1959): 107–117.

Gleick, James. *Genius: The Life and Science of Richard Feynman*. New York: Pantheon Books, 1992.

Glick, Thomas F., ed. *The Comparative Reception of Relativity*. Dordrecht: D. Reidel, 1987.

A Glossary of Terms in Nuclear Science and Technology. Technical report 110-1957. New York: ASME, 1957.

Goenner, Hubert. "On the History of Unified Field Theories." *Living Reviews in Relativity* 7, no. 1 (2004): 2. http://www.livingreviews.org/lrr-2004-2.

Goenner, Hubert. "On the History of Unified Field Theories. Part II (ca. 1930–ca. 1965)." *Living Reviews in Relativity* 17, no. 1 (2014): 5–241. doi:10.12942/lrr-2014-5.

Goenner, Hubert. "The Reaction to Relativity Theory I: The Anti-Einstein Campaign in Germany in 1920." *Science in Context* 6, no. 1 (1993): 107–133.

Goenner, Hubert, and Giuseppe Castagnetti. "Albert Einstein as Pacifist and Democrat during World War I." *Science in Context* 9, no. 4 (1996): 325–386.

Goldberg, Stanley. "In Defense of Ether: The British Response to Einstein's Special Theory of Relativity, 1905–1911." *HSPS* 2 (1970): 89–125.

Goldberg, Stanley. "Putting New Wine in Old Bottles: The Assimilation of Relativity in America." In Glick, *The Comparative Reception of Relativity*, 1–26.

Goldstein, Herbert, Charles Poole, and John Safko. *Classical Mechanics*. 3rd ed. San Francisco: Addison-Wesley, 2001.

Goldstone, Jeffrey. "Derivation of the Brueckner Many-Body Theory." *Proceedings of the Physical Society. A* 239, no. 1217 (1957): 267–279.

Goldstone, Jeffrey. "Field Theories with « Superconductor » Solutions." *INC* 19, no. 1 (1961): 154–164.

Goldstone, Jeffrey, Abdus Salam, and Steven Weinberg. "Broken Symmetries." *PR* 127, no. 3 (1962): 965–970.

Gooday, Graeme, and Daniel Mitchell. "Rethinking 'Classical Physics'." In Buchwald and Fox, *The Oxford Handbook of the History of Physics*, 721–764.

Gordin, Michael D. "Book Review: The Tory Interpretation of History." *HSNS* 44, no. 4 (2014): 413–423.

Gottfried, Kurt. "Quantum Electrodynamics: Matter All in the Mind." *Nature* 419, no. 6903 (2002): 117.

Gowers, Timothy. "Vividness in Mathematics and Narrative." In Doxiadis and Mazur, *Circles Disturbed*, 211–231.

Graham, Loren. "Epilogue." In Graham, Lepenies, and Weingart, *Disciplinary Histories*, 291–295.

Graham, Loren. *Lysenko's Ghost: Epigenetics and Russia*. Cambridge, MA: Harvard University Press, 2016.

Graham, Loren. *Science, Philosophy, and Human Behavior in the Soviet Union*. New York: Columbia University Press, 1987.

Graham, Loren, Wolf Lepenies, and Peter Weingart, eds. *Functions and Uses of Disciplinary Histories*. Dordecht: D. Reidel, 1983.

Grant, Edward. *Much Ado about Nothing: Theories of Space and Vacuum from the Middle Ages to the Scientific Revolution*. Cambridge: Cambridge University Press, 1981.

Gray, Jeremy. *Plato's Ghost: The Modernist Transformation of Mathematics*. Princeton, NJ: Princeton University Press, 2008.

Gray, Louis Harald, and G. T. P. Tarrant. "Phenomena Associated with the Anomalous Absorption of High Energy Gamma Radiation. II." *PRSLA* 143, no. 850 (1934): 681–706.

Gross, Ari. "Pictures and Pedagogy: The Role of Diagrams in Feynman's Early Lectures." *SHPMP* 43, no. 3 (2012): 184–194.

Guicciardini, Niccolò. *Isaac Newton on Mathematical Certainty and Method*. Cambridge, MA: MIT Press, 2009.

Guralnik, Gerald, C. Richard Hagen, and Tom W. B. Kibble. "Global Conservation Laws and Massless Particles." *PRL* 13, no. 20 (1964): 585–587.

Guth, Alan H. "Inflationary Universe: A Possible Solution to the Horizon and Flatness Problems." *PRD* 23, no. 2 (1981): 347–356.

Guth, Alan H., and David Kaiser. "Inflationary Cosmology: Exploring the Universe from the Smallest to the Largest Scales." *Science* 307, no. 5711 (2005): 884–890.

Haar, D. Ter. "Foundations of Statistical Mechanics." *RMP* 27, no. 3 (1955): 289–338.

Hacking, Ian. "Experimentation and Scientific Realism." *Philosophical Topics* 13, no. 1 (1982): 71–87.

Hacking, Ian. "Historical Ontology." In *Historical Ontology*, 1–26.

Hacking, Ian. *Historical Ontology.* Cambridge, MA: Harvard University Press, 2002.

Hacking, Ian. "Language, Truth, and Reason." In *Historical Ontology*, 159–177.

Hacking, Ian. "Michel Foucault's Immature Science." *Noûs* 13, no. 1 (1979): 39–51.

Hacking, Ian. *Representing and Intervening: Introductory Topics in the Philosophy of Natural Science.* Cambridge: Cambridge University Press, 1983.

Hacking, Ian. "The Disunities of Sciences." In *The Disunity of Science: Boundaries, Contexts, and Power*, edited by Peter Galison and David Stump, 37–74. Stanford, CA: Stanford University Press, 1996.

Hacking, Ian. *The Social Construction of What?* Cambridge, MA: Harvard University Press, 1999.

Hacking, Ian. *Why Is There Philosophy of Mathematics at All?* Cambridge: Cambridge University Press, 2014.

Hacking, Ian. "Wittgenstein Rules." *SSS* 14, no. 3 (1984): 469–476.

Hakim, Rémi. "Remarks on Relativistic Statistical Mechanics. I." *JMP* 8, no. 6 (1967): 1315–1344.

Hall, Harvey, and J. Robert Oppenheimer. "Relativistic Theory of the Photoelectric Effect. Part I. Theory of the K-Absorption of X-rays. Part II. Photoelectric Absorption of Ultragamma Radiation." *PR* 38, no. 1 (1931): 57–79.

Hall, Karl Philip. "Purely Practical Revolutionaries: A History of Stalinist Theoretical Physics." PhD diss., Harvard University, 1999.

Halpern, Leopold. "Observations of Our Two Brightest Stars." In *Differential Geometrical Methods in Theoretical Physics*, edited by K. Bleuler and M. Werner, 463–470. Dordrecht: Kluwer Academic, 1988.

Halpern, Otto. "Scattering Processes Produced by Electrons in Negative Energy States." *PR* 44, no. 10 (1933): 855–856.

Halpern, Paul. "Quantum Humor: The Playful Side of Physics at Bohr's Institute for Theoretical Physics." *Physics in Perspective* 14, no. 3 (2012): 279–299.

Haraway, Donna. *Primate Visions: Gender, Race, and Nature in the World of Modern Science.* New York: Routledge, 1989.

Hartmann, Stephan. "Models and Stories in Hadron Physics." In Morgan and Morrison, *Models as Mediators*, 326–346.

Hartree, Douglas R. "The Distribution of Charge and Current in an Atom Consisting of Many Electrons Obeying Dirac's Equations." *MPCPS* 25, no. 2 (1929): 225–236.

Hartree, Douglas R. "The Wave Mechanics of an Atom with a Non-Coulomb Central Field. Part I. Theory and Methods." *MPCPS* 24, no. 1 (1928): 89–110.

Hartree, Douglas R. "The Wave Mechanics of an Atom with a Non-Coulomb Central Field. Part II. Some Results and Discussion." *MPCPS* 24, no. 1 (1928): 111–132.

Hawking, Stephen W. *A Brief History of Time: From the Big Bang to Black Holes.* Toronto: Bantam Books, 1988.

Hawking, Stephen W. "Black Hole Explosions?" *Nature* 248 (1974): 30–31.

Hawking, Stephen W. "Particle Creation by Black Holes." *CMP* 43, no. 3 (1975): 199–220.

Hawking, Stephen W. "Singularities and the Geometry of Spacetime." Orig. pub. 1965. *European Physical Journal H* 39 (2014): 413–503.

Hawking, Stephen W. "Space-Time Foam." *NP* B144 (1978): 349–362.

Hawking, Stephen W. "Wormholes in Spacetime." *PRD* 37 (1988): 904–910.

Hawking, Stephen W., and G. F. R. Ellis. *The Large-scale Structure of Space-time.* Cambridge: Cambridge University Press, 1973.

Hawking, Stephen W., and Roger Penrose. "The Singularities of Gravitational Collapse and Cosmology." *PRSLA* 314, no. 1519 (1970): 529–548.

Heilbron, John L. "Fin-de-siècle Physics." In *Science, Technology and Society in the Time of Alfred Nobel*, edited by C. G. Bernhard, E. Crawford, and P. Sörbom, 51–73. Oxford: Oxford University Press, 1982.

Heilbron, John L. *Physics: A Short History from Quintessence to Quarks.* Oxford: Oxford University Press, 2015.

Heilbron, John L. "The Kossel-Sommerfeld Theory and the Ring Atom." *Isis* 58, no. 4 (1967): 450–485.

Heilbron, John L. "Weighing Imponderables and Other Quantitative Science around 1800." *HSPBS* 24, no. 1. Supplement (1993): 1–33, 35–277, 279–337.

Heilbron, John L., and Robert W. Seidel. *Lawrence and His Laboratory: A History of the Lawrence Berkeley Laboratory.* Berkeley: University of California Press, 1989.

Heisenberg, Werner. "Bemerkungen zur Diracschen Theorie des Positrons." *ZP* 90, nos. 3–4 (1934): 209–231.

Heisenberg, Werner. "On the Physical Content of Quantum Theoretical Kinematics and Mechanics." In Zurek and Wheeler, *Quantum Theory and Measurement*, 62–84.

Heisenberg, Werner. *Physics and Philosophy: The Revolution in Modern Science.* London: George Allen & Unwin, 1959.

Heisenberg, Werner. "Remarks on the Dirac Theory of the Positron." In Miller, *Early Quantum Electrodynamics*, 169–187.

Heisenberg, Werner. "Schwankungserscheinungen und Quantenmechanik." In *Collected Works: Original Scientific Papers*, edited by Walter Blum, Hans-Peter Dürr, and Helmut Rechenberg, vol. A pt. I, 472–477. 1927. Reprint, Berlin: Springer, 1985.

Heisenberg, Werner. *The Physical Principles of the Quantum Theory.* Translated by Carl Eckart and Frank Clark Hoyt. Chicago: University of Chicago Press, 1930.

Heisenberg, Werner. "Über den anschaulichen Inhalt der quantentheoretischen Kinematik und Mechanik." *ZP* 43, no. 3 (1927): 172–198.

Heisenberg, Werner. "Über Energieschwankungen in einem Strahlungsfeld." In *Original Scientific Papers*. Vol. A pt. II of *Werner Heisenberg: Collected Works*, edited by Walter Blum, Hans-Peter Dürr, and Helmut Rechenberg, 116–122. 1931. Berlin: Springer, 1989.

Heisenberg, Werner. "Zum Paulischen Ausschließungsprinzip." *AP* 402, no. 7 (1931): 888–904.

Heisenberg, Werner, and Hans Euler. "Consequences of Dirac Theory of the Positron." Translated by W. Korolevski and H. Kleinert. *ArXiv Physics e-prints*, 2006. eprint: arXiv:physics/0605038. Originally published as "Folgerungen aus der Diracschen Theorie des Positrons," *ZP* 98, nos. 11–12 (1936): 714–732.

Heisenberg, Werner, and Wolfgang Pauli. "Zur Quantendynamik der Wellenfelder." *ZP* 56, nos. 1–2 (1929): 1–61.

Heitler, Walter. "The Quantum Theory of Damping as a Proposal for Heisenberg's S-Matrix." In *Conference on Fundamental Particles and Low Temperatures*, Vol. 1:189–194.

Heitler, Walter. *The Quantum Theory of Radiation*. Corrected ed. Oxford: Oxford University Press, 1944.

Heitler, Walter. *The Quantum Theory of Radiation*. 1936. Oxford: Clarendon Press.

Hendry, John. "Bohr-Kramers-Slater: A Virtual Theory of Virtual Oscillators and Its Role in the History of Quantum Mechanics." *Centaurus* 25, no. 2 (1981): 189–221.

Hentschel, Klaus. *Mapping the Spectrum: Techniques of Visual Representation in Research and Teaching*. Oxford: Oxford University Press, 2002.

Hentschel, Klaus, ed. *Physics and National Socialism: An Anthology of Primary Sources*. Translated by Ann M. Hentschel. Basel: Birkhäser, 1996.

Hentschel, Klaus. "The Conversion of St. John: A Case Study on the Interplay of Theory and Experiment." *Science in Context* 6, no. 1 (1993): 137–194.

Hentschel, Klaus. *Visual Cultures in Science and Technology: A Comparative History*. Oxford: Oxford University Press, 2014.

Hesse, Mary B. *Models and Analogies in Science*. Notre Dame: University of Notre Dame Press, 1966.

Higgs, Peter W. "Broken Symmetries, Massless Particles and Gauge Fields." *PL* 12, no. 2 (1964): 132–133.

Higgs, Peter W. "SBGT and All That." In *Discovery of Weak Neutral Currents: The Weak Interaction Before and After*, 159–163. New York: AIP Conference Proceedings, 1994.

Higgs, Peter W. "Spontaneous Symmetry Breakdown without Massless Bosons." *PR* 145, no. 4 (1966): 1156–1163.

Hirosige, Tetu. "Origins of Lorentz' Theory of Electrons and the Concept of the Electromagnetic Field." *HSPS* 1 (1969): 151–209.

Hirosige, Tetu. "The Ether Problem, the Mechanistic Worldview, and the Origins of the Theory of Relativity." *HSPS* 7 (1976): 3–82.

Hoddeson, Lillian, Gordon Baym, and Michael Eckert. "The Development of the Quantum Mechanical Electron Theory of Metals: 1926–1933." In Hoddeson, Braun, Teichmann, and Weart, *Crystal Maze*, 88–181.

Hoddeson, Lillian, Ernest Braun, Jürgen Teichmann, and Spencer Weart, eds. *Out of the Crystal Maze: Chapters from the History of Solid-state Physics*. New York: Oxford University Press, 1992.

Hoddeson, Lillian, Laurie Brown, Michael Riordan, and Max Dresden, eds. *The Rise of the Standard Model: Particle Physics in the 1960s and 1970s*. Cambridge: Cambridge University Press, 1997.

Hoddeson, Lillian, Helmut Schubert, Steve Heims, and Gordon Baym. "Collective Phenomena." In Hoddeson, Braun, Teichmann, and Weart, *Crystal Maze*, 489–616.

Hoffmann, Christoph. "Processes on Paper: Writing Procedures as Non-material Research Devices." *Science in Context* 26, no. 2 (2013): 279–303.

Hoffmann, Dieter. "Kriegsschicksale: Hans Euler (1909–1941)." *Physikalische Blätter* 45, no. 9 (1989): 382–384.

Hoffmann, Dieter, and Mark Walker, eds. *The German Physical Society in the Third Reich: Physicists between Autonomy and Accommodation*. Translated by Ann M. Hentschel. Cambridge: Cambridge University Press, 2012.

Hofstadter, Douglas R. *Gödel, Escher, Bach: An Eternal Golden Braid*. 20th Anniversary Ed. New York: Basic Books, 1999.

Holmes, Frederic L., Jürgen Renn, and Hans-Jörg Rheinberger, eds. *Reworking the Bench: Research Notebooks in the History of Science*. New York: Kluwer Academic, 2003.

Holton, Gerald. "Mach, Einstein, and the Search for Reality." *Daedalus* 97, no. 2 (1968): 636–673.

Holton, Gerald. "The Birth and Early Days of the Fermi Group in Rome." In *Proceedings of the International Conference "Enrico Fermi and the Universe of Physics"*, edited by C. Bernardini, L. Bonolis, G. Ghisu, D. Savelli, and L. Falera, 53–70. Roma: ENEA / Accademia Nazionale dei Lincei, 2003.

Holton, Gerald. "The Role of Themata in Science." *Foundations of Physics* 26, no. 4 (1996): 453–465.

't Hooft, Gerard. "Computation of the Quantum Effects Due to a Four-dimensional Pseudoparticle." *Physical Review D* 14, no. 12 (1976): 3432–3450.

't Hooft, Gerard. "Renormalizable Lagrangians for Massive Yang-Mills Fields." *NP* B35 (1971): 167–188.

Howard, Don. "Who Invented the 'Copenhagen Interpretation'? A Study in Mythology." *Philosophy of Science* 71, no. 5 (2004): 669–682.

Hoyle, Fred. "Man's Place in the Expanding Universe." *The Listener* 1102 (1950): 419–425.

Hoyle, Fred. *The Nature of the Universe: A Series of Broadcast Lectures*. Oxford: B. Blackwell, 1950.

Hoyle, Fred, and J. V. Narlikar. "Mach's Principle and the Creation of Matter [and Discussion]." *PRSLA* 270, no. 1342 (1962): 334–341.

Huggett, Nick, ed. *Space from Zeno to Einstein: Classic Readings with a Contemporary Commentary*. Cambridge, MA: MIT Press, 1999.

Hund, Friedrich. *Linienspektren und periodisches System der Elemente*. Berlin: Springer, 1927.

Hunt, Bruce J. "[Review] Edmund Whittaker. A History of the Theories of Aether and Electricity. Introduction by Arthur I. Miller." *Isis* 79, no. 3 (1988): 515–516.

Hunt, Bruce J. *The Maxwellians*. Ithaca, NY: Cornell University Press, 1991.

Iggers, Georg G. *Historiography in the Twentieth Century: From Scientific Objectivity to the Postmodern Challenge*. Expanded ed. Middletown, CT: Wesleyan University Press, 2005.

Iliopoulos, John. "Plenary Report on Progress in Gauge Theories." In *High Energy Physics. Proceedings, 17th International Conference, ICHEP 1974, London, England, July 01–July 10, 1974*, edited by J. R. Smith, III:89–116. Oxford: Rutherford Labratory, 1974.

Infeld, Leopold, ed. *Conférence internationale sur les théories relativistes de la gravitation: Sous la direction de L. Infeld*. Paris: Gauthier-Villars, 1964.

Infeld, Leopold, ed. *Proceedings on Theory of Gravitation: Conference in Warszawa and Jabłonna 25–31 July, 1962*. Warsaw: PWN–Editions Scientifiques de Pologne, 1964.

Infeld, Leopold. *Why I Left Canada: Reflections on Science and Politics*. Edited by Lewis Pyenson. Translated by Helen Infeld. Montreal: McGill-Queen's University Press, 1978.

Institut International de Physique Solvay. *Les particules élémentaires: huitième Conseil de physique, tenu à l'Université de Bruxelles du 27 septembre au 2 octobre 1948*. Bruxelles: R. Stoops, 1950.

Institut International de Physique Solvay. *Structure et propriétés des noyaux atomiques: rapports et discussions du septième Conseil de physique tenu à Bruxelles du 22 du 29 octobre 1933*. Paris: Gauthier-Villars, 1934.

Ito, Kenji. "Cultural Difference and Sameness: Historiographic Reflections on Histories of Physics in Modern Japan." In Chemla and Keller, *Cultures without Culturalism*, 49–68.

Ito, Kenji. "Making Sense of Ryôshiron (Quantum Theory): Introduction of Quantum Mechanics into Japan, 1920–1940." PhD diss., Harvard University, 2002.

Ito, Kenji. "Superposing Dynamos and Electrons: Electrical Engineering and Quantum Physics in the Case of Nishina Yoshio." In Katzir, Lehner, and Renn, *Traditions and Transformations in the History of Quantum Physics*, 181–208.

Itô, Daisuke, Zirô Koba, and Sin-itiro Tomonaga. "Corrections Due to the Reaction of 'Cohesive Force Field' to the Elastic Scattering of an Electron. I." *Progress of Theoretical Physics* 3, no. 3 (1948): 276–289.

Itzykson, Claude, and Jean-Bernard Zuber. *Quantum Field Theory*. New York: McGraw-Hill, 1980.

Jackson, John David. *Classical Electrodynamics*. 3rd ed. New York: Wiley, 1998.

Jacobsen, Anja Skaar. *Léon Rosenfeld: Physics, Philosophy, and Politics in the Twentieth Century*. Singapore: World Scientific, 2012.

Jaffe, R. L. "Casimir Effect and the Quantum Vacuum." *PRD* 72, no. 2 (2005): 021301.

James, Jeremiah, and Christian Joas. "Subsequent and Subsidiary? Rethinking the Role of Applications in Establishing Quantum Mechanics." *HSNS* 45, no. 5 (2015): 641–702.

James, Reginald W., Ivar Waller, and Douglas R. Hartree. "An Investigation into the Existence of Zero-point Energy in the Rock-salt Lattice by an X-ray Diffraction Method." *PRSLA* 118, no. 779 (1928): 334–350.

Jammer, Max. *Concepts of Force: A Study in the Foundations of Dynamics*. Cambridge, MA: Harvard University Press, 1957.

Jammer, Max. *Concepts of Space: The History of Theories of Space in Physics*. 3rd ed. New York: Dover, 1993.

Jammer, Max. *The Conceptual Development of Quantum Mechanics*. New York: McGraw-Hill, 1966.

Janssen, Michel. "'No Success like Failure...' Einstein's Quest for General Relativity, 1907–1920." In Janssen and Lehner, *Cambridge Companion to Einstein*, 167–227.

Janssen, Michel, and Christoph Lehner, eds. *The Cambridge Companion to Einstein*. Cambridge: Cambridge University Press, 2014.

Janssen, Michel, John D. Norton, Jürgen Renn, Tilman Sauer, and John Stachel, eds. *The Genesis of General Relativity*. 4 vols. Dordrecht: Springer, 2007.

Jauch, Josef-Maria, and Fritz Rohrlich. *The Theory of Photons and Electrons*. Reading, MA: Addison-Wesley, 1955.

Jeffreys, Bertha Swirles. "John Arthur Gaunt (1904–1944)." *Notes and Records of the Royal Society of London* 44, no. 1 (1990): 73–79.

Jona-Lasinio, Giovanni. "Relativistic Field Theories with Symmetry-breaking Solutions." *INC* 34, no. 6 (1964): 1790–1795.

Josephson, Paul R. *Physics and Politics in Revolutionary Russia*. Berkeley: University of California Press, 1991.

Jost, Res, and J. M. Luttinger. "Vacuumpolarisation und e^4-ladungsrenormalisation für Elektronen." *Helvetica Physica Acta* 23, nos. I–II (1950): 201–213.

Jungnickel, Christa, and Russell McCormmach. *Intellectual Mastery of Nature: Theoretical Physics from Ohm to Einstein*. Chicago: University of Chicago Press, 1986.

Jungnickel, Christa, and Russell McCormmach. *The Second Physicist*. Berlin: Springer, 2017.

Kaiser, David. "A ψ is just a ψ? Pedagogy, Practice, and the Reconstitution of General Relativity, 1942–1975." *SHPMP* 29, no. 3 (1998): 321–338.

Kaiser, David. "A Tale of Two Textbooks: Experiments in Genre." *Isis* 103, no. 1 (2012): 126–138.

Kaiser, David. "Cold War Curvature: Measuring and Modeling Gravitational Systems in Postwar American Physics." 2015.

Kaiser, David. "Cold War Requisitions, Scientific Manpower, and the Production of American Physicists after World War II." *HSPBS* 33, no. 1 (2002): 131–160.

Kaiser, David. *Drawing Theories Apart: The Dispersion of Feynman Diagrams in Postwar Physics*. Chicago: University of Chicago Press, 2005.

Kaiser, David. *How the Hippies Saved Physics: Science, Counterculture, and the Quantum Revival*. New York: W. W. Norton, 2011.

Kaiser, David, ed. *Pedagogy and the Practice of Science: Historical and Contemporary Perspectives*. Cambridge, MA: MIT Press, 2005.

Kaiser, David. "Stick-Figure Realism: Conventions, Reification, and the Persistence of Feynman Diagrams, 1948–1964." *Representations*, no. 70 (2000): 49–86.

Kaiser, David. "Whose Mass Is It Anyway? Particle Cosmology and the Objects of Theory." *SSS* 36, no. 4 (2006): 533–564.

Kaiser, Walter. "Early Theories of the Electron Gas." *HSPBS* 17, no. 2 (1987): 271–297.

Kalckar, Jørgen, ed. "Correspondence Included." In *Foundations of Quantum Physics I (1926–1932)*. Vol. 6 of *Niels Bohr Collected Works*, 412–473. Amsterdam: Elsevier, 1985.

Kalckar, Jørgen, ed. "Correspondence Included." In *Foundations of Quantum Physics II (1933–1958)*. Vol. 7 of *Niels Bohr Collected Works*, 434–515. Amsterdam: Elsevier, 1996.

Kalckar, Jørgen. "Introduction." In *Foundations of Quantum Physics II (1933–1958)*. Vol. 7 of *Niels Bohr Collected Works*, edited by Jørgen Kalckar, 3–51. Amsterdam: Elsevier, 1996.

Kangro, Hans. *Early History of Planck's Radiation Law*. Translated by R. E. W. Maddison. London: Taylor & Francis, 1976.

Kapitza, Peter L. "A Method of Producing Strong Magnetic Fields." *PRSLA* 105, no. 734 (1924): 691–710.

Karplus, Robert, and Maurice Neuman. "Non-linear Interactions between Electromagnetic Fields." *PR* 80, no. 3 (1950): 380–385.

Katzir, Shaul. "Theoretical Challenges by Experimental Physics: Radiation and Its Interaction with Matter." In Katzir, Lehner, and Renn, *Traditions and Transformations in the History of Quantum Physics*, 11–28.

Katzir, Shaul, Christoph Lehner, and Jürgen Renn, eds. *Traditions and Transformations in the History of Quantum Physics*. Berlin: Edition Open Access, 2013.

Keller, Evelyn Fox. "Cognitive Repression in Contemporary Physics." *American Journal of Physics* 47, no. 8 (1979): 718–721.

Keller, Evelyn Fox. *Reflections on Gender and Science*. 10th anniversary ed. New Haven, CT: Yale University Press, 1995.

Keller, Evelyn Fox. "The Anomaly of a Woman in Physics." In *Working It Out*, edited by Sara Ruddick and Pamela Daniels, 77–91. New York: Pantheon Books, 1977.

Kemble, Edwin C. "The General Principles of Quantum Mechanics. Part I." *The Physical Review Supplement* 1, no. 2 (1929): 157–215.

Kemmer, Nicholas. "Über die Lichtstreuung an elektrischen Feldern nach der Theorie des Positrons." *Helvetica Physica Acta* 10, no. 2 (1937): 112–122.

Kemmer, Nicholas, and Victor Weisskopf. "Deviations from the Maxwell Equations Resulting from the Theory of the Positron." *Nature* 137, no. 3468 (1936): 659.

Kennefick, Daniel. "Not Only because of Theory: Dyson, Eddington, and the Competing Myths of the 1919 Eclipse Expedition." In *Einstein and the Changing Worldviews of Physics*, edited by Christoph Lehner, Jürgen Renn, and Matthias Schemmel, 201–232. Boston, MA: Birkhäuser, 2012.

Kennefick, Daniel. "Star Crushing: Theoretical Practice and the Theoreticians' Regress." *SSS* 30, no. 1 (2000): 5–40.

Kennefick, Daniel. *Traveling at the Speed of Thought: Einstein and the Quest for Gravitational Waves*. Princeton, NJ: Princeton University Press, 2007.

Kevles, Daniel J. *In the Name of Eugenics: Genetics and the Uses of Human Heredity*. Cambridge, MA: Harvard University Press, 1995.

Kevles, Daniel J. *The Physicists: The History of a Scientific Community in Modern America*. Cambridge, MA: Harvard University Press, 1995.

Kibble, Tom W. B. "Englert-Brout-Higgs-Guralnik-Hagen-Kibble Mechanism." *Scholarpedia*.

Kibble, Tom W. B. "Symmetry Breaking in Non-Abelian Gauge Theories." *PR* 155, no. 5 (1967): 1554–1561.

Kiefer, Claus. "Quantum Geometrodynamics: Whence, Whither?" *GRG* 41, no. 4 (2009): 877–901.

Kilmister, C. "More about the King and the Vizier." *GRG* 2, no. 1 (1971): 35–36.

Kilpatrick, Franklin P., ed. *Human Behavior from the Transactional Point of View*. Hanover, NH: Institute for Associated Research, 1952.

Kirzhnits, David A., and Andrei D. Linde. "Macroscopic Consequences of the Weinberg Model." *PL* B42, no. 4 (1972): 471–474.

Kitcher, Philip. "Real Realism: The Galilean Strategy." *Philosophical Review* 110, no. 2 (2001): 151–197.

Klein, Martin J. "Fluctuations and Statistical Physics in Einstein's Early Work." In *Albert Einstein: Historical and Cultural Perspectives*, edited by Gerald Holton and Yehuda Elkana, 39–58. Mineola, NY: Dover, 1982.

Klein, Morris. *Mathematical Thought from Ancient to Modern Times*. New York: Oxford University Press, 1972.

Klein, Oskar, and Yoshio Nishina. "On the Scattering of Radiation by Free Electrons According to Dirac's New Relativistic Quantum Dynamics." In *The Oskar Klein Memorial Lectures*, edited by Gösta Ekspong, translated by Lars Bergström, 2:113–129. 1929. Singapore: World Scientific, 1994.

Klein, Ursula. *Experiments, Models, Paper Tools: Cultures of Organic Chemistry in the Nineteenth Century*. Stanford, CA: Stanford University Press, 2003.

Klimchitskaya, G. L., U. Mohideen, and V. M. Mostepanenko. "The Casimir Force between Real Materials: Experiment and Theory." *RMP* 81, no. 4 (2009): 1827–1885.

Knorr Cetina, Karin. "Objectual Practice." In *The Practice Turn in Contemporary Theory*, edited by Karin Knorr Cetina, Theodore Schatzki, and Eike von Savigny, 184–197. London: Routledge, 2001.

Koba, Zirô, and Sin-itiro Tomonaga. "Application of the 'Self-consistent' Subtraction Method to the Elastic Scattering of an Electron." *Progress of Theoretical Physics* 2, no. 4 (1947): 218.

Koba, Zirô, and Sin-itiro Tomonaga. "On Radiation Reactions in Collision Processes. I: Application of the 'Self-consistent' Subtraction Method to the Elastic Scattering of an Electron." *Progress in Theoretical Physics* 3, no. 3 (1948): 290–303.

Kojevnikov, Alexei. "Dirac's Quantum Electrodynamics." In *Einstein Studies in Russia*, edited by Yuri Balashov and Vladimir Vizgin, 229–259. Boston: Birkhäuser, 2002.

Kojevnikov, Alexei. "Einstein's Fluctuation Formula and the Wave-particle Duality." In *Einstein Studies in Russia*, edited by Yuri Balashov and Vladimir Vizgin, 181–228. Boston: Birkhäuser, 2002.

Kojevnikov, Alexei. "Freedom, Collectivism, and Quasiparticles: Social Metaphors in Quantum Physics." *HSPBS* 29, no. 2 (1999): 295–331.

Kojevnikov, Alexei, ed. *Paul Dirac and Igor Tamm Correspondence; 1, 1928-1933*. Munich: Max-Planck Institut für Physik, 1993.

Kojevnikov, Alexei. *Stalin's Great Science: The Times and Adventures of Soviet Physicists*. London: Imperial College Press, 2004.

Korff, S. A., and Gregory Breit. "Optical Dispersion." *RMP* 4, no. 3 (1932): 471–503.

Kragh, Helge. "Bohr-Kramers-Slater Theory." In *Compendium of Quantum Physics*, edited by Daniel Greenberger, Klaus Hentschel, and Friedel Weinert, 62–64. Berlin: Springer, 2009.

Kragh, Helge. *Conceptions of Cosmos: From Myths to the Accelerating Universe : A History of Cosmology*. Oxford: Oxford University Press, 2007.

Kragh, Helge. "Contemporary History of Cosmology and the Controversy over the Multiverse." *AS* 66, no. 4 (2009): 529–551.

Kragh, Helge. "Cosmo-Physics in the Thirties: Towards a History of Dirac Cosmology." *HSPS* 13 (1982): 69–108.

Kragh, Helge. *Cosmology and Controversy: The Historical Development of Two Theories of the Universe.* Princeton, NJ: Princeton University Press, 1996.

Kragh, Helge. *Dirac: A Scientific Biography.* Cambridge: Cambridge University Press, 1990.

Kragh, Helge. *Niels Bohr and the Quantum Atom: The Bohr Model of Atomic Structure 1913–1925.* Oxford: Oxford University Press, 2012.

Kragh, Helge. "Physics and Cosmology." In Buchwald and Fox, *The Oxford Handbook of the History of Physics*, 892–922.

Kragh, Helge. "Preludes to Dark Energy: Zero-point Energy and Vacuum Speculations." *AHES* 66, no. 3 (2012): 199–240.

Kragh, Helge. *Quantum Generations: A History of Physics in the Twentieth Century.* Princeton, NJ: Princeton University Press, 1999.

Kragh, Helge. "Relativistic Collisions: The Work of Christian Møller in the Early 1930s." *AHES* 43, no. 4 (1992): 299–328.

Kragh, Helge. "The Genesis of Dirac's Relativistic Theory of Electrons." *AHES* 24, no. 1 (1981): 31–67.

Kragh, Helge. "Unifying Quanta and Relativity? Schrödinger's Attitude to Relativistic Quantum Mechanics." In *Erwin Schrödinger: Philosophy and the Birth of Quantum Mechanics*, edited by Michel Bitbol and Olivier Darrigol, 315–338. Gif-sur-Yvette, France: Editions Frontiers, 1992.

Kragh, Helge, and Simon Rebsdorf. "Before Cosmophysics: E. A. Milne on Mathematics and Physics." *SHPMP* 33, no. 1 (2002): 35–50.

Kragh, Helge S., and James M. Overduin. *The Weight of the Vacuum: A Scientific History of Dark Energy.* Heidelberg: Springer, 2014.

Kramers, Hendrik A. "Non-relativistic Quantum-electrodynamics and Correspondence Principle." In Institut Solvay, *Particules élémentaires*, 241–265.

Kramers, Hendrik A., and Werner Heisenberg. "On the Dispersion of Radiation by Atoms." In Waerden, *Sources of Quantum Mechanics*, 223–252.

Krige, John. *American Hegemony and the Postwar Reconstruction of Science in Europe.* Cambridge, MA: MIT Press, 2006.

Krige, John. "CERN from the Mid-1960s to the Late 1970s." In *History of CERN*, edited by John Krige, 3:3–38. Amsterdam: Elsevier, 1996.

Krive, I. V., and Andre D. Linde. "On the Vacuum Stability in the σ Model." *NP B* 117, no. 1 (1976): 265–268.

Kronheimer, E. H., and Roger Penrose. "On the Structure of Causal Spaces." *MPCPS* 63, no. 2 (1967): 481–501.

Kruskal, Martin D. "Maximal Extension of Schwarzschild Metric." *PR* 119, no. 5 (1960): 1743–1745.

Kuhn, Thomas S. *Black-Body Theory and the Quantum Discontinuity, 1894–1912*. 1978. Chicago: University of Chicago Press, 1987.

Kuhn, Thomas S. *The Structure of Scientific Revolutions*. 3rd ed. 1962. Chicago: University of Chicago Press, 1996.

Lamb, Willis E. "Anomalous Fine Structure of Hydrogen and Singly Ionized Helium." *Reports on Progress in Physics* 14, no. 1 (1951): 19–63.

Lamb, Willis E., and Robert C. Retherford. "Fine Structure of the Hydrogen Atom by a Microwave Method." *PR* 72, no. 3 (1947): 241–243.

Landau, Lev. "The Damping Problem in Wave Mechanics." In *Collected Papers of L. D. Landau*, edited by D. Ter Haar, 8–18. 1927. Pergamon, 1965.

Landau, Lev, and Rudolf Peierls. "Extension of the Uncertainty Principle to Relativistic Quantum Theory." In *Collected Papers of L. D. Landau*, edited by D. Ter Haar, 40–51. 1931. New York: Gordon and Breach Science Publishers, 1965.

Latour, Bruno. *Science in Action: How to Follow Scientists and Engineers through Society*. Cambridge, MA: Harvard University Press, 1987.

Lauritsen, Charles C., and J. Robert Oppenheimer. "On the Scattering of the Th C" γ-rays." *PR* 46, no. 1 (1934): 80–81.

Law, Jonathan, and Richard Rennie, eds. *A Dictionary of Physics*. 7th ed. Oxford: Oxford University Press, 2015.

Le Guin, Ursula. *The Dispossessed: An Ambiguous Utopia*. 1974. New York: HarperCollins, 2001.

Le Guin, Ursula. "The Stone Ax and the Muskoxen." In *The Language of the Night: Essays on Fantasy and Science Fiction*, Rev. ed., edited by Susan Wood, 225–239. 1975. New York: Putnam, 1993.

Lee, T. D., and Gian Carlo Wick. "Vacuum Stability and Vacuum Excitation in a Spin-0 Field Theory." *PRD* 9, no. 8 (1974): 2291–2316.

Lehmkuhl, Dennis. "General Relativity as a Hybrid Theory: The Genesis of Einstein's Work on the Problem of Motion." *SHPMP*, 2017, 1–15.

Lehmkuhl, Dennis. "Why Einstein Did Not Believe That General Relativity Geometrizes Gravity." *SHPMP* 46 (2014): 316–326.

Leibniz, Gottfried Wilhelm. *Discourse on Metaphysics, Correspondence with Arnauld, and Monadology*. Translated by George R. Montgomery. Chicago: Open Court, 1902.

Leslie, Stuart W. *The Cold War and American Science: The Military-Industrial-Academic Complex at MIT and Stanford*. New York: Columbia University Press, 1993.

Lichnerowicz, André, and Marie-Antoinette Tonnelat, eds. *Fluides et champ gravitationnel en relativité générale Paris, Collège de France, 19–23 juin 1967*. Paris: C.N.R.S., 1969.

Lichnerowicz, André, and Marie-Antoinette Tonnelat, eds. *Les théories relativistes de la gravitation*. Paris: CNRS, 1962.

Lightman, Alan. "Interview of Roger Penrose," 1989. American Institute of Physics. http://www.aip.org/history-programs/niels-bohr-library/oral-histories/34322.

Linde, Andrei. "On the Vacuum Instability and the Higgs Meson Mass." *PL* B70, no. 3 (1977): 306–308.

Linde, Andrei D. "Vacuum Instability, Cosmology and Constraints on Particle Masses in the Weinberg-Salam Model." *PL* B92, nos. 1, 2 (1980): 119–121.

Lippmann, B. A., and Julian Schwinger. "Variational Principles for Scattering Processes. I." *PR* 79, no. 3 (1950): 469–480.

Llewellyn Smith, Christopher Hubert. *An Introduction to Renormalizable Theories of Weak Interactions and Their Experimental Consequences.* Technical report TH-1710. Geneva: CERN, 1973.

London, Fritz, and Edmond Bauer. "The Theory of Observation in Quantum Mechanics." In Zurek and Wheeler, *Quantum Theory and Measurement,* 217–259.

Lorentz, Hendrik A. *The Theory of Electrons and Its Application to the Phenomena of Light and Radiant Heat.* 2nd ed. Leipzig: B. G. Teubner, 1916.

Lyman Spitzer, Jr. *A Propsed Stellarator.* Technical report PM-S-1. Princeton, NJ: Project Matterhorn and Princeton University, 1951.

Ma, Shih-Tsun. "Vacuum Polarization in the Positron Theory." *PM* 40, no. 310 (1949): 1112–1128.

Macagno, Enzo O. "Historico-critical Review of Dimensional Analysis." *Journal of the Franklin Institute* 292, no. 6 (1971): 391–402.

Magnus, P. D., and Craig Callender. "Realist Ennui and the Base Rate Fallacy." *Philosophy of Science* 71, no. 3 (2004): 320–338.

Maiocchi, Roberto. "The Case of Brownian Motion." *BJHS* 23, no. 3 (1990): 257–283.

Mancosu, Paolo. "Visualization in Logic and Mathematics." In Mancosu, Jørgensen, and Pedersen, *Visualization, Explanation, and Reasoning,* 13–29.

Mancosu, Paolo, Klaus Frovin Jørgensen, and Stig Andur Pedersen, eds. *Visualization, Explanation and Reasoning Styles in Mathematics.* Dordecht: Springer, 2005.

Margenau, Henry, and William W. Watson. "Pressure Effects on Spectral Lines." *RMP* 8, no. 1 (1936): 22–53.

Martin, Joseph D. "Fundamental Disputations: The Philosophical Debates That Governed American Physics, 1939–1993." *HSNS* 45, no. 5 (2015): 703–757.

Martin, Paul C., and Julian Schwinger. "Theory of Many-Particle Systems. I." *PR* 115, no. 6 (1959): 1342–1373.

Masani, P. Rustom. *Norbert Wiener, 1894–1964.* Basel: Birkhäuser, 1990.

Massimi, Michela. "Three Tales of Scientific Success." *Philosophy of Science* 83, no. 5 (2016): 757–767.

Mathews, George Ballard. *Projective Geometry.* London: Longmans, Green, & Co., 1914.

Maxwell, James Clerk. "A Dynamical Theory of the Electromagnetic Field." *Philosophical Transactions of the Royal Society. London* 155 (1865): 459–512.

Maxwell, James Clerk. *A Treatise on Electricity and Magnetism.* Unabridged 3rd ed. Edited by W. D. Niven. 2 vols. 1891. Reprint, New York: Dover Publications, 1954.

Maxwell, James Clerk. "Ether." Chap. 97 in *Scientific Papers,* edited by W. D. Niven, 2:763–775. 1878. New York: Dover, 1890.

McAllister, J. W. "Dirac and the Aesthetic Evaluation of Theories." *Methodology and Science* 23 (1990): 87–102.

Mehra, Jagdish. *The Beat of a Different Drum: The Life and Science of Richard Feynman.* Oxford: Oxford University Press, 1994.

Mehra, Jagdish, ed. *The Physicist's Conception of Nature.* Dordrecht: Reidel, 1973.

Mehra, Jagdish, and Kimball Milton. *Climbing the Mountain: The Scientific Biography of Julian Schwinger.* New York: Oxford University Press, 2003.

Mehra, Jagdish, and Helmut Rechenberg. "Planck's Half-quanta: A History of the Concept of Zero-point Energy." *Foundations of Physics* 29, no. 1 (1999): 91–132.

Mehra, Jagdish, and Helmut Rechenberg. *The Historical Development of Quantum Theory.* 6 vols. New York: Springer, 1982–2001.

Meitner, Lise. "Die Streuung Harter γ-strahlen." *Naturwissenschaften* 22, no. 11 (1934): 174.

Meitner, Lise, and H. Kösters. "Über die Streuung Kurzwelliger γ-strahlen." *ZP* 84, nos. 3–4 (1933): 137–144.

Mermin, N. David. "Could Feynman Have Said This?" *PT* 57, no. 5 (2004): 10–11.

Meyenn, Karl von. "The Origins of Nuclear Physics and Carl Friedrich von Weizsäcker's Semi-empirical Mass Formula." In *Time, Quantum and Information,* edited by Lutz Castell and Otfried Ischebeck, 83–114. Berlin: Springer, 2003.

Mialet, Hélène. *Hawking Incorporated: Stephen Hawking and the Anthropology of the Knowing Subject.* Chicago: University of Chicago Press, 2012.

Mielenz, Klaus D., Robert D. Saunders Jr., and John B. Shumaker. "Spectroradiometric Determination of the Freezing Temperature of Gold." *Journal of Research of the National Institute of Standards and Technology* 95, no. 1 (1990): 49–67.

Milam, Erika Lorraine, and Robert A. Nye. "An Introduction to Scientific Masculinities." *Osiris* 30 (2015): 1–14.

Milburn, Colin. "Ahead of Time: Gerald Feinberg and the Governance of Futurity." *Osiris* 34, no. 1 (2019): 216–237.

Milburn, Colin. "Modifiable Futures: Science Fiction at the Bench." *Isis* 101, no. 3 (2010): 560–569.

Miller, Arthur I. *Early Quantum Electrodynamics: A Source Book.* Translated by Walter Grant. Cambridge: Cambridge University Press, 1994.

Miller, Michael E. "The Origins of Schwinger's Euclidean Green's Functions." *SHPMP* 50 (2015): 5–12.

Milne, E. A. "Radiative Equilibrium in the Outer Layers of a Star." *Monthly Notices of the Royal Astronomical Society* 81 (1921): 361–375.

Milne, E. A. *Relativity, Gravitation, and World-structure.* Oxford: Clarendon Press, 1935.

Milonni, Peter W. *The Quantum Vacuum: An Introduction to Quantum Electrodynamics.* San Diego, CA: Academic Press, 1994.

Milonni, Peter W. "Why Spontaneous Emission?" *American Journal of Physics* 52, no. 4 (1984): 340–343.

Minkowski, Hermann. *Raum und Zeit.* Leipzig: B. G. Teubner, 1909.

Minkowski, Hermann. "Raum und Zeit / Space and Time." In Petkov, *Minkowski Spacetime*, xiv–xlii.

Minkowski, Hermann. "Time and Space." Translated by Edward H. Carus. *The Monist* 28, no. 2 (1918): 288–302.

Misner, Charles W. "Feynman Quantization of General Relativity." *RMP* 29, no. 3 (1957): 497–509.

Misner, Charles W. "John Wheeler and the Recertification of General Relativity as True Physics." In *General Relativity and John Archibald Wheeler*, edited by Ignazio Ciufolini and Richard A. Matzner, 9–27. Springer, 2010.

Misner, Charles W. "Waves, Newtonian Fields, and Coordinate Functions." In Infeld, *Proceedings on Theory of Gravitation*, 189–205.

Misner, Charles W., Kip S. Thorne, and John Archibald Wheeler. *Gravitation*. San Francisco: W. H. Freeman and Company, 1973.

Misner, Charles W., and John Archibald Wheeler. "Classical Physics as Geometry." *Annals of Physics* 2, no. 6 (1957): 525–603.

Mody, Cyrus C. M. "What Do Scientists and Engineers Do All Day? On the Structure of Scientific Normalcy." In *Kuhn's Structure of Scientific Revolutions—50 Years On*, edited by William J. Devlin and Alisa Bokulich, 91–104. Heidelberg: Springer, 2015.

Monk, Ray. *Inside the Center: The Life of J. Robert Oppenheimer*. London: Jonathan Cape, 2012.

Morette-DeWitt, Cécile. *The Pursuit of Quantum Gravity: Memoirs of Bryce DeWitt from 1946 to 2004*. Heidelberg: Springer, 2011.

Morgan, Mary, and Margaret Morrison, eds. *Models as Mediators: Perspectives on Natural and Social Science*. Cambridge: Cambridge University Press, 1999.

Morrison, Margaret. *Reconstructing Reality: Models, Mathematics, and Simulations*. Oxford: Oxford University Press, 2015.

Morrison, Margaret. "Where Have All the Theories Gone?" *Philosophy of Science* 74, no. 2 (2007): 195–228.

Morrison, Philip. "Energy Fluctuations in the Electromagnetic Field." *PR* 56, no. 9 (1939): 937–940.

Mott, Neville F., and H. S. W. Massey. *The Theory of Atomic Collisions*. Oxford: Clarendon Press, 1933.

Moyer, Donald Franklin. "Evaluations of Dirac's Electron, 1928–1932." *American Journal of Physics* 49, no. 11 (1981): 1055–1062.

Moyer, Donald Franklin. "Origins of Dirac's Electron, 1925–1928." *American Journal of Physics* 49, no. 10 (1981): 944–949.

Moyer, Donald Franklin. "Vindications of Dirac's Electron, 1932–1934." *American Journal of Physics* 49, no. 12 (1981): 1120–1125.

Museums Correspondent. "Piltdown Man Forgery." *Times* [London, England], November 21st, 1953, 6.

Myrvold, Wayne. "Philosophical Issues in Quantum Theory." In *The Stanford Encyclopedia of Philosophy*, Fall 2018, edited by Edward N. Zalta. Metaphysics Research Lab, Stanford University, 2018. https://plato.stanford.edu/archives/fall2018/entries/qt-issues/.

Nambu, Yoichiro. "Dynamical Theory of Elementary Particles Suggested by Superconductivity." In *Proceedings, 10th International Conference on High-Energy Physics (ICHEP 60): Rochester, NY, USA, 25 Aug–1 Sep 1960*, edited by Ennackal Chandy George Sudarshan, John H. Tinlot, and Adrian Constantin Melissinos, 858–866. Rochester, NY: Rochester University, 1960.

Nambu, Yoichiro. "Gauge Principle, Vector-meson Dominance, and Spontaneous Symmetry Breaking." In Brown, Dresden, and Hoddeson, *Pions to Quarks*, 639–644.

Nambu, Yoichiro. "Nobel Lecture: Spontaneous Symmetry Breaking in Particle Physics: A Case of Cross Fertilization." *RMP* 81, no. 3 (2009): 1015–1018.

Nambu, Yoichiro. "Quasi-Particles and Gauge Invariance in the Theory of Superconductivity." *PR* 117, no. 3 (1960): 648–663.

Nambu, Yoichiro, and Giovanni Jona-Lasinio. "Dynamical Model of Elementary Particles Based on an Analogy with Superconductivity. I." *PR* 122, no. 1 (1961): 345–358.

Nambu, Yoichiro, and Giovanni Jona-Lasinio. "Dynamical Model of Elementary Particles Based on an Analogy with Superconductivity. II." *PR* 124, no. 1 (1961): 246–254.

Nasim, Omar W. *Observing by Hand: Sketching Nebulae in the Nineteenth Century*. Chicago: University of Chicago Press, 2014.

Navarro, Jaume. "'A Dedicated Missionary.' Charles Galton Darwin and the New Quantum Mechanics in Britain." *SHPMP* 40, no. 4 (2009): 316–326.

Navarro, Jaume, ed. *Ether and Modernity: The Recalcitrance of an Epistemic Object in the Early Twentieth Century*. Oxford: Oxford University Press, 2018.

Netz, Reviel. "The Aesthetics of Mathematics: A Study." In Mancosu, Jørgensen, and Pedersen, *Visualization, Explanation, and Reasoning*, 251–194.

Neumann, John von. *Mathematische Gundlagen der Quantenmechanik*. Berlin: Julius Springer, 1932.

Newman, Ezra T. "A Biased and Personal Description of GR at Syracuse University, 1951–1961." In *The Universe of General Relativity*, edited by Anne J. Kox and Jean Eisenstaedt, 373–383. Boston: Birkhäser Boston, 2005.

Newman, Ezra T., and Roger Penrose. "10 Exact Gravitationally-conserved Quantities." *PRL* 15, no. 6 (1965): 231–233.

Newman, Ezra T., and Roger Penrose. "New Conservation Laws for Zero Rest-mass Fields in Asymptotically Flat Space-time." *PRSLA* 305, no. 1481 (1968): 175–204.

Newman, William R. *Newton the Alchemist: Science, Enigma, and the Quest for Nature's "Secret Fire"*. Prnceton, NJ: Princeton University Press, 2018.

Newton, Isaac. *The Principia: Mathematical Principles of Natural Philosophy*. Edited and translated by I. Bernard Cohen and Anne Miller Whitman. 1687. Berkeley: University of California Press, 1999.

Noakes, Richard. "Making Space for the Soul: Oliver Lodge, Maxwellian Psychics and the Etherial Body." In Navarro, *Ether and Modernity*, 88–106.

Norton, John D. "'Nature Is the Realisation of the Simplest Conceivable Mathematical Ideas': Einstein and the Canon of Mathematical Simplicity." *SHPMP* 31, no. 2 (2000): 135–170.

Noyes, H. P., E. M. Hafner, J. Klarmann, and A. E. Woodruff, eds. *Proceedings of the Fourth Annual Rochester Conference on High Energy Nuclear Physics, January 25–27, 1954*. Rochester, NY: University of Rochester, 1954.

Nye, Mary Jo. *Before Big Science: The Pursuit of Modern Chemistry and Physics, 1800–1940*. Cambridge, MA: Harvard University Press, 1996.

Nye, Mary Jo. *From Chemical Philosophy to Theoretical Chemistry: Dynamics of Matter and Dynamics of Disciplines, 1800–1950*. Berkeley: University of California Press, 1994.

Nye, Mary Jo, ed. *The Modern Physical and Mathematical Sciences*. Vol. 5 of *The Cambridge History of Science*. Cambridge: Cambridge University Press, 2002.

Olesko, Kathryn M. *Physics as a Calling: Discipline and Practice in the Königsberg Seminar for Physics*. Ithaca, NY: Cornell University Press, 1991.

Olesko, Kathryn M. "Science Pedagogy as a Category of Historical Analysis: Past, Present, and Future." *Science & Education* 15, no. 7 (2006): 863–880.

Oppenheimer, J. Robert. "[abstract:] The Theory of the Electron and Positive." *PR* 45, no. 4 (1934): 290.

Oppenheimer, J. Robert. "Are the Formulae for the Absorption of High Energy Radiations Valid?" *PR* 47, no. 1 (1935): 44–52.

Oppenheimer, J. Robert. "Electron Theory [and Discussion]." In Institut Solvay, *Particules élémentaires*, 269–286.

Oppenheimer, J. Robert. "Note on Charge and Field Fluctuations." *PR* 47, no. 2 (1935): 144–145.

Oppenheimer, J. Robert. "Note on Light Quanta and the Electromagnetic Field." *PR* 38, no. 4 (1931): 725–746.

Oppenheimer, J. Robert. "Note on the Theory of the Interaction of Field and Matter." *PR* 35, no. 5 (1930): 461–477.

Oppenheimer, J. Robert. *Science and the Common Understanding*. New York: Simon & Schuster, 1954.

Oppenheimer, J. Robert, and Milton S. Plesset. "On the Production of the Positive Electron." *PR* 44, no. 1 (1933): 53–55.

Oppenheimer, J. Robert, and Julian Schwinger. "On Pair Emission in the Proton Bombardment of Fluorine." *PR* 56, no. 10 (1939): 1066–1067.

Oppenheimer, J. Robert, and Hartland Snyder. "On Continued Gravitational Contraction." *PR* 56, no. 5 (1939): 455–459.

Oreskes, Naomi. "Objectivity or Heroism? On the Invisibility of Women in Science." *Osiris* 11 (1996): 87–113.

Pais, Abraham. *Inward Bound: Of Matter and Forces in the Physical World*. Oxford: Clarendon Press, 1986.

Pais, Abraham. *J. Robert Oppenheimer: A Life*. Oxford: Oxford University Press, 2006.

Pais, Abraham. *Niels Bohr's Times, in Physics, Philosophy, and Polity*. Oxford: Clarendon Press, 1991.

Pais, Abraham. "On the Dirac Theory of the Electron (1930–1936)." In *Werner Heisenberg: Collected Works*, edited by W. Blum, H.-P. Dürr, and H. Rechenberg, vol. A pt. II, 95–105. Berlin: Springer, 1989.

Pais, Abraham. "Playing with Equations, the Dirac Way." In *Paul Dirac: The Man and His Work*, edited by Peter Goddard, 93–116. Cambridge: Cambridge University Press, 1998.

Pais, Abraham. *Subtle Is the Lord: The Science and Life of Albert Einstein*. Oxford: Oxford University Press, 1982.

Pais, Abraham. *The Development of the Theory of the Electron*. Princeton, NJ: IAS / Princeton University Press, 1948.

Papatzacos, Paul, and Kjell Mork. "Delbrück Scattering." *Physics Reports* 21, no. 2 (1975): 81–118.

Pauli, Wolfgang. "Difficulties of Field Theories and of Field Quantization." In *Conference on Fundamental Particles and Low Temperatures*, 1:5–10.

Pauli, Wolfgang. *General Principles of Quantum Mechanics*. Translated by P. Achuthan and K. Venkatesan. 1933. Berlin: Springer, 1980.

Pauli, Wolfgang. "Les théories quantiques du magnétism. L' electron magnétique." In *Le magnétisme: rapports et discussions du sixième Conseil de physique, tenu à Bruxelles du 20 au 25 octobre 1930*, 175–280. Paris: Gauthier-Villars, 1932.

Pauli, Wolfgang. "On Dirac's New Method of Field Quantization." *RMP* 15, no. 3 (1943): 175–207.

Pauli, Wolfgang. "Quantentheorie." In *Handbuch der Physik*, edited by Hans Geiger, XXIII:1–278. Berlin: Springer, 1926.

Pauli, Wolfgang. "Relativistic Field Theories of Elementary Particles." *RMP* 13, no. 3 (1941): 203–232.

Pauli, Wolfgang. "Über den Zusammenhang des Abschlusses der Elektronengruppen im Atom mit der Komplexstruktur der Spektren." *ZP* 31, no. 1 (1925): 765–783.

Pauli, Wolfgang. *Wissenschaftlicher Briefwechsel mit Bohr, Einstein, Heisenberg*. Edited by Armin Hermann, Karl von Meyenn, and Victor Frederick Weisskopf. New York: Springer, 1979–2005.

Pauli, Wolfgang. "Zur Quantenmechanik des magnetischen Elektrons." *ZP* 43, no. 9 (1927): 601–623.

Pauli, Wolfgang, and Victor Weisskopf. "The Quantization of the Scalar Relativistic Wave Equation." In Miller, *Early Quantum Electrodynamics*, 188–205.

Pechenkin, Alexander. "The Early Statistical Interpretations of Quantum Mechanics in the USA and USSR." *SHPMP* 43, no. 1 (2012): 25–34.

Peebles, P. J. E. *Principles of Physical Cosmology*. Princeton, NJ: Princeton University Press, 1993.

Peierls, Rudolf. *Bird of Passage: Recollections of a Physicist*. Princeton, NJ: Princeton University Press, 1986.

Peierls, Rudolf. "Elektronentheorie der Metalle." In *Ergebnisse der exakten Naturwissenschaften*, 264–322. Berlin: Springer, 1932.

Peierls, Rudolf. "On the Theory of Galvano-magnetic Effects." In *Selected Scientific Papers*, 1–10.

Peierls, Rudolf. "On the Theory of the Hall Effect." In *Selected Scientific Papers*, 11.

Peierls, Rudolf. *Selected Scientific Papers of Sir Rudolf Peierls (with Commentary)*. Edited by R. H. Dalitz. Singapore: World Scientific, 1997.

Peierls, Rudolf. *Sir Rudolf Peierls: Selected Private and Scientific Correspondence*. Edited by Sabine Lee. 2 vols. Singapore: World Scientific, 2007–2009.

Peierls, Rudolf. "The Vacuum in Dirac's Theory of the Positive Electron." *PRSLA* 146, no. 857 (1934): 420–441.

Penrose, Lionel S., and Roger Penrose. "Impossible Objects: A Special Type of Visual Illusion." *British Journal of Psychology* 49, no. 1 (1958): 31–34.

Penrose, Lionel S., and Roger Penrose. "Puzzles for Christmas." *New Scientist* 4, no. 110 (1958): 1580–1581, 1597.

Penrose, Roger. "An Analysis of the Structure of Space-time." Chap. 28 in *Roger Penrose: collected works*, 1:579–729. 1965. Oxford: Oxford University Press, 2011.

Penrose, Roger. "Asymptotic Properties of Fields and Space-times." *PRL* 10, no. 2 (1963): 66–68.

Penrose, Roger. "Conformal Treatment of Infinity." In DeWitt and DeWitt, *Relativity, Groups, and Topology*, 565–586.

Penrose, Roger. "Cosmological Boundary Conditions for Zero Rest-mass Fields." In *The Nature of Time*, edited by Thomas Gold and Hermann Bondi, 42–54. Ithaca, NY: Cornell University Press, 1967.

Penrose, Roger. "Escher and the Visual Representation of Mathematical Ideas." In Coxeter, Emmer, Penrose, and Teuber, *M. C. Escher*, 143–158.

Penrose, Roger. "Gravitational Collapse and Space-time Singularities." *PRL* 14, no. 3 (1965): 57–59.

Penrose, Roger. Set of tiles for covering a surface. British patent 1,548,164, filed 1979.

Penrose, Roger. "Structure of Space-time." In DeWitt-Morette and Wheeler, *Battelle Rencontres*, 121–235.

Penrose, Roger. "The Apparent Shape of a Relativistically Moving Sphere." *MPCPS* 55, no. 1 (1959): 137–139.

Penrose, Roger. "The Light Cone at Infinity." In Infeld, *Proceedings on Theory of Gravitation*, 369–373.

Penrose, Roger. "Zero Rest-mass Fields Including Gravitation: Asymptotic Behaviour." *PRSLA* 284, no. 1397 (1965): 159–203.

Penrose, Roger, and Wolfgang Rindler. *Spinors and Space-time*. Cambridge: Cambridge University Press, 1986.

Peskin, Michael E., and Daniel V. Schroeder. *An Introduction to Quantum Field Theory*. Reading, MA: Westview Press, 1995.

Petkov, Vesselin, ed. *Minkowski Spacetime: A Hundred Years Later*. Dordecht: Springer, 2010.

Physics Survey Committee. *Physics: Survey and Outlook—Reports of the Subfields of Physics*. Technical report 1295A. Washington, DC: National Academy of Sciences and National Research Council, 1966.

Pickering, Andrew. *Constructing Quarks: A Sociological History of Particle Physics*. Chicago: University of Chicago Press, 1984.

Pickering, Andrew, ed. *Science as Practice and Culture*. Chicago: University of Chicago Press, 1992.

Pickering, Andrew. *The Mangle of Practice: Time, Agency, and Science*. Chicago: University of Chicago Press, 1995.

Plass, Gilbert N., and John Archibald Wheeler. "Blackbody Radiation in the Theory of Action at a Distance." *PR* 70, nos. 9-10 (1946): 793.

Plato, Jan von. *Creating Modern Probability: Its Mathematics, Physics and Philosophy in Historical Perspective*. Cambridge: Cambridge University Press, 1994.

Plesset, Milton S., and John Archibald Wheeler. "Inelastic Scattering of Quanta with Production of Pairs." *PR* 48, no. 4 (1935): 302–306.

Polyakov, Alexander M. "Compact Gauge Fields and the Infrared Catastrophe." *PL* B59, no. 1 (1975): 82–84.

Polyakov, Alexander M. "Quark Confinement and Topology of Gauge Groups." *NP* B120 (1977): 429–458.

Porter, Theodore M. "Statistics and Physical Theories." In Nye, *The Modern Physical and Mathematical Sciences*, 488–504.

Putnam, Hilary. "From Quantum Mechanics to Ethics and Back Again." In *Reading Putnam*, edited by Maria Baghramian, 19–36. London: Routledge, 2013.

Pyenson, Lewis, and Douglas Skopp. "Educating Physicists in Germany circa 1900." *SSS* 7, no. 3 (1977): 329–366.

Rainich, George Yuri. "Electrodynamics in the General Relativity Theory." *Transactions of the American Mathematical Society* 27, no. 1 (1925): 106–136.

Randall, Lisa. *Higgs Discovery: The Power of Empty Space*. New York: Ecco Press, 2013.

Rasmussen, Nicolas, and Alan Chalmers. "The Role of Theory in the Use of Instruments; or, How Much Do We Need to Know about Electrons to Do Science with an Electron Microscope?" In Buchwald and Warwick, *Histories of the Electron*, 467–502.

Rebsdorf, Simon, and Helge Kragh. "Edward Arthur Milne—The Relations of Mathematics to Science." *SHPMP* 33, no. 1 (2002): 51–64.

Rees, Martin J., Remo Ruffini, and John Archibald Wheeler. *Black Holes, Gravitational Waves, and Cosmology: An Introduction to Current Research*. New York: Gordon and Breach, 1974.

Reeves, Barbara J. "Einstein Politicized: The Early Reception of Relativity in Italy." In Glick, *The Comparative Reception of Relativity*, 189–230.

Regt, Henk W. de. "Erwin Schrödinger, Anschaulichkeit, and Quantum Theory." *SHPMP* 28, no. 4 (1997): 461–481.

Reid, Constance. *Courant*. New York: Springer, 1996.

Reines, Fred, and Cowan Jr. "The Reines-Cowan Experiments: Detecting the Poltergeist." *Los Alamos Science* 25 (1997): 4–27.

Renn, Jürgen. "Einstein's Invention of Brownian Motion." *AP* 14, no. S1 (2005): 23–37.

Renn, Jürgen. "The Third Way to General Relativity: Einstein and Mach in Context." In Janssen, Norton, Renn, Sauer, and Stachel, *Genesis of General Relativity*, 945–1000.

Renn, Jürgen, and Tilman Sauer. "Pathways out of Classical Physics." In Janssen, Norton, Renn, Sauer, and Stachel, *Genesis of General Relativity*, 113–312.

Report of an International Conference on Fundamental Particles and Low Temperatures held at the Cavendish Laboratory, Cambridge on 22–27 July 1946. London: The Physical Society, 1947.

Rheinberger, Hans-Jörg. "Cultures of Experimentation." In Chemla and Keller, *Cultures without Culturalism*, 278–295.

Rheinberger, Hans-Jörg. "Cytoplasmic Particles: The Trajectory of a Scientific Object." In Daston, *Biographies of Scientific Objects*, 270–194.

Rheinberger, Hans-Jörg. "Scrips and Scribbles." *MLN* 118, no. 3 (2003): 622–636.

Rheinberger, Hans-Jörg. *Toward a History of Epistemic Things: Synthesizing Proteins in the Test Tube*. Stanford, CA: Stanford University Press, 1997.

Rhodes, Richard. *The Making of the Atomic Bomb*. 1986. New York: Touchstone / Simon & Schuster, 1988.

Rickles, Dean. *A Brief History of String Theory: From Dual Models to M-theory*. Berlin: Springer, 2014.

Rickles, Dean. "Geon Wheeler: From Nuclear to Spacetime Physicist." *European Physical Journal H* 43, no. 3 (2019): 243–265.

Rickles, Dean. "Pourparlers for Amalgamation: Some Early Sources of Quantum Gravity Research." In Katzir, Lehner, and Renn, *Traditions and Transformations in the History of Quantum Physics*, 149–180.

Rickles, Dean, and Cécile M. DeWitt, eds. *The Role of Gravitation in Physics: Report from the 1957 Chapel Hill Conference*. Berlin: Edition Open Access, 2011.

Rindler, Wolfgang. "Visual Horizons in World Models." *Monthly Notices of the Royal Astronomical Society* 116 (1956): 662–667.

Robinson, Ivor, Alfred Schild, and Engelbert Schücking, eds. *Quasi-stellar Sources and Gravitational Collapse, Including the Proceedings of the First Texas Symposium on Relativistic Astrophysics*. Chicago: University of Chicago Press, 1965.

Robinson, Ivor, Alfred Schild, and Engelbert Schücking, eds. "Relativistic Theories of Gravitation." *PT* 16, no. 8 (1963): 17–20.

Robotti, Nadia. "Quantum Numbers and Electron Spin: The History of a Discovery." *Archives Internationale d'Histoire des Sciences* 40, no. 125 (1990): 305–331.

Rocke, Alan J. *Image and Reality: Kekulé, Kopp, and the Scientific Imagination*. Chicago: University of Chicago Press, 2010.

Rohrlich, Fritz. *Applied Quantum Electrodynamics*. Lecture notes. Princeton, NJ: Princeton University, 1953.

Roland, Alex. "Science, Technology, and War." In Nye, *The Modern Physical and Mathematical Sciences*, 559–578.

Roqué, Xavier. "The Manufacture of the Positron." *SHPMP* 28, no. 1 (1997): 73–129.

Rosenfeld, Léon. "Über die Gravitationswirkungen des Lichtes." *ZP* 65, no. 9 (1930): 589–599.

Rubinin, P. "Peter Kapitza and Paul Dirac. Letters of 1935–1937." *Science in the USSR*, 1989–1990, 94–99, 21–24.

Rueger, Alexander. "Attitudes towards Infinities: Responses to Anomalies in Quantum Electrodynamics, 1927–1947." *HSPBS* 22, no. 2 (1992): 309–337.

Rugh, S. E., and H. Zinkernagel. "The Quantum Vacuum and the Cosmological Constant Problem." *SHPMP* 33, no. 4 (2002): 663–705.

Ryckman, Thomas. *Einstein*. New York: Routledge, 2017.

Ryckman, Thomas. *The Reign of Relativity: Philosophy in Physics 1915–1925*. New York: Oxford University Press, 2005.

Sachs, Rainer K. "Gravitional Radiation." In DeWitt and DeWitt, *Relativity, Groups, and Topology*, 521–562.

Sakurai, J. J. *Modern Quantum Mechanics*. Rev. ed. Edited by San Fu Tuan. Reading, MA: Addison-Wesley, 1994.

Salam, Abdus. "Some Speculations on the New Resonances." *RMP* 33, no. 3 (1961): 426–430.

Sauer, Tilman. "Einstein's Unified Field Theory Program." In Janssen and Lehner, *Cambridge Companion to Einstein*, 281–305.

Saunders, Simon. "Is the Zero-Point Energy Real?" In *Ontological Aspects of Quantum Field Theory*, edited by Meinard Kuhlmann, Holger Lyre, and Andrew Wayne, 313–343. Singapore: World Scientific, 2002.

Schaffer, Simon. "Rayleigh and the Establishment of Electrical Standards." *EJP* 15, no. 6 (1994): 277–285.

Schemmel, Matthias. "The Continuity between Classical and Relativistic Cosmology in the Work of Karl Schwarzschild." In Janssen, Norton, Renn, Sauer, and Stachel, *Genesis of General Relativity*, 1081–1108.

Schirrmacher, Arne. "Lenard's Ether and Its Vortex of Emotions between Accommodating and Fighting Modern Physics with Äther and Uräther in the German Political Context." In Navarro, *Ether and Modernity*, 107–129.

Schmidt, Maarten. "3c 273: A Star-like Object with Large Red-shift." *Nature* 197, no. 4872 (1963): 1040.

Scholz, Erhard. "Introducing Groups into Quantum Theory (1926–1930)." *Historia Mathematica* 33, no. 4 (2006): 440–490.

Schrödinger, Erwin. *Collected Papers of Wave Mechanics*. Translated from the Second German Edition. Translated by J. F. Shearer and W. M. Deans. London: Blackie & Son, 1928.

Schrödinger, Erwin. "Quantisation as a Problem of Proper Values (part III)." In *Collected Papers of Wave Mechanics*, 62–101.

Schrödinger, Erwin. "Sur la théorie relativiste de l'électron et l'interprétation de la mécanique quantique." *Annales de l'Institut Henri Poincaré* 2, no. 4 (1932): 269–310.

Schrödinger, Erwin. "Über die kräftefreie Bewegung in der relativistischen Quantenmechanik." *Sitzungsberichte der Preußischen Akademie der Wissenschaften, physikalisch-mathematischen Klasse*, 1930, 418–428.

Schücking, Engelbert. "The First Texas Symposium on Relativistic Astrophysics." *PT*, 1989, 46–52.

Schupbach, Jonah N. "Robustness Analysis as Explanatory Reasoning." *BJPS* 69, no. 1 (2016): 275–300.

Schwarzschild, Bertram. "Why Is the Cosmological Constant So Very Small?" *PT* 42, no. 3 (1989): 21–24.

Schweber, S. S. "A Historical Perspective on the Rise of the Standard Model." In Hoddeson, Brown, Riordan, and Dresden, *Rise of the Standard Model*, 645–684.

Schweber, S. S. *Einstein and Oppenheimer: The Meaning of Genius.* Cambridge, MA: Harvard University Press, 2008.

Schweber, S. S. "Enrico Fermi and Quantum Electrodynamics, 1929–32." *PT* 55, no. 6 (2002): 31–36.

Schweber, S. S. "Feynman and the Visualization of Space-time Processes." *RMP* 58, no. 2 (1986): 449–508.

Schweber, S. S. "Hacking the Quantum Revolution: 1925–1975." *European Physical Journal H* 40, no. 1 (2015): 53–149.

Schweber, S. S. *In the Shadow of the Bomb: Bethe, Oppenheimer, and the Moral Responsibility of the Scientist.* Princeton, NJ: Princeton University Press, 2000.

Schweber, S. S. *Nuclear Forces: The Making of the Physicist Hans Bethe.* Cambridge, MA: Harvard University Press, 2012.

Schweber, S. S. *QED and the Men Who Made It: Dyson, Feynman, Schwinger, and Tomonaga.* Princeton, NJ: Princeton University Press, 1994.

Schweber, S. S. "Quantum Field Theory: From QED to the Standard Model." In Nye, *The Modern Physical and Mathematical Sciences*, 375–393.

Schweber, S. S. "Shelter Island, Pocono, and Oldstone: The Emergence of American Quantum Electrodynamics after World War II." *Osiris* 2 (1986): 265–302.

Schweber, S. S. "The Empiricist Temper Regnant: Theoretical Physics in the United States, 1920–1950." *HSPBS* 17 (1986): 55–98.

Schweber, S. S. "The Metaphysics of Science at the End of a Heroic Age." In *Experimental Metaphysics*, edited by Robert S. Cohen, Michael Horne, and John Stachel, 171–198. Dordrecht: Kluwer Academic Publishers, 1997.

Schweber, S. S. "The Shelter Island Conferences Revisited: 'Fundamental' Physics in the Decade 1975–1985." *Physics in Perspective* 18 (2016): 58–147.

Schweber, S. S. "The Sources of Schwinger's Green's Functions." *Proceedings of the National Academy of Sciences* 102, no. 22 (2005): 7783–7788.

Schweber, S. S. "The Young John Clarke Slater and the Development of Quantum Chemistry." *HSPBS* 20, no. 2 (1990): 339–406.

Schweber, S. S. "Weimar Physics: Sommerfeld's Seminar and the Causality Principle." *Physics in Perspective* 11 (2009): 261–301.

Schwinger, Julian. "A Report on Quantum Electrodynamics." In Mehra, *Physicist's Conception of Nature*, 413–429.

Schwinger, Julian. "A Theory of the Fundamental Interactions." *Annals of Physics* 2, no. 5 (1957): 407–434.

Schwinger, Julian. "Brownian Motion of a Quantum Oscillator." *JMP* 2, no. 3 (1961): 407–432.

Schwinger, Julian. "Euclidean Quantum Electrodynamics." *PR* 115, no. 3 (1959): 721–731.

Schwinger, Julian. "Gauge Invariance and Mass II." *PR* 128, no. 5 (1962): 2425–2429.

Schwinger, Julian. "On Gauge Invariance and Vacuum Polarization." *PR* 82, no. 5 (1951): 664–679.

Schwinger, Julian. "On Quantum-electrodynamics and the Magnetic Moment of the Electron." *PR* 73, no. 4 (1948): 416–417.

Schwinger, Julian. "On the Green's Functions of Quantized Fields. II." *Proceedings of the National Academy of Sciences* 37, no. 7 (1951): 455–459.

Schwinger, Julian. "Quantum Electrodynamics. I. A Covariant Formulation." *PR* 74, no. 10 (1948): 1439–1461.

Schwinger, Julian. "Quantum Electrodynamics. II. Vacuum Polarization and Self-energy." *PR* 75, no. 4 (1949): 651–679.

Schwinger, Julian. "Quantum Electrodynamics. III. The Electromagnetic Properties of the Electron—Radiative Corrections to Scattering." *PR* 76, no. 6 (1949): 790–817.

Schwinger, Julian. "The Theory of Quantized Fields. I." *PR* 82, no. 6 (1951): 914–927.

Schwinger, Julian. "The Theory of Quantized Fields. II." *PR* 91, no. 3 (1953): 713–728.

Schwinger, Julian. "The Theory of Quantized Fields. IV." *PR* 92, no. 5 (1953): 1283–1299.

Schwinger, Julian. "The Theory of Quantized Fields. V." *PR* 93, no. 3 (1954): 615–628.

Schwinger, Julian. "Tomonaga Sin-Itiro: A Memorial—Two Shakers of Physics." In *Nishina Memorial Lectures: Creators of Modern Physics*, 27–42. 1980. Tokyo: Springer, 2008.

Schwinger, Julian, and Victor Weisskopf. "On the Electromagnetic Shift of Energy Levels." *PR* 73, no. 10 (1949): 1272.

Sciama, Dennis W. "Retarded Potentials and the Expansion of the Universe." *PRSLA* 273, no. 1355 (1963): 484–495.

Sciama, Dennis W. "Retarded Potentials and the Expansion of the Universe." In Infeld, *Proceedings on Theory of Gravitation*, 332–335.

Serber, Robert. "Linear Modifications in the Maxwell Field Equations." *PR* 48, no. 1 (1935): 49–54.

Serwer, Daniel. "*Unmechanischer Zwang*: Pauli, Heisenberg, and the Rejection of the Mechanical Atom, 1923-1925." *HSPS* 8 (1977): 189–256.

Seth, Suman. *Crafting the Quantum: Arnold Sommerfeld and the Practice of Theory, 1890–1926.* Cambridge, MA: MIT Press, 2010.

Seth, Suman. "Crisis and the Construction of Modern Theoretical Physics." *BJHS* 40, no. 1 (2007): 25–51.

Shapin, Steven, and Simon Schaffer. *Leviathan and the Air-Pump: Hobbes, Boyle, and the Experimental Life.* Princeton, NJ: Princeton University Press, 1985.

Silva, Indianara, and Olival Freire Jr. "The Concept of the Photon in Question: The Controversy Surrounding the HBT Effect circa 1956–1958." *HSNS* 43, no. 4 (2013): 453–491.

Simões, Ana. "Chemical Physics and Quantum Chemistry in the Twentieth Century." In Nye, *The Modern Physical and Mathematical Sciences*, 394–412.

Simões, Ana, and Kostas Gavroglu. "Quantum Chemistry in Great Britain: Developing a Mathematical Framework for Quantum Chemistry." *SHPMP* 31, no. 4 (2000): 511–548.

Smeenk, Chris. "False Vacuum: Early Universe Cosmology and the Development of Inflation." In Kox and Eisenstaedt. *The Universe of General Relativity*, 2005, 223–257.

Smeenk, Chris. "Tools without Theories." *Metascience* 15, no. 2 (2006): 333–337.

Smith, Alice Kimball, and Charles Weiner, eds. *Robert Oppenheimer: Letters and Recollections.* Stanford, CA: Stanford University Press, 1980.

Smith, Crosbie, and M. Norton Wise. *Energy and Empire: A Biographical Study of Lord Kelvin.* Cambridge: Cambridge University Press, 1989.

Soler, Léna. "Introduction: The Solidity of Scientific Achievements: Structure of the Problem, Difficulties, Philosophical Implications." In *Characterizing the Robustness of Science: After the Practice Turn in Philosophy of Science,* edited by Léna Soler, Emiliano Trizio, Thomas Nickles, and William Wimsatt, 1–60. Dordrecht: Springer Netherlands, 2012.

Sommerfeld, Arnold. *Atomic Structure and Spectral Lines.* Translated from the Third German Edition. Translated by Henry L. Brose. London: Methuen & Co., 1923.

Sommerfeld, Arnold, and Victor Weisskopf. "Besprechungen." *Naturwissenschaften* 23, no. 4 (1935): 70–72.

Stableford, Brian, ed. *Science Fact and Science Fiction: An Encyclopedia.* New York: Routledge, 2006.

Stachel, John. *Einstein from "B" to "Z".* Boston: Birkhäser, 2002.

Stachel, John. "Introduction." In Einstein, *Einstein's Miraculous Year,* 1–27.

Stachel, John. "Notes on the Andover Conference." In *Foundations of Space-Time Theories,* edited by John Earman, Clark Glymour, and John Stachel, vii–xii. Minneapolis: University of Minnesota Press, 1977.

Stachel, John. "Scientific Discoveries as Historical Artifacts." In *Trends in the Historiography of Science,* edited by Kostas Gavroglu, Jean Christianidis, and Efthymios Nicolaidis, 139–148. Dordrecht: Springer, 1994.

Stachel, John. "The Early History of Quantum Gravity (1916–1940)." In *Black Holes, Gravitational Radiation and the Universe,* edited by B. R. Iyer and B. Bhawal, 525–534. Dordrecht: Kluwer, 1999.

Stachel, John. "The Rise and Fall of Geometrodynamics." In *PSA 1972,* edited by Kenneth F. Schaffner and Robert S. Cohen, 31–56. Dordecht: D. Reidel, 1974.

Staley, Richard. *Einstein's Generation: The Origins of the Relativity Revolution.* Chicago: University of Chicago Press, 2008.

Staley, Richard. "Ether and Aesthetics in the Dialogue between Relativists and Their Critics in the Late Nineteenth and Early Twentieth Centuries." In Navarro, *Ether and Modernity,* 179–199.

Staley, Richard. "Trajectories in the History and Historiography of Physics in the Twentieth Century." *History of Science* 51, no. 2 (2013): 151–177.

Stanley, Matthew. "'An Expedition to Heal the Wounds of War': The 1919 Eclipse and Eddington as Quaker Adventurer." *Isis* 94, no. 1 (2003): 57–89.

Star, Susan Leigh. "This Is Not a Boundary Object: Reflections on the Origin of a Concept." *Science, Technology & Human Values* 35, no. 5 (2010): 601–617.

Stefan, V., ed. *Physics and Society: Essays in Honor of Victor Frederick Weisskopf.* Woodbury, NY: AIP Press, 1997.

Stöltzner, Michael. "Higgs Models and Other Stories about Mass Generation." *Journal for General Philosophy of Science* 45, no. 2 (2014): 369–386.

Stöltzner, Michael. "The Causality Debates of the Interwar Years and Their Preconditions: Revisiting the Forman Thesis from a Broader Perspective." In Carson, Kozhevnikov, and Trischler, *Weimar Culture and Quantum Mechanics*, 505–522.

Stone, Michael. "Lifetime and Decay of 'Excited Vacuum' States of a Field Theory Associated with Nonabsolute Minima of Its Effective Potential." *PRD* 14, no. 12 (1976): 3568–3573.

Stone, Michael. "Semiclassical Methods for Unstable States." *PL* B67, no. 2 (1977): 186–188.

Stoner, Edmund C. "The Distribution of Electrons among Atomic Levels." *PM* 48, no. 286 (1924): 719–736.

Strutt (Baron Rayleigh), John William. *The Theory of Sound*. 2nd ed. London: Macmillan, 1894–1896.

Stuewer, Roger H. "G. N. Lewis on Detailed Balancing, the Symmetry of Time, and the Nature of Light." *HSPS* 6 (1975): 469–511.

Stuewer, Roger H. "Gamow's Theory of Alpha-decay." In *The Kaleidoscope of Science: The Israel Colloquium: Studies in History, Philosophy, and Sociology of Science Volume 1*, edited by Edna Ullmann-Margalit, 147–186. Dordrecht: Springer, 1986.

Stuewer, Roger H. "The Compton Effect: Transition to Quantum Mechanics." *AP* 9, nos. 11–12 (2000): 975–989.

Stuewer, Roger H. *The Compton Effect: Turning Point in Physics*. New York: Science History Publications, 1975.

Stuewer, Roger H. "The Origin of the Liquid-drop Model and the Interpretation of Nuclear Fission." *Perspectives on Science* 2, no. 1 (1994): 76–129.

Suárez, Mauricio, and Nancy Cartwright. "Theories: Tools versus Models." *SHPMP* 39, no. 1 (2008): 62–81.

Synge, John L. *Relativity: The General Theory*. Amsterdam: North-Holland Pub. Co, 1960.

Synge, John L. *Relativity: The Special Theory*. Amsterdam: North-Holland Pub. Co., 1956.

Taltavull, Marta Jordi. "Challenging the Boundaries between Classical and Quantum Physics: The Case of Optical Dispersion." In Katzir, Lehner, and Renn, *Traditions and Transformations in the History of Quantum Physics*, 29–59.

Tanona, Scott. "Uncertainty in Bohr's Response to the Heisenberg Microscope." *SHPMP* 35, no. 3 (2004): 483–507.

Tennant, Neil. "Logicism and Neologicism." In *The Stanford Encyclopedia of Philosophy*, Winter, edited by Edward N. Zalta. Metaphysics Research Lab, Stanford University, 2017. https://plato.stanford.edu/archives/win2017/entries/logicism/.

Thomälen, Adolf. *A Text-book of Electrical Engineering*. Translated by George W. O. Howe. London: Longmans, Green, & Co., 1907.

Thorne, Kip S. *Black Holes & Time Warps: Einstein's Outrageous Legacy*. New York: W. W. Norton, 1994.

Thorne, Kip S. "Gravitational Radiation." In *Three Hundred Years of Gravitation*, edited by S. W. Hawking and W. Israel, 330–458. Cambridge: Cambridge University Press, 1987.

Thorne, Kip S. "John Archibald Wheeler (1911–2008)." *Science* 320, no. 5883 (2008): 1603.

Thorsen, Jens. "Introduction [General Theory of Penetration]." In *The Penetration of Charged Particles through Matter (1912–1954)*. Vol. 8 of *Niels Bohr Collected Works*, edited by Jens Thorsen, 203–265. Amsterdam: Elsevier, 1987.

Toll, John, and John Archibald Wheeler. "[abstract.] Some Pair-Theoretic Applications of the Dispersion Relation." *PR* 81, no. 4 (1951): 654–655.

Tomonaga, Sin-Itiro. "On a Relativistically Invariant Formulation of the Quantum Theory of Wave Fields." *Progress of Theoretical Physics* 1, no. 2 (1946): 27–42.

Tomonaga, Sin-Itiro, and J. Robert Oppenheimer. "On Infinite Field Reactions in Quantum Field Theory." *PR* 74, no. 2 (1948): 224–225.

Touschek, Bruno. "Remarks on the Neutrino Gauge Group." In *Ninth Annual International Conference on High Energy Physics*, 2:117–125. Moscow: Academy of Science USSR, 1960.

Traweek, Sharon. *Beamtimes and Lifetimes: The World of High Energy Physics*. Cambridge, MA: Harvard University Press, 1988.

Tucker, Jennifer. "The Historian, the Picture, and the Archive." *Isis* 97, no. 1 (2006): 111–120.

Turner, Paul T. "Vertex Roundtable Interviews Dr. Sidney Coleman and Dr. Gregory Benford." *eI* 36 (2008). http://efanzines.com/EK/eI36/.

Uehling, Edwin A. "Polarization Effects in the Positron Theory." *PR* 48, no. 1 (1935): 55–63.

Uffink, Jos. "Insuperable Difficulties: Einstein's Statistical Road to Molecular Physics." *SHPMP* 37, no. 1 (2006): 36–70.

Van Vleck, John H. *The Theory of Electric and Magnetic Susceptibilities*. Oxford: Oxford University Press, 1932.

Voloshin, Mikhail B., Igor Yu. Kobzarev, and Lev B. Okun. "Bubbles in Metastable Vacuum." Translated by Meinhard E. Mayer. *Soviet Journal of Nuclear Physics* 20 (1975): 644–646.

Waerden, B. L. van der, ed. *Sources of Quantum Mechanics*. Amsterdam: North-Holland Pub. Co, 1967.

Wald, Robert. *General Relativity*. Chicago: University of Chicago Press, 1984.

Waller, Ivar. "Bemerkungen über die Rolle der Eigenenergie des Elektrons in der Quantenthe-orie der Strahlung." *ZP* 62, no. 9 (1930): 673–676.

Waller, Ivar. "Die Streuung von Strahlung durch Gebundene und freie Elektronen nach der Diracschen relativistischen Mechanik." *ZP* 61, no. 11 (1930): 837–851.

Waller, Ivar. "Memories of My Early Work on Lattice Dynamics and X-ray Diffraction." *PRSLA* 371, no. 1744 (1980): 120–124.

Waller, Ivar. "The Transition from Ordinary Dispersion into Compton Effect." *Nature* 120, no. 3013 (1927): 155–156.

Waller, Ivar. "Über eine Verallgemeinerte Streuungsformel." *ZP* 51, no. 3 (1928): 213–231.

Walls, D. F., and Gerard Milburn. *Quantum Optics*. 2nd ed. Berlin: Springer, 2008.

Walter, Scott. "Breaking in the 4-vectors: The Four-dimensional Movement in Gravitation, 1905–1910." In Janssen, Norton, Renn, Sauer, and Stachel, *Genesis of General Relativity*, 1118–1178.

Walter, Scott. "Minkowski's Modern World." In Petkov, *Minkowski Spacetime*, 43–61.

Wang, Jessica. "'Broken Symmetry': Physics, Aesthetics, and Moral Virtue in Nuclear Age America." In *Epistemic Virtues in the Sciences and the Humanities*, edited by Jeroen van Dongen and Herman Paul, 27–47. Cham: Springer, 2017.

Warwick, Andrew. *Masters of Theory: Cambridge and the Rise of Mathematical Physics*. Chicago: University of Chicago Press, 2003.

Warwick, Andrew. "The Laboratory of Theory or What's Exact about the Exact Sciences?" In *The Values of Precision*, edited by M. Norton Wise, 311–351. Princeton, NJ: Princeton University Press, 1997.

Warwick, Andrew, and David Kaiser. "Conclusion: Kuhn, Foucault, and the Power of Pedagogy." In Kaiser, *Pedagogy and the Practice of Science*, 393–410.

Weatherall, James Owen. *Void: The Strange Physics of Nothing*. New Haven, CT: Yale University Press, 2016.

Weinberg, Steven. *Gravitation and Cosmology: Principles and Applications of the General Theory of Relativity*. New York: Wiley, 1972.

Weinberg, Steven. "The Forces of Nature." *Bulletin of the American Academy of Arts and Sciences* 29, no. 4 (1976): 13–29.

Weinberg, Steven. *The Quantum Theory of Fields*. Vol. I, Fundamentals. Cambridge: Cambridge University Press, 1995.

Weinberg, Steven. *The Quantum Theory of Fields*. Vol. II, Modern Applications. Cambridge: Cambridge University Press, 1996.

Weinberg, Steven. "The Search for Unity: Notes for a History of Quantum Field Theory." *Daedalus* 106, no. 4 (1977): 17–35.

Weisskopf, Victor. "On the Self-energy and the Electromagnetic Field of the Electron." *PR* 56 (1939): 72–85.

Weisskopf, Victor. "On the Self-energy of the Electron." *PR* 55, no. 7 (1939): 678–678.

Weisskopf, Victor. "Probleme der neueren Quantentheorie des Elektrons." *Naturwissenschaften* 23, nos. 37–39 (1935): 631–637, 647–653, 669–674.

Weisskopf, Victor. "Recent Developments in the Theory of the Electron." *RMP* 21, no. 2 (1949): 305–315.

Weisskopf, Victor. "The Electrodynamics of the Vacuum Based on the Quantum Theory of the Electron." In Miller, *Early Quantum Electrodynamics*, 206–226.

Weisskopf, Victor. "The Intensity and Structure of Spectral Lines." *The Observatory* 56, no. 713 (1933): 291–308.

Weisskopf, Victor. *The Joy of Insight: Passions of a Physicist*. New York: Basic Books, 1991.

Weisskopf, Victor. "The Self-energy of the Electron [and Correction]." In Miller, *Early Quantum Electrodynamics*, 157–168.

Weisskopf, Victor. "Über die Elektrodynamik des Vakuums auf Grund der Quantentheorie des Elektrons." *Det Kgl. Danske Videnskabernes Selskab. Mathematisk-fysiske Meddelelser* XIV, no. 6 (1936): 1–39.

Weisskopf, Victor. "Über die Lebensdauer angeregter Atomzustände." *Physikalische Zeitschrift der Sowietunion* 4, no. 1 (1933): 97–113.

Weisskopf, Victor. "Über die Polarisation der Elektronen bei der Streuung an Kristallen." ZP 93, no. 9 (1935): 561–581.

Weisskopf, Victor. "Zur Theorie der Resonanzfluoreszenz." AP 401, no. 1 (1931): 23–66.

Weisskopf, Victor, and Eugene Wigner. "Berechnung der natürlichen Linienbreite auf Grund der Diracschen Lichttheorie." ZP 63, no. 1 (1930): 54–73.

Weizsäcker, Carl Friedrich von. "Ausstrahlung bei Stößen sehr schneller Elektronen." ZP 88, no. 9 (1934): 612–625.

Welton, Theodore A. "Some Observable Effects of the Quantum-mechanical Fluctuations of the Electromagnetic Field." PR 74, no. 9 (1948): 1157–1167.

Wentzel, Gregor. Einführung in die Quantentheorie der Wellenfelder. 1943. Reprint, Ann Arbor, MI: Edwards Brothers, 1946.

Wentzel, Gregor. "Quantum Theory of Fields (until 1947)." In Mehra, Physicist's Conception of Nature, 380–403.

Wentzel, Gregor. "Recent Research in Meson Theory." RMP 19, no. 1 (1947): 1–18.

Wentzel, Gregor. "Wellenmechanik der Stoß- und Strahlungsvorgänge." In Handbuch der Physik, edited by Adolf Smekal, XXIV:695–784. I. Berlin: Springer, 1933.

Wesemael, François. "'Unaffected by Fortune, Good or Bad': Context and Reception of Chandrasekhar's Mass-radius Relationship for White Dwarfs, 1935–1965." AS 67, no. 2 (2010): 205–237.

Westfall, Catherine. "Rethinking Big Science: Modest, Mezzo, Grand Science and the Development of the Bevalac, 1971–1993." Isis 94, no. 1 (2003): 30–56.

Weyl, Hermann. Space-Time-Matter. 4th ed. Translated by Henry L. Brose. 1922. London: Methuen & Co., 1952.

Weyl, Hermann. The Theory of Groups and Quantum Mechanics. Translated by H. P. Robertson. 1931. London: Methuen & Co., 1932.

Wheaton, Bruce R. The Tiger and the Shark: Empirical Roots of Wave-particle Dualism. Cambridge: Cambridge University Press, 1983.

Wheeler, John Archibald. "[abstract:] Field and Charge Measurements in Quantum Theory by Niels Bohr and Leon Rosenfeld." In Pais, Development of the Theory of the Electron, 42–45.

Wheeler, John Archibald. "Assessment of Everett's 'Relative State' Formulation of Quantum Theory." RMP 29, no. 3 (1957): 463–465.

Wheeler, John Archibald. Einsteins Vision: wie steht es heute mit Einsteins Vision, alles als Geometrie aufzufassen? Berlin: Springer, 1968.

Wheeler, John Archibald. Final Report of Project Matterhorn. Technical report PM-B-37. Los Alamos, NM: Princeton University and US Atomic Energy Comission, 1953.

Wheeler, John Archibald. "From Mendeleev's Atom to the Collapsing Star." In Atti del Convegno Mendeleeviano, 189–233. Torino: Acad. delle Scienze di Torino, 1971.

Wheeler, John Archibald. "From Relativity to Mutability." In Mehra, Physicist's Conception of Nature, 202–250.

Wheeler, John Archibald. "Geometrodynamics and the Issue of the Final State." In DeWitt and DeWitt, Relativity, Groups, and Topology, 317–522.

Wheeler, John Archibald. "Geons." PR 97, no. 2 (1955): 511–536.

Wheeler, John Archibald. "Information, Physics, Quantum: The Search for Links." In *Complexity, Entropy, and the Physics of Information: Proceedings of the SFI Workshop*, edited by Wojciech H. Zurek, 3–28. Boulder, CO: Westview Press, 1990.

Wheeler, John Archibald. "Molecular Viewpoints in Nuclear Structure." *PR* 52, no. 11 (1937): 1083–1106.

Wheeler, John Archibald. "Neutrinos, Gravitation and Geometry." In *Geometrodynamics*, edited by John Archibald Wheeler, 1–130. 1960. New York: Academic Press, 1962.

Wheeler, John Archibald. "On the Mathematical Description of Light Nuclei by the Method of Resonating Group Structure." *PR* 52, no. 11 (1937): 1107–1122.

Wheeler, John Archibald. "On the Nature of Quantum Geometrodynamics." *Annals of Physics* 2, no. 6 (1957): 604–614.

Wheeler, John Archibald. "Particles and Geometry." In Carmeli, Fickler, and Witten, *Relativity*, 31–42.

Wheeler, John Archibald. "Pregeometry: Motivations and Prospects." In *Quantum Theory and Gravitation*, edited by A. R. Marlov. New York: Academic Press, 1980.

Wheeler, John Archibald. "Some Men and Moments in the History of Nuclear Physics." In *Nuclear Physics in Retrospect: Proceedings of a Symposium on the 1930s*, edited by Roger H. Stuewer, 217–306. Minneapolis: University of Minnesota Press, 1979.

Wheeler, John Archibald. "Superspace and the Nature of Quantum Geometrodynamics." In DeWitt-Morette and Wheeler, *Battelle Rencontres*, 242–307.

Wheeler, John Archibald. "The Origin of Cosmic Rays." In *Proceedings of the International Conference of Theoretical Physics, Kyoto & Tokyo, September, 1953*, 81–93. Tokyo: Science Council of Japan, 1954.

Wheeler, John Archibald, and Richard P. Feynman. "Classical Electrodynamics in Terms of Direct Interparticle Action." *RMP* 21, no. 3 (1949): 425–433.

Wheeler, John Archibald, and Richard P. Feynman. "Interaction with the Absorber as the Mechanism of Radiation." *RMP* 17, nos. 2–3 (1945): 157–181.

Wheeler, John Archibald, and Kenneth William Ford. *Geons, Black Holes, and Quantum Foam: A Life in Physics*. New York: Norton, 1998.

White, Hayden V. *Metahistory: The Historical Imagination in Nineteenth-Century Europe*. Baltimore: Johns Hopkins University Press, 1973.

Whittaker, Edmund Taylor. *A History of the Theories of Aether and Electricity: The Modern Theories 1900–1926*. London: Thomas Nelson & Sons, 1953.

Whittaker, Edmund Taylor. *A Treatise on the Analytical Dynamics of Particles and Rigid Bodies*. 2nd ed. Cambridge: Cambridge University Press, 1917.

Whittaker, Edmund Taylor. "The Influence of Gravitation on Electromagnetic Phenomena." *Journal of the London Mathematical Society* s1-3, no. 2 (1928): 137–144.

Whitworth, Michael. "Transformations of Knowledge in Oliver Lodge's Ether and Reality." In Navarro, *Ether and Modernity*, 30–44.

Wightman, Arthur S. "Superselection Rules; Old and New." *INC* 110, no. 5 (1995): 751–769.

Will, Clifford. "The Renaissance of General Relativity." In *The New Physics*, edited by P. C. W. Davies, 7–33. Cambridge: Cambridge University Press, 1989.

Williams, Evan J. "Correlation of Certain Collision Problems with Radiation Theory." *Det Kgl. Danske Videnskabernes Selskab. Mathematisk-fysiske Meddelelser* 13, no. 4 (1935): 1–50.

Williams, Evan J. "High Energy Formulae." *PR* 47, no. 7 (1935): 569–570.

Wilson, Curtis A. "Perturbations and Solar Tables from Lacaille to Delambre: The Rapprochement of Observation and Theory, Part I." *AHES* 22, nos. 1/2 (1980): 53–188.

Wilson, Curtis A. "Perturbations and Solar Tables from Lacaille to Delambre: The Rapprochement of Observation and Theory, Part II." *AHES* 22, no. 3 (1980): 189–304.

Winch, Peter. "Introduction." In *Lectures on Philosophy*, by Simone Weil, translated by Huw Price, 1–24. 1959. Cambridge: Cambridge University Press, 1978.

Winther, Rasmus Grønfeldt. "The Structure of Scientific Theories." In *The Stanford Encyclopedia of Philosophy*, Winter 2016, edited by Edward N. Zalta. Metaphysics Research Lab, Stanford University, 2016.

Wise, M. Norton. "Forman Reformed, Again." In Carson, Kozhevnikov, and Trischler, *Weimar Culture and Quantum Mechanics*, 415–431.

Wise, M. Norton. "Science as (Historical) Narrative." *Erkenntnis* 75 (2011): 349–376.

Witten, Edward. "Instability of the Kaluza-Klein Vacuum." *NP B* 195 (1982): 481–492.

Wittgenstein, Ludwig. *Wittgenstein's Lectures on the Foundations of Mathematics, Cambridge, 1939*. Edited by Cora Diamond. Chicago: University of Chicago Press, 1975.

Wright, Aaron Sidney. "A Beautiful Sea: P. A. M. Dirac's Epistemology and Ontology of the Vacuum." *AS* 73, no. 3 (2016): 225–256.

Wright, Aaron Sidney. "An Ether by Any Other Name? Paul Dirac's Æther." In Navarro, *Ether and Modernity*, 225–244.

Wright, Aaron Sidney. "Fresnel's Laws, Ceteris Paribus." *Studies in History and Philosophy of Science* 64 (2017): 38–52.

Wright, Aaron Sidney. "The Advantages of Bringing Infinity to a Finite Place: Penrose Diagrams as Objects of Intuition." *HSNS* 44, no. 2 (2014): 99–139.

Wright, Aaron Sidney. "The Origins of Penrose Diagrams in Physics, Art, and the Psychology of Perception, 1958–62." *Endeavour* 37, no. 3 (2013): 133–139.

Wüthrich, Adrian. *The Genesis of Feynman Diagrams*. Dordrecht: Springer, 2010.

Wyler, John Archibald. "Rasputin, Science, and the Transmogrification of Destiny." *GRG* 5, no. 2 (1974): 175–182.

Yang, Chen N., Geoffrey F. Chew, Robert F. Christy, Marvin L. Goldberger, Robert G. Sachs, Leonard I. Schiff, John Archibald Wheeler, and Eugene P. Wigner. "Theoretical Physics." In PSC, *Physics Survey*, 159–165.

Yeang, Chen-Pang. "Tubes, Randomness, and Brownian Motions: or, How Engineers Learned to Start Worrying about Electronic Noise." *AHES* 65 (2011): 437–470.

Yeang, Chen-Pang. "Two Mathematical Approaches to Random Fluctuations." *Perspectives on Science* 24, no. 1 (2016): 45–72.

Zee, Anthony. *Quantum Field Theory in a Nutshell*. Princeton, NJ: Princeton University Press, 2003.

Zel'dovich, Yakov B. "Vacuum Theory: A Possible Solution to the Singularity Problem of Cosmology." *Soviet Physics Uspekhi* 24 (1981): 216–230.

Zel'dovich, Yakov B., Igor Yu. Kobzarev, and Lev B. Okun. "Cosmological Consequences of a Spontaneous Breakdown of a Discrete Symmetry." Translated by Meinhard E. Mayer. *Journal of Experimental and Theoretical Physics* 40, no. 1 (1975): 1–5.

Zhu, Yuelin. "Chien-Shiung Wu: An Intellectual Biography." PhD diss., Harvard University, 2001.

Zichichi, Antonino, ed. *Hadrons and Their Interactions: Current and Field Algebra, Soft Pions, Supermultiplets, and Related Topics.* New York: Academic Press, 1968.

Zichichi, Antonino, ed. *Properties of the Fundamental Interactions.* Bologna: Editrice Compositori, 1973.

Zurek, Wojciech Hubert, and John Archibald Wheeler, eds. *Quantum Theory and Measurement.* Princeton, NJ: Princeton University Press, 1983.

Archives Consulted

American Institute of Physics,
Niels Bohr Library,
College Park, MD

Records of the American Institute of Physics
Hugh Everett III Papers
Gravity Research Foundation

American Philosophical Society,
Philadelphia, PA

John Archibald Wheeler Papers

California Institute of Technology
Archives,
Pasadena, CA

Papers of Richard Phillips Feynman
Murray Gell-Mann Papers
Papers of Milton S. Plesset
H. P. Robertson Papers

Florida State University,
Special Collections, Tallahassee, FL

P. A. M. Dirac Papers

Harvard University Archives,
Cambridge, MA

Papers of Wendell Hinkle Furry

Massachusetts Institute of Technology,
Institute Archives and Special Collections,
Cambridge, MA

Victor Frederick Weisskopf Papers

Private Collection

Sidney R. Coleman Papers

University of California, Los Angeles,
Department of Special Collections

Julian Seymour Schwinger Papers

University of Maryland Libraries,
Special Collections,
College Park, MD

Charles W. Misner Papers
John Toll Papers

Index